PROBABILITY
THEORY AND EXAMPLES

SECOND EDITION

RICHARD DURRETT
CORNELL UNIVERSITY

Duxbury Press
An Imprint of Wadsworth Publishing Company
I**(T)**P® An International Thomson Publishing Company

Belmont • Albany • Bonn • Boston • Cincinnati • Detroit • London • Madrid • Melbourne
Mexico City • New York • Paris • San Francisco • Singapore
Tokyo • Toronto • Washington

Project Development Editor: Jennie Burger
Production Editor: Sheryl Gilbert
Print Buyer: Barbara Britton
Permissions Editor: Peggy Meehan
Copy Editor: Laren Crawford
Cover: Craig Hanson
Printer: Quebecor Printing Fairfield Inc.

Printed in the United States of America
1 2 3 4 5 6 7 8 9 10—01 00 99 98 97 96

For more information, contact Duxbury Press at Wadsworth Publishing Company.

Wadsworth Publishing Company
10 Davis Drive
Belmont, California 94002, USA

International Thomson Editores
Campos Eliseos 385, Piso 7
Col. Polanco
11560 México D.F. México

International Thomson Publishing Europe
Berkshire House 168-173
High Holborn
London, WC1V 7AA, England

International Thomson Publishing GmbH
Königswinterer Strasse 418
53227 Bonn, Germany

Thomas Nelson Australia
102 Dodds Street
South Melbourne 3205
Victoria, Australia

International Thomson Publishing Asia
221 Henderson Road
#05-10 Henderson Building
Singapore 0315

Nelson Canada
1120 Birchmount Road
Scarborough, Ontario
Canada M1K 5G4

International Thomson Publishing Japan
Hirakawacho Kyowa Building, 3F
2-2-1 Hirakawacho
Chiyoda-ku, Tokyo 102, Japan

Library of Congress Cataloging-in-Publication Data

Durrett , Richard
        Probability : theory and examples / Richard Durrett.—2nd ed.
                    p.   cm.
        Includes references and index.
        ISBN: 0-534-24318-5
        1. Probabilities. I. Title
QA273.D865   1995   95-22544
519.2—dc20

Preface

"Something old, something new, something borrowed, and something blue" is the traditional list of ingredients for a wedding dress. It also applies to our preface. To facilitate skipping to your favorite parts, we have divided the discussion here into boldly labelled subdivisions.

Our Manifesto

The first and most obvious use for this book is as a textbook for a one year graduate course in probability taught to students who are familiar with measure theory. An appendix, which gives complete proofs of the results from measure theory we need, is provided so that the book can be used whether or not the students are assumed to be familiar with measure theory.

The title of the book indicates that as we develop the theory, we will focus our attention on examples. Hoping that the book would be a useful reference for people who apply probability in their work, we have tried to emphasize the results that can be used to solve problems.

Exercises are integrated into the text because they are an integral part of it. In general the exercises embedded in the text can be done immediately using the material just presented and the reader should do these "finger exercises" to check her understanding. Exercises at the end of the section present extensions of the results and various complements.

Changes in the Second Edition

The book has undergone a thorough house cleaning as I taught from it during the academic year 1994-95 (and covered all of the unstarred sections).

(i) More than 500 typographical errors have been corrected.

(ii) More details have been added to many proofs to make them easier to understand. For example, Chapter 1 is now 78 pages instead of 63.

(iii) Some sections have been re-arranged and/or divided into subsections.

(iv) Last and most important, I have now worked all the problems and prepared a solutions manual.

As a result of (iv), there are slightly fewer problems (now 472), but more which are reasonable homework problems, since I have eliminated problems that were (a) too easy (just an excuse to state a definition) or (b) too hard (just an excuse to state a related theorem).

In order to achieve approximate conservation of mass (486 pages vs. 434 for the previous edition), some leaves have been pruned off the tree of knowledge presented here. We hope that these modifications will make the text less overwhelming for the student. With this in my mind we have moved some of the exercises which distract from the flow of the development to the end of their sections.

Acknowledgements

I would like to thank the following individuals who have taught from the first edition and sent me their lists of typos:

David Aldous	U. C. Berkeley
Ken Alexander	U. of Southern California
Daren Cline	Texas A&M U.
Ted Cox	Syracuse U.
Robert Dalang	Tufts U.
David Griffeath	U. of Wisconsin
Joe Glover	U. of Florida
Phil Griffin	Syracuse U.
Joe Horowitz	U. Mass., Amherst
Olav Kallenberg	Auburn U.
Jim Kuelbs	U. of Wisconsin
Robin Pemantle	U. of Wisconsin
Yuval Peres	U. C. Berkeley
Ken Ross	U. of Oregon
Byron Schmuland	U. of Alberta
Steve Samuels	Purdue U.
Jon Wellner	U. of Washington, Seattle
Ruth Williams	U. C. San Diego

In the face of this distinguished list of people who have used this book, you can hardly fail to not adopt it. If you want to see your name listed in the (groan) third edition, send your corrections, random insults, or interesting problems to rtd1@cornell.edu

Less famous, but at least as valuable, were my current and former students: Min-jeong Kang, Susan Lee, and Nikhil Shah who worked as "accuracy checkers" in the editorial process. I am sure you will see the names of these talented and hard working individuals again.

Family Update

Turning to the home front, my children: David (a.k.a. Pooh Bear) and Greg are now 8 1/2 and 6. The video games mentioned in the *The Essentials in Probability* have given way to Super-Nintendo and to various software titles on CD-ROM (e.g., Jack Prelutsky's poetry and two Magic School Bus adventures). David takes after his mother and reads (especially Calvin and Hobbes). Greg inherited his dad's fascination with games and puzzles (along with some of my more difficult personality traits) and is learning probability by playing Yahtzee.

At this point it is *de rigeur* (et peut être *de jure*) to thank my wife Susan for her "patience and understanding." This phrase may have meant different things in each of the four other prefaces in which it appears, but now, like the ivy growing on White Hall, I would be lost without her.

Rocking Finale

As usual, I would like think those who gave me muscial encouragement during the many hours I sat in front of my computer: Nirvana, Pearl Jam, Green Day, Candlebox, Counting Crows, Melissa Etheridge, Sheryl Crow, and especially Live for their *Throwing Copper* which has been on endless repeat during the final stages of the process.

A trip to Stockholm May 1–10, 1995 for a workshop on epidemic models organized by Peter Jagers, Anders Martin-Löf, and Ake Svensson, provided an important rest break because the final assault. It was also encouraging to visit a country where my book has been so enthusiastically received. Though Roxette, mentioned in the first edition, is gone from the line-up of rock stars, I hope the second edition of this book will inspire a new generation of graduate students to utter the only two words of Swedish that I know: "Stor Starköl."

Rick Durrett

Contents

3 Random Walks 173

4 Martingales 219

5 Markov Chains 277

6 Ergodic Theorems 335

7 Brownian Motion 374

Appendix: Measure Theory   440

References   484

Notation   494

Normal Table   497

Index   500

Introductory Lecture

As Breiman should have said in his preface: "Probability theory has a right and a left hand. On the left is the rigorous foundational work using the tools of measure theory. The right hand 'thinks probabilistically,' reduces problems to gambling situations, coin-tossing, and motions of a physical particle." We have interchanged Breiman's hands in the quote because we learned in a high school English class that the left hand is sinister and the right is dextrous. While measure theory does not "threaten harm, evil or misfortune," it is an unfortunate fact that we will need four sections of definitions before we come to the first interesting result. To motivate the reader for this necessary foundational work, we will now give some previews of coming attractions.

For a large part of the first two chapters, we will be concerned with the laws of large numbers and the central limit theorem. To introduce these theorems and to illustrate their use, we will begin by giving their interpretation for a person playing roulette. In doing this we will use some terms (e.g. independent, mean, variance) without explaining them. If some of the words that we use are unfamiliar, don't worry. There will be more than enough definitions when the time comes.

A roulette wheel has 38 slots – 18 red, 18 black, and 2 green ones that are numbered 0 and 00 – so if our gambler bets \$1 on red coming up he wins \$1 with probability 18/38 and loses \$1 with probability 20/38. Let $X_1, X_2, \ldots$ be the outcomes of the first, second, and subsequent bets. If the house and gambler are honest, $X_1, X_2, \ldots$ are independent random variables and each has the same distribution, namely $P(X_i = 1) = 9/19$ and $P(X_i = -1) = 10/19$. One of the first things we will have to do is to construct a probability space and define on it a sequence of independent random variables $X_1, X_2, \ldots$ with this distribution, but our friend the gambler doesn't care about this technicality. He wants to know what we can tell him about the amount he has won at time n: $S_n = X_1 + \cdots + X_n$.

The first facts we can tell him are that (i) the average amount of money he will win on one play (= the mean of X_1 and denoted EX_1) is

$$(9/19) \cdot \$1 + (10/19) \cdot (-\$1) = -\$1/19 = -\$.05263$$

and (ii) on the average after n ways his winnings will be $ES_n = nEX_1 = -\$n/19$. For most values of n the probability of having lost exactly $n/19$ dollars is zero, so the next question to be answered is: How close will his experience be to the average? The first answer is provided by

The weak law of large numbers If $X_1, X_2, \ldots$ are independent and identically distributed random variables with mean $EX_1 = \mu$ then for all $\epsilon > 0$

$$P(|S_n/n - \mu| > \epsilon) \to 0 \text{ as } n \to \infty$$

Less formally, if n is large S_n/n is close to μ with high probability.

This result provides some information but leaves several questions unanswered. The first one is: if our gambler was statistically minded and wrote down the values of S_n/n, would the resulting sequence of numbers converge to $-1/19$? The answer to this question is given by

The strong law of large numbers. If $X_1, X_2, \ldots$ are independent and identically distributed random variables with mean $EX_i = \mu$ then with probability one, S_n/n converges to μ.

An immediate consequence of the last result of interest to our gambler is that with probability one $S_n \to -\infty$ as $n \to \infty$. That is, the gambler will eventually go bankrupt no matter how much money he starts with.

The laws of large numbers tell us what happens in the long run but do not provide much information about what happens over the short run. That gap is filled by

The central limit theorem. If $X_1, X_2, \ldots$ are independent and identically distributed random variables with mean $EX_i = \mu$ and variance $\sigma^2 = E(X_i - \mu)^2$ then for any y

$$P\left(\frac{S_n - n\mu}{\sigma n^{1/2}} \le y\right) \to \mathcal{N}(y)$$

where $\mathcal{N}(y) = \int_{-\infty}^{y} (2\pi)^{-1/2} e^{-x^2/2} \, dx$ is the (standard) normal distribution.

If we let χ denote a random variable with a normal distribution then the last conclusion can be written informally as

$$S_n \approx n\mu + \sigma n^{1/2}\chi$$

In the example we have been considering $\mu = -1/19$ and

$$\sigma^2 = \frac{9}{19}(1 + 1/19)^2 + \frac{10}{19}(-1 + 1/19)^2 = 1 - (1/19)^2 = .9972$$

If we use $\sigma^2 \approx 1$ to simplify the arithmetic then the central limit theorem tells us

$$S_n \approx -n/19 + n^{1/2}\chi$$

or when $n = 100$

$$S_{100} \approx -5.26 + 10\chi$$

If we are interested in the probability $S_{100} \geq 0$, this is

$$P(-5.26 + 10\chi \geq 0) = P(\chi \geq .526) \approx .30$$

from the table of the normal distribution at the back of the book.

The last result shows that after 100 plays the negative drift is not too noticeable. The gambler has lost \$5.26 on the average and has a probability .3 of being ahead. To see why casinos make money suppose there are 100 gamblers playing 100 times and set $n = 10,000$ to get

$$S_{10,000} \approx -526 + 100\chi$$

Now $P(\chi \leq 2.3) = .99$ so with that probability $S_{10,000} \leq -296$, i.e., the casino is slowly but surely making money.

1 Laws of Large Numbers

In the first three sections we will recall some definitions and results from measure theory. Our purpose is not only to review that material but also to introduce the terminology of probability theory, which differs slightly from that of measure theory. In Section 1.4 we introduce the crucial concept of independence and explore its properties. In Section 1.5 we prove the weak law of large numbers and give several applications. In Sections 1.6 we prove some Borel-Cantelli lemmas to prepare for the proof of the strong law of large numbers in Section 1.7. In Section 1.8 we investigate the convergence of random series which leads to estimates on the rate of convergence in the law of large numbers. Finally, in Section 1.9 we show that in nice situations convergence in the weak law occurs exponentially rapidly.

1.1. Basic Definitions

Here, and throughout the book, terms being defined are set in **boldface**. We begin with the most basic quantity. A **probability space** is a triple $(\Omega, \mathcal{F}, P)$ where Ω is a set of "outcomes", $\mathcal{F}$ is a set of "events", and $P : \mathcal{F} \to [0, 1]$ is a function that assigns probabilities to events. We assume that $\mathcal{F}$ is a σ-field (or σ-**algebra**), i.e., a (nonempty) collection of subsets of Ω that satisfy

(i) if $A \in \mathcal{F}$ then $A^c \in \mathcal{F}$, and

(ii) if $A_i \in \mathcal{F}$ is a countable sequence of sets then $\cup_i A_i \in \mathcal{F}$.

Here and in what follows **countable** means finite or countably infinite. Since $\cap_i A_i = (\cup_i A_i^c)^c$, it follows that a σ-field is closed under countable intersections. We omit the last property from the definition to make it easier to check.

Without P, $(\Omega, \mathcal{F})$ is called a **measurable space**, i.e., it is a space on which we can put a measure. A **measure** is a nonnegative countably additive set function. That is, a function $\mu : \mathcal{F} \to \mathbf{R}$ with

(i) $\mu(A) \geq \mu(\emptyset) = 0$ for all $A \in \mathcal{F}$, and

(ii) if $A_i \in \mathcal{F}$ is a countable sequence of disjoint sets then

$$\mu(\cup_i A_i) = \sum_i \mu(A_i)$$

If $\mu(\Omega) = 1$ we call μ a **probability measure**. In this book, probability measures are usually denoted by P. The next exercise gives some consequences of the definition that we will need later. In all cases we assume that the sets we mention are in $\mathcal{F}$. For (i) one needs to know that $B - A = B \cap A^c$. For (iv) it is useful to note that (ii) of the definition with $A_1 = A$ and $A_2 = A^c$ implies $P(A^c) = 1 - P(A)$.

EXERCISE 1.1. Let P be a probability measure on $(\Omega, \mathcal{F})$

(i) **monotonicity.** If $A \subset B$ then $P(B) - P(A) = P(B - A) \geq 0$.

(ii) **subadditivity.** If $A_m \in \mathcal{F}$ for $m \geq 1$ and $A \subset \cup_{m=1}^{\infty} A_m$ then $P(A) \leq \sum_{m=1}^{\infty} P(A_m)$.

(iii) **continuity from below.** If $A_i \uparrow A$ (i.e., $A_1 \subset A_2 \subset \ldots$ and $\cup_i A_i = A$) then $P(A_i) \uparrow P(A)$.

(iv) **continuity from above.** If $A_i \downarrow A$ (i.e., $A_1 \supset A_2 \supset \ldots$ and $\cap_i A_i = A$) then $P(A_i) \downarrow P(A)$.

Some examples of probability measures should help to clarify the concept. We leave it to the reader to check that they are examples, i.e., $\mathcal{F}$ is a σ-field and P is a probability measure.

Example 1.1. Discrete probability spaces. Let $\Omega = $ a countable set, i.e., finite or countably infinite. Let $\mathcal{F} = $ the set of all subsets of Ω. Let

$$P(A) = \sum_{\omega \in A} p(\omega) \text{ where } p(\omega) \geq 0 \text{ and } \sum_{\omega \in \Omega} p(\omega) = 1.$$

A little thought reveals that this is the most general probability measure on this space. In many cases when Ω is a finite set, we have $p(\omega) = 1/|\Omega|$ where $|\Omega| = $ the number of points in Ω. Concrete examples in this category are:

a. flipping a fair coin: $\Omega = \{$ Heads, Tails $\}$

b. rolling a die: $\Omega = \{1, 2, 3, 4, 5, 6\}$

Example 1.2. Real line and unit interval. Let $\mathbf{R} = $ the real line, $\mathcal{R} = $ the **Borel sets** $ = $ the smallest σ-field containing the open sets, $\lambda = $ **Lebesgue measure** $ = $ the only measure on $\mathcal{R}$ with $\lambda((a, b]) = b - a$ for all $a < b$. The construction of Lebesgue measure is carried out in Section 1 of the Appendix. $\lambda(\mathbf{R}) = \infty$. To get a probability space let $\Omega = (0,1)$, $\mathcal{F} = \{A \cap (0,1) : A \in \mathcal{R}\}$ and $P(B) = \lambda(B)$ for $B \in \mathcal{F}$. P is Lebesgue measure restricted to the Borel subsets of $(0,1)$.

EXERCISE 1.2. (i) If $\mathcal{F}_i$, $i \in I$ are σ-fields then $\cap_{i \in I} \mathcal{F}_i$ is. Here $I \neq \emptyset$ is an arbitrary index set (i.e., possibly uncountable). (ii) Use the result in (i) to show if we are given a set Ω and a collection $\mathcal{A}$ of subsets of Ω then there is a smallest σ-field containing $\mathcal{A}$. We will call this the σ-**field generated by** $\mathcal{A}$ and denote it by $\sigma(\mathcal{A})$.

Example 1.3. Product spaces. If $(\Omega_i, \mathcal{F}_i, P_i)$ $i = 1, \ldots, n$ are probability spaces, we can let $\Omega = \Omega_1 \times \cdots \times \Omega_n = \{(\omega_1, \ldots, \omega_n) : \omega_i \in \Omega_i\}$. $\mathcal{F} = \mathcal{F}_1 \times \cdots \times \mathcal{F}_n =$ the σ-field generated by $\{A_1 \times \cdots \times A_n : A_i \in \mathcal{F}_i\}$. Let $P = P_1 \times \cdots \times P_n =$ the measure on $\mathcal{F}$ that has

$$P(A_1 \times \cdots \times A_n) = P_1(A_1) \cdot P_2(A_2) \cdots P_n(A_n)$$

For more details see Section 6 of the Appendix. Concrete examples of product spaces are:

a. Roll two dice. $\Omega = \{1,2,3,4,5,6\} \times \{1,2,3,4,5,6\}$, $\mathcal{F} =$ all subsets of Ω, $P(A) = |A|/36$.

b. Unit cube. If $\Omega_i = (0,1)$, $\mathcal{F}_i =$ the Borel sets, and $P_i =$Lebesgue measure, then the product space defined above is the unit cube $\Omega = (0,1)^n$, $\mathcal{F} =$ the Borel subsets of Ω, and P is n-dimensional Lebesgue measure restricted to $\mathcal{F}$.

EXERCISE 1.3. Let $\mathbf{R}^n = \{(x_1, \ldots, x_n) : x_i \in \mathbf{R}\}$. $\mathcal{R}^n =$ the Borel subsets of $\mathbf{R}^n$ is defined to be the σ-field generated by the open subsets of $\mathbf{R}^n$. Prove this is the same as $\mathcal{R} \times \cdots \times \mathcal{R} =$ the σ-field generated by sets of the form $A_1 \times \cdots \times A_n$. Hint: Show that both σ-fields coincide with the one generated by $(a_1, b_1) \times \cdots \times (a_n, b_n)$.

Probability spaces become a little more interesting when we define random variables on them. A real valued function X defined on Ω is said to be a **random variable** if for every Borel set $B \subset \mathbf{R}$ we have

$$X^{-1}(B) = \{\omega : X(\omega) \in B\} \in \mathcal{F}$$

When we need to emphasize the σ-field we will say that X is $\mathcal{F}$-**measurable** or write $X \in \mathcal{F}$. If Ω is a discrete probability space (see Example 1.1) then any function $X : \Omega \to \mathbf{R}$ is a random variable. A second trivial, but useful, type of example of a random variable is the **indicator function** of a set $A \in \mathcal{F}$:

$$1_A(\omega) = \begin{cases} 1 & \omega \in A \\ 0 & \omega \notin A \end{cases}$$

The notation is supposed to remind you that this function is 1 on A. Analysts call this object the characteristic function of A. In probability that term is used for something quite different. (See Section 2.3.)

If X is a random variable then X induces a probability measure on $\mathbf{R}$ called its **distribution** by setting $\mu(A) = P(X \in A)$ for Borel sets A. Using the notation introduced above, the right hand side can be written as $P(X^{-1}(A))$. In words we pull $A \in \mathcal{R}$ back to $X^{-1}(A) \in \mathcal{F}$ and then take P of that set. For a picture see Figure 1.1.1.

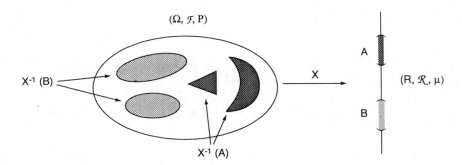

Figure 1.1.1

To check that μ is a probability measure we observe that if the A_i are disjoint then using the definition of μ; the fact that X lands in the union if and only if it lands in one of the A_i; if the sets $A_i \in \mathcal{R}$ are disjoint then the events $\{X \in A_i\}$ are disjoint; and the definition of μ again; we have:

$$\mu\left(\cup_i A_i\right) = P\left(X \in \cup_i A_i\right) = P\left(\cup_i \{X \in A_i\}\right) = \sum_i P(X \in A_i) = \sum_i \mu(A_i)$$

The distribution of a random variable X is usually described by giving its **distribution function**, $F(x) = P(X \leq x)$.

(1.1) Theorem. Any distribution function F has the following properties:

(i) F is nondecreasing

(ii) $\lim_{x \to \infty} F(x) = 1$, $\lim_{x \to -\infty} F(x) = 0$

(iii) F is right continuous, i.e. $\lim_{y \downarrow x} F(y) = F(x)$

(iv) If $F(x-) = \lim_{y \uparrow x} F(y)$ then $F(x-) = P(X < x)$

(v) $P(X = x) = F(x) - F(x-)$

Proof   To prove (i) note that if $x \le y$ then $\{X \le x\} \subset \{X \le y\}$ and then use (i) in Exercise 1.1 to conclude that $P(X \le x) \le P(X \le y)$.

To prove (ii) we observe that if $x \uparrow \infty$ then $\{X \le x\} \uparrow \Omega$ while if $x \downarrow -\infty$ then $\{X \le x\} \downarrow \emptyset$ and then use (iii) and (iv) of Exercise 1.1.

To prove (iii) we observe that if $y \downarrow x$ then $\{X \le y\} \downarrow \{X \le x\}$.

To prove (iv) we observe that if $y \uparrow x$ then $\{X \le y\} \uparrow \{X < x\}$.

For (v) note $P(X = x) = P(X \le x) - P(X < x)$ and use (iii) and (iv). $\square$

The next result shows that we have found more than enough properties to characterize distribution functions.

(1.2) Theorem. If F satisfies (i), (ii), and (iii) in (1.1) then it is the distribution function of some random variable.

Proof   Let $\Omega = (0,1)$, $\mathcal{F} = $ the Borel sets, and $P = $ Lebesgue measure. If $\omega \in (0,1)$ let

$$X(\omega) = \sup\{y : F(y) < \omega\}$$

Once we show that

(⋆) $$\{\omega : X(\omega) \le x\} = \{\omega : \omega \le F(x)\}$$

the desired result follows immediately since $P(\omega : \omega \le F(x)) = F(x)$. (Recall P is Lebesgue measure.) To check (⋆), we observe that if $\omega \le F(x)$ then $X(\omega) \le x$, since $x \notin \{y : F(y) < \omega\}$. On the other hand if $\omega > F(x)$, then since F is right continuous, there is an $\epsilon > 0$ so that $F(x + \epsilon) < \omega$ and $X(\omega) \ge x + \epsilon > x$. $\square$

Remark. To make the reader appreciate the care that went into this definition we note that there are four sensible combinations of sup, inf and inequality to try: $(\sup, <)$, $(\sup, \le)$, $(\inf, >)$, and $(\inf, \ge)$ but only the one chosen will cope correctly with the two trouble spots on F: discontinuities and intervals on which F is constant. (See Figure 1.1.2.)

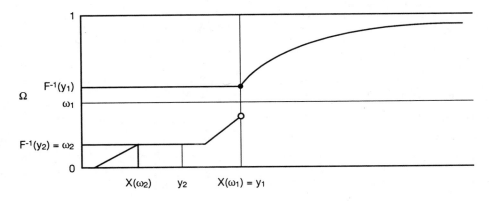

Figure 1.1.2

Even though F may not be 1-1 and onto we will call X the inverse of F and denote it by F^{-1}. The scheme in the proof of (1.2) is useful in generating random variables on a computer. Standard algorithms generate random variables U with a uniform distribution, then one applies the inverse of the distribution function defined in (1.2) to get a random variable $F^{-1}(U)$ with distribution function F.

An immediate consequence of (1.2) is

(1.3) Corollary. If F satisfies (i), (ii), and (iii) in (1.1) there is a unique probability measure μ on $(\mathbf{R}, \mathcal{R})$ that has $\mu((a, b]) = F(b) - F(a)$ for all a, b.

Proof   (1.2) gives the existence of a random variable X with distribution function F. The measure it induces on $(\mathbf{R}, \mathcal{R})$ is the desired μ. There is only one measure associated with a given F because the sets $(a, b]$ are closed under intersection and generate the σ-field. (See (2.2) in the Appendix.)   □

If X and Y induce the same distribution μ on $(\mathbf{R}, \mathcal{R})$ we say X and Y are **equal in distribution**. In view of (1.3) this holds if and only if X and Y have the same distribution function, i.e., $P(X \leq x) = P(Y \leq x)$ for all x. When X and Y have the same distribution, we like to write

$$X \overset{d}{=} Y$$

but this is too tall to use in text, so for typographical reasons we will also use $X =_d Y$.

When the distribution function $F(x) = P(X \leq x)$ has the form

$$(*) \qquad\qquad F(x) = \int_{-\infty}^{x} f(y) \, dy$$

we say that X has **density function** f. In remembering formulas it is often useful to think of $f(x)$ as being $P(X = x)$ although

$$P(X = x) = \lim_{\epsilon \to 0} \int_{x-\epsilon}^{x+\epsilon} f(y) \, dy = 0$$

We can start with f and use $(*)$ to define F. In order to end up with a distribution function it is necessary and sufficient that $f(x) \geq 0$ and $\int f(x) \, dx = 1$. Three examples that will be important in what follows are:

Example 1.4. Uniform distribution on (0,1). $f(x) = 1$ for $x \in (0, 1)$, 0 otherwise. Distribution function

$$F(x) = \begin{cases} 0 & x \leq 0 \\ x & 0 \leq x \leq 1 \\ 1 & x > 1 \end{cases}$$

Example 1.5. Exponential distribution. $f(x) = e^{-x}$ for $x \geq 0$, 0 otherwise. Distribution function

$$F(x) = \begin{cases} 0 & x \leq 0 \\ 1 - e^{-x} & x \geq 0 \end{cases}$$

Example 1.6. Standard normal distribution.

$$f(x) = (2\pi)^{-1/2} \exp(-x^2/2)$$

In this case there is no closed form expression but we have the following bounds that are useful for large x:

(1.4) Theorem. For $x > 0$,

$$(x^{-1} - x^{-3}) \exp(-x^2/2) \leq \int_{x}^{\infty} \exp(-y^2/2) dy \leq x^{-1} \exp(-x^2/2)$$

Proof   Changing variables $y = x + z$ and using $\exp(-z^2/2) \leq 1$ gives

$$\int_{x}^{\infty} \exp(-y^2/2) \, dy \leq \exp(-x^2/2) \int_{0}^{\infty} \exp(-xz) \, dz = x^{-1} \exp(-x^2/2)$$

For the other direction we observe

$$\int_x^\infty (1 - 3y^{-4}) \exp(-y^2/2)\, dy = (x^{-1} - x^{-3}) \exp(-x^2/2) \qquad \square$$

A distribution function on **R** is said to be **absolutely continuous** if it has a density and **singular** if the corresponding measure is singular w.r.t. Lebesgue measure. See Section 8 of the Appendix for more on these notions. An example of a singular distribution is:

Example 1.7. Uniform distribution on the Cantor set. The Cantor set C is defined by removing $(1/3, 2/3)$ from $[0,1]$ and then removing the middle third of each interval that remains. We define an associated distribution function by setting $F(x) = 0$ for $x \leq 0$, $F(x) = 1$ for $x \geq 1$, $F(x) = 1/2$ for $x \in [1/3, 2/3]$, $F(x) = 1/4$ for $x \in [1/9, 2/9]$, $F(x) = 3/4$ for $x \in [7/9, 8/9]$, ... The function F that results is called **Lebesgue's singular function** because there is no f for which $(*)$ holds. From the definition it is immediate that the corresponding measure has $\mu(C^c) = 0$.

A probability measure P (or its associated distribution function) is said to be **discrete** if there is a countable set S with $P(S^c) = 0$. The simplest example of a discrete distribution is

Example 1.8. Pointmass at 0. $F(x) = 1$ for $x \geq 0$, $F(x) = 0$ for $x < 0$.

The next example shows that the distribution function associated with a discrete probability measure can be quite wild.

Example 1.9. Dense discontinuities. Let q_1, q_2, ... be an enumeration of the rationals and let

$$F(x) = \sum_{i=1}^\infty 2^{-i} 1_{[q_i, \infty)}$$

where $1_{[\theta, \infty)}(x) = 1$ if $x \in [\theta, \infty)$, $= 0$ otherwise.

EXERCISES

1.4. Let $\Omega = \mathbf{R}$, $\mathcal{F} = $ all subsets so that A or A^c is countable, $P(A) = 0$ in the first case and $= 1$ in the second. Show that $(\Omega, \mathcal{F}, P)$ is a probability space.

1.5. A σ-field $\mathcal{F}$ is said to be **countably generated** if there is a countable collection $\mathcal{C} \subset \mathcal{F}$ so that $\sigma(\mathcal{C}) = \mathcal{F}$. Show that $\mathcal{R}^d$ is countably generated.

1.6. Suppose X and Y are random variables on $(\Omega, \mathcal{F}, P)$ and let $A \in \mathcal{F}$. Show that if we let $Z(\omega) = X(\omega)$ for $\omega \in A$ and $Z(\omega) = Y(\omega)$ for $\omega \in A^c$, then Z is a random variable.

1.7. Let χ have the standard normal distribution. Use (1.4) to get upper and lower bounds on $P(\chi \geq 4)$.

1.8. Show that a distribution function has at most countably many discontinuities.

1.9. Show that if $F(x) = P(X \leq x)$ is continuous then $Y = F(X)$ has a uniform distribution on (0,1). That is, if $y \in [0, 1]$, $P(Y \leq y) = y$.

1.10. Suppose X has density f, $P(\alpha \leq X \leq \beta) = 1$ and g is a function that is increasing and differentiable on (α, β). Then $g(X)$ has density $f(g^{-1}(x))/g'(g^{-1}(x))$ for $x \in (g(\alpha), g(\beta))$ and 0 otherwise. When $g(x) = ax + b$ with $a > 0$ the answer is $f((y - b)/a)/a$.

1.11. Suppose X has a normal distribution. Use the previous exercise to compute the density of $\exp(X)$. (The answer is called the **lognormal distribution**.)

1.12. (i) Suppose X has density function f. Compute the distribution function of X^2 and then differentiate to find its density function. (ii) Work out the answer when X has a standard normal distribution to find the density of the **chi-square distribution**.

1.2. Random Variables

In this section we will develop some results that will help us later to prove that quantities we define are random variables, i.e., they are measurable. Since most of what we have to say is true for random elements of an arbitrary measurable space $(S, \mathcal{S})$, and the proofs are the same (sometimes easier), we will develop our results in that generality. First we need a definition. A function $X : \Omega \to S$ is said to be a **measurable map** from $(\Omega, \mathcal{F})$ to $(S, \mathcal{S})$ if

$$X^{-1}(B) \equiv \{\omega : X(\omega) \in B\} \in \mathcal{F} \quad \text{for all } B \in \mathcal{S}.$$

If $(S, \mathcal{S}) = (\mathbf{R}^d, \mathcal{R}^d)$ and $d > 1$ then X is called a **random vector**. Of course, if $d = 1$, X is called a random variable.

The next result is useful for proving that maps are measurable.

(2.1) **Theorem.** If $\{\omega : X(\omega) \in A\} \in \mathcal{F}$ for all $A \in \mathcal{A}$ and $\mathcal{A}$ **generates** $\mathcal{S}$ (i.e., $\mathcal{S}$ is the smallest σ-field that contains $\mathcal{A}$), then X is measurable.

Proof   Writing $\{X \in B\}$ as shorthand for $\{\omega : X(\omega) \in B\}$, we have

$$\{X \in \cup_i B_i\} = \cup_i \{X \in B_i\}$$
$$\{X \in B^c\} = \{X \in B\}^c$$

So the class of sets $\mathcal{B} = \{B : \{X \in B\} \in \mathcal{F}\}$ is a σ-field. Since $\mathcal{B} \supset \mathcal{A}$ and $\mathcal{A}$ generates $\mathcal{S}$, $\mathcal{B} \supset \mathcal{S}$.   □

It follows from the two equations displayed in the previous proof that if $\mathcal{S}$ is a σ-field then $\{\{X \in B\} : B \in \mathcal{S}\}$ is a σ-field. It is the smallest σ-field on Ω that makes X a measurable map. It is called the **σ-field generated by** X and denoted $\sigma(X)$.

EXERCISE 2.1. Show that if $\mathcal{A}$ generates $\mathcal{S}$ then $X^{-1}(\mathcal{A}) = \{\{X \in A\} : A \in \mathcal{A}\}$ generates $\sigma(X) = \{\{X \in B\} : B \in \mathcal{S}\}$.

Example 2.1.   If $(S, \mathcal{S}) = (\mathbf{R}, \mathcal{R})$ then possible choices of $\mathcal{A}$ in (2.1) are $\{(-\infty, x] : x \in \mathbf{R}\}$ or $\{(-\infty, x) : x \in \mathbf{Q}\}$ where $\mathbf{Q} =$ the rationals.

Example 2.2.   If $(S, \mathcal{S}) = (\mathbf{R}^d, \mathcal{R}^d)$ a useful choice of $\mathcal{A}$ is

$$\{(a_1, b_1) \times \cdots \times (a_d, b_d) : -\infty < a_i < b_i < \infty\}$$

or occasionally the larger collection of open sets.

(2.2) **Theorem.** If $X : (\Omega, \mathcal{F}) \to (S, \mathcal{S})$ and $f : (S, \mathcal{S}) \to (T, \mathcal{T})$ are measurable maps then $f(X)$ is a measurable map from $(\Omega, \mathcal{F})$ to $(T, \mathcal{T})$.

Proof   See Figure 1.2.1. Let $B \in \mathcal{T}$. $\{\omega : f(X(\omega)) \in B\} = \{\omega : X(\omega) \in f^{-1}(B)\} \in \mathcal{F}$, since by assumption $f^{-1}(B) \in \mathcal{S}$.   □

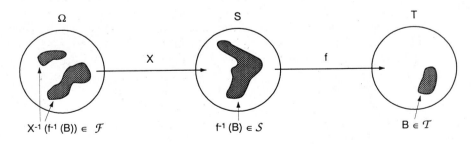

Figure 1.2.1

From (2.2) it follows immediately that if X is a random variable then so is cX for all $c \in \mathbf{R}$, X^2, $\sin(X)$, etc. The next result shows why we wanted to prove (2.2) for measurable maps.

(2.3) Theorem. If $X_1, \ldots, X_n$ are random variables and $f : \mathbf{R}^n \to \mathbf{R}$ is measurable then $f(X_1, \ldots, X_n)$ is a random variable.

Proof In view of (2.2), it suffices to show that $(X_1, \ldots, X_n)$ is a random vector. To do this we observe that if $A_1, \ldots, A_n$ are Borel sets then

$$\{(X_1, \ldots, X_n) \in A_1 \times \cdots \times A_n\} = \cap_i \{X_i \in A_i\} \in \mathcal{F}.$$

Since sets of the form $A_1 \times \cdots \times A_n$ generate $\mathcal{R}^n$, (2.3) follows from (2.1).   □

(2.4) Corollary. If $X_1, \ldots, X_n$ are random variables then $X_1 + \ldots + X_n$ is a random variable.

Proof In view of (2.3) it suffices to show that $f(x_1, \ldots, x_n) = x_1 + \ldots + x_n$ is measurable. To do this we use Example 2.1 and note that $\{x : x_1 + \ldots + x_n < a\}$ is an open set and hence is in $\mathcal{R}^n$.   □

By combining (2.4) with a remark after (2.2) we see that if X and Y are random variables then $X - Y$ is. To get a feeling for the barehands approach to proving measurability, try

EXERCISE 2.2. Prove (2.4) when $n = 2$ by checking $\{X_1 + X_2 < x\} \in \mathcal{F}$.

(2.5) Theorem. If $X_1, X_2, \ldots$ are random variables then so are

$$\inf_n X_n \qquad \sup_n X_n \qquad \limsup_n X_n \qquad \liminf_n X_n$$

Proof Since the infimum of a sequence is $< a$ if and only if some term is $< a$ (if all terms are $\geq a$ then the infimum is), we have

$$\{\inf_n X_n < a\} = \cup_n \{X_n < a\} \in \mathcal{F}$$

A similar argument shows $\{\sup_n X_n > a\} = \cup_n \{X_n > a\} \in \mathcal{F}$. For the last two we observe

$$\liminf_{n \to \infty} X_n = \sup_n \left(\inf_{m \geq n} X_m \right)$$

$$\limsup_{n \to \infty} X_n = \inf_n \left(\sup_{m \geq n} X_m \right)$$

To complete the proof in the first case note that $Y_n = \inf_{m \geq n} X_m$ is a random variable for each n so $\sup_n Y_n$ is as well. □

From (2.5) we see that

$$\Omega_o \equiv \{\omega : \lim_{n \to \infty} X_n \text{ exists }\} = \{\omega : \limsup_{n \to \infty} X_n - \liminf_{n \to \infty} X_n = 0\}$$

is a measurable set. (Here $\equiv$ indicates that the first equality is a definition.) If $P(\Omega_o) = 1$ we say that X_n **converges almost surely** (a type of convergence called almost everywhere in measure theory). To have a limit defined on the whole space it is convenient to let

$$X_\infty = \limsup_{n \to \infty} X_n$$

but this random variable may take the value $+\infty$. To accomodate this and some other headaches, we will generalize the definition of random variable.

A function whose domain is a set $D \in \mathcal{F}$ and whose range is $\mathbf{R}^* \equiv [-\infty, \infty]$ is said to be a **random variable** if for all $B \in \mathcal{R}^*$ we have $X^{-1}(B) = \{\omega : X(\omega) \in B\} \in \mathcal{F}$. Here $\mathcal{R}^* =$ the Borel subsets of $\mathbf{R}^*$ with $\mathbf{R}^*$ given the usual topology, i.e., the one generated by intervals of the form $[-\infty, a)$, (a, b) and $(b, \infty]$ where $a, b \in \mathbf{R}$. The reader should note that the **extended real line** $(\mathbf{R}^*, \mathcal{R}^*)$ is a measurable space, so all the results above generalize immediately.

EXERCISES

2.3. Show that if f is continuous and $X_n \to X$ almost surely then $f(X_n) \to f(X)$ almost surely.

2.4. (i) Show that a continuous function from $\mathbf{R}^d \to \mathbf{R}$ is a measurable map from $(\mathbf{R}^d, \mathcal{R}^d)$ to $(\mathbf{R}, \mathcal{R})$. (ii) Show that $\mathcal{R}^d$ is the smallest σ-field that makes all the continuous functions measurable.

2.5. A function f is said to be **lower semicontinuous** or l.s.c. if

$$\liminf_{y \to x} f(y) \geq f(x)$$

and **upper semicontinuous** (u.s.c.) if $-f$ is l.s.c. Show that f is l.s.c. if and only if $\{x : f(x) \leq a\}$ is closed for each $a \in \mathbf{R}$ and conclude that semicontinuous functions are measurable.

2.6. Let $f : \mathbf{R}^d \to \mathbf{R}$ be an arbitrary function and let $f^\delta(x) = \sup\{f(y) : |y - x| < \delta\}$ and $f_\delta(x) = \inf\{f(y) : |y - x| < \delta\}$ where $|z| = (z_1^2 + \ldots + z_d^2)^{1/2}$. Show that f^δ is l.s.c. and f_δ is u.s.c. Let $f^0 = \lim_{\delta \downarrow 0} f^\delta$, $f_0 = \lim_{\delta \downarrow 0} f_\delta$, and conclude that the set of points at which f is discontinuous $= \{f^0 \neq f_0\}$ is measurable.

2.7. A function $\varphi : \Omega \to \mathbf{R}$ is said to be **simple** if

$$\varphi(\omega) = \sum_{m=1}^{n} c_m 1_{A_m}(\omega)$$

where the c_m are real numbers and $A_m \in \mathcal{F}$. Show that the class of $\mathcal{F}$ measurable functions is the smallest class containing the simple functions and closed under pointwise limits.

2.8. Use the previous exercise to conclude that Y is measurable with respect to $\sigma(X)$ if and only if $Y = f(X)$ where $f : \mathbf{R} \to \mathbf{R}$ is measurable.

2.9. To get a constructive proof of the last result, note that $\{\omega : m2^{-n} \leq Y < (m+1)2^{-n}\} = \{X \in B_{m,n}\}$ for some $B_{m,n} \in \mathcal{R}$ and set $f_n(x) = m2^{-n}$ for $x \in B_{m,n}$ and show that as $n \to \infty$ $f_n(x) \to f(x)$ and $Y = f(X)$.

1.3. Expected Value

If $X \geq 0$ is a random variable on $(\Omega, \mathcal{F}, P)$ then we define its **expected value** to be $EX = \int X \, dP$, which always makes sense, but may be ∞. (The integral is defined in Section 4 of the Appendix.) To reduce the general case to the nonnegative case, let $x^+ = \max\{x, 0\}$ be the **positive part** and let $x^- = \max\{-x, 0\}$ be the **negative part** of x. We declare that EX **exists** and set $EX = EX^+ - EX^-$ whenever the subtraction makes sense, i.e., $EX^+ < \infty$ or $EX^- < \infty$.

EX is often called the **mean** of X and denoted by μ. EX is defined by integrating X, so it has all the properties that integrals do. From (4.5) and (4.7) in the Appendix and the trivial observation that $E(b) = b$ for any real number b, we get the following:

Theorem. Suppose $X, Y \geq 0$ or $E|X|, E|Y| < \infty$

(3.1a) $E(X + Y) = EX + EY$

(3.1b) $E(aX + b) = aE(X) + b$ for any real numbers a, b.

(3.1c) If $X \geq Y$ then $EX \geq EY$.

EXERCISE 3.1. Suppose $E|X|, E|Y| < \infty$. Show that equality holds in (3.1c) if and only if $X = Y$ a.s. Hint: use (3.4) below.

EXERCISE 3.2. Suppose only that EX and EY exist. Show that (3.1c) always holds; (3.1a) holds unless one expected value is ∞ and the other is $-\infty$; (3.1b) holds unless $a = 0$ and EX is infinite.

In this section we will recall some properties of expected value and prove some new ones. To organize things we will divide the developments into three sub-sections.

a. Inequalities

Our first two results are (5.2) and (5.3) from the Appendix.

(3.2) **Jensen's inequality.** Suppose φ is **convex**, that is,

$$\lambda\varphi(x) + (1 - \lambda)\varphi(y) \geq \varphi(\lambda x + (1 - \lambda)y)$$

for all $\lambda \in (0,1)$ and $x, y \in \mathbf{R}$. Then

$$E(\varphi(X)) \geq \varphi(EX)$$

provided both expectations exist, i.e., $E|X|$ and $E|\varphi(X)| < \infty$.

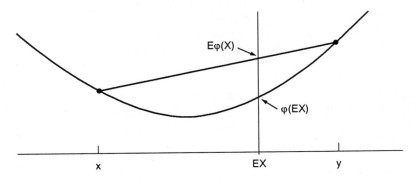

Figure 1.3.1

To recall the direction in which the inequality goes consider the special case in which $P(X = x) = \lambda$, $P(X = y) = 1 - \lambda$, and look at Figure 1.3.1.

EXERCISE 3.3. Suppose φ is strictly convex, i.e., $>$ holds for $\lambda \in (0,1)$. Show that, under the assumptions of (3.2), $\varphi(EX) = E\varphi(X)$ implies $X = EX$ a.s.

EXERCISE 3.4. Suppose $\varphi : \mathbf{R}^n \to \mathbf{R}$ is convex. Imitate the proof of (5.2) in the Appendix to show

$$E\varphi(X_1, \ldots, X_n) \geq \varphi(EX_1, \ldots, EX_n)$$

provided $E|\varphi(X_1, \ldots, X_n)| < \infty$ and $E|X_i| < \infty$ for all i.

(3.3) Hölder's inequality. If $p, q \in [1, \infty]$ with $1/p + 1/q = 1$ then

$$E|XY| \leq \|X\|_p \|Y\|_q$$

Here $\|X\|_r = (E|X|^r)^{1/r}$ for $r \in [1, \infty)$; $\|X\|_\infty = \inf\{M : P(|X| > M) = 0\}$. The special case $p = q = 2$ is called the **Cauchy-Schwarz inequality**:

$$E|XY| \leq \left(EX^2 EY^2\right)^{1/2}$$

To state our next result we need some notation. If we only integrate over $A \subset \Omega$ we write

$$E(X; A) = \int_A X\, dP$$

(3.4) Chebyshev's inequality. Suppose $\varphi : \mathbf{R} \to \mathbf{R}$ has $\varphi \geq 0$, let $A \in \mathcal{R}$ and let $i_A = \inf\{\varphi(y) : y \in A\}$.

$$i_A P(X \in A) \leq E(\varphi(X); X \in A) \leq E\varphi(X)$$

Proof The definition of i_A and the fact that $\varphi \geq 0$ imply that

$$i_A 1_{(X \in A)} \leq \varphi(X) 1_{(X \in A)} \leq \varphi(X)$$

So taking expected values and using (3.1.c) gives the desired result.

Remark. Some authors call (3.4) **Markov's inequality** and use the name Chebyshev's inequality for the special case $\varphi(x) = x^2$, $A = \{x : |x| \geq a\}$

$$(*) \qquad\qquad a^2 P(|X| \geq a) \leq EX^2$$

Our next four exercises are concerned with how good $(*)$ is and with complements and converses. These constitute a digression from the main story and can be skipped without much loss.

EXERCISE 3.5. $(*)$ **is and is not sharp.** (i) Show that $(*)$ is sharp by showing that if $0 < b \leq a$ are fixed there is an X with $EX^2 = b^2$ for which equality holds. (ii) Show that (∗) is not sharp by showing that if X has $0 < EX^2 < \infty$ then

$$\lim_{a \to \infty} a^2 P(|X| \geq a)/EX^2 = 0$$

EXERCISE 3.6. **One sided Chebyshev bound.** (i) Let $0 < p < 1$, and let X have $P(X = a) = p$ and $P(X = -b) = 1-p$. Apply (3.4) to $\varphi(x) = (x+b)^2$ and conclude that if Y is any random variable with $EY = EX$ and var$(Y) = $ var(X) then $P(Y \geq a) \leq p$ and equality holds when $Y = X$.
(ii) Suppose $EY = 0$, var$(Y) = \sigma^2$, and $a > 0$. Show that $P(Y \geq a) \leq \sigma^2/(a^2 + \sigma^2)$ and there is a Y for which equality holds.

EXERCISE 3.7. **Two nonexistent lower bounds.**
Show that: (i) if $\epsilon > 0$, $\inf\{P(|X| > \epsilon) : EX = 0, \text{var}(X) = 1\} = 0$.
(ii) if $y \geq 1$, $\sigma^2 \in (0,\infty)$, $\inf\{P(|X| > y) : EX = 1, \text{var}(X) = \sigma^2\} = 0$.

EXERCISE 3.8. **A useful lower bound.** Let $Y \geq 0$ with $EY^2 < \infty$ and let $a < EY$. Apply the Cauchy-Schwarz inequality to $Y1_{(Y>a)}$ and conclude

$$P(Y > a) \geq (EY - a)^2/EY^2$$

This is often applied with $a = 0$.

b. Integration to the limit

There are three classic real analysis results, (5.5)–(5.7) in the Appendix, about what happens when we interchange limits and integrals.

(3.5) **Fatou's lemma.** If $X_n \geq 0$ then $\liminf_{n\to\infty} EX_n \geq E(\liminf_{n\to\infty} X_n)$.

To recall the direction of the inequality think of the special case $X_n = n1_{(0,1/n)}$ (on the unit interval equipped with the Borel sets and Lebesgue measure). Here $X_n \to 0$ a.s. but $EX_n = 1$ for all n.

(3.6) **Monotone convergence theorem.** If $0 \leq X_n \uparrow X$ then $EX_n \uparrow EX$.

This follows immediately from (3.5) since $X_n \uparrow X$ and (1.3c) imply

$$\limsup_{n\to\infty} EX_n \leq EX$$

(3.7) **Dominated convergence theorem.** If $X_n \to X$ a.s., $|X_n| \leq Y$ for all n, and $EY < \infty$, then $EX_n \to EX$.

The special case of (3.7) in which Y is constant is called the **bounded convergence theorem**.

In the developments below we will need another result on integration to the limit. Perhaps the most important special case of this result occurs when $g(x) = |x|^p$ with $p > 1$ and $h(x) = x$.

(3.8) Theorem. Suppose $X_n \to X$ a.s. and there are continuous functions $g, h \geq 0$ with $g(x) > 0$ for large x and $|h(x)|/g(x) \to 0$ as $|x| \to \infty$ and $Eg(X_n) \leq K < \infty$ for all n. Then $Eh(X_n) \to Eh(X)$.

Proof Pick M large and so that $P(|X| = M) = 0$. Let $\bar{X}_n = X_n 1_{(|X_n| \leq M)}$. Since $P(|X| = M) = 0$, $\bar{X}_n \to \bar{X}$ a.s. Since $h(\bar{X}_n)$ is bounded, it follows from the bounded convergence theorem that

(a)
$$Eh(\bar{X}_n) \to Eh(\bar{X})$$

To control the effect of the truncation we note that

(b)
$$E|h(\bar{Y}) - h(Y)| \leq E(|h(Y)|; Y > M) \leq \epsilon_M Eg(Y)$$

where $\epsilon_M = \sup\{|h(x)|/g(x) : |x| \geq M\}$
Taking $Y = X_n$ in (b), it follows that

(c)
$$|Eh(\bar{X}_n) - Eh(X_n)| \leq K\epsilon_M$$

To estimate $|Eh(\bar{X}) - Eh(X)|$, we observe since g is bounded below, Fatou's lemma implies
$$Eg(X) \leq \liminf_{n \to \infty} Eg(X_n) \leq K$$

Taking $Y = X$ in (b) gives

(d)
$$|Eh(\bar{X}) - Eh(X)| \leq K\epsilon_M$$

The triangle inequality implies

$$|Eh(X_n) - Eh(X)| \leq |Eh(X_n) - Eh(\bar{X}_n)|$$
$$+ |Eh(\bar{X}_n) - Eh(\bar{X})| + |Eh(\bar{X}) - Eh(X)|$$

Taking limits and using (a), (c), (d) we have

$$\limsup_{n \to \infty} |Eh(X_n) - Eh(X)| \leq 2K\epsilon_M$$

which proves the desired result since $K < \infty$ and $\epsilon_M \to 0$ as $M \to \infty$.   □

A simple example shows that (3.8) can sometimes be applied when (3.7) cannot.

EXERCISE 3.9. Let $\Omega = (0,1)$ equipped with the Borel sets and Lebesgue measure. Let $\alpha \in (1,2)$ and $X_n = n^\alpha 1_{(1/(n+1),1/n)} \to 0$ a.s. Show that (3.8) can be applied with $h(x) = x$ and $g(x) = |x|^{2/\alpha}$ but the X_n are not dominated by an integrable function.

c. Computing expected values

Integrating over $(\Omega, \mathcal{F}, P)$ is nice in theory, but to do computations we have to shift to a space on which we can do calculus. In most cases we will apply the next result with $S = \mathbf{R}^d$.

(3.9) Change of variables formula. Let X be a random element of $(S, \mathcal{S})$ with distribution μ, i.e., $\mu(A) = P(X \in A)$. If f is a measurable function from $(S, \mathcal{S})$ to $(\mathbf{R}, \mathcal{R})$ so that $f \geq 0$ or $E|f(X)| < \infty$ then

$$Ef(X) = \int_S f(y)\, \mu(dy)$$

Remark. To explain the name, write h for X and $P \circ h^{-1}$ for μ to get

$$\int_\Omega f(h(\omega))\, dP = \int_S f(y)\, d(P \circ h^{-1}).$$

Proof   We will prove this result by verifying it in four increasingly more general special cases. The reader should note the method employed, since it will be used several times below.

CASE 1: INDICATOR FUNCTIONS. If $B \in \mathcal{S}$ and $f = 1_B$ then recalling the relevant definitions shows

$$E1_B(X) = P(X \in B) = \mu(B) = \int_S 1_B(y)\, \mu(dy)$$

CASE 2: SIMPLE FUNCTIONS. Let $f(x) = \sum_{m=1}^n c_m 1_{B_m}$ where $c_m \in \mathbf{R}$, $B_m \in \mathcal{S}$. The linearity of expected value, the result of Case 1, and the linearity of integration imply

$$Ef(X) = \sum_{m=1}^n c_m E1_{B_m}(X)$$

$$= \sum_{m=1}^n c_m \int_S 1_{B_m}(y)\, \mu(dy) = \int_S f(y)\, \mu(dy)$$

CASE 3: NONNEGATIVE FUNCTIONS. Now if $f \geq 0$ and we let

$$f_n(x) = ([2^n f(x)]/2^n) \wedge n$$

where $[x]$ = the largest integer $\leq x$ and $a \wedge b = \min\{a, b\}$, then the f_n are simple and $f_n \uparrow f$, so using the result for simple functions and the monotone convergence theorem

$$Ef(X) = \lim_n Ef_n(X) = \lim_n \int_S f_n(y)\, \mu(dy) = \int_S f(y)\, \mu(dy)$$

CASE 4: INTEGRABLE FUNCTIONS. The general case now follows by writing $f(x) = f(x)^+ - f(x)^-$. The condition $E|f(X)| < \infty$ guarantees that $Ef(X)^+$ and $Ef(X)^-$ are finite. So using the result for nonnegative functions and linearity of expected value and integration

$$Ef(X) = Ef(X)^+ - Ef(X)^- = \int_S f(y)^+ \mu(dy) - \int_S f(y)^- \mu(dy)$$

$$= \int_S f(y)\, \mu(dy) \qquad\qquad \square$$

For practice with the proof technique of (3.9) do

EXERCISE 3.10. Suppose that the probability measure μ has $\mu(A) = \int_A f(x)\, dx$ for all $A \in \mathcal{R}$. Then for any g with $g \geq 0$ or $\int |g(x)|\, \mu(dx) < \infty$ we have

$$\int g(x)\, \mu(dx) = \int g(x) f(x)\, dx$$

A consequence of (3.9) is that we can compute expected values of functions of random variables by performing integrals on the real line. Before we can do some examples we need to introduce the terminology for what we are about to compute. If k is a positive integer then EX^k is called the **kth moment** of X. The first moment EX is usually called the **mean** and denoted by μ. If $EX^2 < \infty$ then the **variance** of X is defined to be $\mathrm{var}(X) = E(X - \mu)^2$. To compute the variance the following formula is useful

(3.10a) $$\mathrm{var}(X) = E(X - \mu)^2 = EX^2 - 2\mu EX + \mu^2 = EX^2 - \mu^2$$

From this it is immediate that

(3.10b) $$\mathrm{var}(X) \leq EX^2$$

Here EX^2 is the expected value of X^2. When we want the square of EX we will write $(EX)^2$. Since $E(aX + b) = aEX + b$ by (3.1b), it follows easily from the definition that

$$(3.10c) \qquad \begin{aligned} \text{var}(aX + b) &= E(aX + b - E(aX + b))^2 \\ &= a^2 E(X - EX)^2 = a^2 \text{var}(X) \end{aligned}$$

We turn now to concrete examples and leave the calculus in the first two examples to the reader. (Integrate by parts.)

Example 3.1. If X has an **exponential distribution** then

$$EX^k = \int_0^\infty x^k e^{-x} dx = k!$$

So the mean of X is 1 and the variance is $EX^2 - (EX)^2 = 2 - 1^2 = 1$. If we let $Y = X/\lambda$ then by Exercise 1.10, Y has density $\lambda e^{-\lambda y}$ for $y \geq 0$, the **exponential density** with parameter λ. From (3.1b) and (3.10c) it follows that Y has mean $1/\lambda$ and variance $1/\lambda^2$.

Example 3.2. If X has a standard normal distribution,

$$EX = \int x(2\pi)^{-1/2} \exp(-x^2/2) \, dx = 0 \quad \text{(by symmetry)}$$

$$\text{var}(X) = EX^2 = \int x^2 (2\pi)^{-1/2} \exp(-x^2/2) \, dx = 1$$

If we let $\sigma > 0$, $\mu \in \mathbf{R}$, and $Y = \sigma X + \mu$ then (3.1b) and (3.10c) imply $EY = \mu$ and $\text{var}(Y) = \sigma^2$. By Exercise 1.10, Y has density

$$(2\pi\sigma^2)^{-1/2} \exp(-(y - \mu)^2/2\sigma^2)$$

the **normal distribution** with mean μ and variance σ^2.

We will next consider some discrete distributions. The first is ridiculously simple but we will need the result several times below, so we record it here.

Example 3.3. We say that X has a **Bernoulli distribution** with parameter p if $P(X = 1) = p$ and $P(X = 0) = 1 - p$. Clearly,

$$EX = p \cdot 1 + (1 - p) \cdot 0 = p$$

Since $X^2 = X$ we have $EX^2 = EX = p$

$$\text{var}(X) = EX^2 - (EX)^2 = p - p^2 = p(1 - p)$$

Example 3.4. We say that X has a **Poisson distribution** with parameter λ if

$$P(X = k) = e^{-\lambda}\lambda^k/k! \text{ for } k = 0, 1, 2, \ldots$$

To evaluate the moments of the Poisson random variable we use a little inspiration to observe that for $k \geq 1$

$$E(X(X-1)\cdots(X-k+1)) = \sum_{j=k}^{\infty} j(j-1)\cdots(j-k+1)e^{-\lambda}\frac{\lambda^j}{j!}$$

$$= \lambda^k \sum_{j=k}^{\infty} e^{-\lambda}\frac{\lambda^{j-k}}{(j-k)!} = \lambda^k$$

where the equalities follow from (i) the fact that $j(j-1)\cdots(j-k+1) = 0$ when $j < k$, (ii) cancelling part of the factorial, (iii) the fact that the Poisson distribution has total mass 1. Using the last formula it follows that $EX = \lambda$ while

$$\text{var}(X) = EX^2 - (EX)^2 = E(X(X-1)) + EX - \lambda^2 = \lambda$$

Example 3.5. N is said to have a **geometric distribution** with success probability $p \in (0,1)$ if

$$P(N = k) = p(1-p)^{k-1} \text{ for } k = 1, 2, \ldots$$

N is the number of independent trials needed to observe an event with probability p. Differentiating the identity

$$\sum_{k=0}^{\infty}(1-p)^k = 1/p$$

and referring to Example 9.2 in the Appendix for the justification gives

$$-\sum_{k=1}^{\infty} k(1-p)^{k-1} = -1/p^2$$

$$\sum_{k=2}^{\infty} k(k-1)(1-p)^{k-2} = 2/p^3$$

From this it follows that

$$EN = \sum_{k=1}^{\infty} kp(1-p)^{k-1} = 1/p$$

$$EN(N-1) = \sum_{k=2}^{\infty} k(k-1)p(1-p)^{k-1} = 2(1-p)/p^2$$

$$\text{var}(N) = EN^2 - (EN)^2 = EN(N-1) + EN - (EN)^2$$

$$= \frac{2(1-p)}{p^2} + \frac{p}{p^2} - \frac{1}{p^2} = \frac{1-p}{p^2}$$

EXERCISES

3.11. Inclusion exclusion formula. Let $A_1, A_2, \ldots, A_n$ be events and $A = \cup_{i=1}^n A_i$. Prove that $1_A = 1 - \prod_{i=1}^n (1 - 1_{A_i})$. Expand out the right hand side, then take expected value to conclude

$$P\left(\cup_{i=1}^n A_i\right) = \sum_{i=1}^n P(A_i) - \sum_{i<j} P(A_i \cap A_j)$$
$$+ \sum_{i<j<k} P(A_i \cap A_j \cap A_k) - \ldots + (-1)^{n-1} P(\cap_{i=1}^n A_i)$$

3.12. Bonferroni inequalities. Let $A_1, A_2, \ldots, A_n$ be events and $A = \cup_{i=1}^n A_i$. Show that $1_A \leq \sum_{i=1}^n 1_{A_i}$, etc. and then take expected values to conclude

$$P\left(\cup_{i=1}^n A_i\right) \leq \sum_{i=1}^n P(A_i)$$

$$P\left(\cup_{i=1}^n A_i\right) \geq \sum_{i=1}^n P(A_i) - \sum_{i<j} P(A_i \cap A_j)$$

$$P\left(\cup_{i=1}^n A_i\right) \leq \sum_{i=1}^n P(A_i) - \sum_{i<j} P(A_i \cap A_j) + \sum_{i<j<k} P(A_i \cap A_j \cap A_k)$$

In general if we stop the inclusion exclusion formula after an even (odd) number of sums we get a lower (an upper) bound.

3.13. If $E|X|^k < \infty$ then for $0 < j < k$, $E|X|^j < \infty$ and furthermore

$$E|X|^j \leq (E|X|^k)^{j/k}$$

3.14. Apply Jensen's inequality with $\varphi(x) = e^x$ and $P(X = \log\ y_m) = p(m)$ to conclude that if $\sum_{m=1}^n p(m) = 1$ and $p(m), y_m > 0$ then

$$\sum_{m=1}^n p(m) y_m \geq \prod_{m=1}^n y_m^{p(m)}$$

When $p(m) = 1/n$ this says the arithmetic mean exceeds the geometric mean.

3.15. If $EX_1^- < \infty$ and $X_n \uparrow X$ then $EX_n \uparrow EX$.

3.16. Let $X \geq 0$ but do NOT assume $E(1/X) < \infty$. Show

$$\lim_{y \to \infty} yE(1/X; X > y) = 0, \qquad \lim_{y \to 0} yE(1/X; X > y) = 0.$$

3.17. If $X_n \geq 0$ then $E(\sum_{n=0}^{\infty} X_n) = \sum_{n=0}^{\infty} EX_n$.

3.18. If X is integrable and A_n are disjoint sets with union A

$$\sum_{n=0}^{\infty} E(X; A_n) = E(X; A)$$

i.e., the sum converges absolutely and has the value on the right.

1.4. Independence

We begin with what is hopefully a familiar definition and then work our way up to a definition that is appropriate for our current setting.

Two events A and B are **independent** if $P(A \cap B) = P(A)P(B)$.

Two random variables X and Y are **independent** if for all $C, D \in \mathcal{R}$,

$$P(X \in C, Y \in D) = P(X \in C)P(Y \in D)$$

i.e., the events $A = \{X \in C\}$ and $B = \{Y \in D\}$ are independent.

Two σ-fields $\mathcal{F}$ and $\mathcal{G}$ are **independent** if for all $A \in \mathcal{F}$ and $B \in \mathcal{G}$ the events A and B are independent.

As the next exercise shows, the second definition is a special case of the third.

EXERCISE 4.1. (i) Show that if X and Y are independent then $\sigma(X)$ and $\sigma(Y)$ are. (ii) Conversely if $\mathcal{F}$ and $\mathcal{G}$ are independent, $X \in \mathcal{F}$, and $Y \in \mathcal{G}$, then X and Y are independent.

The second definition above is, in turn, a special case of the first.

EXERCISE 4.2. (i) Show that if A and B are independent then so are A^c and B, A and B^c, and A^c and B^c. (ii) Conclude that events A and B are independent if and only if their indicator random variables 1_A and 1_B are independent.

In view of the fact that the first definition is a special case of the second which is a special case of the third, we take things in the opposite order when we say what it means for several things to be independent. We begin by reducing to the case of finitely many objects. An infinite collection of objects (σ-fields, random variables, or sets) is said to be independent if every finite subcollection is.

σ-fields $\mathcal{F}_1, \mathcal{F}_2, \ldots, \mathcal{F}_n$ are **independent** if whenever $A_i \in \mathcal{F}_i$ for $i = 1, \ldots, n$ we have

$$P\left(\cap_{i=1}^n A_i\right) = \prod_{i=1}^n P(A_i)$$

Random variables $X_1, \ldots, X_n$ are **independent** if whenever $B_i \in \mathcal{R}$ for $i = 1, \ldots, n$ we have

$$P\left(\cap_{i=1}^n \{X_i \in B_i\}\right) = \prod_{i=1}^n P(X_i \in B_i)$$

Sets $A_1, \ldots, A_n$ are **independent** if whenever $I \subset \{1, \ldots n\}$ we have

$$P\left(\cap_{i \in I} A_i\right) = \prod_{i \in I} P(A_i)$$

At first glance it might seem that the last definition does not match the other two. However, if you think about it for a minute, you will see that if the indicator variables 1_{A_i}, $1 \le i \le n$ are independent and we take $B_i = \{1\}$ for $i \in I$ $B_i = \mathbf{R}$ for $i \notin I$ then the condition in the definition results. Conversely,

EXERCISE 4.3. Let $A_1, A_2, \ldots, A_n$ be independent. Show (i) $A_1^c, A_2, \ldots, A_n$ are independent; (ii) $1_{A_1}, \ldots, 1_{A_n}$ are independent.

One of the first things to understand about the definition of independent events is that it is not enough to assume $P(A_i \cap A_j) = P(A_i)P(A_j)$ for all $i \ne j$. A sequence of events $A_1, \ldots, A_n$ with the last property is called **pairwise independent**. It is clear that independent events are pairwise independent. The next example shows that the converse is not true.

Example 4.1. Let X_1, X_2, X_3 be independent random variables with

$$P(X_i = 0) = P(X_i = 1) = 1/2$$

Let $A_1 = \{X_2 = X_3\}$, $A_2 = \{X_3 = X_1\}$ and $A_3 = \{X_1 = X_2\}$. These events are pairwise independent since if $i \ne j$ then

$$P(A_i \cap A_j) = P(X_1 = X_2 = X_3) = 1/4 = P(A_i)P(A_j)$$

but they are not independent since

$$P(A_1 \cap A_2 \cap A_3) = 1/4 \ne 1/8 = P(A_1)P(A_2)P(A_3)$$

In order to show that random variables X and Y are independent we have to check that $P(X \in A, Y \in B) = P(X \in A)P(Y \in B)$ for all Borel sets A and B. Since there are a lot of Borel sets, our next topic is

a. Sufficient conditions for independence

Our main result is (4.2). To state that result we need a definition that generalizes all our earlier definitions.

Collections of sets $\mathcal{A}_1, \mathcal{A}_2, \ldots, \mathcal{A}_n \subset \mathcal{F}$ are said to be **independent** if whenever $A_i \in \mathcal{A}_i$ and $I \subset \{1, \ldots, n\}$ we have

$$P\left(\cap_{i \in I} A_i\right) = \prod_{i \in I} P(A_i)$$

If each collection is a single set i.e., $\mathcal{A}_i = \{A_i\}$ this reduces to the definition for sets. If each $\mathcal{A}_i$ contains Ω, e.g., $\mathcal{A}_i$ is a σ-field the condition is equivalent to

$$P\left(\cap_{i=1}^n A_i\right) = \prod_{i=1}^n P(A_i) \quad \text{whenever } A_i \in \mathcal{A}_i$$

since we can set $A_i = \Omega$ for $i \notin I$. Conversely, if $\mathcal{A}_1, \mathcal{A}_2, \ldots, \mathcal{A}_n$ are independent and $\bar{\mathcal{A}}_i = \mathcal{A}_i \cup \{\Omega\}$ so there is no loss of generality in supposing $\Omega \in \mathcal{A}_i$.

The proof of (4.2) is based on Dynkin's $\pi - \lambda$ theorem ((2.1) in the Appendix). To state this result we need two definitions. We say that $\mathcal{A}$ is a **π-system** if it is closed under intersection, i.e., if $A, B \in \mathcal{A}$ then $A \cap B \in \mathcal{A}$. We say that $\mathcal{L}$ is a **λ-system** if: (i) $\Omega \in \mathcal{L}$. (ii) If $A, B \in \mathcal{L}$ and $A \subset B$ then $B - A \in \mathcal{L}$. (iii) If $A_n \in \mathcal{L}$ and $A_n \uparrow A$ then $A \in \mathcal{L}$.

(4.1) $\pi - \lambda$ **Theorem.** If $\mathcal{P}$ is a π-system and $\mathcal{L}$ is a λ-system that contains $\mathcal{P}$ then $\sigma(\mathcal{P}) \subset \mathcal{L}$.

(4.2) **Theorem.** Suppose $\mathcal{A}_1, \mathcal{A}_2, \ldots, \mathcal{A}_n$ are independent and each $\mathcal{A}_i$ is a π-system. Then $\sigma(\mathcal{A}_1), \sigma(\mathcal{A}_2), \ldots, \sigma(\mathcal{A}_n)$ are independent.

Proof Let $A_2, \ldots, A_n$ be sets with $A_i \in \mathcal{A}_i$, let $F = A_2 \cap \cdots \cap A_n$ and let $\mathcal{L} = \{A : P(A \cap F) = P(A)P(F)\}$. As noted after the definition, we can without loss of generality suppose $\Omega \in \mathcal{A}_1$. So we have $P(F) = \prod_{i=2}^n P(A_i)$ and (i) $\Omega \in \mathcal{L}$. To check (ii), we note that if $A, B \in \mathcal{L}$ with $A \subset B$ then $(B - A) \cap F = (B \cap F) - (A \cap F)$. So using (i) in Exercise 1.1, the fact $A, B \in \mathcal{L}$ and then (i) in Exercise 1.1 again:

$$P((B - A) \cap F) = P(B \cap F) - P(A \cap F) = P(B)P(F) - P(A)P(F)$$
$$= \{P(B) - P(A)\}P(F) = P(B - A)P(F)$$

and we have $B - A \in \mathcal{L}$. To check (iii) let $B_k \in \mathcal{L}$ with $B_k \uparrow B$ and note that $(B_k \cap F) \uparrow (B \cap F)$ so using (iii) in Exercise 1.1, then the fact $A, B \in \mathcal{L}$ and then (iii) in Exercise 1.1 again:

$$P(B \cap F) = \lim_k P(B_k \cap F) = \lim_k P(B_k)P(F) = P(B)P(F)$$

Applying the $\pi - \lambda$ theorem now gives $\mathcal{L} \supset \sigma(\mathcal{A}_1)$ and since $\mathcal{A}_2, \ldots, \mathcal{A}_n$ are arbitrary members of $\mathcal{A}_2, \ldots, \mathcal{A}_n$, we have

(4.2′) If $\mathcal{A}_1, \mathcal{A}_2, \ldots, \mathcal{A}_n$ are independent then $\sigma(\mathcal{A}_1), \mathcal{A}_2, \ldots, \mathcal{A}_n$ are independent.

Applying (4.2′) to $\mathcal{A}_2, \ldots, \mathcal{A}_n, \sigma(\mathcal{A}_1)$ (which are independent since the definition is unchanged by permuting the order of the collections) shows that $\sigma(\mathcal{A}_2), \mathcal{A}_3, \ldots, \mathcal{A}_n, \sigma(\mathcal{A}_1)$ are independent and after n iterations we have the desired result. □

Remark. The reader should note that it is not easy to show that if $A, B \in \mathcal{L}$ then $A \cap B \in \mathcal{L}$, or $A \cup B \in \mathcal{L}$, but it is easy to check that if $A, B \in \mathcal{L}$ with $A \subset B$ then $B - A \in \mathcal{L}$.

Having worked to establish (4.2) we get several corollaries.

(4.3) **Corollary.** In order for $X_1, \ldots, X_n$ to be independent it is sufficient that for all $x_1, \ldots, x_n \in (-\infty, \infty]$

$$P(X_1 \leq x_1, \ldots, X_n \leq x_n) = \prod_{i=1}^{n} P(X_i \leq x_i)$$

Proof   Let $\mathcal{A}_i = $ the sets of the form $\{X_i \leq x_i\}$. Since $\{X_i \leq x\} \cap \{X_i \leq y\} = \{X_i \leq x \wedge y\}$, $\mathcal{A}_i$ is a π-system. Since we have allowed $x_i = \infty$, $\Omega \in \mathcal{A}_i$. Exercise 2.1 implies $\sigma(\mathcal{A}_i) = \sigma(X_i)$, so the result follows from (4.2). □

The last result expresses independence of random variables in terms of their distribution functions. The next two exercises treat density functions and discrete random variables.

EXERCISE 4.4. Suppose $(X_1, \ldots, X_n)$ has density $f(x_1, x_2, \ldots, x_n)$, that is

$$P((X_1, X_2, \ldots, X_n) \in A) = \int_A f(x)\, dx \text{ for } A \in \mathcal{R}^n$$

If $f(x)$ can be written as $g_1(x_1) \cdots g_n(x_n)$ where the $g_m \geq 0$ are measurable then $X_1, X_2, \ldots, X_n$ are independent. Note that the g_m are not assumed to be probability densities.

EXERCISE 4.5. Suppose $X_1, \ldots, X_n$ are random variables that take values in countable sets $S_1, \ldots, S_n$. Then in order for $X_1, \ldots, X_n$ to be independent it is sufficient that whenever $x_i \in S_i$

$$P(X_1 = x_1, \ldots, X_n = x_n) = \prod_{i=1}^{n} P(X_i = x_i)$$

Our next goal is to prove that functions of disjoint collections of independent random variables are independent. See (4.5) for the precise statement. First we will prove an analogous result for σ-fields.

(4.4) Corollary. Suppose $\mathcal{F}_{i,j}, 1 \leq i \leq n, 1 \leq j \leq m(i)$ are independent and let $\mathcal{G}_i = \sigma(\cup_j \mathcal{F}_{i,j})$. Then $\mathcal{G}_1, \ldots, \mathcal{G}_n$ are independent.

Proof Let $\mathcal{A}_i$ be the collection of sets of the form $\cap_j A_{i,j}$ where $A_{i,j} \in \mathcal{F}_{i,j}$. $\mathcal{A}_i$ is a π-system that contains Ω and contains $\cup_j \mathcal{F}_{i,j}$ so (4.2) implies $\sigma(\mathcal{A}_i) \supset \mathcal{G}_i$ are independent. □

(4.5) Corollary. If for $1 \leq i \leq n$, $1 \leq j \leq m(i)$, $X_{i,j}$ are independent and $f_i : \mathbf{R}^{m(i)} \to \mathbf{R}$ are measurable then $f_i(X_{i,1}, \ldots, X_{i,m(i)})$ are independent.

Proof Let $\mathcal{F}_{i,j} = \sigma(X_{i,j})$ and $\mathcal{G}_i = \sigma(\cup_j \mathcal{F}_{i,j})$. Since $f_i(X_{i,1}, \ldots, X_{i,m(i)}) \in \mathcal{G}_i$, the desired result follows from Exercise 4.1. □

A concrete special case of (4.5) that we will use in a minute is: if $X_1, \ldots, X_n$ are independent then $X = X_1$ and $Y = X_2 \cdots X_n$ are independent. Later when we study sums $S_m = X_1 + \cdots + X_m$ of independent random variables $X_1, \ldots, X_n$ we will use (4.5) to conclude that if $m < n$ then $S_n - S_m$ is independent of the indicator function of the event $\{\max_{1 \leq k \leq m} S_k > x\}$.

b. Independence, distribution, and expectation

Our next goal is to obtain formulas for the distribution and expectation of independent random variables.

(4.6) Theorem. Suppose $X_1, \ldots, X_n$ are independent random variables and X_i has distribution μ_i, then $(X_1, \ldots, X_n)$ has distribution $\mu_1 \times \cdots \times \mu_n$.

Proof Using the definitions of (i) $A_1 \times \cdots \times A_n$, (ii) independence, (iii) μ_i, and (iv) $\mu_1 \times \cdots \times \mu_n$

$$P((X_1, \ldots, X_n) \in A_1 \times \cdots \times A_n) = P(X_1 \in A_1, \ldots, X_n \in A_n)$$

$$= \prod_{i=1}^{n} P(X_i \in A_i) = \prod_{i=1}^{n} \mu_i(A_i) = \mu_1 \times \cdots \times \mu_n(A_1 \times \cdots \times A_n)$$

The last formula shows that the distribution of $(X_1, \ldots, X_n)$ and the measure $\mu_1 \times \cdots \times \mu_n$ agree on sets of the form $A_1 \times \cdots \times A_n$, a π-system that generates $\mathcal{R}^n$. So (2.2) in the Appendix implies they must agree. $\square$

(4.7) **Theorem.** Suppose X and Y are independent and have distributions μ and ν. If $h : \mathbf{R}^2 \to \mathbf{R}$ is a measurable function with $h \geq 0$ or $E|h(X,Y)| < \infty$ then

$$Eh(X,Y) = \iint h(x,y)\, \mu(dx)\, \nu(dy)$$

In particular, if $h(x,y) = f(x)g(y)$ where $f, g : \mathbf{R} \to \mathbf{R}$ are measurable functions with $f, g \geq 0$ or $E|f(X)|$ and $E|g(Y)| < \infty$ then

$$Ef(X)g(Y) = Ef(X) \cdot Eg(Y)$$

Proof   Using (3.9) and then Fubini's theorem ((6.2) in the Appendix) we have

$$Eh(X,Y) = \int_{\mathbf{R}^2} h\, d(\mu \times \nu) = \iint h(x,y)\, \mu(dx)\, \nu(dy)$$

To prove the second result we start with the result when $f, g \geq 0$. In this case, using the first result, the fact that $g(y)$ does not depend on x, and then (3.9) twice we get

$$Ef(X)g(Y) = \iint f(x)g(y)\, \mu(dx)\, \nu(dy) = \int g(y) \int f(x)\, \mu(dx)\, \nu(dy)$$

$$= \int Ef(X)g(y)\, \nu(dy) = Ef(X)Eg(Y)$$

Applying the result for nonnegative f and g to $|f|$ and $|g|$ we get $E|f(X)g(Y)| = E|f(X)|E|g(Y)| < \infty$ and we can repeat the last argument to prove the desired result. $\square$

From (4.7) it is only a small step to

(4.8) **Theorem.** If $X_1, \ldots, X_n$ are independent and have $X_i \geq 0$ or $E|X_i| < \infty$ then

$$E\left(\prod_{i=1}^{n} X_i\right) = \prod_{i=1}^{n} EX_i$$

i.e., the expectation on the left exists and has the value given on the right.

Proof $X = X_1$ and $Y = X_2 \cdots X_n$ are independent by (4.5) so taking $f(x) = |x|$ and $g(y) = |y|$ we have $E|X_1 \cdots X_n| = E|X_1| E|X_2 \cdots X_n|$ and it follows by induction that if $1 \le m \le n$

$$E|X_m \cdots X_n| = \prod_{i=m}^{n} E|X_k|$$

If the $X_i \ge 0$, then $|X_i| = X_i$ and the desired result follows from the special case $m = 1$. To prove the result in general note that the special case $m = 2$ implies $E|Y| = E|X_2 \cdots X_n| < \infty$ so using (4.7) with $f(x) = x$ and $g(y) = y$ shows $E(X_1 \cdots X_n) = EX_1 \cdot E(X_2 \cdots X_n)$ and the desired result follows by induction. □

Example 4.2. It can happen that $E(XY) = EX \cdot EY$ without the variables being independent. Suppose the joint distribution of X and Y is given by the following table

$X\backslash Y$	1	0	-1
1	0	a	0
0	b	c	b
-1	0	a	0

where $a, b > 0$, $c \ge 0$, and $2a + 2b + c = 1$. Things are arranged so that $XY \equiv 0$. Symmetry implies $EX = 0$, $EY = 0$ so $E(XY) = 0 = EXEY$. The random variables are not independent since

$$P(X = 1, Y = 1) = 0 < ab = P(X = 1)P(Y = 1)$$

Two random variables X and Y with $EX^2, EY^2 < \infty$ that have $EXY = EXEY$ are said to be **uncorrelated**. The finite second moments are needed so that we know $E|XY| < \infty$ by the Cauchy-Schwarz inequality.

EXERCISE 4.6. Let $\Omega = (0,1)$, $\mathcal{F}$ = Borel sets, P = Lebesgue measure. $X_n(\omega) = \sin(2\pi n\omega)$, $n = 1, 2, \ldots$ are uncorrelated but not independent.

We turn now to the distribution of the sum of two independent r.v.'s

(4.9) Theorem. If X and Y are independent, $F(x) = P(X \le x)$ and $G(y) = P(Y \le y)$ then

$$P(X + Y \le z) = \int F(z - y)\, dG(y)$$

The integral on the right hand side is called the **convolution** of F and G, and is denoted $F * G(z)$. The meaning of $dG(y)$ will be explained in the proof.

Proof   Let $h(x, y) = 1_{(x+y \le z)}$. Let μ and ν be the probability measures with distribution functions F and G. Since for fixed y

$$\int h(x, y) \, \mu(dx) = \int 1_{(-\infty, z-y]}(x) \, \mu(dx) = F(z - y)$$

using (4.7) gives

$$P(X + Y \le z) = \iint 1_{(x+y \le z)} \, \mu(dx) \, \nu(dy)$$

$$= \int F(z - y) \, \nu(dy) = \int F(z - y) \, dG(y)$$

The last equality is just change of notation – we regard $dG(y)$ as a shorthand for "integrate with respect to the measure ν with distribution function G".   □

EXERCISE 4.7. (i) Show that if X and Y are independent with distributions μ and ν then

$$P(X + Y = 0) = \sum_y \mu(\{-y\})\nu(\{y\})$$

(ii) Conclude that if X has continuous distribution $P(X = Y) = 0$.

To treat concrete examples we need a special case of (4.9).

(4.10) **Theorem.** Suppose that X with density f, and Y with distribution function G are independent. Then $X + Y$ has density

$$h(x) = \int f(x - y) \, dG(y)$$

When Y has density g, the last formula can be written as

$$h(x) = \int f(x - y) \, g(y) \, dy$$

Proof   From (4.9), the definition of density function, and Fubini's theorem ((6.2) in the Appendix) which is justified since everything is nonnegative, we get

$$P(X + Y \le z) = \int F(z - y) \, dG(y) = \int \int_{-\infty}^{z} f(x - y) \, dx \, dG(y)$$

$$= \int_{-\infty}^{z} \int f(x - y) \, dG(y) \, dx$$

The last equation says that $X + Y$ has density $h(x) = \int f(x - y)dG(y)$. The second formula follows from the first when we recall the meaning of $dG(y)$ given in (4.9) and use Exercise 3.10. $\qquad\qquad\qquad\qquad\qquad\qquad\qquad\qquad\qquad\qquad\square$

(4.10) plus some ugly calculus allows us to treat two standard examples. These facts should be familiar from undergraduate probability. We give one calculation and leave the other to the reader.

Example 4.3. The **gamma density** with parameters α and λ, is given by

$$f(x) = \begin{cases} \lambda^\alpha x^{\alpha-1} e^{-\lambda x}/\Gamma(\alpha) & \text{for } x \geq 0 \\ 0 & \text{for } x < 0 \end{cases}$$

where $\Gamma(\alpha) = \int_0^\infty x^{\alpha-1} e^{-x} \, dx$. We will now show

If $X = \text{gamma}(\alpha, \lambda)$ and $Y = \text{gamma}(\beta, \lambda)$ are independent then $X + Y$ is gamma$(\alpha + \beta, \lambda)$.

Proof   Writing $f_{X+Y}(z)$ for the density funtion of $X + Y$ and using (4.10)

$$\begin{aligned} f_{X+Y}(x) &= \int_0^x \frac{\lambda^\alpha (x-y)^{\alpha-1}}{\Gamma(\alpha)} e^{-\lambda(x-y)} \frac{\lambda^\beta y^{\beta-1}}{\Gamma(\beta)} e^{-\lambda y} \, dy \\ &= \frac{\lambda^{\alpha+\beta} e^{-\lambda x}}{\Gamma(\alpha)\Gamma(\beta)} \int_0^x (x-y)^{\alpha-1} y^{\beta-1} \, dy \end{aligned}$$

so it suffices to show the integral is $x^{\alpha+\beta-1}\Gamma(\alpha)\Gamma(\beta)/\Gamma(\alpha + \beta)$. To do this we begin by changing variables $y = xu$, $dy = x \, du$ to get

$$x^{\alpha+\beta-1} \int_0^1 (1 - u)^{\alpha-1} u^{\beta-1} \, du = \int_0^x (x - y)^{\alpha-1} y^{\beta-1} \, dy$$

Multiplying each side by e^{-x}, integrating from 0 to ∞, and then using Fubini's theorem on the right we have

$$\begin{aligned} \Gamma(\alpha + \beta) &\int_0^1 (1 - u)^{\alpha-1} u^{\beta-1} \, du \\ &= \int_0^\infty \int_0^x y^{\beta-1} e^{-y} (x - y)^{\alpha-1} e^{-(x-y)} \, dy \, dx \\ &= \int_0^\infty y^{\beta-1} e^{-y} \int_x^\infty (x - y)^{\alpha-1} e^{-(x-y)} \, dx \, dy = \Gamma(\alpha)\Gamma(\beta) \end{aligned}$$

which gives the desired result.

EXERCISE 4.8. Use the fact that a gamma$(1, \lambda)$ is an exponential with parameter λ, and induction to show that the sum of n independent exponential(λ) r.v.'s, $X_1 + \cdots + X_n$, has a gamma(n, λ) distribution.

EXERCISE 4.9. In Example 3.2 we introduced the normal density with mean μ and variance a, $(2\pi a)^{-1/2} \exp(-(x - \mu)^2/2a)$. Show that if $X = \text{normal}(\mu, a)$ and $Y = \text{normal}(\nu, b)$ are independent then $X + Y = \text{normal}(\mu + \nu, a + b)$. To simplify this tedious calculation notice that it is enough to prove the result for $\mu = \nu = 0$. In Exercise 3.4 of Chapter 2 you will give a simpler proof of this result.

c. Constructing independent random variables

The last question that we have to address before we can study independent random variables is: Do they exist? (If they don't exist then there is no point in studying them!) If we are given a finite number of distribution functions $F_i, 1 \le i \le n$, it is easy to construct independent random variables $X_1, \ldots, X_n$ with $P(X_i \le x) = F_i(x)$. Let $\Omega = \mathbf{R}^n$, $\mathcal{F} = \mathcal{R}^n$, $X_i(\omega_1, \ldots, \omega_n) = \omega_i$ (the ith coordinate of $\omega \in \mathbf{R}^n$), and let P be the measure on $\mathcal{R}^n$ that has

$$P((a_1, b_1] \times \cdots \times (a_n, b_n]) = (F_1(b_1) - F_1(a_1)) \cdots (F_n(b_n) - F_n(a_n))$$

If μ_i is the measure with distribution function F_i then $P = \mu_1 \times \cdots \times \mu_n$.

To construct an infinite sequence $X_1, X_2, \ldots$ of independent random variables with given distribution functions we want to perform the last construction on the infinite product space

$$\mathbf{R}^{\mathbf{N}} = \{(\omega_1, \omega_2, \ldots) : \omega_i \in \mathbf{R}\} = \{\text{functions } \omega : \mathbf{N} \to \mathbf{R}\}$$

where $\mathbf{N} = \{1, 2, \ldots\}$ and $\mathbf{N}$ stands for **natural numbers**. We define $X_i(\omega) = \omega_i$ and we equip $\mathbf{R}^{\mathbf{N}}$ with the product σ-field $\mathcal{R}^{\mathbf{N}}$, which is generated by the **finite dimensional sets** = sets of the form $\{\omega : \omega_i \in B_i, 1 \le i \le n\}$ where $B_i \in \mathcal{R}$. It is clear how we want to define P for finite dimensional sets. To assert the existence of a unique extension to $\mathcal{R}^{\mathbf{N}}$ we use (7.1) from the Appendix:

(4.11) **Kolmogorov's extension theorem.** Suppose we are given probability measures μ_n on $(\mathbf{R}^n, \mathcal{R}^n)$ that are consistent, that is,

$$\mu_{n+1}((a_1, b_1] \times \cdots \times (a_n, b_n] \times \mathbf{R}) = \mu_n((a_1, b_1] \times \cdots \times (a_n, b_n])$$

Then there is a unique probability measure P on $(\mathbf{R}^{\mathbf{N}}, \mathcal{R}^{\mathbf{N}})$ with

$$P(\omega : \omega_i \in (a_i, b_i], 1 \le i \le n) = \mu_n((a_1, b_1] \times \cdots \times (a_n, b_n])$$

In what follows we will need to construct sequences of random variables that take values in other measurable spaces $(S, \mathcal{S})$. Unfortunately, (4.11) is not valid for arbitrary measurable spaces. The first example (on an infinite product of different spaces $\Omega_1 \times \Omega_2 \times \ldots$) was due to Andersen and Jessen (1948). (See Halmos p. 214 or Neveu p. 84.) For an example in which all the spaces Ω_i are the same see Wegner (1973). Fortunately, there is a class of spaces that is adequate for all of our results and for which the generalization of Kolmogorov's theorem is trivial.

$(S, \mathcal{S})$ is said to be **nice** if there is a 1-1 map φ from S into **R** so that φ and φ^{-1} are both measurable.

Such spaces are often called **standard Borel spaces**, but we already have too many things named after Borel. The next result shows that most spaces arising in applications are nice.

(4.12) Theorem. If S is a Borel subset of a complete separable metric space M, and $\mathcal{S}$ is the collection of Borel subsets of S, then $(S, \mathcal{S})$ is nice.

Proof   We begin with the special case $S = [0, 1)^{\mathbf{N}}$ with metric

$$\rho(x, y) = \sum_{n=1}^{\infty} |x_n - y_n|/2^n$$

If $x = (x^1, x^2, x^3, \ldots)$, expand each component in binary $x^j = .x_1^j x_2^j x_3^j \ldots$ (taking the expansion with an infinite number of 0's). Let

$$\varphi_o(x) = .x_1^1 x_2^1 x_1^2 x_3^1 x_2^2 x_1^3 x_4^1 x_3^2 x_2^3 x_1^4 \ldots$$

To treat the general case we observe that by letting

$$d(x, y) = \rho(x, y)/(1 + \rho(x, y))$$

(for more details see Exercise 4.11) we can suppose that the metric has $d(x, y) < 1$ for all x, y. Let $q_1, q_2, \ldots$ be a countable dense set in S. Let

$$\psi(x) = (d(x, q_1), d(x, q_2), \ldots).$$

$\psi : S \to [0, 1)^{\mathbf{N}}$ is continuous and 1-1. $\varphi_o \circ \psi$ gives the desired mapping. $\square$

EXERCISE 4.10. Let $\rho(x, y)$ be a metric. (i) Suppose h is differentiable with $h(0) = 0$, $h'(x) > 0$ for $x > 0$ and $h'(x)$ decreasing on $[0, \infty)$. Then $h(\rho(x, y))$ is a metric. (ii) $h(x) = x/(x + 1)$ satisfies the hypotheses in (i).

EXERCISES

4.11. (i) Prove directly from the definition that if X and Y are independent and f and g are measurable functions then $f(X)$ and $g(Y)$ are independent.

4.12. Let $K \geq 3$ be a prime and let X and Y be independent random variables that are uniformly distributed on $\{0, 1, \ldots, K-1\}$. For $0 \leq n < K$ let $Z_n = X + nY$ mod K. Show that $Z_0, Z_1, \ldots, Z_{K-1}$ are **pairwise independent**, i.e., each pair is independent, but if we know the values of two of the variables then we know the values of all the variables.

4.13. Find four random variables taking values in $\{-1, 1\}$ so that any three are independent but all four are not. Hint: consider products of independent random variables.

4.14. Let $\Omega = \{1, 2, 3, 4\}$, $\mathcal{F}$ = all subsets of Ω, and $P(\{i\}) = 1/4$. Give an example of two collections of sets $\mathcal{A}_1$ and $\mathcal{A}_2$ that are independent but the generated σ-fields are not.

4.15. Show that if X and Y are independent integer valued random variables then
$$P(X + Y = n) = \sum_m P(X = m)P(Y = n - m)$$

4.16. In Example 3.4 we introduced the Poisson distribution with parameter λ which is given by $P(Z = k) = e^{-\lambda}\lambda^k/k!$ for $k = 0, 1, 2, \ldots$. Use the previous exercise to show that if $X = \text{Poisson}(\lambda)$ and $Y = \text{Poisson}(\mu)$ are independent then $X + Y = \text{Poisson}(\lambda + \mu)$.

4.17. X is said to have a Binomial(n, p) distribution if
$$P(X = m) = \binom{n}{m} p^m (1 - p)^{n-m}$$

(i) Show that if $X = \text{Binomial}(n, p)$ and $Y = \text{Binomial}(m, p)$ are independent then $X + Y = \text{Binomial}(n + m, p)$. (ii) Look at Example 3.3 and use induction to conclude that the sum of n independent Bernoulli(p) random variables is Binomial(n, p).

4.18. It should not be surprising that the distribution of $X + Y$ can be $F * G$ without the random variables being independent. Suppose $X, Y \in \{0, 1, 2\}$ and take each value with probability $1/3$. (a) Find the distribution of $X + Y$ assuming X and Y are independent. (b) Find all the joint distributions (X, Y) so that the distribution of $X + Y$ is the same as the answer to (a).

4.19. Let $X, Y \geq 0$ be independent with distribution functions F and G. Find the distribution function of XY.

4.20. If we want an infinite sequence of coin tossings we do not have to use Kolmogorov's theorem. Let Ω be the unit interval $(0,1)$ equipped with the Borel sets $\mathcal{F}$ and Lebesgue measure P. Let $Y_n(\omega) = 1$ if $[2^n\omega]$ is odd and 0 if $[2^n\omega]$ is even. Show that $Y_1, Y_2, \ldots$ are independent with $P(Y_k = 0) = P(Y_k = 1) = 1/2$.

1.5. Weak Laws of Large Numbers

In this section we will prove several "weak laws of large numbers." The first order of business is to define the mode of convergence that appears in the conclusion of the theorems. We say that Y_n converges to Y **in probability** if for all $\epsilon > 0$, $P(|Y_n - Y| > \epsilon) \to 0$ as $n \to \infty$.

a. L^2 weak laws

Our first set of weak laws come from computing variances and using Chebyshev's inequalities. Extending a definition given in Example 4.2 for two random variables, a family of random variables X_i, $i \in I$ with $E(X_i^2) < \infty$ is said to be **uncorrelated** if we have

$$E(X_i X_j) = E X_i E X_j \quad \text{whenever } i \neq j$$

The key to our weak law for uncorrelated random variables, (5.2), is:

(5.1) Lemma. Let $X_1, \ldots, X_n$ have $E(X_i^2) < \infty$ and be uncorrelated. Then

$$\text{var}(X_1 + \cdots + X_n) = \text{var}(X_1) + \cdots + \text{var}(X_n)$$

where $\text{var}(Y) =$ the variance of Y.

Proof Let $\mu_i = EX_i$ and $S_n = \sum_{i=1}^n X_i$. Since $ES_n = \sum_{i=1}^n \mu_i$, using the definition of the variance, writing the square of the sum as the product of two copies of the sum, and then expanding we have

$$\text{var}(S_n) = E(S_n - ES_n)^2 = E\left(\sum_{i=1}^n (X_i - \mu_i)\right)^2$$

$$= E\left(\sum_{i=1}^n \sum_{j=1}^n (X_i - \mu_i)(X_j - \mu_j)\right)$$

$$= \sum_{i=1}^n E(X_i - \mu_i)^2 + 2\sum_{i=1}^n \sum_{j=1}^{i-1} E((X_i - \mu_i)(X_j - \mu_j))$$

where in the last equality we have separated out the diagonal terms $i = j$ and used the fact that the sum over $1 \leq i < j \leq n$ is the same as the sum over $1 \leq j < i \leq n$.

The first sum is $\text{var}(X_1) + \ldots + \text{var}(X_n)$ so we want to show that the second sum is zero. To do this we observe

$$E((X_i - \mu_i)(X_j - \mu_j)) = EX_iX_j - \mu_iEX_j - \mu_jEX_i + \mu_i\mu_j$$
$$= EX_iX_j - \mu_i\mu_j = 0$$

since X_i and X_j are uncorrelated. $\square$

In words, (5.1) says that for uncorrelated random variables the variance of the sum is the sum of the variances. The second ingredient in our proof of (5.2) is the following consequence of (3.10c)

$$\text{var}(cY) = c^2\text{var}(Y)$$

This result and (5.1) lead easily to

(5.2) L^2 **weak law.** Let $X_1, X_2, \ldots$ be uncorrelated random variables with $E(X_i) = \mu$ and $\text{var}(X_i) \leq C < \infty$. If $S_n = X_1 + \ldots + X_n$ then as $n \to \infty$, $S_n/n \to \mu$ in L^2 and in probability.

Proof   To prove L^2 convergence observe that $E(S_n/n) = \mu$, so

$$E(S_n/n - \mu)^2 = \text{var}(S_n/n) = \frac{1}{n^2}(\text{var}(X_1) + \cdots + \text{var}(X_n)) \leq \frac{Cn}{n^2} \to 0$$

To conclude there is also convergence in probability we apply the next result to $Z = S_n/n - \mu$.

(5.3) **Lemma.** If $p > 0$ and $E|Z_n|^p \to 0$ then $Z_n \to 0$ in probability.

Proof   Chebyshev's inequality, (3.4), with $\varphi(x) = x^p$ and $X = |Z_n|$ implies that if $\epsilon > 0$ then $P(|Z_n| \geq \epsilon) \leq \epsilon^{-p}E|Z_n|^p \to 0$. $\square$

The most important special case of (5.2) occurs when $X_1, X_2, \ldots$ are independent random variables that all have the same distribution. In the jargon, they are **independent and identically distributed** or **i.i.d.**, for short. The L^2 weak law (5.2) tells us that if $EX_i^2 < \infty$ then S_n/n converges to $\mu = EX_i$ in probability as $n \to \infty$. In (5.8) below, we will see that $E|X_i| < \infty$ is sufficient for the last conclusion, but for the moment, we will concern ourselves with consequences of the weaker result.

Our first application is to a situation that on the surface has nothing to do with randomness.

Example 5.1. Polynomial approximation. Let f be a continuous function on [0,1], and let

$$f_n(x) = \sum_{m=0}^{n} \binom{n}{m} x^m (1-x)^{n-m} f(m/n) \quad \text{where} \quad \binom{n}{m} = \frac{n!}{m!(n-m)!}$$

be the **Bernstein polynomial of degree** n associated with f. Then as $n \to \infty$

$$\sup_{x \in [0,1]} |f_n(x) - f(x)| \to 0$$

Proof First observe that if S_n is the sum of n independent random variables with $P(X_i = 1) = p$ and $P(X_i = 0) = 1 - p$ then $EX_i = p$, $\text{var}(X_i) = p(1-p)$ and

$$P(S_n = m) = \binom{n}{m} p^m (1-p)^{n-m}$$

so $Ef(S_n/n) = f_n(p)$. (5.2) tells us that as $n \to \infty$, $S_n/n \to p$ in probability. The last two observations motivate the definition of $f_n(p)$ but to prove the desired conclusion we have to use the proof of (5.2) rather than the result itself.

Combining the proof of (5.2) with our formula for the variance of X_i and the fact that $p(1-p) \le 1/4$ when $p \in [0,1]$ we have

$$P(|S_n/n - p| > \delta) \le \text{var}(S_n/n)/\delta^2 = p(1-p)/n\delta^2 \le 1/4n\delta^2$$

To conclude now that $Ef(S_n/n) \to f(p)$, let $M = \sup_{x \in [0,1]} |f(x)|$, let $\epsilon > 0$, and pick $\delta > 0$ so that if $|x - y| < \delta$ then $|f(x) - f(y)| < \epsilon$. (This is possible since a continuous function is uniformly continuous on each bounded interval.) Now, using Jensen's inequality gives

$$|Ef(S_n/n) - f(p)| \le E|f(S_n/n) - f(p)| \le \epsilon + 2MP(|S_n/n - p| > \delta)$$

Letting $n \to \infty$ we have $\limsup_{n \to \infty} |Ef(S_n/n) - f(p)| \le \epsilon$, but ϵ is arbitrary so this gives the desired result. □

Our next result is for comic relief.

Example 5.2. A high dimensional cube is almost the boundary of a ball. Let $X_1, X_2, \ldots$ be independent and uniformly distributed on $(-1, 1)$.

Let $Y_i = X_i^2$, which are independent since they are functions of independent random variables. $EY_i = 1/3$ and $\text{var}(Y_i) \leq EY_i^2 \leq 1$, so (5.2) implies

$$(X_1^2 + \ldots + X_n^2)/n \to 1/3 \quad \text{in probability as } n \to \infty$$

Let $A_{n,\epsilon} = \{x \in \mathbf{R}^n : (1-\epsilon)\sqrt{n/3} < |x| < (1+\epsilon)\sqrt{n/3}\}$ where $|x| = (x_1^2 + \cdots + x_n^2)^{1/2}$. If we let $|S|$ denote the Lebesgue measure of S then the last conclusion implies that for any $\epsilon > 0$, $|A_{n,\epsilon} \cap (-1,1)^n|/2^n \to 1$ or in words, most of the volume of the cube $(-1,1)^n$ comes from $A_{n,\epsilon}$ which is almost the boundary of the ball of radius $\sqrt{n/3}$.

   To try to visualize the relationship between the sphere and the cube note (i) the cube intersects the x_1 axis at 1, while the sphere does at $\sqrt{n/3}$, (ii) the cube interesects the "diagonal" where all the coordinates are equal when they are all 1, while the sphere cuts it where they all are $1/\sqrt{3}$.

b. Triangular arrays

Many classical limit theorems in probability concern arrays $X_{n,k}$, $1 \leq k \leq n$ of random variables and investigate the limiting behavior of their row sums $S_n = X_{n,1} + \cdots + X_{n,n}$. In most cases we assume that the random variables on each row are independent but for the next trivial (but useful) result we do not need that assumption. Indeed, here S_n can be any sequence of random variables.

(5.4) Theorem. Let $\mu_n = ES_n$, $\sigma_n^2 = \text{var}(S_n)$. If $\sigma_n^2/b_n^2 \to 0$ then

$$\frac{S_n - \mu_n}{b_n} \to 0 \quad \text{in probability}$$

Proof Our assumptions imply $E((S_n - \mu_n)/b_n)^2 = b_n^{-2}\text{var}(S_n) \to 0$, so the desired conclusion follows from (5.3). $\qquad\square$

We will now give three applications of (5.4). For these three examples the following calculation is useful

$$\sum_{m=1}^{n} \frac{1}{m} \geq \int_1^n \frac{dx}{x} \geq \sum_{m=2}^{n} \frac{1}{m}$$

(*) $$\log n \leq \sum_{m=1}^{n} \frac{1}{m} \leq 1 + \log n$$

Example 5.3. Coupon collector's problem. Let $X_1, X_2, \ldots$ be i.i.d. uniform on $\{1, 2, \ldots, n\}$. To motivate the name, think of collecting baseball cards (or coupons). Suppose that the ith item we collect is chosen at random from the set of possibilities and is independent of the previous choices. Let $\tau_k^n = \inf\{m : |\{X_1, \ldots, X_m\}| = k\}$ be the first time we have k different items. In this problem we are interested in the asymptotic behavior of $T_n = \tau_n^n$, the time to collect a complete set. It is easy to see that $\tau_1^n = 1$. To make later formulas work out nicely we will set $\tau_0^n = 0$. For $1 \leq k \leq n$, $X_{n,k} \equiv \tau_k^n - \tau_{k-1}^n$ represents the time to get a choice different from our first $k - 1$, so $X_{n,k}$ has a geometric distribution with parameter $1 - (k - 1)/n$, and is independent of the earlier waiting times $X_{n,j}$, $1 \leq j < k$. Example 3.5 tells us that if X has a geometric distribution with parameter p then $EX = 1/p$ and $\text{var}(X) \leq 1/p^2$. Using the linearity of expected value, (∗) and (5.1) we see that

$$ET_n = \sum_{k=1}^{n} \left(1 - \frac{k-1}{n}\right)^{-1} = n \sum_{m=1}^{n} m^{-1} \sim n \log n$$

$$\text{var}(T_n) \leq \sum_{k=1}^{n} \left(1 - \frac{k-1}{n}\right)^{-2} = n^2 \sum_{m=1}^{n} m^{-2} \leq n^2 \sum_{m=1}^{\infty} m^{-2}$$

Taking $b_n = n \log n$ and using (5.4) it follows that

$$\frac{T_n - n \sum_{m=1}^{n} m^{-1}}{n \log n} \to 0 \quad \text{in probability}$$

and hence $T_n/(n \log n) \to 1$ in probability. For a concrete example take $n = 500$. In this case the limit theorem says it will take about $500 \log 500 = 3107$ tries to get a complete set.

Example 5.4. Random permutations. Let Ω_n consist of the $n!$ permutations (i.e., one to one mappings from $\{1, \ldots, n\}$ onto $\{1, \ldots, n\}$) and make this into a probability space by assuming all the permutations are equally likely. This application of the weak law concerns the cycle structure of a random permutation π, so we begin by describing the decompostion of a permutation into cycles. Consider the sequence $1, \pi(1), \pi(\pi(1)), \ldots$ Eventually $\pi^k(1) = 1$. When it does we say the first cycle is completed and has length k. To start the second cycle we pick the smallest integer i not in the first cycle and look at $i, \pi(i), \pi(\pi(i)), \ldots$ until we come back to i. We repeat the construction until all the elements are accounted for. For example if the permutation is

i	1	2	3	4	5	6	7	8	9
$\pi(i)$	3	9	6	8	2	1	5	4	7

then the cycle decomposition is $(136)(2975)(48)$.

Let $X_{n,k} = 1$ if a right parenthesis occurs after the kth number in the decomposition, and let $S_n = X_{n,1} + \ldots + X_{n,n} = $ the number of cycles. (In the example $X_{9,3} = X_{9,7} = X_{9,9} = 1$ and the other $X_{9,m} = 0$.) I claim that

Lemma. $X_{n,1}, \ldots, X_{n,n}$ are independent and $P(X_{n,j} = 1) = 1/(n - j + 1)$.

Intuitively, this is true since, independent of what has happened so far, there are $n - j + 1$ values that have not appeared in the range, and only 1 of them will complete the cycle.

Proof To prove this, it is useful to generate the permutation in a special way. Let $i_1 = 1$. Pick j_1 at random from $\{1, \ldots, n\}$ and let $\pi(i_1) = j_1$. If $j_1 \neq 1$ let $i_2 = j_1$. If $j_1 = 1$ let $i_2 = 2$. In either case pick j_2 at random from $\{1, \ldots, n\} - \{j_1\}$. In general, if $i_1, j_1, \ldots, i_{k-1}, j_{k-1}$ have been selected and we have set $\pi(i_\ell) = j_\ell$ for $1 \leq \ell < k$ then (a) if $j_{k-1} \in \{i_1, \ldots, i_{k-1}\}$ so a cycle has just been completed, we let $i_k = \inf(\{1, \ldots, n\} - \{i_1, \ldots, i_k\})$ and (b) if $j_{k-1} \notin \{i_1, \ldots, i_{k-1}\}$ we let $i_k = j_{k-1}$. In either case we pick j_k at random from $\{1, \ldots, n\} - \{j_1, \ldots, j_{k-1}\}$ and let $\pi(i_k) = j_k$.

The construction above is tedious to write out, or read, but now I can claim with a clear conscience that $X_{n,1}, \ldots, X_{n,n}$ are independent and $P(X_{n,k} = 1) = 1/(n - j + 1)$ since when we pick j_k there are $n - j + 1$ values in $\{1, \ldots, n\} - \{j_1, \ldots, j_{k-1}\}$ and only 1 of them will complete the cycle.   □

To check the conditions of (5.4) now note

$$ES_n = 1/n + 1/(n - 1) + \cdots + 1/2 + 1$$

$$\mathrm{var}(S_n) = \sum_{k=1}^n \mathrm{var}(X_{n,k}) \leq \sum_{k=1}^n E(X_{n,k}^2) = \sum_{k=1}^n E(X_{n,k}) = ES_n$$

where the results on the second line follow from (5.1), (3.10b), and $X_{n,k}^2 = X_{n,k}$. Now $ES_n \sim \log n$, so if $b_n = (\log n)^{.5+\epsilon}$, the conditions of (5.4) are satisfied and it follows that

(∗)
$$\frac{S_n - \sum_{m=1}^n m^{-1}}{(\log n)^{.5+\epsilon}} \to 0$$

Taking $\epsilon = 0.5$ we have that $S_n / \log n \to 1$ in probability but (∗) says more. We will see in Example 4.7 of Chapter 2 that (∗) is false if $\epsilon = 0$.

Example 5.5. An occupancy problem. Suppose we put r balls at random in n boxes, i.e., all n^r assignments of balls to boxes have equal probability. Let A_i be the event that the ith box is empty, and $N_n =$ the number of empty boxes. It is easy to see that

$$P(A_i) = (1 - 1/n)^r \qquad \text{and} \qquad EN_n = n(1 - 1/n)^r$$

A little calculus (take logarithms) shows that if $r/n \to c$, $EN_n/n \to e^{-c}$. (For a proof see (1.3) in Chapter 2.) To compute the variance of N_n we observe that

$$EN_n^2 = E\left(\sum_{m=1}^{n} 1_{A_m}\right)^2 = \sum_{1 \le k,m \le n} P(A_k \cap A_m)$$

$$\text{var}(N_n) = EN_n^2 - (EN_n)^2 = \sum_{1 \le k,m \le n} P(A_k \cap A_m) - P(A_k)P(A_m)$$

$$= n(n-1)\{(1-2/n)^r - (1-1/n)^{2r}\} + n\{(1-1/n)^r - (1-1/n)^{2r}\}$$

The first term coming from $k \ne m$ and the second from $k = m$. Since $(1-2/n)^r \to e^{-2c}$ and $(1-1/n)^r \to e^{-c}$, it follows easily from the last formula that $\text{var}(N_n/n) = \text{var}(N_n)/n^2 \to 0$. Taking $b_n = n$ in (5.4) now we have

$$N_n/n \to e^{-c} \quad \text{in probability}$$

c. Truncation

To truncate a random variable X at level M means to consider

$$\bar{X} = X1_{(|X| \le M)} = \begin{cases} X & \text{if } |X| \le M \\ 0 & \text{if } |X| > M \end{cases}$$

To extend the weak law to random variables without a finite second moment, we will truncate and then use Chebyshev's inequality. We begin with a very general but also very useful result. Its proof is easy because we have assumed what we need for the proof. Later we will have to work a little to verify the assumptions in special cases but the general result serves to identify the essential ingredients in the proof.

(5.5) **Weak law for triangular arrays.** For each n let $X_{n,k}$, $1 \le k \le n$, be independent. Let $b_n > 0$ with $b_n \to \infty$, and let $\bar{X}_{n,k} = X_{n,k}1_{(|X_{n,k}| \le b_n)}$. Suppose that

(i) $\sum_{k=1}^{n} P(|X_{n,k}| > b_n) \to 0$, and

(ii) $b_n^{-2} \sum_{k=1}^{n} E\bar{X}_{n,k}^2 \to 0$ as $n \to \infty$.

If we let $S_n = X_{n,1} + \ldots + X_{n,n}$ and put $a_n = \sum_{k=1}^{n} E\bar{X}_{n,k}$ then

$$(S_n - a_n)/b_n \to 0 \quad \text{in probability}$$

Proof   Let $\bar{S}_n = \bar{X}_{n,1} + \cdots + \bar{X}_{n,n}$. Clearly,

$$P\left(\left|\frac{S_n - a_n}{b_n}\right| > \epsilon\right) \le P(S_n \ne \bar{S}_n) + P\left(\left|\frac{\bar{S}_n - a_n}{b_n}\right| > \epsilon\right)$$

To estimate the first term we note that

$$P(S_n \ne \bar{S}_n) \le P\left(\cup_{k=1}^n \{\bar{X}_{n,k} \ne X_{n,k}\}\right) \le \sum_{k=1}^n P(|X_{n,k}| > b_n) \to 0$$

by (i). For the second term we note that Chebyshev's inequality, $a_n = E\bar{S}_n$, (5.1), and $\text{var}(X) \le EX^2$ imply

$$P\left(\left|\frac{\bar{S}_n - a_n}{b_n}\right| > \epsilon\right) \le \epsilon^{-2} E\left|\frac{\bar{S}_n - a_n}{b_n}\right|^2 = \epsilon^{-2} b_n^{-2} \text{var}(\bar{S}_n)$$

$$= (b_n \epsilon)^{-2} \sum_{k=1}^n \text{var}(\bar{X}_{n,k}) \le (b_n \epsilon)^{-2} \sum_{k=1}^n E(\bar{X}_{n,k})^2 \to 0$$

by (ii) and the proof is complete. $\square$

From (5.5) we get the following result for a single sequence.

(5.6) **Weak law of large numbers.** Let $X_1, X_2, \ldots$ be i.i.d. with

$$xP(|X_i| > x) \to 0 \quad \text{as } x \to \infty$$

Let $S_n = X_1 + \cdots + X_n$ and let $\mu_n = E(X_1 1_{(|X_1| \le n)})$. Then $S_n/n - \mu_n \to 0$ in probability.

Remark.   The assumption in the theorem is necessary for the existence of constants a_n so that $S_n/n - a_n \to 0$. See Feller, Vol. II (1971) p. 234-236 for a proof.

Proof   We will apply (5.5) with $X_{n,k} = X_k$ and $b_n = n$. To check (i) we note

$$\sum_{k=1}^n P(|X_{n,k}| > n) = nP(|X_i| > n) \to 0$$

by assumption. To check (ii) we need to show $n^{-2} \cdot nE\bar{X}_{n,1}^2 \to 0$. To do this we need the following result, which will be useful several times below.

(5.7) **Lemma.** If $Y \geq 0$ and $p > 0$ then $E(Y^p) = \int_0^\infty py^{p-1} P(Y > y) \, dy$.

Proof Using the definition of expected value, Fubini's theorem (for nonnegative random variables), and then calculating the resulting integrals gives

$$
\int_0^\infty py^{p-1} P(Y > y) \, dy = \int_0^\infty \int_\Omega py^{p-1} 1_{(Y>y)} \, dP \, dy
$$

$$
= \int_\Omega \int_0^\infty py^{p-1} 1_{(Y>y)} \, dy \, dP
$$

$$
= \int_\Omega \int_0^Y py^{p-1} \, dy \, dP = \int_\Omega Y^p \, dP = EY^p \qquad \square
$$

Returning to the proof of (5.6), we observe that (5.7) and the fact that $\bar{X}_{n,1} = X_1 1_{(|X_1| \leq n)}$ imply

$$
E(\bar{X}_{n,1}^2) = \int_0^\infty 2y P(|\bar{X}_{n,1}| > y) \, dy \leq \int_0^n 2y P(|X_1| > y) \, dy
$$

since $P(|\bar{X}_{n,1}| > y) = 0$ for $y \geq n$ and $= P(|X_1| > y) - P(|X_1| > n)$ for $y \leq n$. We claim that $yP(|X_1| > y) \to 0$ implies

$$
E(X_{n,1}^2)/n = \frac{1}{n} \int_0^n 2y P(|X_1| > y) \, dy \to 0
$$

as $n \to \infty$. Intuitively this holds since the right-hand side is the average of $2yP(|X_1| > y)$ over $[0, n]$. To prove this, let $g(y) = 2yP(|X_1| > y)$. Since $0 \leq g(y) \leq 2y$ and $g(y) \to 0$ as $y \to \infty$ we must have $M = \sup g(y) < \infty$. If we let $\epsilon_K = \sup\{g(y) : y > K\}$ then by considering the integrals over $[0, K]$ and $[K, N]$ separately

$$
\int_0^n 2y P(|X_1| > y) \, dy \leq KM + (n - K)\epsilon_K
$$

Dividing by n and letting $n \to \infty$ we have

$$
\limsup_{n \to \infty} \frac{1}{n} \int_0^n 2y P(|X_1| > y) \, dy \leq \epsilon_K
$$

Since K is arbitrary and $\epsilon_K \to 0$ as $K \to \infty$ the desired result follows. $\square$

Finally, we have the weak law in its most familiar form.

(5.8) **Corollary.** Let $X_1, X_2, \ldots$ be i.i.d. with $E|X_i| < \infty$. Let $S_n = X_1 + \cdots + X_n$ and let $\mu = EX_1$. Then $S_n/n \to \mu$ in probability.

Remark. Applying (5.7) with $p = 1 - \epsilon$ and $\epsilon > 0$, we see that $xP(|X_1| > x) \to 0$ implies $E|X_1|^{1-\epsilon} < \infty$ so the assumption in (5.6) is not much weaker than finite mean.

Proof   Two applications of the dominated convergence theorem imply

$$xP(|X_1| > x) \leq E(|X_1|1_{(|X_1|>x)}) \to 0 \quad \text{as } x \to \infty$$
$$\mu_n = E(X_1 1_{(|X_1|\leq n)}) \to E(X_1) = \mu \quad \text{as } n \to \infty$$

Using (5.6) we see that if $\epsilon > 0$ then $P(|S_n/n - \mu_n| > \epsilon/2) \to 0$. Since $\mu_n \to \mu$ it follows that $P(|S_n/n - \mu_n| > \epsilon) \to 0$. □

Example 5.6.   For an example where the weak law does not hold, suppose $X_1, X_2, \ldots$ are independent and have a **Cauchy distribution**:

$$P(X_i \leq x) = \int_{-\infty}^{x} \frac{dt}{\pi(1 + t^2)}$$

As $x \to \infty$

$$P(|X_1| > x) = 2\int_{x}^{\infty} \frac{dt}{\pi(1 + t^2)} \sim \frac{2}{\pi} \int_{x}^{\infty} t^{-2}dt = \frac{2}{\pi}x^{-1}$$

From the necessity of the condition above we can conclude that there is no sequence of constants μ_n so that $S_n/n - \mu_n \to 0$. We will see later that S_n/n always has the same distribution as X_1. (See Exercise 3.10 in Chapter 2.)

As the next example shows we can have a weak law in some situations in which $E|X| = \infty$.

Example 5.7. The "St. Petersburg paradox".   Let $X_1, X_2, \ldots$ be independent random variables with

$$P(X_i = 2^j) = 2^{-j} \quad \text{for } j \geq 1.$$

In words, you win 2^j dollars if it takes j tosses to get a heads. The paradox here is that $EX_1 = \infty$ but you clearly wouldn't pay an infinite amount to play this game. An application of (5.5) will tell us how much we should pay to play the game n times.

In this example $X_{n,k} = X_k$. To apply (5.5) we have to pick b_n. To do this, we are guided by the principle that in checking (ii) we want to take b_n as small as we can and have (i) hold. With this in mind, we observe that if m is an integer

$$P(X_1 \geq 2^m) = \sum_{j=m}^{\infty} 2^{-j} = 2^{-m+1}$$

Let $m(n) = \log_2 n + K(n)$ where $K(n) \to \infty$ and is chosen so that $m(n)$ is an integer (and hence the displayed formula is valid). Letting $b_n = 2^{m(n)}$ we have

$$nP(X_1 \geq b_n) = n2^{-m(n)+1} = 2^{-K(n)+1} \to 0$$

proving (i). To check (ii) we observe that if $\bar{X}_{n,k} = X_k 1_{(|X_k| \leq b_n)}$ then

$$E\bar{X}_{n,k}^2 = \sum_{j=1}^{m(n)} 2^{2j} \cdot 2^{-j} \leq 2^{m(n)} \sum_{k=0}^{\infty} 2^{-k} = 2b_n$$

So the expression in (ii) is smaller than $2n/b_n$ which $\to 0$ since

$$b_n = 2^{m(n)} = n2^{K(n)} \quad \text{and} \quad K(n) \to \infty$$

The last step is to evaluate a_n. $E\bar{X}_{n,k} = \sum_{n=1}^{m(n)} 2^n 2^{-n} = m(n)$, so $a_n = nm(n)$. We have $m(n) = \log n + K(n)$ (here and until the end of the example all logs are base 2), so if we pick $K(n)/\log n \to 0$ then $a_n/n\log n \to 1$ as $n \to \infty$. If we pick

$$K(n) = \sup\{k \leq \log\log n \text{ such that } \log n + k \text{ is an integer}\}$$

then $m(n) \leq \log n + \log\log n$ so $b_n = 2^{m(n)} \leq n\log n$ and (5.5) implies

$$(S_n - n\log n)/n\log n \to 0 \quad \text{in probability}$$

That is, $S_n/(n\log n) \to 1$ in probability.

Returning to our original question, we see that a fair price for playing n times is \$ $\log_2 n$ per play. When $n = 1024$ this is \$10 per play. Nicolas Bernoulli wrote in 1713, "there ought not to exist any even halfway sensible person who would not sell the right of playing the game for 40 ducates." If the wager were 1 ducat, one would need $2^{40} \approx 10^{12}$ plays to start to break even.

EXERCISES

5.1. Let $X_1, X_2, \ldots$ be uncorrelated random variables with $EX_i = \mu_i$ and $\text{var}(X_i)/i \to 0$ as $i \to \infty$. Let $S_n = X_1 + \ldots + X_n$ and $\nu_n = ES_n/n$ then as $n \to \infty$, $S_n/n - \nu_n \to 0$ in L^2 and in probability.

5.2. The L^2 weak law generalizes immediately to certain dependent sequences. Suppose $EX_n = 0$ and $EX_n X_m \le r(n-m)$ for $m \le n$ (no absolute value on the left-hand side!) with $r(k) \to 0$ as $k \to \infty$. Show that $(X_1 + \ldots + X_n)/n \to 0$ in probability.

5.3. Monte Carlo integration. (i) Let f be a measurable function on $[0,1]$ with $\int_0^1 |f(x)| \, dx < \infty$. Let $U_1, U_2, \ldots$ be independent and uniformly distributed on $[0,1]$ and let

$$I_n = n^{-1}(f(U_1) + \ldots + f(U_n))$$

Show that $I_n \to I \equiv \int_0^1 f \, dx$ in probability. (ii) Suppose $\int_0^1 |f(x)|^2 \, dx < \infty$. Use Chebyshev's inequality to estimate $P(|I_n - I| > a/n^{1/2})$.

5.4. Let $X_1, X_2, \ldots$ be i.i.d. with $P(X_i = (-1)^k k) = C/k^2 \log k$ for $k \ge 2$ where C is chosen to make the sum of the probabilities $= 1$. Show that $E|X_i| = \infty$ but there is a finite constant μ so that $S_n/n \to \mu$ in probability.

5.5. Let $X_1, X_2, \ldots$ be i.i.d. with $P(X_i > x) = e/x \log x$ for $x \ge e$. Show that $E|X_i| = \infty$ but there is a sequence of constants $\mu_n \to \infty$ so that $S_n/n - \mu_n \to 0$ in probability.

5.6. (i) Show that if $X \ge 0$ is integer valued $EX = \sum_{n \ge 1} P(X \ge n)$. (ii) Find a similar expression for EX^2.

5.7. Generalize (5.6) to conclude that if $H(x) = \int_{(-\infty, x]} h(y) \, dy$ with $h(y) \ge 0$ then

$$E\, H(X) = \int_{-\infty}^{\infty} h(y) P(X \ge y) \, dy$$

An important special case is $H(x) = \exp(\theta x)$ with $\theta > 0$.

5.8. An unfair "fair game". Let $p_k = 1/2^k k(k+1)$, $k = 1, 2, \ldots$ and $p_0 = 1 - \sum_{k \ge 1} p_k$.

$$\sum_{k=1}^{\infty} 2^k p_k = \left(1 - \frac{1}{2}\right) + \left(\frac{1}{2} - \frac{1}{3}\right) + \ldots = 1$$

so if we let $X_1, X_2, \ldots$ be i.i.d. with $P(X_n = -1) = p_0$ and

$$P(X_n = 2^k - 1) = p_k \quad \text{for } k \ge 1$$

then $EX_n = 0$. Let $S_n = X_1 + \ldots + X_n$. Use (5.5) with $b_n = 2^{m(n)}$ where $m(n) = \min\{m : 2^{-m}m^{-3/2} \le n^{-1}\}$ to conclude that

$$S_n/(n/\log_2 n) \to -1 \text{ in probability}$$

5.9. Weak law for positive variables. Suppose $X_1, X_2, \ldots$ are i.i.d. and ≥ 0. Let $\mu(s) = \int_0^s x\, dF(x)$ and $\nu(s) = \mu(s)/s(1 - F(s))$. It is known that there exist constants a_n so that $S_n/a_n \to 1$ in probability, if and only if $\nu(s) \to \infty$ as $s \to \infty$. Pick $b_n \ge 1$ so that $n\mu(b_n) = b_n$ (this works for large n) and use (5.5) to prove that the condition is sufficient.

1.6. Borel-Cantelli Lemmas

If A_n is a sequence of subsets of Ω we let

$$\limsup_{m \to \infty} A_n = \lim_{m \to \infty} \cup_{n=m}^{\infty} A_n = \{\omega \text{ that are in infinitely many } A_n\}$$

(the limit exists since the sequence is decreasing in m) and let

$$\liminf A_n = \lim_{m \to \infty} \cap_{n=m}^{\infty} A_n = \{\omega \text{ that are in all but finitely many } A_n\}$$

(the limit exists since the sequence is increasing in m). The names limsup and liminf can be explained by noting that

$$\limsup_{n \to \infty} 1_{A_n} = 1_{(\limsup A_n)} \qquad \liminf_{n \to \infty} 1_{A_n} = 1_{(\liminf A_n)}$$

It is common to write $\limsup A_n = \{\omega : \omega \in A_n \text{ i.o.}\}$ where i.o. stands for infinitely often. An example which illustrates the use of this notation is: "$X_n \to 0$ a.s. if and only if for all $\epsilon > 0$, $P(|X_n| > \epsilon \text{ i.o.}) = 0$." The reader will see many other examples below. The next result should be familiar from measure theory even though its name may not be.

(6.1) Borel-Cantelli Lemma. If $\sum_{n=1}^{\infty} P(A_n) < \infty$ then $P(A_n \text{ i.o.}) = 0$.

Proof   Let $N = \sum_k 1_{A_k}$ be the number of events that occur. Fubini's theorem implies $EN = \sum_k P(A_k) < \infty$, so we must have $N < \infty$ a.s.   □

The next result is a typical application of the Borel-Cantelli lemma.

(6.2) Theorem. $X_n \to X$ in probability if and only if for every subsequence $X_{n(m)}$ there is a further subsequence $X_{n(m_k)}$ that converges almost surely to X.

Proof Let ϵ_k be a sequence of positive numbers that $\downarrow 0$. For each k there is an $n(m_k) > n(m_{k-1})$ so that $P(|X_{n(m_k)} - X| > \epsilon_k) \leq 2^{-k}$. Since

$$\sum_{k=1}^{\infty} P(|X_{n(m_k)} - X| > \epsilon_k) < \infty$$

the Borel-Cantelli lemma implies $P(|X_{n(m_k)} - X| > \epsilon_k \text{ i.o.}) = 0$, i.e., $X_{n(m_k)} \to X$ a.s. To prove the second conclusion we note that if for every subsequence $X_{n(m)}$ there is a further subsequence $X_{n(m_k)}$ that converges almost surely to X then we can apply the next lemma to the sequence of numbers $y_n = P(|X_n - X| > \delta)$ for any $\delta > 0$ to get the desired result

(6.3) Lemma. Let y_n be a sequence of elements of a topological space. If every subsequence $y_{n(m)}$ has a further subseqeunce $y_{n(m_k)}$ that converges to y then $y_n \to y$

Proof If $y_n \not\to y$ then there is an open set G containing y and a subsequence $y_{n(m)}$ with $y_{n(m)} \notin G$ for all m, but clearly no subsequence of $y_{n(m)}$ converges to y.   □

Remark. Since there is a sequence of random variables which converges in probability but not a.s. (for an example see Exercises 6.13 or 6.14), it follows from (6.3) that a.s. convergence does not come from a metric, or even from a topology. Exercises 6.4 and 6.5 will give a metric for convergence in probability, and show that the space of random variables is a complete space under this metric.

An example of the usefulness of (6.2) is

(6.4) Corollary. If f is continuous and $X_n \to X$ in probability then $f(X_n) \to f(X)$ in probability. If in addition f is bounded then $Ef(X_n) \to Ef(X)$.

Proof If $X_{n(m)}$ is a subsequence then (6.2) implies there is a further subsequence $X_{n(m_k)} \to X$ almost surely. Since f is continuous, Exercise 2.3 implies $f(X_{n(m_k)}) \to f(X)$ almost surely and (6.2) implies $f(X_n) \to f(X)$ in probability. If f is bounded then the bounded convergenece theorem implies $Ef(X_{n(m_k)}) \to Ef(X)$ and applying (6.3) to $y_n = Ef(X_n)$ gives the desired result.   □

EXERCISE 6.1. Prove the first result in (6.4) directly from the definition.

EXERCISE 6.2. **Fatou's lemma.** Suppose $X_n \geq 0$ and $X_n \to X$ in probability. Show that $\liminf_{n\to\infty} EX_n \geq EX$.

EXERCISE 6.3. **Dominated convergence.** Suppose $X_n \to X$ in probability and (a) $|X_n| \le Y$ with $EY < \infty$ or (b) there is a continuous function g with $g(x) > 0$ for large x with $|x|/g(x) \to \infty$ as $|x| \to \infty$ so that $Eg(X_n) \le C < \infty$ for all n. Show that $EX_n \to EX$.

EXERCISE 6.4. Show (a) that $d(X,Y) = E(|X - Y|/(1 + |X - Y|))$ defines a metric on the set of random variables, i.e., (i) $d(X,Y) = 0$ if and only if $X = Y$ a.s., (ii) $d(X,Y) = d(Y,X)$, (iii) $d(X,Z) \le d(X,Y) + d(Y,Z)$ and (b) that $d(X_n, X) \to 0$ as $n \to \infty$ if and only if $X_n \to X$ in probability.

EXERCISE 6.5. Show that random variables are a complete space under the metric defined in the previous exercise, i.e., if $d(X_m, X_n) \to 0$ whenever m, $n \to \infty$ then there is a r.v. X_∞ so that $X_n \to X_\infty$ in probability.

As our second application of the Borel-Cantelli lemma we get our first strong law of large numbers:

(6.5) Theorem. Let $X_1, X_2, \ldots$ be i.i.d. with $EX_i = \mu$ and $EX_i^4 < \infty$. If $S_n = X_1 + \cdots + X_n$ then $S_n/n \to \mu$ a.s.

Proof By letting $X_i' = X_i - \mu$ we can suppose without loss of generality that $\mu = 0$. Now

$$ES_n^4 = E\left(\sum_{i=1}^{n} X_i\right)^4 = E \sum_{1 \le i,j,k,\ell \le n} X_i X_j X_k X_\ell$$

Terms in the sum of the form $E(X_i^3 X_j)$, $E(X_i^2 X_j X_k)$, and $E(X_i X_j X_k X_\ell)$ are 0 (if i, j, k, ℓ are distinct) since the expectation of the product is the product of the expectations, and in each case one of the terms has expectation 0. The only terms that do not vanish are those of the form EX_i^4 and $EX_i^2 X_j^2 = (EX_i^2)^2$. There are n and $3n(n-1)$ of these terms respectively. (In the second case we can pick the two indices in $n(n-1)/2$ ways and with the indices fixed the term can arise in a total of 6 ways.) The last observation implies

$$ES_n^4 = nEX_1^4 + 3(n^2 - n)(EX_1^2)^2 \le Cn^2$$

where $C < \infty$. Chebyshev's inequality gives us

$$P(|S_n| > n\epsilon) \le E(S_n^4)/(n\epsilon)^4 \le C/(n^2 \epsilon^4)$$

Summing on n and using the Borel-Cantelli lemma gives $P(|S_n| > n\epsilon \text{ i.o.}) = 0$. Since ϵ is arbitrary, the proof is complete. $\square$

The converse of the Borel-Cantelli lemma is trivially false:

Example 6.1. Let $\Omega = (0,1)$, $\mathcal{F}$ = Borel sets, P = Lebesgue measure. If $A_n = (0, a_n)$ where $a_n \to 0$ as $n \to \infty$ then $\limsup A_n = \emptyset$ but if $a_n \geq 1/n$ we have $\sum a_n = \infty$.

The example just given suggests that for general sets we cannot say much more than the next result.

EXERCISE 6.6. Prove that $P(\limsup A_n) \geq \limsup P(A_n)$ and
$\qquad P(\liminf A_n) \leq \liminf P(A_n)$

For independent events, however, the necessary condition for $P(\limsup A_n) > 0$ is sufficient for $P(\limsup A_n) = 1$.

(6.6) The second Borel-Cantelli lemma. If the events A_n are independent then $\sum P(A_n) = \infty$ implies $P(A_n \text{ i.o.}) = 1$.

Proof   Let $M < N < \infty$. Independence and $1 - x \leq e^{-x}$ imply

$$P\left(\cap_{n=M}^N A_n^c\right) = \prod_{n=M}^N (1 - P(A_n)) \leq \prod_{n=M}^N \exp(-P(A_n))$$

$$= \exp\left(-\sum_{n=M}^N P(A_n)\right) \to 0 \quad \text{as } N \to \infty$$

So $P(\cup_{n=M}^\infty A_n) = 1$ for all M, and since $\cup_{n=M}^\infty A_n \downarrow \limsup A_n$ it follows that $P(\limsup A_n) = 1$. $\square$

A typical application of the second Borel-Cantelli lemma is:

(6.7) Theorem. If $X_1, X_2, \ldots$ are i.i.d. with $E|X_i| = \infty$, then $P(|X_n| \geq n \text{ i.o.}) = 1$. So if $S_n = X_1 + \cdots + X_n$ then $P(\lim S_n/n \text{ exists} \in (-\infty, \infty)) = 0$.

Proof   From (5.7) we get

$$E|X_1| = \int_0^\infty P(|X_1| > x)\,dx \leq \sum_{n=0}^\infty P(|X_1| > n)$$

Since $E|X_1| = \infty$ and $X_1, X_2, \ldots$ are i.i.d., it follows from the second Borel-Cantelli lemma that $P(|X_n| \geq n \text{ i.o.}) = 1$. To prove the second claim observe that

$$\frac{S_n}{n} - \frac{S_{n+1}}{n+1} = \frac{S_n}{n(n+1)} - \frac{X_{n+1}}{n+1}$$

and on $C \equiv \{\omega : \lim_{n\to\infty} S_n/n$ exists $\in (-\infty, \infty)\}$, $S_n/(n(n+1)) \to 0$. So on $C \cap \{\omega : |X_n| \geq n$ i.o.$\}$ we have

$$\left| \frac{S_n}{n} - \frac{S_{n+1}}{n+1} \right| > 2/3 \quad \text{i.o.}$$

contradicting the fact that $\omega \in C$. From the last observation we conclude that

$$\{\omega : |X_n| \geq n \text{ i.o.}\} \cap C = \emptyset$$

and since $P(|X_n| \geq n$ i.o.$) = 1$, it follows that $P(C) = 0$. $\qquad\square$

(6.7) shows that $E|X_i| < \infty$ is necessary for the strong law of large numbers. The reader will have to wait until (7.1) to see that condition is also sufficient. The next result extends the second Borel-Cantelli lemma and sharpens its conclusion.

(6.8) Theorem. If $A_1, A_2, \ldots$ are pairwise independent and $\sum_{n=1}^{\infty} P(A_n) = \infty$ then as $n \to \infty$

$$\sum_{m=1}^{n} 1_{A_m} \Big/ \sum_{m=1}^{n} P(A_m) \to 1 \quad \text{a.s.}$$

Proof Let $X_m = 1_{A_m}$ and let $S_n = X_1 + \cdots + X_n$. Since the A_m are pairwise independent, the X_m are uncorrelated and hence (5.1) implies

$$\text{var}(S_n) = \text{var}(X_1) + \cdots + \text{var}(X_n)$$

(3.10b) and the fact $X_m \in \{0, 1\}$ imply $\text{var}(X_m) \leq E(X_m^2) = E(X_m)$, so $\text{var}(S_n) \leq E(S_n)$. Chebyshev's inequality implies

$$(*) \qquad P(|S_n - ES_n| > \delta ES_n) \leq \text{var}(S_n)/(\delta ES_n)^2 \leq 1/(\delta^2 ES_n) \to 0$$

as $n \to \infty$. (Since we have assumed $ES_n \to \infty$.)

The last computation shows that $S_n/ES_n \to 1$ in probability. To get almost sure convergence we have to take subsequences. Let $n_k = \inf\{n : ES_n \geq k^2\}$. Let $T_k = S_{n_k}$ and note that the defintion and $EX_m \leq 1$ imply $k^2 \leq ET_k \leq k^2 + 1$. Replacing n by n_k in $(*)$ and using $ET_k \geq k^2$ shows

$$P(|T_k - ET_k| > \delta ET_k) \leq 1/(\delta^2 k^2)$$

So $\sum_{k=1}^{\infty} P(|T_k - ET_k| > \delta ET_k) < \infty$, and the Borel-Cantelli lemma implies $P(|T_k - ET_k| > \delta ET_k$ i.o.$) = 0$. Since δ is arbitrary, it follows that $T_k/ET_k \to 1$

a.s. To show $S_n/ES_n \to 1$ a.s., pick an ω so that $T_k(\omega)/ET_k \to 1$ and observe that if $n_k \leq n < n_{k+1}$ then

$$\frac{T_k(\omega)}{ET_{k+1}} \leq \frac{S_n(\omega)}{ES_n} \leq \frac{T_{k+1}(\omega)}{ET_k}$$

To show that the terms at the left and right ends $\to 1$, we rewrite the last inequalities as

$$\frac{ET_k}{ET_{k+1}} \cdot \frac{T_k(\omega)}{ET_k} \leq \frac{S_n(\omega)}{ES_n} \leq \frac{T_{k+1}(\omega)}{ET_{k+1}} \cdot \frac{ET_{k+1}}{ET_k}$$

From this we see it is enough to show $ET_{k+1}/ET_k \to 1$ but this follows from

$$k^2 \leq ET_k \leq ET_{k+1} \leq (k+1)^2 + 1$$

and the fact that $\{(k+1)^2 + 1\}/k^2 = 1 + 2/k + 2/k^2 \to 1$. $\qquad\square$

The moral of the proof of (6.8) is that if you want to show that $X_n/c_n \to 1$ a.s. for sequences $c_n, X_n \geq 0$ that are increasing, it is enough to prove the result for a subsequence $n(k)$ that has $c_{n(k+1)}/c_{n(k)} \to 1$. For practice with this technique try the following.

EXERCISE 6.7. Let $0 \leq X_1 \leq X_2 \ldots$ be random variables with $EX_n \sim an^\alpha$ with $a, \alpha > 0$, and $\text{var}(X_n) \leq Bn^\beta$ with $\beta < 2\alpha$. Show that $X_n/n^\alpha \to a$ a.s.

EXERCISE 6.8. Let X_n be independent Poisson r.v.'s with $EX_n = \lambda_n$ and let $S_n = X_1 + \cdots + X_n$. Show that if $\sum \lambda_n = \infty$ then $S_n/ES_n \to 1$ a.s.

Example 6.2. Record values. Let $X_1, X_2, \ldots$ be a sequence of random variables and think of X_k as the distance for an individual's kth high jump or shot put toss so that $A_k = \{X_k > \sup_{j<k} X_j\}$ is the event that a record occurs at time k. Ignoring the fact that an athelete's performance may get better with more experience or that injuries may occur we will suppose that $X_1, X_2, \ldots$ are i.i.d. with a distribution $F(x)$ that is continuous. Even though it may seem that the occurrence of a record at time k will make it less likely that one will occur at time $k + 1$, we

Claim. The A_k are independent with $P(A_k) = 1/k$.

To prove this we start by observing that since F is continuous $P(X_j = X_k) = 0$ for any $j \neq k$ (see Exercise 4.8) so we can let $Y_1^n > Y_2^n > \cdots > Y_n^n$ be the random variables $X_1, \ldots, X_n$ put into decreasing order and define a random permutation of $\{1, \ldots, n\}$ by $\pi_n(i) = j$ if $X_i = Y_j^n$, i.e., if the ith random variable has rank j. Since the distribution of $(X_1, \ldots, X_n)$ is not affected by changing the order of the random variables, it is easy to see

(a) The permutation π_n is uniformly distributed over the set of $n!$ possibilities.

Proof of (a) This is "obvious" by symmetry, but if one wants to hear more, we can argue as follows. Let π_n be the permutation induced by $(X_1, \ldots, X_n)$, and let σ_n be a randomly chosen permutation of $\{1, \ldots, n\}$ independent of the X sequence. Then we can say two things about the permutation induced by $(X_{\sigma(1)}, \ldots, X_{\sigma(n)})$: (i) it is $\pi_n \circ \sigma_n$ and (ii) it has the same distribution as π_n. The desired result follows now by noting that if π is any permutation, $\pi \circ \sigma_n$, is uniform over the $n!$ possibilities. $\square$

Once you believe (a) the rest is easy:

(b) $P(A_n) = P(\pi_n(n) = 1) = 1/n$.

(c) If $m < n$ and $i_{m+1}, \ldots i_n$ are distinct elements of $\{1, \ldots, n\}$ then

$$P(A_m | \pi_n(j) = i_j \text{ for } m + 1 \le j \le n) = 1/m$$

Intuitively this is true since if we condition on the ranks of $X_{m+1}, \ldots, X_n$ then this determines the set of ranks available for $X_1, \ldots, X_m$, but all possible orderings of the ranks are equally likely and hence there is probability $1/m$ that the smallest rank will end up at m.

Proof of (c) If we let σ_m be a randomly chosen permutation of $\{1, \ldots, m\}$ then (i) $\pi_n \circ \sigma_m$ has the same distribution as π_n and (ii) since the application of σ_m randomly rearranges $\pi_n(1), \ldots, \pi_n(m)$ the desired result follows. $\square$

If we let $m_1 < m_2 \ldots < m_k$ then it follows from (c) that

$$P(A_{m_1} | A_{m_2} \cap \ldots \cap A_{m_k}) = P(A_{m_1}) P(A_{m_2} \cap \ldots \cap A_{m_k})$$

and the claim follows by induction.

From (6.8) and the by now familiar fact that $\sum_{m=1}^n 1/m \sim \log n$ it follows that if $R_n = \sum_{m=1}^n 1_{A_m}$ is the number of records at time n then as $n \to \infty$,

$$(6.9) \qquad\qquad R_n / \log n \to 1 \quad \text{a.s.}$$

The reader should note that the last result is independent of the distribution F (as long as it is continuous).

Remark. Let $X_1, X_2, \ldots$ be i.i.d. with a distribution that is continuous. Let Y_i be the number of $j \leq i$ with $X_j > X_i$. It follows from (a) that Y_i are independent random variables with $P(Y_i = j) = 1/i$ for $0 \leq j < i - 1$.

Comic relief. Let $X_0, X_1, \ldots$ be i.i.d. and imagine they are the offers you get for a car you are going to sell. Let $N = \inf\{n \geq 1 : X_n > X_0\}$. Symmetry implies $P(N > n) \geq 1/(n+1)$. (When the distribution is continuous this probability is exactly $1/(n+1)$, but our distribution now is general and ties go to the first person who calls.) Using Exercise 5.6 now

$$EN = \sum_{n=0}^{\infty} P(N > n) = \sum_{n=0}^{\infty} \frac{1}{n+1} = \infty$$

so the expected time you have to wait until you get an offer better than the first one is ∞. To avoid lawsuits let me hasten to add that I am not suggesting that you should take the first offer you get!

Example 6.3.   Head runs. Let X_n, $n \in \mathbf{Z}$, be i.i.d. with $P(X_n = 1) = P(X_n = -1) = 1/2$. Let $\ell_n = \max\{m : X_{n-m+1} = \ldots = X_n = 1\}$ be the length of the run of $+1$'s at time n, and let $L_n = \max_{1 \leq m \leq n} \ell_m$ be the longest run at time n. We use a two sided sequence so that for all n, $P(\ell_n = k) = (1/2)^{k+1}$ for $k \geq 0$. Since $\ell_1 < \infty$ the result we are going to prove

(6.10) $L_n / \log_2 n \to 1$ a.s.

is also true for a one sided sequence. To prove (6.10) we begin by observing

$$P(\ell_n \geq (1+\epsilon) \log_2 n) \leq n^{-(1+\epsilon)}$$

for any $\epsilon > 0$ so it follows from the Borel-Cantelli lemma that $\ell_n \leq (1+\epsilon) \log_2 n$ for $n \geq N_\epsilon$. Since ϵ is arbitrary it follows that

$$\limsup_{n \to \infty} L_n / \log_2 n \leq 1 \quad \text{a.s.}$$

To get a result in the other direction we break the first n trials into disjoint blocks of length $[(1 - \epsilon) \log_2 n] + 1$, on which the variables are all 1 with probability $2^{-[(1-\epsilon)\log_2 n]-1} \geq n^{-(1-\epsilon)}/2$ to conclude that if n is large enough so that $[n/\{[(1 - \epsilon) \log_2 n] + 1\}] \geq n/\log_2 n$

$$P(L_n \leq (1 - \epsilon) \log_2 n) \leq (1 - n^{-(1-\epsilon)}/2)^{n/(\log_2 n)} \leq \exp(-n^\epsilon/2\log_2 n)$$

which is summable, so the Borel-Cantelli lemma implies

$$\liminf_{n \to \infty} L_n / \log_2 n \geq 1 \quad \text{a.s.}$$

EXERCISE 6.9. Show that $\limsup_{n\to\infty} \ell_n/\log_2 n = 1$, $\liminf_{n\to\infty} \ell_n = 0$ a.s.

EXERCISES

6.10. If X_n is any sequence of random variables, there are constants $c_n \to \infty$ so that $X_n/c_n \to 0$ a.s.

6.11. (i) If $P(A_n) \to 0$ and $\sum_{n=1}^{\infty} P(A_n^c \cap A_{n+1}) < \infty$ then $P(A_n \text{ i.o.}) = 0$. (ii) Find an example of a sequence A_n to which the result in (i) can be applied but the Borel-Cantelli lemma cannot.

6.12. Let A_n be a sequence of independent events with $P(A_n) < 1$ for all n. Show that $P(\cup A_n) = 1$ implies $P(A_n \text{ i.o.}) = 1$.

6.13. Let $X_1, X_2, \ldots$ be independent. Show that $\sup X_n < \infty$ if and only if $\sum_n P(X_n > A) < \infty$ for some A.

6.14. Let $X_1, X_2, \ldots$ be independent with $P(X_n = 1) = p_n$ and $P(X_n = 0) = 1 - p_n$. Show that (i) $X_n \to 0$ in probability if and only if $p_n \to 0$, and (ii) $X_n \to 0$ a.s. if and only if $\sum p_n < \infty$.

6.15. Let $Y_1, Y_2, \ldots$ be i.i.d. Find necessary and sufficient conditions for (i) $Y_n/n \to 0$ almost surely, (ii) $(\max_{m \le n} Y_m)/n \to 0$ almost surely, (iii) $(\max_{m \le n} Y_m)/n \to 0$ in probability, (iv) $Y_n/n \to 0$ in probability.

6.16. The last two exercises give examples with $X_n \to X$ in probability without $X_n \to X$ a.s. There is one situation in which the two notions are equivalent. Let $X_1, X_2, \ldots$ be a sequence of r.v.'s on $(\Omega, \mathcal{F}, P)$ where Ω is a countable set. Show that $X_n \to X$ in probability implies $X_n \to X$ a.s.

6.17. Show that if X_n is the outcome of the nth play of the St. Petersburg game (Example 5.7) then $\limsup_{n\to\infty} X_n/(n \log_2 n) = \infty$ a.s. and hence the same result holds for S_n. This shows that the convergence $S_n/(n \log_2 n) \to 1$ in probability proved in Section 5 does not occur a.s.

6.18. Let $X_1, X_2, \ldots$ be i.i.d. with $P(X_i > x) = e^{-x}$, let $M_n = \max_{1 \le m \le n} X_m$. Show that (i) $\limsup_{n\to\infty} X_n/\log n = 1$ and (ii) $M_n/\log n \to 1$ a.s.

6.19. Let $X_1, X_2, \ldots$ be i.i.d. with distribution F, let $\lambda_n \uparrow \infty$, and let $A_n = \{\max_{1 \le m \le n} X_m > \lambda_n\}$. Show that $P(A_n \text{ i.o.}) = 0$ or 1 according as $\sum_{n \ge 1}(1 - F(\lambda_n)) < \infty$ or $= \infty$.

6.20. **Kochen-Stone lemma.** Suppose $\sum P(A_k) = \infty$. Use Exercises 3.8 and 6.6 to show that if

$$\limsup_{n\to\infty} \left(\sum_{k=1}^{n} P(A_k) \right)^2 \bigg/ \left(\sum_{1 \le j,k \le n} P(A_j \cap A_k) \right) = \alpha > 0.$$

then $P(A_n \text{ i.o.}) \geq \alpha$. The case $\alpha = 1$ contains (6.6).

1.7. Strong Law Of Large Numbers

We are now ready to give Etemadi's proof of

(7.1) Strong law of large numbers. Let $X_1, X_2, \ldots$ be pairwise independent identically distributed random variables with $E|X_i| < \infty$. Let $EX_i = \mu$ and $S_n = X_1 + \ldots + X_n$. Then $S_n/n \to \mu$ a.s. as $n \to \infty$.

Proof As in the proof of weak law of large numbers, we begin by truncating.

(a) Lemma. Let $Y_k = X_k 1_{(|X_k| \leq k)}$ and $T_n = Y_1 + \cdots + Y_n$. It is sufficient to prove that $T_n/n \to \mu$ a.s.

Proof $\sum_{k=1}^{\infty} P(|X_k| > k) \leq \int_0^\infty P(|X_1| > t)\, dt = E|X_1| < \infty$ so $P(X_k \neq Y_k \text{ i.o.}) = 0$. This shows that $|S_n(\omega) - T_n(\omega)| \leq R(\omega) < \infty$ a.s. for all n, from which the desired result follows. $\qquad\square$

The second step is not so intuitive but it is an important part of this proof and the one given in Section 1.8.

(b) Lemma. $\sum_{k=1}^{\infty} \operatorname{var}(Y_k)/k^2 \leq 4E|X_1| < \infty$.

Proof To bound the sum, we observe

$$\operatorname{var}(Y_k) \leq E(Y_k^2) = \int_0^\infty 2y P(|Y_k| > y)\, dy \leq \int_0^k 2y P(|X_1| > y)\, dy$$

so using Fubini's theorem (since everything is ≥ 0 and the sum is just an integral with respect to counting measure on $\{1, 2, \ldots\}$)

$$\sum_{k=1}^{\infty} E(Y_k^2)/k^2 \leq \sum_{k=1}^{\infty} k^{-2} \int_0^\infty 1_{(y<k)}\, 2y\, P(|X_1| > y)\, dy$$

$$= \int_0^\infty \left\{ \sum_{k=1}^{\infty} k^{-2} 1_{(y<k)} \right\} 2y P(|X_1| > y)\, dy$$

Since $E|X_1| = \int_0^\infty P(|X_1| > y)\, dy$, we can complete the proof by showing

(c) Lemma. If $y \geq 0$ then $2y \sum_{k>y} k^{-2} \leq 4$.

Proof We being with the observation that if $m \geq 2$ then

$$\sum_{k \geq m} k^{-2} \leq \int_{m-1}^{\infty} x^{-2} dx = (m-1)^{-1}$$

When $y \geq 1$ the sum starts with $k = [y] + 1 \geq 2$ so

$$2y \sum_{k > y} k^{-2} \leq 2y/[y] \leq 4$$

since $y/[y] \leq 2$ for $y \geq 1$ (the worst case being y close to 2). To cover $0 \leq y < 1$ we note that in this case

$$2y \sum_{k > y} k^{-2} \leq 2 \left(1 + \sum_{k=2}^{\infty} k^{-2} \right) \leq 4 \qquad \square$$

The first two steps, (a) and (b) above, are standard. Etemadi's inspiration was that since X_n^+, $n \geq 1$, and X_n^-, $n \geq 1$, satisfy the assumptions of the theorem and $X_n = X_n^+ - X_n^-$, we can without loss of generality suppose $X_n \geq 0$. As in the proof of (6.8) we will prove the result first for a subsequence and then use monotonicity to control the values in between. This time however, we let $\alpha > 1$, and $k(n) = [\alpha^n]$. Chebyshev's inequality implies that if $\epsilon > 0$

$$\sum_{n=1}^{\infty} P(|T_{k(n)} - ET_{k(n)}| > \epsilon k(n)) \leq \epsilon^{-2} \sum_{n=1}^{\infty} \text{var}(T_{k(n)})/k(n)^2$$

$$= \epsilon^{-2} \sum_{n=1}^{\infty} k(n)^{-2} \sum_{m=1}^{k(n)} \text{var}(Y_m)$$

$$= \epsilon^{-2} \sum_{m=1}^{\infty} \text{var}(Y_m) \sum_{n:k(n) \geq m} k(n)^{-2}$$

where we have used Fubini's theorem to interchange the two summations (everything is ≥ 0). Now $k(n) = [\alpha^n]$ and $[\alpha^n] \geq \alpha^n/2$ for $n \geq 1$, so summing the geometric series and noting that the first term is $\leq m^{-2}$

$$\sum_{n:\alpha^n \geq m} [\alpha^n]^{-2} \leq 4 \sum_{n:\alpha^n \geq m} \alpha^{-2n} \leq 4(1 - \alpha^{-2})^{-1} m^{-2}$$

Combining our computations shows

$$\sum_{n=1}^{\infty} P(|T_{k(n)} - ET_{k(n)}| > \epsilon k(n)) \leq 4(1 - \alpha^{-2})^{-1} \epsilon^{-2} \sum_{m=1}^{\infty} E(Y_m^2) m^{-2} < \infty$$

by (b). Since ϵ is arbitrary $(T_{k(n)} - ET_{k(n)})/k(n) \to 0$. The dominated convergence theorem implies $EY_k \to EX_1$ as $k \to \infty$, so $ET_{k(n)}/k(n) \to EX_1$ and we have shown $T_{k(n)}/k(n) \to EX_1$ a.s. To handle the intermediate values, we observe that if $k(n) \le m < k(n+1)$

$$\frac{T_{k(n)}}{k(n+1)} \le \frac{T_m}{m} \le \frac{T_{k(n+1)}}{k(n)}$$

(here we use $Y_i \ge 0$), so recalling $k(n) = [\alpha^n]$ we have $k(n+1)/k(n) \to \alpha$ and

$$\frac{1}{\alpha}EX_1 \le \liminf_{n \to \infty} T_m/m \le \limsup_{m \to \infty} T_m/m \le \alpha EX_1$$

Since $\alpha > 1$ is arbitrary the proof is complete. $\square$

The next result shows that the strong law holds whenever EX_i exists.

(7.2) Theorem. Let $X_1, X_2, \ldots$ be i.i.d. with $EX_i^+ = \infty$ and $EX_i^- < \infty$. If $S_n = X_1 + \cdots + X_n$ then $S_n/n \to \infty$ a.s.

Proof   Let $M > 0$ and $X_i^M = X_i \wedge M$. The X_i^M are i.i.d. with $E|X_i^M| < \infty$ so if $S_i^M = X_1^M + \cdots + X_n^M$ then (7.1) implies $S_n^M/n \to EX_i^M$. Since $X_i \ge X_i^M$ it follows that

$$\liminf_{n \to \infty} S_n/n \ge \lim_{n \to \infty} S_n^M/n = EX_i^M$$

The monotone convergence theorem implies $E(X_i^M)^+ \uparrow EX_i^+ = \infty$ as $M \uparrow \infty$, so $EX_i^M = E(X_i^M)^+ - E(X_i^M)^- \uparrow \infty$ and we have $\liminf_{n \to \infty} S_n/n \ge \infty$ which implies the desired result. $\square$

The rest of this section is devoted to applications of the strong law of large numbers.

Example 7.1.   Renewal theory. Let $X_1, X_2, \ldots$ be i.i.d. with $0 < X_i < \infty$. Let $T_n = X_1 + \ldots + X_n$ and think of T_n as the time of nth occurrence of some event. For a concrete situation consider a diligent janitor who replaces a light bulb the instant it burns out. Suppose the first bulb is put in at time 0 and let X_i be the lifetime of the ith lightbulb. In this interpretation T_n is the time the nth light bulb burns out and $N_t = \sup\{n : T_n \le t\}$ is the number of light bulbs that have burnt out by time t.

(7.3) Theorem. If $EX_1 = \mu \le \infty$ then as $t \to \infty$, $N_t/t \to 1/\mu$ a.s. $(1/\infty = 0)$

Proof By (7.2), $T_n/n \to \mu$ a.s. From the definition of N_t it follows that $T(N_t) \le t < T(N_t + 1)$ so dividing through by N_t gives

$$\frac{T(N_t)}{N_t} \le \frac{t}{N_t} \le \frac{T(N_t + 1)}{N_t + 1} \cdot \frac{N_t + 1}{N_t}$$

To take the limit we note that since $T_n < \infty$ for all n we have $N_t \uparrow \infty$ as $t \to \infty$. The strong law of large numbers implies that for $\omega \in \Omega_0$ with $P(\Omega_0) = 1$ we have $T_n(\omega)/n \to \mu$, $N_t(\omega) \uparrow \infty$, and hence

$$\frac{T_{N_t(\omega)}(\omega)}{N_t(\omega)} \to \mu \qquad \frac{N_t(\omega) + 1}{N_t(\omega)} \to 1$$

From this it follows that for $\omega \in \Omega_0$ that $t/N_t(\omega) \to \mu$ a.s. □

The last argument shows that if $X_n \to X_\infty$ a.s. and $N(n) \to \infty$ a.s. then $X_{N(n)} \to X_\infty$ a.s. We have written this out with care because the analogous result for convergence in probability is false.

EXERCISE 7.1. Give an example with $X_n \in \{0,1\}$, $X_n \to 0$ in probability, $N(n) \uparrow \infty$ a.s., and $X_{N(n)} \to 1$ a.s.

EXERCISE 7.2. **Lazy janitor.** Suppose the ith light bulb burns for an amount of time X_i and then remains burned out for time Y_i before being replaced. Suppose the X_i, Y_i are positive and independent with the X's having distribution F and the Y's having distribution G, both of which have finite mean. Let R_t be the amount of time in $[0, t]$ that we have a working light bulb. Show that $R_t/t \to EX_i/(EX_i + EY_i)$ almost surely.

Example 7.2. Empirical distribution functions. Let $X_1, X_2, \ldots$ be i.i.d. with distribution F and let

$$F_n(x) = n^{-1} \sum_{m=1}^{n} 1_{(X_m \le x)}$$

$F_n(x) = $ the observed frequency of values that are $\le x$, hence the name given above. The next result shows that F_n converges uniformly to F as $n \to \infty$.

(7.4) **The Glivenko-Cantelli theorem.** As $n \to \infty$,

$$\sup_x |F_n(x) - F(x)| \to 0 \quad \text{a.s.}$$

Proof Fix x and let $Y_n = 1_{(X_n \le x)}$. Since the Y_n are i.i.d. with $EY_n = P(X_n \le x) = F(x)$ the strong law of large numbers implies that $F_n(x) = $

$n^{-1} \sum_{m=1}^{n} Y_m \to F(x)$ a.s. In general if F_n is a sequence of nondecreasing functions which converges pointwise to a bounded and continuous limit F then $\sup_x |F_n(x) - F(x)| \to 0$. However, the distribution function $F(x)$ may have jumps so we have to work a little harder.

Again fix x and let $Z_n = 1_{(X_n < x)}$. Since the Z_n are i.i.d. with $EZ_n = P(X_n < x) = F(x-) = \lim_{y \uparrow x} F(y)$ the strong law of large numbers implies that $F_n(x-) = n^{-1} \sum_{m=1}^{n} Z_m \to F(x-)$ a.s. For $1 \le j \le k - 1$ let $x_{j,k} = \inf\{y : F(y) \ge j/k\}$. The pointwise convergence of $F_n(x)$ and $F_n(x-)$ imply that we can pick $N_k(\omega)$ so that if $n \ge N_k(\omega)$ then

$$|F_n(x_{j,k}) - F(x_{j,k})| < k^{-1} \quad \text{and} \quad |F_n(x_{j,k}-) - F(x_{j,k}-)| < k^{-1}$$

for $1 \le j \le k - 1$. If we let $x_{0,k} = -\infty$ and $x_{k,k} = \infty$ then the last two inequalities hold for $j = 0$ or k. If $x \in (x_{j-1,k}, x_{j,k})$ with $1 \le j \le k$ and $n \ge N_k(\omega)$ then using the monotonicity of F_n and F, and $F(x_{j,k}-) - F(x_{j-1,k}) \le k^{-1}$ (see Figure 1.7.1), we have

$$F_n(x) \le F_n(x_{j,k}-) \le F(x_{j,k}-) + k^{-1} \le F(x_{j-1,k}) + 2k^{-1} \le F(x) + 2k^{-1}$$
$$F_n(x) \ge F_n(x_{j-1,k}) \ge F(x_{j-1,k}) - k^{-1} \ge F(x_{j,k}-) - 2k^{-1} \ge F(x) - 2k^{-1}$$

so $\sup_x |F_n(x) - F(x)| \le 2k^{-1}$ and we have proved the result.   □

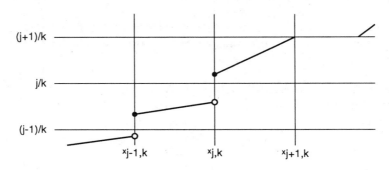

Figure 1.7.1

Example 7.3. Shannon's theorem. Let $X_1, X_2, \ldots \in \{1, \ldots, r\}$ be independent with $P(X_i = k) = p(k) > 0$ for $1 \le k \le r$. Here we are thinking of $1, \ldots, r$ as the letters of an alphabet, and $X_1, X_2, \ldots$ are the successive letters produced by an information source, in this i.i.d. case the proverbial monkey at a typewriter. Let $\pi_n(\omega) = p(X_1(\omega)) \cdots p(X_n(\omega))$ be the probability of the realization we observed in the first n trials. Since $\log \pi_n(\omega)$ is a sum of independent random variables it follows from the strong law of large numbers that

$$-n^{-1} \log \pi_n(\omega) \to H \equiv -\sum_{k=1}^{r} p(k) \log p(k) \text{ a.s.}$$

The constant H is called the **entropy** of the source and is a meaure of how random it is. The last result is the **asymptotic equipartition property**: if $\epsilon > 0$ then as $n \to \infty$

$$P\left(\exp(-n(H+\epsilon)) \le \pi_n(\omega) \le \exp(-n(H-\epsilon))\right) \to 1$$

We will give a more general version of this result in (5.1) of Chapter 6.

EXERCISES

7.3. Let $X_0 = (1,0)$ and define $X_n \in \mathbf{R}^2$ inductively by declaring that X_{n+1} is chosen at random from the ball of radius $|X_n|$ centered at the origin, i.e., $X_{n+1}/|X_n|$ is uniformly distributed on the ball of radius 1 and independent of $X_1, \ldots, X_n$. Prove that $n^{-1} \log X_n \to c$ a.s. and compute c.

7.4. Investment problem. We assume that at the beginning of each year you can buy bonds for \$1 that are worth \$ a at the end of the year or stocks that are worth a random amount $V \ge 0$. If you always invest a fixed proportion p of your wealth in bonds then your wealth at the end of year $n+1$ is $W_{n+1} = (ap + (1-p)V_n)W_n$. Suppose $V_1, V_2, \ldots$ are i.i.d. with $EV_n^2 < \infty$ and $E(V_n^{-2}) < \infty$. (i) Show that $n^{-1} \log W_n \to c(p)$ a.s. (ii) Show that $c(p)$ is concave. [Use (9.1) in the Appendix to justify differentiating under the expected value.] (iii) By investigating $c'(0)$ and $c'(1)$, give conditions on V that guarantee that the optimal choice of p is in (0,1). (iv) Suppose $P(V = 1) = P(V = 4) = 1/2$. Find the optimal p as a function of a.

*1.8.   Convergence of Random Series

In this section we will pursue a second approach to the strong law of large numbers based on the convergence of random series. This approach has the advantage that it leads to estimates on the rate of convergence under moment assumptions, (8.7) and (8.8), and to a negative result for the infinite mean case, (8.9). The first two results in this section are of considerable interest in their own right, although we will see more general versions in (1.1) of Chapter 3 and (4.2) of Chapter 4.

   To state the first result we need some notation. Let $\mathcal{F}_n' = \sigma(X_n, X_{n+1}, \ldots)$ = the future after time n = the smallest σ-field with respect to which all the

X_m, $m \geq n$ are measurable. Let $\mathcal{T} = \cap_n \mathcal{F}'_n =$ the remote future, or **tail** σ-field. Intuitively, $A \in \mathcal{T}$ if and only if changing a finite number of values does not affect the occurrence of the event. As usual, we turn to examples to help explain the definition.

Example 8.1. If $B_n \in \mathcal{R}$ then $\{X_n \in B_n \text{ i.o.}\} \in \mathcal{T}$. If we let $X_n = 1_{A_n}$ and $B_n = \{1\}$ this example becomes $\{A_n \text{ i.o.}\}$.

Example 8.2. Let $S_n = X_1 + \ldots + X_n$. It is easy to check that
$\{\lim_{n \to \infty} S_n \text{ exists }\} \in \mathcal{T}$,
$\{\limsup_{n \to \infty} S_n > 0\} \notin \mathcal{T}$,
$\{\limsup_{n \to \infty} S_n/c_n > x\} \in \mathcal{T}$ if $c_n \to \infty$.

The next result shows that all examples are trivial.

(8.1) Kolmogorov's 0-1 law. If $X_1, X_2, \ldots$ are independent and $A \in \mathcal{T}$ then $P(A) = 0$ or 1.

Proof We will show that A is independent of itself, that is, $P(A \cap A) = P(A)P(A)$, so $P(A) = P(A)^2$, and hence $P(A) = 0$ or 1. We will sneak up on this conclusion in two steps:

(a) $A \in \sigma(X_1, \ldots, X_k)$ and $B \in \sigma(X_{k+1}, X_{k+2}, \ldots)$ are independent.

Proof of (a) If $B \in \sigma(X_{k+1}, \ldots, X_{k+j})$ for some j this follows from (4.5). Since $\sigma(X_1, \ldots, X_k)$ and $\cup_j \sigma(X_{k+1}, \ldots, X_{k+j})$ are π-systems that contain Ω (a) follows from (4.2).

(b) $A \in \sigma(X_1, X_2, \ldots)$ and $B \in \mathcal{T}$ are independent.

Proof of (b) Since $\mathcal{T} \subset \sigma(X_{k+1}, X_{k+2}, \ldots)$, if $A \in \sigma(X_1, \ldots, X_k)$ for some k, this follows from (a). $\cup_k \sigma(X_1, \ldots, X_k)$ and $\mathcal{T}$ are π-systems that contain Ω, so (b) follows from (4.2).

Since $\mathcal{T} \subset \sigma(X_1, X_2, \ldots)$, (b) implies an $A \in \mathcal{T}$ is independent of itself and (8.1) follows. $\square$

If $A_1, A_2, \ldots$ are independent then (8.1) implies $P(A_n \text{ i.o.}) = 0$ or 1. Applying (8.1) to Example 8.2 gives $P(\lim_{n \to \infty} S_n \text{ exists}) = 0$ or 1. The next result will help us prove the probability is 1 in certain situations.

(8.2) Kolmogorov's inequality. Suppose $X_1, \ldots, X_n$ are independent with

$EX_i = 0$ and $\text{var}(X_i) < \infty$. If $S_n = X_1 + \cdots + X_n$ then

$$P\left(\max_{1\leq k\leq n} |S_k| \geq x\right) \leq x^{-2}\text{var}(S_n)$$

Remark. Under the same hypotheses, Chebyshev's inequality (3.4) gives only

$$P(|S_n| \geq x) \leq x^{-2}\text{var}(S_n)$$

Proof Let $A_k = \{|S_k| \geq x$ but $|S_j| < x$ for $j < k\}$, i.e., we break things down according to the time that $|S_k|$ first exceeds x. Since the A_k are disjoint and $(S_n - S_k)^2 \geq 0$

$$ES_n^2 \geq \sum_{k=1}^{n} \int_{A_k} S_n^2 \, dP = \sum_{k=1}^{n} \int_{A_k} S_k^2 + 2S_k(S_n - S_k) + (S_n - S_k)^2 \, dP$$

$$\geq \sum_{k=1}^{n} \int_{A_k} S_k^2 \, dP + \sum_{k=1}^{n} \int 2S_k 1_{A_k} \cdot (S_n - S_k) \, dP$$

$S_k 1_{A_k} \in \sigma(X_1, \ldots, X_k)$ and $S_n - S_k \in \sigma(X_{k+1}, \ldots, X_n)$ are independent by (4.5), so using (4.8) and $E(S_n - S_k) = 0$ shows

$$\int 2S_k 1_{A_k} \cdot (S_n - S_k) \, dP = E(2S_k 1_{A_k}) \cdot E(S_n - S_k) = 0$$

Using now the fact that $|S_k| \geq x$ on A_k and the A_k are disjoint

$$ES_n^2 \geq \sum_{k=1}^{n} \int_{A_k} S_k^2 \, dP \geq \sum_{k=1}^{n} x^2 P(A_k) = x^2 P\left(\max_{1\leq k\leq n} |S_k| \geq x\right) \qquad \square$$

EXERCISE 8.1. Suppose $X_1, X_2, \ldots$ are i.i.d. with $EX_i = 0$, $\text{var}(X_i) = C < \infty$. Use (8.2) with $n = m^\alpha$ where $\alpha(2p-1) > 1$ to conclude that if $S_n = X_1 + \cdots + X_n$ and $p > 1/2$ then $S_n/n^p \to 0$ almost surely.

We turn now to our results on convergence of series. To state them we need a definition. We say that $\sum_{n=1}^{\infty} a_n$ converges, if $\lim_{N\to\infty} \sum_{n=1}^{N} a_n$ exists.

(8.3) **Theorem.** Suppose $X_1, X_2, \ldots$ are independent with $EX_n = 0$. If $\sum_{n=1}^{\infty} \text{var}(X_n) < \infty$ then with probability one $\sum_{n=1}^{\infty} X_n(\omega)$ converges.

Proof Let $S_N = \sum_{n=1}^{N} X_n$. From (8.2) we get

$$P\left(\max_{M\leq m\leq N} |S_m - S_M| > \epsilon\right) \leq \epsilon^{-2}\text{var}(S_N - S_M) = \epsilon^{-2} \sum_{n=M+1}^{N} \text{var}(X_n)$$

Letting $N \to \infty$ in the last result we get

$$P\left(\max_{m \geq M} |S_m - S_M| > \epsilon\right) \leq \epsilon^{-2} \sum_{n=M+1}^{\infty} \text{var}(X_n) \to 0 \quad \text{as } M \to \infty$$

If we let $w_M = \sup_{m,n \geq M} |S_m - S_n|$ then $w_M \downarrow$ as $M \uparrow$ and

$$P(w_M > 2\epsilon) \leq P\left(\max_{m \leq M} |S_m - S_M| > \epsilon\right) \to 0$$

as $M \to \infty$ so $w_M \downarrow 0$ almost surely. But $w_M(\omega) \downarrow 0$ implies $S_n(\omega)$ is a Cauchy sequence and hence $\lim_{n \to \infty} S_n(\omega)$ exists, so we have proved (8.3). $\quad\square$

Example 8.3. Let $X_1, X_2, \ldots$ be independent with

$$P(X_n = n^{-\alpha}) = P(X_n = -n^{-\alpha}) = 1/2$$

$EX_n = 0$ and $\text{var}(X_n) = n^{-2\alpha}$ so if $\alpha > 1/2$ it follows from (8.3) that $\sum X_n$ converges. (8.4) shows that $\alpha > 1/2$ is also necessary for this conclusion. Notice that there is absolute convergence, i.e., $\sum |X_n| < \infty$, if and only if $\alpha > 1$.

(8.3) is sufficient for all of our applications but our treatment would not be complete if we did not mention the last word on convergence of random series.

(8.4) Kolmogorov's three series theorem. Let $X_1, X_2, \ldots$ be independent. Let $A > 0$ and let $Y_i = X_i 1_{(|X_i| \leq A)}$. In order that $\sum_{n=1}^{\infty} X_n$ converges a.s. it is necessary and sufficient that

(i) $\sum_{n=1}^{\infty} P(|X_n| > A) < \infty$, (ii) $\sum_{n=1}^{\infty} EY_n$ converges, and (iii) $\sum_{n=1}^{\infty} \text{var}(Y_n) < \infty$

Proof We will prove the necessity in Example 4.8 of Chapter 2 as an application of the central limit theorem. To prove the sufficiency, let $\mu_n = EY_n$. (iii) and (8.3) imply that $\sum_{n=1}^{\infty}(Y_n - \mu_n)$ converges a.s. Using (ii) now gives that $\sum_{n=1}^{\infty} Y_n$ converges a.s. (i) and the Borel-Cantelli lemma imply $P(X_n \neq Y_n \text{ i.o.}) = 0$ so $\sum_{n=1}^{\infty} X_n$ converges a.s. $\quad\square$

The link between convergence of series and the strong law of large numbers is provided by

(8.5) Kronecker's lemma. If $a_n \uparrow \infty$ and $\sum_{n=1}^{\infty} x_n/a_n$ converges then

$$a_n^{-1} \sum_{m=1}^{n} x_m \to 0$$

Proof Let $a_0 = 0$, $b_0 = 0$, and for $m \geq 1$, let $b_m = \sum_{k=1}^{m} x_k/a_k$. Then $x_m = a_m(b_m - b_{m-1})$ and so

$$a_n^{-1} \sum_{m=1}^{n} x_m = a_n^{-1} \left\{ \sum_{m=1}^{n} a_m b_m - \sum_{m=1}^{n} a_m b_{m-1} \right\}$$

$$= a_n^{-1} \left\{ a_n b_n + \sum_{m=2}^{n} a_{m-1} b_{m-1} - \sum_{m=1}^{n} a_m b_{m-1} \right\}$$

$$= b_n - \sum_{m=1}^{n} \frac{(a_m - a_{m-1})}{a_n} b_{m-1}$$

(Recall $a_0 = 0$.) By hypothesis $b_n \to b_\infty$ as $n \to \infty$. Since $a_m - a_{m-1} \geq 0$, the last sum is an average of $b_0, \dots, b_n$. Intuitively, if $\epsilon > 0$ and $M < \infty$ are fixed and n is large, the average assigns mass $\geq 1 - \epsilon$ to the b_m with $m \geq M$, so

$$\sum_{m=1}^{n} \frac{(a_m - a_{m-1})}{a_n} b_{m-1} \to b_\infty$$

To argue formally, let $B = \sup |b_n|$, pick M so that $|b_m - b_\infty| < \epsilon/2$ for $m \geq M$, then pick N so that $a_M/a_n < \epsilon/4B$ for $n \geq N$. Now if $n \geq N$ we have

$$\left| \sum_{m=1}^{n} \frac{(a_m - a_{m-1})}{a_n} b_{m-1} - b_\infty \right| \leq \sum_{m=1}^{n} \frac{(a_m - a_{m-1})}{a_n} |b_{m-1} - b_\infty| \cdot$$

$$\leq \frac{a_M}{a_n} \cdot 2B + \frac{a_n - a_M}{a_n} \cdot \frac{\epsilon}{2} < \epsilon$$

proving the desired result since ϵ is arbitrary. $\qquad \square$

(8.6) **The strong law of large numbers.** Let $X_1, X_2, \dots$ be i.i.d. random variables with $E|X_i| < \infty$. Let $EX_i = \mu$ and $S_n = X_1 + \dots + X_n$. Then $S_n/n \to \mu$ a.s. as $n \to \infty$.

Proof Let $Y_k = X_k 1_{(|X_k| \leq k)}$ and $T_n = Y_1 + \dots + Y_n$. By (a) in the proof of (7.1) it suffices to show that $T_n/n \to \mu$. Let $Z_k = Y_k - EY_k$ and note that (3.10b) and (b) in the proof of (7.1) imply

$$\sum_{k=1}^{\infty} \text{var}(Z_k)/k^2 \leq \sum_{k=1}^{\infty} EY_k^2/k^2 < \infty$$

Applying (8.3) now we conclude that $\sum_{k=1}^{\infty} Z_k/k$ converges a.s. so (8.5) implies

$$n^{-1} \sum_{k=1}^{n} (Y_k - EY_k) \to 0 \quad \text{and hence} \quad \frac{T_n}{n} - n^{-1} \sum_{k=1}^{n} EY_k \to 0 \text{ a.s.}$$

The dominated convergence theorem implies $EY_k \to \mu$ as $k \to \infty$. From this it follows easily that $n^{-1} \sum_{k=1}^{n} EY_k \to \mu$ and hence $T_n/n \to \mu$. □

Rates of convergence. As mentioned earlier, one of the advantages of the random series proof is that it provides estimates on the rate of convergence of $S_n/n \to \mu$. By subtracting μ from each random variable we can and will suppose without loss of generality that $\mu = 0$.

(8.7) Theorem. Let $X_1, X_2, \ldots$ be i.i.d. random variables with $EX_i = 0$ and $EX_i^2 = \sigma^2 < \infty$. Let $S_n = X_1 + \ldots + X_n$. If $\epsilon > 0$ then

$$S_n/n^{1/2}(\log n)^{1/2+\epsilon} \to 0 \quad \text{a.s.}$$

Remark. (9.6) in Chapter 7 will show that

$$\limsup_{n \to \infty} S_n/n^{1/2}(\log \log n)^{1/2} = \sigma\sqrt{2} \quad \text{a.s.}$$

so the last result is not far from the best possible.

Proof   Let $a_n = n^{1/2}(\log n)^{1/2+\epsilon}$ for $n \geq 2$ and $a_1 > 0$.

$$\sum_{n=1}^{\infty} \text{var}(X_n/a_n) = \sigma^2 \left(\frac{1}{a_1^2} + \sum_{n=2}^{\infty} \frac{1}{n(\log n)^{1+2\epsilon}} \right) < \infty$$

so applying (8.3) we get $\sum_{n=1}^{\infty} X_n/a_n$ converges a.s. and the indicated result follows from (8.5). □

The next result due to Marcinkiewicz and Zygmund treats the situation in which $EX_i^2 = \infty$ but $E|X_i|^p < \infty$ for some $1 < p < 2$.

(8.8) Theorem. Let $X_1, X_2, \ldots$ be i.i.d. with $EX_1 = 0$ and $E|X_1|^p < \infty$ where $1 < p < 2$. If $S_n = X_1 + \ldots + X_n$ then $S_n/n^{1/p} \to 0$ a.s.

Proof   Let $Y_k = X_k 1_{(|X_k| \leq k^{1/p})}$ and $T_n = Y_1 + \cdots + Y_n$.

$$\sum_{k=1}^{\infty} P(Y_k \neq X_k) = \sum_{k=1}^{\infty} P(|X_k|^p > k) \leq E|X_k|^p < \infty$$

so the Borel-Cantelli lemma implies $P(Y_k \neq X_k \text{ i.o.}) = 0$ and it suffices to show $T_n/n^{1/p} \to 0$. Using (3.10b), (5.7) with $p = 2$, $P(|Y_m| > y) \leq P(|X_1| > y)$ and

Fubini's theorem (everything is ≥ 0) we have

$$\sum_{m=1}^{\infty} \text{var}(Y_m/m^{1/p}) \leq \sum_{m=1}^{\infty} EY_m^2/m^{2/p}$$

$$\leq \sum_{m=1}^{\infty} \sum_{n=1}^{m} \int_{(n-1)^{1/p}}^{n^{1/p}} \frac{2y}{m^{2/p}} P(|X_1| > y)\, dy$$

$$= \sum_{n=1}^{\infty} \int_{(n-1)^{1/p}}^{n^{1/p}} \sum_{m=n}^{\infty} \frac{2y}{m^{2/p}} P(|X_1| > y)\, dy$$

To bound the integral we note that for $n \geq 2$ comparing the sum with the integral of $x^{-2/p}$

$$\sum_{m=n}^{\infty} m^{-2/p} \leq \frac{p}{2-p}(n-1)^{(p-2)/p} \leq Cy^{p-2}$$

when $y \in [(n-1)^{1/p}, n^{1/p}]$. Since $E|X_i|^p = \int_0^{\infty} px^{p-1} P(|X_i| > x)\, dx < \infty$ it follows that

$$\sum_{m=1}^{\infty} \text{var}(Y_m/m^{1/p}) < \infty$$

If we let $\mu_m = EY_m$ and apply (8.3) and (8.5) it follows that

$$n^{-1/p} \sum_{m=1}^{n} (Y_m - \mu_m) \to 0 \quad \text{a.s.}$$

To estimate μ_m we note that since $EX_m = 0$, $\mu_m = -E(X_i; |X_i| > m^{1/p})$ so

$$|\mu_m| \leq \int_{m^{1/p}}^{\infty} P(|X_i| > x)\, dx$$

$$\leq p^{-1} m^{-(p-1)/p} \int_{m^{1/p}}^{\infty} px^{p-1} P(|X_i| > x)\, dx$$

$$\leq m^{-1+1/p} p^{-1} E(|X_i|^p; |X_i| > m^{1/p})$$

Now $\sum_{m=1}^{n} m^{-1+1/p} \leq Cn^{1/p}$ and $E(|X_i|^p; |X_i| > m^{1/p}) \to 0$ as $m \to \infty$ so $n^{-1/p} \sum_{m=1}^{n} \mu_m \to 0$ and the desired result follows. $\square$

EXERCISE 8.2. The converse of the last exercise is much easier. Let $p > 0$. If $S_n/n^{1/p} \to 0$ a.s. then $E|X_1|^p < \infty$.

Infinite Mean. The St. Petersburg game, discussed in Example 5.7 and Exercise 6.17 is a situation in which $EX_i = \infty$, $S_n/n \log_2 n \to 1$ in probability but

$$\limsup_{n\to\infty} S_n/(n \log_2 n) = \infty \text{ a.s.}$$

The next result, due to Feller (1946), shows that when $E|X_1| = \infty$, S_n/a_n cannot converge almost surely to a nonzero limit. In (6.7) we considered the special case $a_n = n$.

(8.9) Theorem. Let $X_1, X_2, \ldots$ be i.i.d. with $E|X_1| = \infty$ and let $S_n = X_1 + \cdots + X_n$. Let a_n be a sequence of positive numbers with a_n/n increasing. Then $\limsup_{n\to\infty} |S_n|/a_n = 0$ or ∞ according as $\sum_n P(|X_1| \geq a_n) < \infty$ or $= \infty$.

Proof Since $a_n/n \uparrow$, $a_{kn} \geq ka_n$ for any integer k and hence

$$\sum_{n=1}^{\infty} P(|X_1| \geq ka_n) \geq \sum_{n=1}^{\infty} P(|X_1| \geq a_{kn}) \geq \frac{1}{k} \sum_{m=k}^{\infty} P(|X_1| \geq a_m) = \infty$$

The last observation shows that if the sum is infinite $\limsup_{n\to\infty} |X_n|/a_n = \infty$. Since $\max\{|S_{n-1}|, |S_n|\} \geq |X_n|/2$, it follows that $\limsup_{n\to\infty} |S_n|/a_n = \infty$.

To prove the other half, we begin with the mysterious but useful

$$(*) \qquad \sum_{m=1}^{\infty} mP(a_{m-1} \leq |X_i| < a_m) = \sum_{n=1}^{\infty} P(|X_i| \geq a_{n-1})$$

To see this write $m = \sum_{n=1}^{m} 1$ and then use Fubini's theorem. Turning to more routine details, let $Y_n = X_n 1_{(|X_n|<a_n)}$, and $T_n = Y_1 + \ldots + Y_n$. When the sum is finite $P(Y_n \neq X_n \text{ i.o.}) = 0$ and it suffices to investigate the behavior of the T_n. To do this we let $a_0 = 0$ and compute

$$\sum_{n=1}^{\infty} \text{var}(Y_n/a_n)^2 \leq \sum_{n=1}^{\infty} EY_n^2/a_n^2$$

$$= \sum_{n=1}^{\infty} a_n^{-2} \sum_{m=1}^{n} \int_{[a_{m-1}, a_m)} y^2 \, dF(y)$$

$$= \sum_{m=1}^{\infty} \int_{[a_{m-1}, a_m)} y^2 \, dF(y) \sum_{n=m}^{\infty} a_n^{-2}$$

Since $a_n \geq na_m/m$ we have $\sum_{n=m}^{\infty} a_n^{-2} \leq (m^2/a_m^2) \sum_{n=m}^{\infty} n^{-2} \leq Cma_m^{-2}$ and using $(*)$ it follows that

$$\leq C \sum_{m=1}^{\infty} m \int_{[a_{m-1}, a_m)} dF(y) < \infty$$

The last step is to show $ET_n/a_n \to 0$. To begin we note that if $E|X_i| = \infty$, $\sum_{n=1}^{\infty} P(|X_i| > a_n) < \infty$, and $a_n/n \uparrow$ we must have $a_n/n \uparrow \infty$. To estimate ET_n/a_n now, we observe that

$$\left| a_n^{-1} \sum_{m=1}^{n} EY_m \right| \leq a_n^{-1} \sum_{m=1}^{n} E(|X_m|; |X_m| < a_m)$$

$$\leq \frac{na_N}{a_n} + \frac{n}{a_n} E(|X_i|; a_N \leq |X_i| < a_n)$$

where the last inequality holds for any fixed N. Since $a_n/n \to \infty$, the first term converges to 0. Since $m/a_m \downarrow$ the second is

$$\leq \sum_{m=N+1}^{n} \frac{m}{a_m} E(|X_i|; a_{m-1} \leq |X_i| < a_m)$$

$$\leq \sum_{m=N+1}^{\infty} m P(a_{m-1} \leq |X_i| < a_m)$$

$(*)$ shows that the sum is finite so it is small if N is large and the desired result follows. $\qquad\qquad\qquad\qquad\qquad\qquad\qquad\qquad\qquad\qquad\qquad\qquad\qquad\square$

EXERCISES

8.3. Let $X_1, X_2, \ldots$ be i.i.d. standard normals. Show that for any t

$$\sum_{n=1}^{\infty} X_n \cdot \frac{\sin(n\pi t)}{n} \qquad \text{converges a.s.}$$

We will see this series again at the end of Section 1 of Chapter 7.

8.4. Let $X_1, X_2, \ldots$ be independent with $EX_n = 0$, $\text{var}(X_n) = \sigma_n^2$. (i) Show that if $\sum_n \sigma_n^2/n^2 < \infty$ then $\sum_n X_n/n < \infty$ and hence $n^{-1} \sum_{m=1}^{n} X_m \to 0$. (ii) Suppose $\sum \sigma_n^2/n^2 = \infty$ and without loss of generality that $\sigma_n^2 \leq n^2$ for all n. Show that there are independent random variables X_n with $EX_n = 0$ and $\text{var}(X_n) \leq \sigma_n^2$ so that X_n/n and hence $n^{-1} \sum_{m \leq n} X_m$ does not converge to 0 a.s.

8.5. Let $X_n \geq 0$ be independent for $n \geq 1$. The following are equivalent: (i) $\sum_{n=1}^{\infty} X_n < \infty$ a.s. (ii) $\sum_{n=1}^{\infty} P(X_n > 1) + E(X_n 1_{(X_n \leq 1)}) < \infty$ (iii) $\sum_{n=1}^{\infty} E(X_n/(1 + X_n)) < \infty$.

8.6. Let $\psi(x) = x^2$ when $|x| \leq 1$ and $= |x|$ when $|x| \geq 1$. Show that if $X_1, X_2, \ldots$ are independent with $EX_n = 0$ and $\sum_{n=1}^{\infty} E\psi(X_n) < \infty$ then $\sum_{n=1}^{\infty} X_n$ converges a.s.

8.7. Suppose $\sum_{n=1}^{\infty} E|X_n|^{p(n)} < \infty$ where $0 < p(n) \le 2$ for all n and $EX_n = 0$ when $p(n) > 1$. Show that $\sum_{n=1}^{\infty} X_n$ converges a.s.

8.8. Let $X_1, X_2, \ldots$ be i.i.d. and not $\equiv 0$. Then the radius of convergence of the power series $\sum_{n \ge 1} X_n(\omega) z^n$ (i.e., $r(\omega) = \sup\{c : \sum |X_n(\omega)| c^n < \infty\}$) is 1 a.s. or 0 a.s. according as $E \log^+ |X_1| < \infty$ or $= \infty$ where $\log^+ x = \max(\log x, 0)$.

8.9. Let $X_1, X_2, \ldots$ be independent and let $S_{m,n} = X_{m+1} + \ldots + X_n$. Then

$$(\star) \qquad P\left(\max_{m < j \le n} |S_{m,j}| > 2a\right) \min_{m < k \le n} P(|S_{k,n}| \le a) \le P(|S_{m,n}| > a)$$

8.10. Use $(\star)$ to prove a theorem of P. Lévy: Let $X_1, X_2, \ldots$ be independent and let $S_n = X_1 + \ldots + X_n$. If $\lim_{n \to \infty} S_n$ exists in probability then it also exists a.s.

8.11. Let $X_1, X_2, \ldots$ be i.i.d. and $S_n = X_1 + \ldots + X_n$. Use $(\star)$ to conclude that if $S_n/n \to 0$ in probability then $(\max_{1 \le m \le n} S_m)/n \to 0$ in probability.

8.12. Let $X_1, X_2, \ldots$ be i.i.d. and $S_n = X_1 + \ldots + X_n$. Suppose $a_n \uparrow \infty$ and $a(2^n)/a(2^{n-1})$ is bounded. Use $(\star)$ to show that if $S_n/a(n) \to 0$ in probability and $S_{2^n}/a(2^n) \to 0$ a.s. then $S_n/a(n) \to 0$ a.s.
(ii) Suppose in addition that $EX_1 = 0$ and $EX_1^2 < \infty$. Use the previous exercise and Chebyshev's inequality to conclude that $S_n/n^{1/2}(\log_2 n)^{1/2+\epsilon} \to 0$ a.s.

*1.9. Large Deviations

Let $X_1, X_2, \ldots$ be i.i.d. and let $S_n = X_1 + \cdots + X_n$. In this section we will investigate the rate at which $P(S_n > na) \to 0$ for $a > \mu = EX_i$. We will ultimately conclude that if the **moment generating function** $\varphi(\theta) = E \exp(\theta X_i) < \infty$ for some $\theta > 0$, $P(S_n \ge na) \to 0$ exponentially rapidly and we will identify

$$\gamma(a) = \lim_{n \to \infty} \frac{1}{n} \log P(S_n \ge na).$$

Our first step is to prove that the limit exists. This is based on an observation that will be useful several times below. Let $\pi_n = P(S_n \ge na)$.

$$\pi_{m+n} \ge P(S_m \ge ma, S_{n+m} - S_m \ge na) = \pi_m \pi_n$$

since S_m and $S_{n+m} - S_m$ are independent. Letting $\gamma_n = \log \pi_n$ transforms multiplication into addition.

(9.1) Lemma. If $\gamma_{m+n} \ge \gamma_m + \gamma_n$ then as $n \to \infty$, $\gamma_n/n \to \sup_m \gamma_m/m$.

Proof Clearly $\limsup \gamma_n/n \le \sup \gamma_m/m$. To complete the proof, it suffices to prove that for any m $\liminf \gamma_n/n \ge \gamma_m/m$. Writing $n = km + \ell$ with $0 \le \ell < m$ and making repeated use of the hypothesis gives $\gamma_n \ge k\gamma_m + \gamma_\ell$. Dividing by $n = km + \ell$ gives

$$\frac{\gamma(n)}{n} \ge \left(\frac{km}{km + \ell}\right) \frac{\gamma(m)}{m} + \frac{\gamma(\ell)}{n}$$

Letting $n \to \infty$ and recalling $n = km + \ell$ with $0 \le \ell < m$ gives the desired result. $\square$

(9.1) implies that $\lim_{n \to \infty} \frac{1}{n} \log P(S_n \ge na) = \gamma(a)$ exists ≤ 0. It follows from the formula for the limit that

(9.2) $$P(S_n \ge na) \le e^{n\gamma(a)}$$

The last two observations give us some useful information about $\gamma(a)$.

EXERCISE 9.1. The following are equivalent: (a) $\gamma(a) = -\infty$, (b) $P(X_1 \ge a) = 0$, and (c) $P(S_n \ge na) = 0$ for all n.

EXERCISE 9.2. Use the definition to conclude that if $\lambda \in [0,1]$ is rational then $\gamma(\lambda a + (1 - \lambda)b) \ge \lambda\gamma(a) + (1 - \lambda)\gamma(b))/2$. Use monotonicity to conclude that the last relationship holds for all $\lambda \in [0,1]$ so γ is concave and hence Lipschitz continuous on compact subsets of $\gamma(a) > -\infty$.

The conclusions above are valid for any distribution. For the rest of this section we will suppose:

(H1) $\quad \varphi(\theta) = E \exp(\theta X_i) < \infty$ for some $\theta > 0$

Let $\theta_+ = \sup\{\theta : \varphi(\theta) < \infty\}$, $\theta_- = \inf\{\theta : \varphi(\theta) < \infty\}$ and note that $\varphi(\theta) < \infty$ for $\theta \in (\theta_-, \theta_+)$. (H1) implies that $EX_i^+ < \infty$ so $\mu = EX^+ - EX^- \in [-\infty, \infty)$. If $\theta > 0$ Chebyshev's inequality implies

$$e^{\theta na} P(S_n \ge na) \le E \exp(\theta S_n) = \varphi(\theta)^n$$

or letting $\kappa(\theta) = \log \varphi(\theta)$

(9.3) $$P(S_n \ge na) \le \exp(-n\{a\theta - \kappa(\theta)\})$$

Our first goal is to show:

(9.4) **Lemma.** If $a > \mu$ and θ is small $a\theta - \kappa(\theta) > 0$.

Proof $\kappa(0) = \log \varphi(0) = 0$ so it suffices to show that (i) κ is continuous at 0, (ii) differentiable on $(0, \theta_+)$, and (iii) $\kappa'(\theta) \to \mu$ as $\theta \to 0$. For then

$$a\theta - \kappa(\theta) = \int_0^\theta a - \kappa'(x)\, dx < 0$$

for small θ. The first step is to show that the derivative exists. Let $F(x) = P(X_i \leq x)$. Since

$$|e^{hx} - 1| = \left| \int_0^{hx} e^y\, dy \right| \leq |hx|(e^{hx} + 1)$$

(the two terms are to cover the cases $h > 0$ and $h < 0$), an application of the dominated convergence theorem shows that

$$\begin{aligned}
\varphi'(\theta) &= \lim_{h \to 0} \frac{\varphi(\theta + h) - \varphi(\theta)}{h} \\
&= \lim_{h \to 0} \int \frac{e^{hx} - 1}{h} e^{\theta x}\, dF(x) \\
&= \int x e^{\theta x}\, dF(x) \quad \text{for } \theta \in (\theta_-, \theta_+)
\end{aligned}$$

From the last equation it follows that $\kappa(\theta) = \log \varphi(\theta)$ has $\kappa'(\theta) = \varphi'(\theta)/\varphi(\theta)$. To take the limit as $\theta \downarrow 0$, we note that using the monotone convergence theorem for $x \leq 0$ and the dominated convergence theorem for $x \geq 0$ shows that $\varphi'(\theta) \to \mu$ as $\theta \to 0$. A similar but simpler argument show that $\varphi(\theta) \to 1$ as $\theta \downarrow 0$, so we have shown (i)–(iii) and proved (9.4). $\square$

Having found an upper bound on $P(S_n \geq na)$, it is natural to optimize it by finding the maximum of $\theta a - \kappa(\theta)$.

$$\frac{d}{d\theta}\{\theta a - \log \varphi(\theta)\} = a - \varphi'(\theta)/\varphi(\theta)$$

so (assuming things are nice) the maximum occurs when $a = \varphi'(\theta)/\varphi(\theta)$. To turn the parenthetical clause into a mathematical hypothesis we begin by defining

$$F_\theta(x) = \frac{1}{\varphi(\theta)} \int_{-\infty}^x e^{\theta y}\, dF(y)$$

whenever $\varphi(\theta) < \infty$. It follows from the proof of (9.4) that if $\theta \in (\theta_-, \theta_+)$, F_θ is a distribution function with mean

$$\int x\, dF_\theta(x) = \frac{1}{\varphi(\theta)} \int_{-\infty}^\infty x e^{\theta x}\, dF(x) = \varphi'(\theta)/\varphi(\theta)$$

Repeating the proof in (9.4) it is easy to see that if $\theta \in (\theta_-, \theta_+)$ then

$$\varphi''(\theta) = \int_{-\infty}^{\infty} x^2 e^{\theta x} \, dF(x)$$

So we have

$$\frac{d}{d\theta} \frac{\varphi'(\theta)}{\varphi(\theta)} = \frac{\varphi''(\theta)}{\varphi(\theta)} - \left(\frac{\varphi'(\theta)}{\varphi(\theta)}\right)^2 = \int x^2 \, dF_\theta(x) - \left(\int x \, dF_\theta(x)\right)^2 \geq 0$$

since the last expression is the variance of F_θ. If we assume

(H2)   the distribution F is not a point mass at μ

then $\varphi'(\theta)/\varphi(\theta)$ is strictly increasing and $a\theta - \log \varphi(\theta)$ is concave. Since we have $\varphi'(0)/\varphi(0) = \mu$ this shows that for each $a > \mu$ there is at most one $\theta_a \geq 0$ that solves $a = \varphi'(\theta_a)/\varphi(\theta_a)$ and this value of θ maximizes $a\theta - \log \varphi(\theta)$. Before discussing the existence of θ_a we will consider some examples:

Example 9.1. Normal distribution.

$$\int e^{\theta x} (2\pi)^{-1/2} \exp(-x^2/2) \, dx = \exp(\theta^2/2) \int (2\pi)^{-1/2} \exp(-(x-\theta)^2/2) \, dx$$

The integrand in the last integral is the density of a normal distribution with mean θ and variance 1, so $\varphi(\theta) = \exp(\theta^2/2)$, $\theta \in (-\infty, \infty)$. In this case $\varphi'(\theta)/\varphi(\theta) = \theta$ and F_θ is a normal distribution with mean θ and variance 1.

Example 9.2. Exponential distribution with parameter λ. If $\theta < \lambda$

$$\int_0^{\infty} e^{\theta x} \lambda e^{-\lambda x} \, dx = \lambda/(\lambda - \theta)$$

$\varphi'(\theta)/\varphi(\theta) = 1/(\lambda - \theta)$ and F_θ is an exponential distribution with parameter $\lambda - \theta$ and hence mean $1/(\lambda - \theta)$.

Example 9.3. Coin flips. $P(X_i = 1) = P(X_i = -1) = 1/2$

$$\varphi(\theta) = (e^\theta + e^{-\theta})/2$$
$$\varphi'(\theta)/\varphi(\theta) = (e^\theta - e^{-\theta})/(e^\theta + e^{-\theta})$$

$F_\theta(\{1\}) = e^\theta/(e^\theta + e^{-\theta})$ and $F_\theta(\{-1\}) = e^{-\theta}/(e^\theta + e^{-\theta})$.

Example 9.4. Perverted exponential. Let $g(x) = Cx^{-3}e^{-x}$ for $x \geq 1$, $g(x) = 0$ otherwise, and choose C so that g is a probability density. In this case

$$\varphi(\theta) = \int e^{\theta x} g(x) dx < \infty$$

if and only if $\theta \leq 1$ and when $\theta \leq 1$ we have

$$\varphi'(\theta)/\varphi(\theta) \leq \varphi'(1)/\varphi(1) = \int_1^\infty Cx^{-2}\, dx \bigg/ \int_1^\infty Cx^{-3} dx = 2$$

Recall $\theta_+ = \sup\{\theta : \varphi(\theta) < \infty\}$. In Examples 9.1 and 9.2, we have $\varphi'(\theta)/\varphi(\theta) \uparrow \infty$ as $\theta \uparrow \theta_+$ so we can solve $a = \varphi'(\theta)/\theta)$ for any $a > \mu$. In Example 9.3 $\varphi'(\theta)/\varphi(\theta) \uparrow 1$ as $\theta \to \infty$, but we cannot hope for much more since F and hence F_θ is supported on $\{-1, 1\}$.

EXERCISE 9.3.   Let $x_o = \sup\{x : F(x) < 1\}$. Show that if $x_o < \infty$ then $\varphi(\theta) < \infty$ for all $\theta > 0$ and $\varphi'(\theta)/\varphi(\theta) \to x_o$ as $\theta \uparrow \infty$.

Example 9.4 presents a problem since we cannot solve $a = \varphi'(\theta)/\varphi(\theta)$ when $a > 2$. (9.6) will cover this problem case but first we will treat the cases in which we can solve the equation.

(9.5) Theorem. Suppose in addition to (H1) and (H2) that there is a $\theta_a \in (0, \theta_+)$ so that $a = \varphi'(\theta_a)/\varphi(\theta_a)$. Then as $n \to \infty$

$$n^{-1} \log P(S_n \geq na) \to -a\theta_a + \log \varphi(\theta_a).$$

Proof   The fact that the limsup of the left-hand side $\leq$ the right-hand side follows from (9.3). To prove the other inequality pick $\lambda \in (\theta_a, \theta_+)$, let $X_1^\lambda, X_2^\lambda, \ldots$ be i.i.d. with distribution F_λ and let $S_n^\lambda = X_1^\lambda + \cdots + X_n^\lambda$. Writing dF/dF_λ for the Radon-Nikodym derivative of the associated measures it is immediate from the definition that $dF/dF_\lambda = e^{-\lambda x}\varphi(\lambda)$. If we let F_λ^n and F^n denote the distributions of S_n^λ and S_n, then

Lemma. $dF^n/dF_\lambda^n = e^{-\lambda x}\varphi(\lambda)^n$.

Proof   We will prove this by induction. The result holds when $n = 1$. For

$n > 1$ we note that

$$F^n = F^{n-1} * F(z) = \int_{-\infty}^{\infty} dF^{n-1}(x) \int_{-\infty}^{z-x} dF(y)$$

$$= \int dF_\lambda^{n-1}(x) \int dF_\lambda(y) \, 1_{(x+y \le z)} e^{-\lambda(x+y)} \varphi(\lambda)^n$$

$$= E\left(1_{(S_{n-1}^\lambda + X_n^\lambda \le z)} e^{-\lambda(S_{n-1}^\lambda + X_n^\lambda)} \varphi(\lambda)^n\right)$$

$$= \int_{-\infty}^{z} dF^n(u) e^{-\lambda u} \varphi(\lambda)^n$$

where in the last two equalities we have used (3.9) for $(S_{n-1}^\lambda, X_n^\lambda)$ and S_n^λ. □

If $\nu > a$ then the lemma and monotonicity imply

$$(*) \quad P(S_n \ge na) \ge \int_{na}^{n\nu} e^{-\lambda x} \varphi(\lambda)^n \, dF_\lambda^n(x) \ge \varphi(\lambda)^n e^{-\lambda n\nu}(F_\lambda^n(n\nu) - F_\lambda^n(na))$$

F_λ has mean $\varphi'(\lambda)/\varphi(\lambda)$, so if we have $a < \varphi'(\lambda)/\varphi(\lambda) < \nu$, then the weak law of large numbers implies

$$F_\lambda^n(n\nu) - F_\lambda^n(na) \to 1 \text{ as } n \to \infty$$

From the last conclusion and $(*)$ it follows that

$$\liminf_{n \to \infty} n^{-1} \log P(S_n > na) \ge -\lambda \nu + \log \varphi(\lambda)$$

Since $\lambda > \theta_a$ and $\nu > \varphi'(\lambda)/\varphi(\lambda)$ are arbitrary, the proof is complete. □

To get a feel for what the answers look like we consider our examples. To prepare for the computations we recall some important information

$$\kappa(\theta) = \log \varphi(\theta) \quad \kappa'(\theta) = \varphi'(\theta)/\varphi(\theta) \quad \theta_a \text{ solves } \kappa'(\theta_a) = a$$
$$\gamma(a) = \lim_{n \to \infty} (1/n) \log P(S_n \ge na) = -a\theta_a + \kappa(\theta_a)$$

Normal distribution (Example 9.1)

$$\kappa(\theta) = \theta^2/2 \quad \kappa'(\theta) = \theta \quad \theta_a = a$$
$$\gamma(a) = -a\theta_a + \kappa(\theta_a) = -a^2/2$$

EXERCISE 9.4. Check the last result by observing that S_n has a normal distribution with mean 0 and variance n, and using (1.4).

Exponential distribution (Example 9.2) with $\lambda = 1$

$$\kappa(\theta) = -\log(1-\theta) \qquad \kappa'(\theta) = 1/(1-\theta) \qquad \theta_a = 1 - 1/a$$
$$\gamma(a) = -a\theta_a + \kappa(\theta_a) = -a + 1 + \log a$$

With these two examples as models the reader should be able to do

EXERCISE 9.5. Let $X_1, X_2, \ldots$ be i.i.d. Poisson with mean 1, and let $S_n = X_1 + \cdots + X_n$. Find $\lim_{n\to\infty}(1/n)\log P(S_n \geq na)$ for $a > 1$. The answer and another proof can be found in Exercise 1.4 of Chapter 2.

Coin flips (Example 9.3). Here we take a different approach. To find the θ which makes the mean of $F_\theta = a$ we set $F_\theta(\{1\}) = e^\theta/(e^\theta + e^{-\theta}) = (1+a)/2$. Letting $x = e^\theta$ gives

$$2x = (1+a)(x + x^{-1}) \qquad (a-1)x^2 + (1+a) = 0$$

So $x = \sqrt{(1+a)/(1-a)}$, $\theta_a = \log x = \{\log(1+a) - \log(1-a)\}/2$.

$$\varphi(\theta_a) = \frac{e^{\theta_a} + e^{-\theta_a}}{2} = \frac{e^{\theta_a}}{1+a} = \frac{1}{\sqrt{(1+a)(1-a)}}$$
$$\gamma(a) = -a\theta_a + \kappa(\theta_a) = -\{(1+a)\log(1+a) + (1-a)\log(1-a)\}/2$$

In Exercise 1.3 of Chapter 2, this result will be proved by a direct computation. Since the formula for $\gamma(a)$ is rather ugly, the following simpler bound is useful.

EXERCISE 9.6. Show that for coin flips $\varphi(\theta) \leq \exp(\varphi(\theta) - 1) \leq \exp(\beta\theta^2)$ for $\theta \leq 1$ where $\beta = \sum_{n=1}^{\infty} 1/(2n)! \approx .586$, and use (9.3) to conclude that $P(S_n \geq an) \leq \exp(-na^2/4\beta)$ for all $a \in [0,1]$. It is customary to simplify this further by using $\beta \leq \sum_{n=1}^{\infty} 2^{-n} = 1$.

Turning now to the problematic values for which we cannot solve $a = \varphi'(\theta_a)/\varphi(\theta)$, we begin by observing that if $x_o = \sup\{x : F(x) < 1\}$ and F is not a point mass at x_o then $\varphi'(\theta)/\varphi(\theta) \uparrow x_0$ as $\theta \uparrow \infty$ but $\varphi'(\theta)/\varphi(\theta) < x_0$ for all $\theta < \infty$. However, the result for $a = x_o$ is trivial:

$$\frac{1}{n}\log P(S_n \geq nx_o) = \log P(X_i = x_o) \qquad \text{for all } n$$

EXERCISE 9.7. Show that as $a \uparrow x_o$, $\gamma(a) \downarrow \log P(X_i = x_o)$.

When $x_o = \infty$, $\varphi'(\theta)/\varphi(\theta) \uparrow \infty$ as $\theta \uparrow \infty$ so the only case that remains is covered by

(9.6) Theorem. Suppose $x_o = \infty$, $\theta_+ < \infty$, and $\varphi'(\theta)/\varphi(\theta)$ increases to a finite limit a_0 as $\theta \uparrow \theta_+$. If $a_0 \leq a < \infty$

$$n^{-1} \log P(S_n \geq na) \rightarrow -a\theta_+ + \log \varphi(\theta_+)$$

i.e., $\gamma(a)$ is linear for $a \geq a_0$.

Proof Since $(\log \varphi(\theta))' = \varphi'(\theta)/\varphi(\theta)$, integrating from 0 to θ_+ shows that $\log(\varphi(\theta_+)) < \infty$. Letting $\theta = \theta_+$ in (9.3) shows that the limsup of the left-hand side $\leq$ the right-hand side. To get the other direction we will use the transformed distribution F_λ, for $\lambda = \theta_+$. Letting $\theta \uparrow \theta_+$ and using the dominated convergence theorem for $x \leq 0$ and the monotone convergence theorem for $x \geq 0$ we see that F_λ has mean a_0. From $(*)$ in the proof of (9.5), we see that if $a_0 \leq a < \nu = a + 3\epsilon$

$$P(S_n \geq na) \geq \varphi(\lambda)^n e^{-n\lambda\nu} (F_\lambda^n(n\nu) - F_\lambda^n(na))$$

and hence

$$\frac{1}{n} \log P(S_n \geq na) \geq \varphi(\lambda) - \lambda\nu + \frac{1}{n} \log P(S_n^\lambda \in (na, n\nu])$$

Letting $X_1^\lambda, X_2^\lambda, \ldots$ be i.i.d. with distribution F_λ and $S_n^\lambda = X_1^\lambda + \cdots + X_n^\lambda$ we have

$$P(S_n^\lambda \in (na, n\nu]) \geq P\{S_{n-1}^\lambda \in ((a_0 - \epsilon)n, (a_0 + \epsilon)n]\}$$
$$\cdot P\{X_n^\lambda \in ((a - a_0 + \epsilon)n, (a - a_0 + 2\epsilon)n]\}$$
$$\geq \frac{1}{2} P\{X_n^\lambda \in ((a - a_0 + \epsilon)n, (a - a_0 + \epsilon)(n + 1)]\}$$

for large n by the weak law of large numbers. To get a lower bound on the right-hand side of the last equation we observe that

$$\limsup_{n \to \infty} \frac{1}{n} \log P(X_1^\lambda \in ((a - a_0 + \epsilon)n, (a - a_0 + \epsilon)(n + 1)]) = 0$$

for if the limsup was < 0 we would have $E \exp(\eta X_1^\lambda) < \infty$ for some $\eta > 0$ and hence $E \exp((\lambda + \eta) X_1) < \infty$, contradicting the definition of $\lambda = \theta_+$. To finish the argument now we recall that (9.1) implies that

$$\lim_{n \to \infty} \frac{1}{n} \log P(S_n \geq na) = \gamma(a)$$

exists so our lower bound on the limsup is good enough. $\square$

By adapting the proof of the last result, you can show that (H1) is necessary for exponential convergence:

EXERCISE 9.8. Suppose $EX_i = 0$ and $E\exp(\theta X_i) = \infty$ for all $\theta > 0$. Then

$$\frac{1}{n}\log P(S_n \geq na) \to 0 \text{ for all } a > 0$$

EXERCISE 9.9. Suppose $EX_i = 0$. Show that if $\epsilon > 0$ then

$$\liminf_{n\to\infty} P(S_n \geq na)/nP(X_1 \geq n(a+\epsilon)) \geq 1$$

Hint: Let $F_n = \{X_i \geq n(a+\epsilon) \text{ for exactly one } i \leq n\}$.

2 Central Limit Theorems

The first four sections of this chapter develop the central limit theorem. The last five treat various extensions and complements. We begin this chapter by considering special cases of these results that can be treated by elementary computations.

2.1. The De Moivre-Laplace Theorem

Let $X_1, X_2, \ldots$ be i.i.d. with $P(X_1 = 1) = P(X_1 = -1) = 1/2$ and let $S_n = X_1 + \cdots + X_n$. In words, we are betting \$1 on the flipping of a fair coin and S_n is our winnings at time n. If n and k are integers

$$P(S_{2n} = 2k) = \binom{2n}{n+k} 2^{-2n}$$

since $S_{2n} = 2k$ if and only if there are $n + k$ flips that are $+1$ and $n - k$ flips that are -1 in the first $2n$. The first factor gives the number of such outcomes and the second the probability of each one. **Stirling's formula** (see Feller, Vol. I. (1968) p.52) tells us

(1.1)
$$n! \sim n^n e^{-n} \sqrt{2\pi n} \quad \text{as } n \to \infty$$

where $a_n \sim b_n$ means $a_n/b_n \to 1$ as $n \to \infty$, so

$$\binom{2n}{n+k} = \frac{(2n)!}{(n+k)!(n-k)!}$$

$$\sim \frac{(2n)^{2n}}{(n+k)^{n+k}(n-k)^{n-k}} \cdot \frac{(2\pi(2n))^{1/2}}{(2\pi(n+k))^{1/2}(2\pi(n-k))^{1/2}}$$

and we have

(1.2)
$$\binom{2n}{n+k} 2^{-2n} \sim \left(1 + \frac{k}{n}\right)^{-n-k} \cdot \left(1 - \frac{k}{n}\right)^{-n+k}$$

$$\cdot (\pi n)^{-1/2} \cdot \left(1 + \frac{k}{n}\right)^{-1/2} \cdot \left(1 - \frac{k}{n}\right)^{-1/2}$$

The first two terms on the right are

$$= \left(1 - \frac{k^2}{n^2}\right)^{-n} \cdot \left(1 + \frac{k}{n}\right)^{-k} \cdot \left(1 - \frac{k}{n}\right)^{k}$$

A little calculus shows that

(1.3) Lemma. If $a_j \to \infty$ and $a_j c_j \to \lambda$ then $(1 + c_j)^{a_j} \to e^{\lambda}$.

Proof   As $x \to 0$, $\log(1 + x)/x \to 1$, so $a_j \log(1 + c_j) \to \lambda$ and the desired result follows. □

EXERCISE 1.1. Generalize the last proof to conclude that if $\max_{1 \le j \le n} c_{j,n} \to 0$ and $\sum_{j=1}^{n} c_{j,n} \to \lambda$ then $\prod_{j=1}^{n}(1 + c_{j,n}) \to e^{\lambda}$.

Using (1.3) now we see that if $2k = x\sqrt{2n}$, i.e., $k = x\sqrt{n/2}$, then

$$\left(1 - \frac{k^2}{n^2}\right)^{-n} = \left(1 - x^2/2n\right)^{-n} \to e^{x^2/2}$$

$$\left(1 + \frac{k}{n}\right)^{-k} = \left(1 + x/\sqrt{2n}\right)^{-x\sqrt{n/2}} \to e^{-x^2/2}$$

$$\left(1 - \frac{k}{n}\right)^{k} = \left(1 - x/\sqrt{2n}\right)^{x\sqrt{n/2}} \to e^{-x^2/2}$$

For this choice of k, $k/n \to 0$, so

$$\left(1 + \frac{k}{n}\right)^{-1/2} \cdot \left(1 - \frac{k}{n}\right)^{-1/2} \to 1$$

and putting things together gives:

(1.4) Theorem. If $2k/\sqrt{2n} \to x$ then $P(S_{2n} = 2k) \sim (\pi n)^{-1/2} e^{-x^2/2}$.

Our next step is to compute

$$P(a\sqrt{2n} \le S_{2n} \le b\sqrt{2n}) = \sum_{m \in [a\sqrt{2n}, b\sqrt{2n}] \cap 2\mathbf{Z}} P(S_{2n} = m)$$

Changing variables $m = x\sqrt{2n}$ we have that the above is

$$\approx \sum_{x \in [a,b] \cap (2\mathbf{Z}/\sqrt{2n})} (2\pi)^{-1/2} e^{-x^2/2} \cdot (2/n)^{1/2}$$

where $2\mathbf{Z}/\sqrt{2n} = \{2z/\sqrt{2n} : z \in \mathbf{Z}\}$. We have multiplied and divided by $\sqrt{2}$ since the space between points in the sum is $(2/n)^{1/2}$, so if n is large the sum above is

$$\approx \int_a^b (2\pi)^{-1/2} e^{-x^2/2} dx$$

The integrand is the density of the (standard) normal distribution, so changing notation we can write the last quantity as $P(a \leq \chi \leq b)$ where χ is a random variable with that distribution.

It is not hard to fill in the details to get

(1.5) The De Moivre-Laplace Theorem. If $a < b$ then as $m \to \infty$

$$P(a \leq S_m/\sqrt{m} \leq b) \to \int_a^b (2\pi)^{-1/2} e^{-x^2/2} dx$$

(To remove the restriction to even integers observe $S_{2n+1} = S_{2n} \pm 1$.) The last result is a special case of the central limit theorem given in Section 2.4, so further details are left to the reader.

Another special case that can be treated with Stirling's formula is

EXERCISE 1.2. Let $X_1, X_2, \ldots$ be independent and have a Poisson distribution with mean 1. Then $S_n = X_1 + \cdots + X_n$ has a Poisson distribution with mean n, i.e., $P(S_n = k) = e^{-n} n^k /k!$ Use Stirling's formula to show that if $(k-n)/\sqrt{n} \to x$ then

$$\sqrt{2\pi n} P(S_n = k) \to \exp(-x^2/2)$$

As in the case of coin flips it follows that

$$P(a \leq (S_n - n)/\sqrt{n} \leq b) \to \int_a^b (2\pi)^{-1/2} e^{-x^2/2} \, dx$$

but proving the last conclusion is not part of the exercise.

Stirling's formula can also be used to compute some large deviations probabilities considered in Section 1.9. In the next two exercises, $X_1, X_2, \ldots$ are i.i.d. and $S_n = X_1 + \cdots + X_n$. In each case you should begin by considering $P(S_n = k)$ when $k/n \to a$ and then relate $P(S_n = j+1)$ to $P(S_n = j)$ to show $P(S_n \geq k) \leq CP(S_n = k)$.

EXERCISE 1.3. Suppose $P(X_i = 1) = P(X_i = -1) = 1/2$. Show that if $a \in (0, 1)$

$$\frac{1}{2n} \log P(S_{2n} \geq 2na) \to -\gamma(a)$$

where $\gamma(a) = \frac{1}{2}\{(1 + a)\log(1 + a) + (1 - a)\log(1 - a)\}$.

EXERCISE 1.4. Suppose $P(X_i = k) = e^{-1}/k!$ for $k = 0, 1, \ldots$ Show that if $a > 1$

$$\frac{1}{n}\log P(S_n \geq na) \to a - 1 - a\log a$$

2.2. Weak Convergence

In this section we will define the type of convergence that appears in the central limit theorem and explore some of its properties. A sequence of distribution functions is said to **converge weakly** to a limit F (written $F_n \Rightarrow F$) if $F_n(y) \to F(y)$ for all y that are continuity points of F. A sequence of random variables X_n is said to **converge weakly** or **converge in distribution** to a limit X_∞ (written $X_n \Rightarrow X_\infty$) if their distribution functions $F_n(x) = P(X_n \leq x)$ converge weakly. To see that convergence at continuity points is enough to identify the limit, observe that F is right continuous and by Exercise 1.8 in Chapter 1, the discontinuities of F are at most a countable set.

a. Examples

Two examples of weak convergence that we have seen earlier are:

Example 2.1. Let $X_1, X_2, \ldots$ be i.i.d. with $P(X_i = 1) = P(X_i = -1) = 1/2$ and let $S_n = X_1 + \cdots + X_n$. Then (1.5) implies

$$F_n(y) = P(S_n/\sqrt{n} \leq y) \to \int_{-\infty}^{y} (2\pi)^{-1/2} e^{-x^2/2} \, dx$$

Example 2.2. Let $X_1, X_2, \ldots$ be i.i.d. with distribution F. The Glivenko-Cantelli theorem ((7.4) in Chapter 1) implies that for almost every ω,

$$F_n(y) = n^{-1} \sum_{m=1}^{n} 1_{(X_m(\omega) \leq y)} \to F(y) \text{ for all } y$$

In the last two examples convergence occurred for all y, even though in the second case the distribution function could have discontinuities. The next example shows why we restrict our attention to continuity points.

Example 2.3. Let X have distribution F. Then $X + 1/n$ has distribution

$$F_n(x) = P(X + 1/n \leq x) = F(x - 1/n)$$

As $n \to \infty$, $F_n(x) \to F(x-) = \lim_{y \uparrow x} F(y)$ so convergence only occurs at continuity points.

Example 2.4. Waiting for rare events. Let X_p be the number of trials needed to get a success in a sequence of independent trials with success probability p. Then $P(X_p \geq n) = (1-p)^{n-1}$ for $n = 1, 2, 3, \ldots$ and it follows from (1.3) that as $p \to 0$,

$$P(pX_p > x) \to e^{-x} \quad \text{for all } x \geq 0$$

In words, pX_p converges weakly to an exponential distribution.

Example 2.5. Birthday problem. Let $X_1, X_2, \ldots$ be independent and uniformly distributed on $\{1, \ldots, N\}$, and let $T_N = \min\{n : X_n = X_m \text{ for some } m < n\}$.

$$P(T_N > n) = \prod_{m=2}^{n} \left(1 - \frac{m-1}{N}\right)$$

When $N = 365$ this is the probability that two people in a group of size n do not have the same birthday (assuming all birthdays are equally likely). Using Exercise 1.1 it is easy to see that

$$P(T_N/N^{1/2} > x) \to \exp(-x^2/2) \text{ for all } x \geq 0$$

Taking $N = 365$ and noting $22/\sqrt{365} = 1.1515$ and $(1.1515)^2/2 = 0.6630$ this says that

$$P(T_{365} > 22) \approx e^{-.6630} \approx .515$$

This answer is 2% smaller than the true probability .524.

Before giving our sixth example we need a simple result called **Scheffé's Theorem.** Suppose we have probability densities f_n and $f_n \to f_\infty$ pointwise as $n \to \infty$. Then for all Borel sets B

$$\left| \int_B f_n(x)dx - \int_B f_\infty(x)dx \right| \leq \int |f_n(x) - f_\infty(x)|dx$$

$$= 2 \int (f_\infty(x) - f_n(x))^+ \, dx \to 0$$

by the dominated convergence theorem, the equality following from the fact that the $f_n \geq 0$ and have integral $= 1$. Writing μ_n for the corresponding measures, we have shown that the **total variation norm**

$$\|\mu_n - \mu_\infty\| \equiv \sup_B |\mu_n(B) - \mu_\infty(B)| \to 0$$

a conclusion stronger than weak convergence. (Take $B = (-\infty, x]$.) The example $\mu_n = a$ point mass at $1/n$ (with $1/\infty = 0$) shows that we may have $\mu_n \Rightarrow \mu_\infty$ with $\|\mu_n - \mu_\infty\| = 1$ for all n.

EXERCISE 2.1. Give an example of random variables X_n with densities f_n so that $X_n \Rightarrow$ a uniform distribution on $(0,1)$ but $f_n(x)$ does not converge to 1 for any $x \in [0, 1]$.

Example 2.6. Central order statistic. Put $(2n + 1)$ points at random in $(0,1)$, i.e., with locations that are independent and uniformly distributed. Let V_{n+1} be the $(n + 1)$th largest point. It is easy to see that

Lemma. V_{n+1} has density

$$(2n + 1)\binom{2n}{n} x^n (1 - x)^n$$

Proof   There are $2n + 1$ ways to pick the observation that falls at x, then we have to pick n indices for observations $< x$, which can be done in $\binom{2n}{n}$ ways. Once we have decided on the indices that will land $< x$ and $> x$, the probability the corresponding random variables will do what we want is $x^n(1 - x)^n$, and the probability density that the remaining one will land at x is 1. If you don't like the previous sentence compute the probability $X_1 < x - \epsilon, \ldots, X_n < x - \epsilon$, $x - \epsilon < X_{n+1} < x + \epsilon$, $X_{n+2} > x + \epsilon, \ldots X_{2n+1} > x + \epsilon$ then let $\epsilon \to 0$.   □

To compute the density function of $Y_n = 2(V_{n+1} - 1/2)\sqrt{2n}$, we use Exercise 1.10 in Chapter 1, or simply change variables $x = 1/2 + y/2\sqrt{2n}$, $dx = dy/2\sqrt{2n}$ to get

$$(2n + 1)\binom{2n}{n}\left(\frac{1}{2} + \frac{y}{2\sqrt{2n}}\right)^n \left(\frac{1}{2} - \frac{y}{2\sqrt{2n}}\right)^n \frac{1}{2\sqrt{2n}}$$

$$= \binom{2n}{n} 2^{-2n} \cdot (1 - y^2/2n)^n \cdot \frac{2n + 1}{2n} \sqrt{\frac{n}{2}}$$

The first factor is $P(S_{2n} = 0)$ for a simple random walk so (1.4) and (1.3) imply that

$$P(Y_n = y) \to (2\pi)^{-1/2} \exp(-y^2/2) \quad \text{as} \quad n \to \infty$$

Here and in what follows we write $P(Y_n = y)$ for the density function of Y_n. Using Scheffe's theorem now, we conclude that Y_n converges weakly to a standard normal distribution.

EXERCISE 2.2. **Convergence of maxima.** Let $X_1, X_2, \ldots$ be independent with distribution F, and let $M_n = \max_{m \leq n} X_m$. Then $P(M_n \leq x) = F(x)^n$. Prove the following limit laws for M_n:

(i) If $F(x) = 1 - x^{-\alpha}$ for $x \geq 1$ where $\alpha > 0$ then for $y > 0$

$$P(M_n/n^{1/\alpha} \leq y) \to \exp(-y^{-\alpha})$$

(ii) If $F(x) = 1 - |x|^\beta$ for $-1 \leq x \leq 0$ where $\beta > 0$ then for $y < 0$

$$P(n^{1/\beta} M_n \leq y) \to \exp(-|y|^\beta)$$

(iii) If $F(x) = 1 - e^{-x}$ for $x \geq 0$ then for all $y \in (-\infty, \infty)$

$$P(M_n - \log n \leq y) \to \exp(-e^{-y})$$

The limits that appear above are called the **extreme value distributions.** The last one is called the **double exponential** or **Gumbel distribution.** Necessary and sufficient conditions for $(M_n - b_n)/a_n$ to converge to these limits were obtained by Gnedenko (1943). For a recent treatment, see Resnick (1987).

EXERCISE 2.3. Let $X_1, X_2, \ldots$ be i.i.d. and have the standard normal distribution. (i) From (1.4) in Chapter 1 we know

$$P(X_i > x) \sim \frac{1}{x} e^{-x^2/2} \quad \text{as } x \to \infty$$

Use this to conclude that for any real number θ

$$P(X_i > x + (\theta/x))/P(X_i > x) \to e^{-\theta}$$

(ii) Show that if we define b_n by $P(X_i > b_n) = 1/n$

$$P(b_n(M_n - b_n) \leq x) \to \exp(-e^{-x})$$

(iii) Show that $b_n \sim (2 \log n)^{1/2}$ and conclude $M_n/(2 \log n)^{1/2} \to 1$ in probability.

b. Theory

The next result is useful for proving things about weak convergence.

(2.1) Theorem. If $F_n \Rightarrow F_\infty$ then there are random variables Y_n, $1 \leq n \leq \infty$, with distribution F_n so that $Y_n \to Y_\infty$ a.s.

Proof Let $\Omega = (0,1)$, $\mathcal{F} =$ Borel sets, $P =$ Lebesgue measure, and let $Y_n(x) = \sup\{y : F_n(y) < x\}$. By (1.1) in Chapter 1, Y_n has distribution F_n. We will now show that $Y_n(x) \to Y_\infty(x)$ for all but a countable number of x. To do this it is convenient to write $Y_n(x)$ as $F_n^{-1}(x)$ and drop the subscript when $n = \infty$. We begin by identifying the exceptional set. Let $a_x = \sup\{y : F(y) < x\}$, $b_x = \inf\{y : F(y) > x\}$, and $\Omega_0 = \{x : (a_x, b_x) = \emptyset\}$ where (a_x, b_x) is the open interval with the indicated endpoints. $\Omega - \Omega_0$ is countable since the (a_x, b_x) are disjoint and each nonempty interval contains a different rational number. If $x \in \Omega_0$ then $F(y) < x$ for $y < F^{-1}(x)$ and $F(z) > x$ for $z > F^{-1}(x)$. To prove that $F_n^{-1}(x) \to F^{-1}(x)$ for $x \in \Omega_0$ there are two things to show

(a) $\liminf_{n \to \infty} F_n^{-1}(x) \geq F^{-1}(x)$

Proof of (a) Let $y < F^{-1}(x)$ be such that F is continuous at y. Since $x \in \Omega_0$, $F(y) < x$ and if n is sufficiently large $F_n(y) < x$, i.e., $F_n^{-1}(x) \geq y$. Since this holds for all y satisfying the indicated restrictions, the result follows.

(b) $\limsup_{n \to \infty} F_n^{-1}(x) \leq F^{-1}(x)$

Proof of (b) Let $y > F^{-1}(x)$ be such that F is continuous at y. Since $x \in \Omega_0$, $F(y) > x$ and if n is sufficiently large $F_n(y) > x$, i.e., $F_n^{-1}(x) \leq y$. Since this holds for all y satisfying the indicated restrictions, the result follows and we have completed the proof of (2.1). $\qquad\square$

(2.1) allows us to immediately generalize some of our earlier results.

EXERCISE 2.4. **Fatou's lemma.** Let $g \geq 0$ be continuous. If $X_n \Rightarrow X_\infty$ then

$$\liminf_{n \to \infty} Eg(X_n) \geq Eg(X_\infty)$$

EXERCISE 2.5. **Integration to the limit.** Suppose g, h are continuous with $g(x) > 0$, and $|h(x)|/g(x) \to 0$ as $|x| \to \infty$. If $F_n \Rightarrow F$ and $\int g(x)\, dF_n(x) \leq C < \infty$ then

$$\int h(x)\, dF_n(x) \to \int h(x) dF(x)$$

The next result illustrates the usefulness of (2.1) and gives an equivalent definition of weak convergence that makes sense in any topological space.

(2.2) **Theorem.** $X_n \Rightarrow X_\infty$ if and only if for every bounded continuous function g we have $Eg(X_n) \to Eg(X_\infty)$.

Proof   Let Y_n have the same distribution as X_n and converge a.s. Since g is continuous $g(Y_n) \to g(Y_\infty)$ a.s. and the bounded convergence theorem implies

$$Eg(X_n) = Eg(Y_n) \to Eg(Y_\infty) = Eg(X_\infty)$$

To prove the converse let

$$g_{x,\epsilon}(y) = \begin{cases} 1 & y \le x \\ 0 & y \ge x + \epsilon \\ \text{linear} & x \le y \le x + \epsilon \end{cases}$$

Since $g_{x,\epsilon}(y) = 1$ for $y \le x$, $g_{x,\epsilon}$ is continuous, and $g_{x,\epsilon}(y) = 0$ for $y > x + \epsilon$,

$$\limsup_{n \to \infty} P(X_n \le x) \le \limsup_{n \to \infty} Eg_{x,\epsilon}(X_n) = Eg_{x,\epsilon}(X_\infty) \le P(X_\infty \le x + \epsilon)$$

Letting $\epsilon \to 0$ gives $\limsup_{n \to \infty} P(X_n \le x) \le P(X_\infty \le x)$. The last conclusion is valid for any x. To get the other direction we observe

$$\liminf_{n \to \infty} P(X_n \le x) \ge \liminf_{n \to \infty} Eg_{x-\epsilon,\epsilon}(X_n) = Eg_{x-\epsilon,\epsilon}(X_\infty) \ge P(X_\infty \le x - \epsilon)$$

Letting $\epsilon \to 0$ gives $\liminf_{n \to \infty} P(X_n \le x) \ge P(X_\infty < x) = P(X_\infty \le x)$ if x is a continuity point. The results for the $\limsup$ and the $\liminf$ combine to give the desired result. $\square$

The next result is a trivial but useful generalization of (2.2).

(2.3) Continuous mapping theorem. Let g be a measurable function and $D_g = \{x : g$ is discontinuous at $x\}$. If $X_n \Rightarrow X_\infty$ and $P(X_\infty \in D_g) = 0$ then $g(X_n) \Rightarrow g(X)$. If in addition g is bounded then $Eg(X_n) \to Eg(X_\infty)$.

Remark. D_g is always a Borel set. See Exercise 2.6 in Chapter 1.

Proof   Let $Y_n =_d X_n$ with $Y_n \to Y_\infty$ a.s. If f is continuous then $D_{f \circ g} \subset D_g$ so $P(Y_\infty \in D_{f \circ g}) = 0$ and it follows that $f(g(Y_n)) \to f(g(Y_\infty))$ a.s. Since $f \circ g$ is bounded the bounded convergence theorem implies $Ef(g(Y_n)) \to Ef(g(Y_\infty))$. Since this holds for all bounded continuous functions it follows from (2.2) that $g(X_n) \Rightarrow g(X_\infty)$. The second conclusion can be proved by taking $f(x) = x$ and repeating the first three lines. $\square$

The next result provides a number of useful alternative defintions of weak convergence.

(2.4) Theorem. The following statements are equivalent: (i) $X_n \Rightarrow X_\infty$

(ii) For all open sets G, $\liminf_{n\to\infty} P(X_n \in G) \geq P(X_\infty \in G)$.
(iii) For all closed sets K, $\limsup_{n\to\infty} P(X_n \in K) \leq P(X_\infty \in K)$.
(iv) For all sets A with $P(X_\infty \in \partial A) = 0$, $\lim_{n\to\infty} P(X_n \in A) = P(X_\infty \in A)$.

Remark. To help remember the directions of the inequalities in (ii) and (iii) consider the special case in which $P(X_n = x_n) = 1$. In this case if $x_n \in G$ and $x_n \to x_\infty \in \partial G$ then $P(X_n \in G) = 1$ for all n but $P(X_\infty \in G) = 0$. Letting $K = G^c$ gives an example for (iii).

Proof We will prove four things and leave it to the reader to check that we have proved the result given above.

(i) implies (ii): Let Y_n have the same distribution as X_n and $Y_n \to Y_\infty$ a.s. Since G is open

$$\liminf_{n\to\infty} 1_G(Y_n) \geq 1_G(Y_\infty)$$

so Fatou's Lemma implies

$$\liminf_{n\to\infty} P(Y_n \in G) \geq P(Y_\infty \in G)$$

(ii) is equivalent to (iii): This follows easily from: A is open if and only if A^c is closed and $P(A) + P(A^c) = 1$.

(ii) and (iii) imply (iv): Let $K = \bar{A}$ and $G = A^o$ be the closure and interior of A respectively. The boundary of A, $\partial A = \bar{A} - A^o$ and $P(X_\infty \in \partial A) = 0$ so

$$P(X_\infty \in K) = P(X_\infty \in A) = P(X_\infty \in G)$$

Using (ii) and (iii) now

$$\limsup_{n\to\infty} P(X_n \in A) \leq \limsup_{n\to\infty} P(X_n \in K) \leq P(X_\infty \in K) = P(X_\infty \in A)$$
$$\liminf_{n\to\infty} P(X_n \in A) \geq \liminf_{n\to\infty} P(X_n \in G) \geq P(X_\infty \in G) = P(X_\infty \in A)$$

(iv) implies (i): Let x be such that $P(X_\infty = x) = 0$ and let $A = (-\infty, x]$.   □

The next result is useful in studying limits of sequences of distributions.

(2.5) Helly's selection theorem. For every sequence F_n of distribution functions there is a subsequence $F_{n(k)}$ and a right continuous nondecreasing function F so that $\lim_{k\to\infty} F_{n(k)}(y) = F(y)$ at all continuity points y of F.

Remark. The limit may not be a distribution function. For example if $a + b + c = 1$ and $F_n(x) = a\, 1_{(x \geq n)} + b\, 1_{(x \geq -n)} + c\, G(x)$ where G is a distribution function then $F_n(x) \to F(x) = b + cG(x)$,

$$\lim_{x \downarrow -\infty} F(x) = b \quad \text{and} \quad \lim_{x \uparrow \infty} F(x) = b + c = 1 - a$$

In words, an amount of mass a escapes to $+\infty$, and mass b escapes to $-\infty$. The type of convergence that occurs in (2.4) is sometimes called **vague convergence**, and will be denoted here by $\Rightarrow_v$.

Proof of (2.5) Let $q_1, q_2, \ldots$ be an enumeration of the rationals. Since for each k, $F_m(q_k) \in [0, 1]$ for all m, there is a sequence $m_k(i) \to \infty$ that is a subsequence of $m_{k-1}(j)$ (let $m_0(j) \equiv j$) so that

$$F_{m_k(i)}(q_k) \text{ converges to } G(q_k) \text{ as } i \to \infty$$

Let $F_{n(k)} = F_{m_k(k)}$. By construction $F_{n(k)}(q) \to G(q)$ for all rational q. The function G may not be right continuous but $F(x) = \inf\{G(q) : q \in \mathbf{Q}, q > x\}$ is since

$$\lim_{x_n \downarrow x} F(x_n) = \inf\{G(q) : q \in \mathbf{Q}, q > x_n \text{ for some } n\}$$

$$= \inf\{G(q) : q \in \mathbf{Q}, q > x\} = F(x)$$

To complete the proof, let x be a continuity point of F. Pick rationals r_1, r_2, s with $r_1 < r_2 < x < s$ so that

$$F(x) - \epsilon < F(r_1) \leq F(r_2) \leq F(x) \leq F(s) < F(x) + \epsilon$$

Since $F_n(r_2) \to G(r_2) \geq F(r_1)$, and $F_n(s) \to G(s) \leq F(s)$ it follows that if n is large

$$F(x) - \epsilon < F_n(r_2) \leq F_n(x) \leq F_n(s) < F(x) + \epsilon$$

which is the desired conclusion. $\square$

The last result raises a question: When can we conclude that no mass is lost in the limit in (2.5)?

(2.6) Theorem. Every subsequential limit is the distribution function of a probability measure if and only if the sequence F_n is **tight**, i.e., for all $\epsilon > 0$ there is an M_ϵ so that

$$\limsup_{n \to \infty} 1 - F_n(M_\epsilon) + F_n(-M_\epsilon) \leq \epsilon$$

Proof   Suppose the sequence is tight and $F_{n(k)} \Rightarrow_v F$. Let $r < -M_\epsilon$ and $s > M_\epsilon$ be continuity points of F. Since $F_n(r) \to F(r)$ and $F_n(s) \to F(s)$, we have

$$1 - F(s) + F(r) = \lim_{k \to \infty} 1 - F_{n(k)}(s) + F_{n(k)}(r)$$

$$\leq \limsup_{n \to \infty} 1 - F_n(M_\epsilon) + F_n(-M_\epsilon) \leq \epsilon$$

The last result implies $\limsup_{x \to \infty} 1 - F(x) + F(-x) \leq \epsilon$ and since ϵ is arbitrary it follows that F is the distribution function of a probability measure.

To prove the converse now suppose F_n is not tight. In this case there is an $\epsilon > 0$ and a subsequence $n(k) \to \infty$ so that

$$1 - F_{n(k)}(k) + F_{n(k)}(-k) \geq \epsilon$$

for all k. By passing to a further subsequence $F_{n(k_j)}$ we can suppose that $F_{n(k_j)} \Rightarrow_v F$. Let $r < 0 < s$ be continuity points of F.

$$1 - F(s) + F(r) = \lim_{j \to \infty} 1 - F_{n(k_j)}(s) + F_{n(k_j)}(r)$$

$$\geq \liminf_{j \to \infty} 1 - F_{n(k_j)}(k_j) + F_{n(k_j)}(-k_j) \geq \epsilon$$

Letting $s \to \infty$ and $r \to -\infty$ we see that F is not the distribution function of a probability measure.   □

The following sufficient condition for tightness is often useful.

(2.7) Theorem. If there is a $\varphi \geq 0$ so that $\varphi(x) \to \infty$ as $|x| \to \infty$ and

$$C = \sup_n \int \varphi(x) dF_n(x) < \infty$$

then F_n is tight.

Proof $1 - F_n(M) + F_n(-M) \leq C / \inf_{|x| \geq M} \varphi(x)$.   □

EXERCISES

2.6. If $F_n \Rightarrow F$ and F is continuous then $\sup_x |F_n(x) - F(x)| \to 0$.

2.7. If F is any distribution function there is a sequence of distribution functions of the form $\sum_{m=1}^n a_{n,m} 1_{(x_{n,m} \leq x)}$ with $F_n \Rightarrow F$. Hint: use (7.4) in Chapter 1.

2.8. Let X_n, $1 \leq n \leq \infty$, be integer valued. Show that $X_n \Rightarrow X_\infty$ if and only if $P(X_n = m) \to P(X_\infty = m)$ for all m.

2.9. Show that if $X_n \to X$ in probability then $X_n \Rightarrow X$ and that conversely, if $X_n \Rightarrow c$, where c is a constant then $X_n \to c$ in probability.

2.10. **Converging together lemma.** Suppose $X_n \Rightarrow X$ and $Y_n \Rightarrow c$, where c is a constant then $X_n + Y_n \Rightarrow X + c$. A useful consequence of this result is that if $X_n \Rightarrow X$ and $Z_n - X_n \Rightarrow 0$ then $Z_n \Rightarrow X$.

2.11. Suppose $X_n \Rightarrow X$, $Y_n \geq 0$, and $Y_n \Rightarrow c$, where $c > 0$ is a constant then $X_n Y_n \Rightarrow cX$. This result is true without the assumptions $Y_n \geq 0$ and $c > 0$. We have imposed these only to make the proof less tedious.

2.12. Show that if $X_n = (X_n^1, \ldots, X_n^n)$ is uniformly distributed over the surface of the sphere of radius $\sqrt{n}$ in $\mathbf{R}^n$ then $X_n^1 \Rightarrow$ a standard normal. Hint: Let $Y_1, Y_2, \ldots$ be i.i.d. standard normals and let $X_n^i = Y_i(n/\sum_{m=1}^n Y_m^2)^{1/2}$.

2.13. Suppose $Y_n \geq 0$, $EY_n^\alpha \to 1$ and $EY_n^\beta \to 1$ for some $0 < \alpha < \beta$. Show that $Y_n \to 1$ in probability.

2.14. For each $K < \infty$ and $y < 1$ there is a $c_{y,K} > 0$ so that $EX^2 = 1$ and $EX^4 \leq K$ implies $P(|X| > y) \geq c_{y,K}$.

2.15. **The Lévy Metric.** Show that

$$\rho(F,G) = \inf\{\epsilon : F(x - \epsilon) - \epsilon \leq G(x) \leq F(x + \epsilon) + \epsilon \text{ for all } x\}$$

defines a metric on the space of distributions and $\rho(F_n, F) \to 0$ if and only if $F_n \Rightarrow F$.

2.16. **The Ky Fan metric** on random variables is defined by

$$\alpha(X,Y) = \inf\{\epsilon \geq 0 : P(|X - Y| > \epsilon) \leq \epsilon\}$$

Show that if $\alpha(X,Y) = \alpha$ then the corresponding distributions have Lévy distance $\rho(F,G) \leq \alpha$.

2.17. Let $\alpha(X,Y)$ be the metric in the previous exercise and let $\beta(X,Y) = E(|X - Y|/(1 + |X - Y|))$ be the metric of Exercise 6.3 in Chapter 1. If $\alpha(X,Y) = a$ then

$$a^2/(1 + a) \leq \beta(X,Y) \leq a + (1 - a)a/(1 + a)$$

2.3. Characteristic Functions

This long section is divided into five parts. The first three are required reading, the last two are optional. In part a we show that the characteristic function

$\varphi(t) = E \exp(itX)$ determines $F(x) = P(X \leq x)$, and we give recipes for computing F from φ. In part b we relate weak convergence of distributions to the behavior of the corresponding characteristic functions. In part c we relate the behavior of $\varphi(t)$ at 0 to the moments of X. In part d we prove Polya's criterion and use it to construct some famous and some strange examples of characteristic functions. Finally in part e we consider the moment problem, i.e., when is a distribution characterized by its moments.

a. Definition, inversion formula

If X is a random variable we define its **characteristic function (ch.f.)** by

$$\varphi(t) = Ee^{itX} = E \cos tX + iE \sin tX$$

The last formula requires taking the expected value of a complex valued random variable but as the second equality may suggest no new theory is required. If Z is complex valued we define $EZ = E(\mathrm{Re}\, Z) + iE(\mathrm{Im}\, Z)$ where $\mathrm{Re}\,(a + bi) = a$ is the **real part** and $\mathrm{Im}\,(a + bi) = b$ is the **imaginary part**. Some properties are immediate from the definition:

(3.1a) $\varphi(0) = 1$

(3.1b) $\varphi(-t) = E(\cos(-tX) + i \sin(-tX)) = \overline{\varphi(t)}$

where $\bar{z}$ denotes the **complex conjugate** of z, $\overline{a + bi} = a - bi$.

(3.1c) $|\varphi(t)| = |Ee^{itX}| \leq E|e^{itX}| = 1$

Here $|z|$ denotes the **modulus** of the complex number z, $|a + bi| = (a^2 + b^2)^{1/2}$. The inequality follows from Exercise 3.4 in Chapter 1 since $\varphi(x, y) = (x^2 + y^2)^{1/2}$ is convex.

$$(3.1d) \qquad |\varphi(t + h) - \varphi(t)| = |E(e^{i(t+h)X} - e^{itX})|$$
$$\leq E|e^{i(t+h)X} - e^{itX}| = E|e^{ihX} - 1|$$

since $|zw| = |z| \cdot |w|$. The last quantity $\to 0$ as $h \to 0$ by the bounded convergence theorem, so $\varphi(t)$ is uniformly continuous on $(-\infty, \infty)$.

(3.1e) $Ee^{it(aX+b)} = e^{itb} Ee^{i(ta)X} = e^{itb} \varphi(at)$

(3.1e) and (3.1b) imply that if X has ch.f. $\varphi(t)$ then $-X$ has ch.f. $\varphi(-t) = \overline{\varphi(t)}$.

(3.1f) If X_1 and X_2 are independent and have ch.f.'s φ_1 and φ_2 then $X + Y$ has ch.f. $\varphi_1(t)\varphi_2(t)$.

Proof $Ee^{it(X_1+X_2)} = E(e^{itX_1}e^{itX_2}) = Ee^{itX_1}Ee^{itX_2}$. $\square$

The next order of business is to give some examples.

Example 3.1. Coin flips. If $P(X = 1) = P(X = -1) = 1/2$ then

$$Ee^{itX} = (e^{it} + e^{-it})/2 = \cos t$$

Example 3.2. Poisson distribution. If $P(X = k) = e^{-\lambda}\lambda^k/k!$ for $k = 0, 1, 2, \ldots$ then

$$Ee^{itX} = \sum_{k=0}^{\infty} e^{-\lambda}\frac{\lambda^k e^{itk}}{k!} = \exp(\lambda(e^{it} - 1))$$

Example 3.3. Normal distribution

Density $(2\pi)^{-1/2}\exp(-x^2/2)$
Ch.f. $\exp(-t^2/2)$

Combining this result with (3.1e), we see that a normal distribution with mean μ and variance σ^2 has ch.f. $\exp(i\mu t - \sigma^2 t^2/2)$. Similar scalings can be applied to other examples so we will often just give the ch.f. for one member of the family.

Physics Proof

$$\int e^{itx}(2\pi)^{-1/2}e^{-x^2/2}\,dx = e^{-t^2/2}\int (2\pi)^{-1/2}e^{-(x-it)^2/2}\,dx$$

The integral is 1 since the integrand is the normal density with mean it and variance 1. $\square$

Math Proof Now that we have cheated and figured out the answer we can verify it by a formal calculation that gives very little insight into why it is true. Let

$$\varphi(t) = \int e^{itx}(2\pi)^{-1/2}e^{-x^2/2}dx = \int \cos tx\,(2\pi)^{-1/2}e^{-x^2/2}dx$$

since $i\sin tx$ is an odd function. Differentiating with respect to t (referring to Example 9.1 in the Appendix for the justification) and then integrating by parts gives

$$\varphi'(t) = \int -x \sin tx \, (2\pi)^{-1/2} e^{-x^2/2} dx$$

$$= -\int t \cos tx \, (2\pi)^{-1/2} e^{-x^2/2} dx = -t\varphi(t)$$

This implies $\frac{d}{dt}\{\varphi(t)\exp(t^2/2)\} = 0$ so $\varphi(t)\exp(t^2/2) = \varphi(0) = 1$.   □

In the next three examples the density is 0 outside the indicated range.

Example 3.4. Uniform distribution on (a, b)

Density $1/(b-a)$ $x \in (a, b)$
Ch.f. $(e^{itb} - e^{ita})/it(b-a)$

In the special case $a = -c$, $b = c$ the ch.f. is $(e^{itc} - e^{-itc})/2cit = (\sin ct)/ct$.

Proof Once you recall that $\int_a^b e^{\lambda x}\, dx = (e^{\lambda b} - e^{\lambda a})/\lambda$ holds for complex λ this is immediate.   □

Example 3.5. Triangular distribution

Density $1 - |x|$ $x \in (-1, 1)$
Ch.f. $2(1 - \cos t)/t^2$

Proof To see this notice that if X and Y are independent and uniform on $(-1/2, 1/2)$ then $X + Y$ has a triangular distribution. Using Example 3.4 now and (3.1f) it follows that the desired ch.f. is

$$\{(e^{it/2} - e^{-it/2})/it\}^2 = \{2\sin(t/2)/t\}^2$$

Using the trig identity $\cos 2\theta = 1 - 2\sin^2\theta$ with $\theta = t/2$ converts the answer into the form given above.   □

Example 3.6. Exponential distribution

Density e^{-x} $x \in (0, \infty)$
Ch.f. $1/(1 - it)$

Proof Integrating gives

$$\int_0^\infty e^{itx} e^{-x}\, dx = \frac{e^{(it-1)x}}{it - 1}\Big|_0^\infty = \frac{1}{1 - it}$$

since $\exp((it - 1)x) \to 0$ as $x \to \infty$.   □

Example 3.7. Bilateral exponential

Density $\quad \frac{1}{2}e^{-|x|} \quad x \in (-\infty, \infty)$

Ch.f. $\quad 1/(1+t^2)$

Proof This follows from a more general fact:

(3.1g) If $F_1, \ldots, F_n$ have ch.f. $\varphi_1, \ldots, \varphi_n$ and $\lambda_i \geq 0$ have $\lambda_1 + \ldots + \lambda_n = 1$ then $\sum_{i=1}^{n} \lambda_i F_i$ has ch.f. $\sum_{i=1}^{n} \lambda_i \varphi_i$.

Applying (3.1g) with F_1 the distribution of an exponential random variable X, F_2 the distribution of $-X$, and $\lambda_1 = \lambda_2 = 1/2$ then using (3.1b) we see the desired ch.f. is

$$\frac{1}{2(1-it)} + \frac{1}{2(1+it)} = \frac{(1+it) + (1-it)}{2(1+t^2)} = \frac{1}{(1+t^2)} \qquad \square$$

EXERCISE 3.1. Show that if φ is a ch.f. then $\mathrm{Re}\,\varphi$ and $|\varphi|^2$ are also.

The first issue to be settled is that the characteristic function uniquely determines the distribution. This and more is provided by

(3.2) **The inversion formula.** Let $\varphi(t) = \int e^{itx} \mu(dx)$ where μ is a probability measure. If $a < b$ then

$$\lim_{T \to \infty} (2\pi)^{-1} \int_{-T}^{T} \frac{e^{-ita} - e^{-itb}}{it} \varphi(t)\, dt = \mu(a, b) + \frac{1}{2}\mu(\{a, b\})$$

Remark. The existence of the limit is part of the conclusion. If $\mu = \delta_0$, a point mass at 0, $\varphi(t) \equiv 1$. In this case if $a = -1$ and $b = 1$, the integrand is $(2\sin t)/t$ and the integral does not converge absolutely.

Proof Let

$$I_T = \int_{-T}^{T} \frac{e^{-ita} - e^{-itb}}{it} \varphi(t)\, dt = \int_{-T}^{T} \int \frac{e^{-ita} - e^{-itb}}{it} e^{itx} \mu(dx)\, dt$$

The integrand may look bad near $t = 0$ but if we observe that

$$\frac{e^{-ita} - e^{-itb}}{it} = \int_{a}^{b} e^{-ity}\, dy$$

we see that the modulus of the integrand is bounded by $b - a$. Since μ is a probability measure and $[-T, T]$ is a finite interval it follows from Fubini's theorem, $\cos(-x) = \cos x$, and $\sin(-x) = -\sin x$ that

$$
I_T = \int \int_{-T}^{T} \frac{e^{-ita} - e^{-itb}}{it} e^{itx}\, dt\, \mu(dx)
$$

$$
= \int \left\{ \int_{-T}^{T} \frac{\sin(t(x - a))}{t}\, dt - \int_{-T}^{T} \frac{\sin(t(x - b))}{t}\, dt \right\} \mu(dx)
$$

Introducing $R(\theta, T) = \int_{-T}^{T} (\sin \theta t)/t\, dt$, we can write the last result as

(∗) $$I_T = \int \{ R(x - a, T) - R(x - b, T) \} \mu(dx)$$

If we let $S(T) = \int_0^T (\sin x)/x\, dx$ then for $\theta > 0$ changing variables $t = x/\theta$ shows that

$$
R(\theta, T) = 2 \int_0^{T\theta} \frac{\sin x}{x}\, dx = 2S(T\theta)
$$

while for $\theta < 0$, $R(\theta, T) = -R(|\theta|, T)$. Introducing the function sgn x which is 1 if $x > 0$, -1 if $x < 0$, and 0 if $x = 0$, we can write the last two formulas together as

$$
R(\theta, T) = 2(\text{sgn}\,\theta) S(T|\theta|)
$$

As $T \to \infty$, $S(T) \to \pi/2$ (see Exercise 6.7 in the Appendix), so we have $R(\theta, T) \to \pi \,\text{sgn}\, \theta$ and

$$
R(x - a, T) - R(x - b, T) \to
\begin{cases}
2\pi & a < x < b \\
\pi & x = a \text{ or } x = b \\
0 & x < a \text{ or } x > b
\end{cases}
$$

$R(\theta, T) \leq 2 \sup S(y) < \infty$, so using the bounded convergence theorem with (∗) implies

$$
(2\pi)^{-1} I_T \to \mu(a, b) + \frac{1}{2}\mu(\{a, b\})
$$

proving (3.2).   □

EXERCISE 3.2. (i) Imitate the proof of (3.2) to show that

$$
\mu(\{a\}) = \lim_{T \to \infty} \frac{1}{2T} \int_{-T}^{T} e^{-ita} \varphi(t)\, dt
$$

(ii) If $P(X \in h\mathbf{Z}) = 1$ where $h > 0$ then its ch.f. has $\varphi(2\pi/h + t) = \varphi(t)$ so

$$P(X = x) = \frac{h}{2\pi} \int_{-\pi/h}^{\pi/h} e^{-itx} \varphi(t)\, dt \quad \text{for } x \in h\mathbf{Z}$$

(iii) If $X = Y + b$ then $E\exp(itX) = e^{itb}E\exp(itY)$. So if $P(X \in b + h\mathbf{Z}) = 1$ the inversion formula in (ii) is valid for $x \in b + h\mathbf{Z}$.

Two trivial consequences of the inversion formula are:

EXERCISE 3.3. If φ is real then X and $-X$ have the same distribution.

EXERCISE 3.4. If X_i, $i = 1, 2$ are independent and have normal distributions with mean 0 and variance σ_i^2 then $X_1 + X_2$ has a normal distribution with mean 0 and variance $\sigma_1^2 + \sigma_2^2$.

The inversion formula is simpler when φ is integrable but as the next result shows this only happens when the underlying measure is nice.

(3.3) **Theorem.** If $\int |\varphi(t)|\, dt < \infty$ then μ has bounded continuous density

$$f(y) = \frac{1}{2\pi} \int e^{-ity} \varphi(t)\, dt$$

Proof As we observed in the proof of (3.2)

$$\left| \frac{e^{-ita} - e^{-itb}}{it} \right| = \left| \int_a^b e^{-ity}\, dy \right| \leq |b - a|$$

so the integral in (3.2) converges absolutely in this case and

$$\mu(a, b) + \frac{1}{2}\mu(\{a, b\}) = \frac{1}{2\pi} \int_{-\infty}^{\infty} \frac{e^{-ita} - e^{-itb}}{it} \varphi(t)\, dt \leq \frac{(b - a)}{2\pi} \int_{-\infty}^{\infty} |\varphi(t)|\, dt$$

The last result implies μ has no point masses and

$$\mu(x, x + h) = \frac{1}{2\pi} \int \frac{e^{-itx} - e^{-it(x+h)}}{it} \varphi(t)\, dt$$

$$= \frac{1}{2\pi} \int \left(\int_x^{x+h} e^{-ity}\, dy \right) \varphi(t)\, dt$$

$$= \int_x^{x+h} \left(\frac{1}{2\pi} \int e^{-ity} \varphi(t)\, dt \right) dy$$

by Fubini's theorem, so the distribution μ has density function

$$f(y) = \frac{1}{2\pi} \int e^{-ity} \varphi(t) \, dt$$

The dominated convergence theorem implies f is continuous and the proof is complete.   □

EXERCISE 3.5. Give an example of a measure μ with a density but for which $\int |\varphi(t)| dt = \infty$. Hint: Two of the examples above have this property.

EXERCISE 3.6. Show that if $X_1, \ldots, X_n$ are independent and uniformly distributed on $(-1, 1)$ then for $n \geq 2$, $X_1 + \cdots + X_n$ has density

$$f(x) = \frac{1}{\pi} \int_0^\infty (\sin t/t)^n \cos tx \, dt$$

Although it is not obvious from the formula, f is a polynomial in each interval $(k, k+1)$, $k \in \mathbf{Z}$ and vanishes on $[-n, n]^c$.

(3.3) and the next result show that the behavior of φ at infinity is related to the smoothness of the underlying measure.

EXERCISE 3.7. Suppose X and Y are independent and have ch.f. φ and distribution μ. Apply Exercise 3.2 to $X - Y$ and use Exercise 4.7 in Chapter 1 to get

$$\lim_{T \to \infty} \frac{1}{2T} \int_{-T}^T |\varphi(t)|^2 \, dt = P(X - Y = 0) = \sum_x \mu(\{x\})^2$$

Remark. The last result implies that if $\varphi(t) \to 0$ as $t \to \infty$, μ has no point masses. Exercise 3.13 gives an example to show that the converse is false. The Riemann-Lebesgue Lemma (Exercise 4.5 in the Appendix) shows that if μ has a density, $\varphi(t) \to 0$ as $t \to \infty$.

Applying the inversion formula (3.3) to the ch.f. in Examples 3.5 and 3.7 gives us two more examples of ch.f. The first one does not have an official name so we gave it one to honor its role in the proof of Polya's criterion (see (3.8)).

Example 3.8. Polya's distribution

Density	$(1 - \cos x)/\pi x^2$
Ch.f.	$(1 - |t|)^+$

Proof (3.3) implies

$$\frac{1}{2\pi}\int \frac{2(1-\cos s)}{s^2}e^{-isy}\,ds = (1-|y|)^+$$

Now let $s = x$, $y = -t$.   □

Example 3.9. The Cauchy distribution

Density $1/\pi(1+x^2)$

Ch.f. $\exp(-|t|)$

Proof (3.3) implies

$$\frac{1}{2\pi}\int \frac{1}{1+s^2}e^{-isy}\,ds = \frac{1}{2}e^{-|y|}$$

Now let $s = x$, $y = -t$ and multiply each side by 2.   □

EXERCISE 3.8. Use the last result to conclude that if $X_1, X_2, \ldots$ are independent and have the Cauchy distribution then $(X_1 + \cdots + X_n)/n$ has the same distribution as X_1.

b. Weak convergence

Our next step toward the central limit theorem is to relate convergence of characteristic functions to weak convergence.

(3.4) **Continuity theorem.** Let μ_n, $1 \le n \le \infty$ be probability measures with ch.f. φ_n. (i) If $\mu_n \Rightarrow \mu_\infty$ then $\varphi_n(t) \to \varphi_\infty(t)$ for all t. (ii) If $\varphi_n(t)$ converges pointwise to a limit $\varphi(t)$ that is continuous at 0, then the associated sequence of distributions μ_n is tight and converges weakly to the measure μ with characteristic function φ.

Remark. To see why continuity of the limit at 0 is needed in (ii), let μ_n have a normal distribution with mean 0 and variance n. In this case $\varphi_n(t) = \exp(-nt^2/2) \to 0$ for $t \ne 0$, and $\varphi_n(0) = 1$ for all n, but the measures do not converge weakly since $\mu_n((-\infty, x]) \to 1/2$ for all x.

Proof (i) is easy. e^{itx} is bounded and continuous so if $\mu_n \Rightarrow \mu_\infty$ then (2.2) implies $\varphi_n(t) \to \varphi_\infty(t)$. To prove (ii), our first goal is to prove tightness. We begin with some calculations that may look mysterious but will prove to be very useful.

$$\int_{-u}^{u} 1 - e^{itx}\,dt = 2u - \int_{-u}^{u}(\cos tx + i\sin tx)\,dt = 2u - \frac{2\sin ux}{x}$$

Dividing both sides by u, integrating $\mu_n(dx)$, and using Fubini's theorem on the left-hand side gives

$$u^{-1} \int_{-u}^{u} (1 - \varphi_n(t))\, dt = 2 \int \left(1 - \frac{\sin ux}{ux}\right) \mu_n(dx)$$

To bound the right-hand side we note that $|\sin x| \le |x|$ for all x so we have $1 - (\sin ux / ux) \ge 0$. Discarding the integral over $(-2/u, 2/u)$ and using $|\sin ux| \le 1$ on the rest, the right-hand side is

$$\ge 2 \int_{|x| \ge 2/u} \left(1 - \frac{1}{|ux|}\right) \mu_n(dx) \ge \mu_n(\{x : |x| > 2/u\})$$

Since $\varphi(t) \to 1$ as $t \to 0$,

$$u^{-1} \int_{-u}^{u} (1 - \varphi(t))\, dt \to 0 \text{ as } u \to 0$$

Pick u so that the integral is $< \epsilon$. Since $\varphi_n(t) \to \varphi(t)$ for each t, it follows from the dominated convergence theorem that for $n \ge N$

$$2\epsilon \ge u^{-1} \int_{-u}^{u} (1 - \varphi_n(t))\, dt \ge \mu_n\{x : |x| > 2/u\}$$

Since ϵ is arbitrary, the sequnce μ_n is tight.

To complete the proof now we observe that if $\mu_{n(k)} \Rightarrow \mu$ then it follows from the first sentence of the proof that μ has ch.f. φ. The last observation and tightness imply that every subsequence has a further subsequence that converges to μ. I claim that this implies the whole sequence converges to μ. To see this, observe that we have shown that if f is bounded and continuous then every subsequence of $\int f \, d\mu_n$ has a further subsequence that converges to $\int f \, d\mu$, so (6.3) in Chapter 1 implies that the whole sequence converges to that limit. This shows $\int f \, d\mu_n \to \int f \, d\mu$ for all bounded continuous functions f so the desired result follows from (2.2). $\square$

EXERCISE 3.9. Suppose that $X_n \Rightarrow X$ and X_n has a normal distribution with mean 0 and variance σ_n^2. Prove that $\sigma_n^2 \to \sigma^2 \in [0, \infty)$.

EXERCISE 3.10. Show that if X_n and Y_n are independent for $1 \le n \le \infty$, $X_n \Rightarrow X_\infty$, and $Y_n \Rightarrow Y_\infty$, then $X_n + Y_n \Rightarrow X_\infty + Y_\infty$.

EXERCISE 3.11. Let $X_1, X_2, \ldots$ be independent and let $S_n = X_1 + \cdots + X_n$. Let φ_j be the ch.f. of X_j and suppose that $S_n \to S_\infty$ a.s. Then S_∞ has ch.f. $\prod_{j=1}^{\infty} \varphi_j(t)$.

EXERCISE 3.12. Using the identity $\sin t = 2\sin(t/2)\cos(t/2)$ repeatedly leads to $(\sin t)/t = \prod_{m=1}^{\infty}\cos(t/2^m)$. Prove the last identity by interpreting each side as a characteristic function.

EXERCISE 3.13. Let $X_1, X_2, \ldots$ be independent taking vaules 0 and 1 with probability 1/2 each. $X = 2\sum_{j\geq 1} X_j/3^j$ has the Cantor distribution. Compute the ch.f. φ of X and notice that φ has the same value at all points $t = 3^k\pi$.

c. Moments and derivatives

In the proof of (3.4) we derived the inequality

$$(3.5) \qquad \mu\{x : |x| > 2/u\} \leq u^{-1}\int_{-u}^{u}(1 - \varphi(t))\,dt$$

which shows that the smoothness of the characteristic function at 0 is related to the decay of the measure at ∞. The next result continues this theme. We leave the proof to the reader. (Use (9.1) in the Appendix.)

EXERCISE 3.14. If $\int|x|^n\mu(dx) < \infty$ then its characteristic function φ has a continuous derivative of order n given by $\varphi^{(n)}(t) = \int(ix)^n e^{itx}\mu(dx)$.

EXERCISE 3.15. Use the last exercise and the series expansion for $e^{-t^2/2}$ to show that the standard normal distribution has

$$EX^{2n} = (2n)!/2^n n! = (2n-1)(2n-3)\cdots 3\cdot 1 \equiv (2n-1)!!$$

The result in Exercise 3.14 shows that if $E|X|^n < \infty$, then its characteristic function is n times differentiable at 0, and $\varphi^n(0) = E(iX)^n$. Expanding φ in a Taylor series about 0, leads to

$$\varphi(t) = \sum_{m=0}^{n}\frac{E(itX)^m}{m!} + o(t^n)$$

where $o(t^n)$ indicates a quantity $g(t)$ that has $g(t)/t^n \to 0$ as $t \to 0$. For our purposes below it will be important to have a good estimate on the error term, so we will now derive the last result. The starting point is a little calculus.

(3.6) Lemma. $\qquad \left| e^{ix} - \sum_{m=0}^{n}\frac{(ix)^m}{m!}\right| \leq \min\left(\frac{|x|^{n+1}}{(n+1)!}, \frac{2|x|^n}{n!}\right)$

Proof   Integrating by parts gives

$$\int_0^x (x-s)^n e^{is}\, ds = \frac{x^{n+1}}{n+1} + \frac{i}{n+1}\int_0^x (x-s)^{n+1} e^{is}\, ds$$

When $n = 0$ this says

$$\int_0^x e^{is}\, ds = x + i\int_0^x (x-s)e^{is}\, ds$$

The left-hand side is $(e^{ix} - 1)/i$, so rearranging gives

$$e^{ix} = 1 + ix + i^2 \int_0^x (x-s)e^{is}\, ds$$

Using the result for $n = 1$ now gives

$$e^{ix} = 1 + ix + \frac{i^2 x^2}{2} + \frac{i^3}{2}\int_0^x (x-s)^2 e^{is}\, ds$$

and iterating we arrive at

(a) $$e^{ix} - \sum_{m=0}^n \frac{(ix)^m}{m!} = \frac{i^{n+1}}{n!}\int_0^x (x-s)^n e^{is}\, ds$$

To prove (3.6) now it only remains to estimate the "error term" on the right-hand side. Since $|e^{is}| \le 1$ for all s,

(b) $$\left| \frac{i^{n+1}}{n!}\int_0^x (x-s)^n e^{is}\, ds \right| \le |x|^{n+1}/(n+1)!$$

The last estimate is good when x is small. The next is designed for large x. Integrating by parts

$$\frac{i}{n}\int_0^x (x-s)^n e^{is}\, ds = -\frac{x^n}{n} + \int_0^x (x-s)^{n-1} e^{is}\, ds$$

Noticing $x^n/n = \int_0^x (x-s)^{n-1} ds$ now gives

$$\frac{i^{n+1}}{n!}\int_0^x (x-s)^n e^{is}\, ds = \frac{i^n}{(n-1)!}\int_0^x (x-s)^{n-1}(e^{is} - 1)\, ds$$

and since $|e^{ix} - 1| \le 2$, it follows that

(c) $$\left| \frac{i^{n+1}}{n!}\int_0^x (x-s)^n e^{is}\, ds \right| \le \left| \frac{2}{(n-1)!}\int_0^x (x-s)^{n-1}\, ds \right| \le 2|x|^n/n!$$

Combining (a), (b), and (c) we have (3.6). $\square$

Taking expected values, using Jensen's inequality, applying (3.6) to $x = tX$, gives

(3.7)
$$\left| Ee^{itX} - \sum_{m=0}^{n} E\frac{(itX)^m}{m!} \right| \le E\left| e^{itX} - \sum_{m=0}^{n} \frac{(itX)^m}{m!} \right|$$
$$\le E\min\left(\frac{|tX|^{n+1}}{(n+1)!}, \frac{2|tX|^n}{n!} \right)$$

In the next section the following special case will be useful.

(3.8) Theorem. If $EX = \mu$ and $E|X|^2 = \sigma^2 < \infty$ then

$$\varphi(t) = 1 + it\mu - t^2\sigma^2/2 + o(t^2)$$

Proof The error term is $\le t^2 E(|t| \cdot |X|^3 \wedge 6|X|^2)/3!$. The variable in parentheses is smaller than $6|X|^2$ and converges to 0 as $t \to 0$, so the desired conclusion follows from the dominated convergence theorem. $\square$

Remark. The point of the estimate in (3.7) which involves the minimum of two terms rather than just the first one which would result from a naive application of Taylor series, is that we get the conclusion in (3.8) under the assumption $E|X|^2 < \infty$, i.e., we do not have to assume $E|X|^3 < \infty$.

EXERCISE 3.16. (i) Suppose that the family of measures $\{\mu_i, i \in I\}$ is tight, i.e., $\sup_i \mu_i([-M, M]^c) \to 0$ as $M \to \infty$. Use (3.1d) and (3.7) with $n = 0$ to show that their ch.f.'s φ_i are equicontinuous, i.e., if $\epsilon > 0$ we can pick $\delta > 0$ so that if $|h| < \delta$ then $|\varphi_i(t + h) - \varphi_i(t)| < \epsilon$. (ii) Suppose $\mu_n \Rightarrow \mu_\infty$. Use (3.4) and equicontinuity to conclude that the ch.f.'s $\varphi_n \to \varphi_\infty$ uniformly on compact sets. [Argue directly, you don't need to go to AA.] (iii) Give an example to show that the convergence need not be uniform on the whole real line.

EXERCISE 3.17. Let $X_1, X_2, \ldots$ be i.i.d. with characteristic function φ. (i) If $\varphi'(0) = ia$ and $S_n = X_1 + \cdots + X_n$ then $S_n/n \to a$ in probability. (ii) If $S_n/n \to a$ in probability then $\varphi(t/n)^n \to e^{iat}$. Use this to conclude that $\varphi'(0) = ia$ so the weak law holds if and only if $\varphi'(0)$ exists. This is due to E.J.G. Pitman (1956).

The last exercise in combination with Exercise 5.6 from Chapter 1 shows that $\varphi'(0)$ may exist when $E|X| = \infty$.

EXERCISE 3.18. $2\int_0^\infty (1 - \operatorname{Re}\varphi(t))/(\pi t^2)\,dt = \int |y|dF(y)$. Hint: change variables $x = |y|t$ in the density function of Example 3.8, which integrates to 1.

The next result shows that the existence of second derivatives implies the existence of second moments.

(3.9) **Theorem.** If $\limsup_{h\downarrow 0}\{\varphi(h) - 2\varphi(0) + \varphi(-h)\}/h^2 > -\infty$, then $E|X|^2 < \infty$.

Proof $(e^{ihx} - 2 + e^{-ihx})/h^2 = -2(1 - \cos hx)/h^2 \le 0$ and $2(1 - \cos hx)/h^2 \to x^2$ as $h \to 0$ so Fatou's lemma and Fubini's theorem imply

$$\int x^2\,dF(x) \le 2\liminf_{h\to 0}\int \frac{1 - \cos hx}{h^2}\,dF(x)$$

$$= -\limsup_{h\to 0}\frac{\varphi(h) - 2\varphi(0) + \varphi(-h)}{h^2} < \infty \qquad \square$$

EXERCISE 3.19. Show that if $\lim_{t\downarrow 0}(\varphi(t) - 1)/t^2 = c > -\infty$ then $EX = 0$ and $E|X|^2 = -2c < \infty$. In particular, if $\varphi(t) = 1 + o(t^2)$ then $\varphi(t) \equiv 1$.

EXERCISE 3.20. If Y_n are r.v.'s with ch.f.'s φ_n then $Y_n \Rightarrow 0$ if and only if there is a $\delta > 0$ so that $\varphi_n(t) \to 1$ for $|t| \le \delta$.

EXERCISE 3.21. Let $X_1, X_2, \ldots$ be independent. If $S_n = \sum_{m\le n} X_m$ converges in distribution then it converges in probability (and hence a.s. by Exercise 8.11 in Chapter 1). Hint: The last exercise implies that if $m, n \to \infty$ then $S_m - S_n \to 0$ in probability. Now use Exercise 6.4 in Chapter 1.

*d. Polya's criterion

The next result is useful for constructing examples of ch.f.'s.

(3.10) **Polya's criterion.** Let $\varphi(t)$ be real nonnegative and have $\varphi(0) = 1$, $\varphi(t) = \varphi(-t)$, and φ is decreasing and convex on $(0, \infty)$ with

$$\lim_{t\downarrow 0}\varphi(t) = 1, \qquad \lim_{t\uparrow\infty}\varphi(t) = 0$$

Then there is a probability measure ν on $(0, \infty)$, so that

$$(*) \qquad \varphi(t) = \int_0^\infty \left(1 - \left|\frac{t}{s}\right|\right)^+ \nu(ds)$$

and hence φ is a characteristic function.

Remark. Before we get lost in the details of the proof, the reader should note that $(*)$ displays φ as a convex combination of ch.f.'s of the form given in Example 3.8, so an extension of (3.1g) (to be proved below) implies that this is a ch.f.

The assumption that $\lim_{t\to 0}\varphi(t) = 0$ is necessary because the function $\varphi(t) = 1_{\{0\}}(t)$ which is 1 at 0 and 0 otherwise satisfies all the other hypotheses. We could allow $\lim_{t\to\infty}\varphi(t) = c > 0$ by having a pointmass of size c at 0 but we leave this extension to the reader.

Proof Let φ' be the right derivative of φ, i.e.,

$$\varphi'(t) = \lim_{h\downarrow 0} \frac{\varphi(t+h) - \varphi(t)}{h}$$

Since φ is convex this exists and is right continuous and increasing. So we can let μ be the measure on $(0, \infty)$ with $\mu(a, b] = \varphi'(b) - \varphi'(a)$ for all $0 \le a < b < \infty$, and let ν be the measure on $(0, \infty)$ with $d\nu/d\mu = s$.

Now $\varphi'(t) \to 0$ as $t \to \infty$ (for if $\varphi'(t) \downarrow -\epsilon$ we would have $\varphi(t) \le 1 - \epsilon t$ for all t) so Exercise 8.7 in the Appendix implies

$$-\varphi'(s) = \int_s^\infty r^{-1}\nu(dr)$$

Integrating again and using Fubini's theorem we have for $t \ge 0$

$$\varphi(t) = \int_t^\infty \int_s^\infty r^{-1}\nu(dr)\,ds = \int_t^\infty r^{-1}\int_t^r ds\,\nu(dr)$$

$$= \int_t^\infty \left(1 - \frac{t}{r}\right)\nu(dr) = \int_0^\infty \left(1 - \frac{t}{r}\right)^+ \nu(dr)$$

Using $\varphi(-t) = \varphi(t)$ to extend the formula to $t \le 0$ we have $(*)$. Setting $t = 0$ in $(*)$ shows ν has total mass 1.

If φ is piecewise linear, ν has a finite number of atoms and the result follows from Example 3.8 and (3.1g). To prove the general result, let ν_n be a sequence of measures on $(0, \infty)$ with a finite number of atoms that converges weakly to ν (see Exercise 2.7) and let

$$\varphi_n(t) = \int_0^\infty \left(1 - \left|\frac{t}{s}\right|\right)^+ \nu_n(ds)$$

Since $s \to (1 - |t/s|)^+$ is bounded and continuous, $\varphi_n(t) \to \varphi(t)$ and the desired result follows from part (ii) of (3.4). $\square$

A classic application of Polya's criterion is

EXERCISE 3.22. Show that $\exp(-|t|^{\alpha})$ is a characteristic function for $0 < \alpha \leq 1$.

(The case $\alpha = 1$ corresponds to the Cauchy distribution). The next argument, which we learned from Frank Spitzer, proves that this is true for $0 < \alpha \leq 2$. The case $\alpha = 2$ corresponds to a normal distribution, so that case can be safely ignored in the proof.

Example 3.10. $\exp(-|t|^{\alpha})$ is a characteristic function for $0 < \alpha < 2$.

Proof   A little calculus shows that for any β and $|x| < 1$

$$(1 - x)^{\beta} = \sum_{n=0}^{\infty} \binom{\beta}{n}(-x)^n$$

where

$$\binom{\beta}{n} = \frac{\beta(\beta - 1)\cdots(\beta - n + 1)}{1 \cdot 2 \cdots n}$$

Let $\psi(t) = 1 - (1 - \cos t)^{\alpha/2} = \sum_{n=1}^{\infty} c_n(\cos t)^n$ where

$$c_n = \binom{\alpha/2}{n}(-1)^{n+1}$$

$c_n \geq 0$ (here we use $\alpha < 2$), and $\sum_{n=1}^{\infty} c_n = 1$ (take $t = 0$ in the definition of ψ). $\cos t$ is a characteristic function (see Example 3.1) so an easy extension of (3.1g) shows that ψ is a ch.f. We have $1 - \cos t \sim t^2/2$ as $t \to 0$ so

$$1 - \cos(t \cdot 2^{1/2} \cdot n^{-1/\alpha}) \sim n^{-2/\alpha}t^2$$

Using (1.2) and (ii) of (3.4) now it follows that

$$\exp(-|t|^{\alpha}) = \lim_{n \to \infty} \{\psi(t \cdot 2^{1/2} \cdot n^{-1/\alpha})\}^n$$

is a ch.f. $\square$

Exercise 3.19 shows that $\exp(-|t|^{\alpha})$ is not a ch.f. when $\alpha > 2$. A reason for interest in these characteristic functions is explained by the following generalization of Exercise 3.8.

EXERCISE 3.23. If $X_1, X_2, \ldots$ are independent and have characteristic function $\exp(-|t|^{\alpha})$ then $(X_1 + \cdots + X_n)/n^{1/\alpha}$ has the same distribution as X_1.

We will return to this topic in Section 2.7. Polya's criterion can also be used to construct some "pathological examples."

EXERCISE 3.24. Let φ_1 and φ_2 be ch.f.'s. Show that $A = \{t : \varphi_1(t) = \varphi_2(t)\}$ is closed, contains 0, and is symmetric about 0. Show that if A is a set with these properties and $\varphi_1(t) = e^{-|t|}$ there is a φ_2 so that $\{t : \varphi_1(t) = \varphi_2(t)\} = A$.

Example 3.11. For some purposes it is nice to have an explicit example of two ch.f.'s that agree on $[-1, 1]$. From Example 3.8 we know that $(1 - |t|)^+$ is the ch.f. of the density $(1 - \cos x)/\pi x^2$. Define $\psi(t)$ to be equal to φ on $[-1, 1]$ and periodic with period 2, i.e., $\psi(t) = \psi(t + 2)$. The Fourier series for ψ is

$$\psi(u) = \frac{1}{2} + \sum_{n=-\infty}^{\infty} \frac{2}{\pi^2(2n - 1)^2} \exp(i(2n - 1)\pi u)$$

The right-hand side is the ch.f. of a discrete distribution with

$$P(X = 0) = 1/2 \quad \text{and} \quad P(X = (2n - 1)\pi) = 2\pi^{-2}(2n - 1)^{-2} \quad n \in \mathbf{Z}.$$

EXERCISE 3.25. Find independent r.v.'s X, Y, and Z so that Y and Z do not have the same distribution but $X + Y$ and $X + Z$ do.

EXERCISE 3.26. Show that if X and Y are independent and $X + Y$ and X have the same distribution then $Y \equiv 0$.

For more curiosities see Feller, Vol. II (1971), Section XV.2a.

*e. The moment problem

Suppose $\int x^k dF_n(x)$ has a limit μ_k for each k. Then the sequence of distributions is tight by (2.6) and every subsequential limit has the moments μ_k by Exercise 2.5, so we can conclude the sequence converges weakly if there is only one distribution with these moments. It is easy to see that this is true if F is concentrated on a finite interval $[-M, M]$ since every continuous function can be approximated uniformly on $[-M, M]$ by polynomials. The result is false in general.

Counterexample 1. Heyde (1963) Consider the **lognormal density**

$$f_0(x) = (2\pi)^{-1/2} x^{-1} \exp(-(\log x)^2/2) \qquad x \geq 0$$

and for $-1 \le a \le 1$ let

$$f_a(x) = f_0(x)\{1 + a\sin(2\pi \log x)\}$$

To see that f_a is a density and has the same moments as f_0, it suffices to show that

$$\int_0^\infty x^r f_0(x) \sin(2\pi \log x)\,dx = 0 \text{ for } r = 0,1,2,\dots$$

Changing variables $x = \exp(s+r)$, $s = \log x - r$, $ds = dx/x$ the integral becomes

$$(2\pi)^{-1/2} \int_{-\infty}^\infty \exp(rs + r^2)\exp(-(s+r)^2/2)\sin(2\pi(s+r))\,ds$$

$$= (2\pi)^{-1/2}\exp(r^2/2)\int_{-\infty}^\infty \exp(-s^2/2)\sin(2\pi s)\,ds = 0$$

The two equalities holding because r is an integer and the integrand is odd. From the proof it should be clear that we could let

$$g(x) = f_0(x)\left\{1 + \sum_{k=1}^\infty a_k \sin(k\pi \log x)\right\} \quad \text{if } \sum_{k=1}^\infty |a_k| \le 1$$

to get a large family of densities having the same moments as the lognormal.

The moments of the lognormal are easy to compute. Recall that if χ has the standard normal distribution then Exercise 1.11 in Chapter 1 implies $\exp(\chi)$ has the lognormal distribution.

$$EX^n = E\exp(n\chi) = \int e^{nx}(2\pi)^{-1/2}e^{-x^2/2}\,dx$$

$$= e^{n^2/2}\int (2\pi)^{-1/2}e^{-(x-n)^2/2}\,dx = \exp(n^2/2)$$

since the last integrand is the density of the normal with mean n and variance 1. Somewhat remarkably there is a family of discrete random variables with these moments. Let $a > 0$ and

$$P(Y_a = ae^k) = a^{-k}\exp(-k^2/2)/c_a \quad \text{for } k \in \mathbf{Z}$$

where c_a is chosen to make the total mass 1.

$$\exp(-n^2/2)EY_a^n = \exp(-n^2/2)\sum_k (ae^k)^n a^{-k}\exp(-k^2/2)/c_a$$

$$= \sum_k a^{-(k-n)}\exp(-(k-n)^2/2)/c_a = 1$$

by the definition of c_a. This example is due to Liepnik (1981).

The lognormal density decays like $\exp(-(\log x)^2/2)$ as $|x| \to \infty$. The next counterexample has more rapid decay. Since the exponential distribution, e^{-x} for $x \geq 0$, is determined by its moments (see Exercise 3.28 below) we cannot hope to do much better than this.

Counterexample 2. Let $\lambda \in (0,1)$ and for $-1 \leq a \leq 1$ let

$$f_{a,\lambda}(x) = c_\lambda \exp(-|x|^\lambda)\{1 + a\sin(\beta|x|^\lambda \text{sgn}(x))\}$$

where $\beta = \tan(\lambda\pi/2)$ and $1/c_\lambda = \int \exp(-|x|^\lambda)\,dx$. To prove that these are density functions and that for a fixed value of λ they have the same moments, it suffices to show

$$\int x^n \exp(-|x|^\lambda)\sin(\beta|x|^\lambda \text{sgn}(x))\,dx = 0 \quad \text{for } n = 0, 1, 2, \ldots$$

This is clear for even n since the integrand is odd. To prove the result for odd n, it suffices to integrate over $[0, \infty)$. Using the identity

$$\int_0^\infty t^{p-1}e^{-qt}dt = \Gamma(p)/q^p \quad \text{when } \text{Re}\,q > 0$$

with $p = (n+1)/\lambda$, $q = 1 + \beta i$, and changing variables $t = x^\lambda$ we get

$$\Gamma((n+1)/\lambda)/(1 + \beta\,i)^{(n+1)/\lambda}$$
$$= \int_0^\infty x^{\lambda\{(n+1)/\lambda - 1\}} \exp(-(1+\beta i)x^\lambda)\lambda\, x^{\lambda-1}\,dx$$
$$= \lambda \int_0^\infty x^n \exp(-x^\lambda)\cos(\beta x^\lambda)dx - i\lambda \int_0^\infty x^n \exp(-x^\lambda)\sin(\beta x^\lambda)\,dx$$

Since $\beta = \tan(\lambda\pi/2)$

$$(1 + \beta i)^{(n+1)/\lambda} = (\cos\lambda\pi/2)^{-(n+1)/\lambda}(\exp(i\lambda\pi/2))^{(n+1)/\lambda}$$

The right-hand side is real since $\lambda < 1$ and $(n+1)$ is even so

$$\int_0^\infty x^n \exp(-x^\lambda)\sin(\beta x^\lambda)\,dx = 0$$

A useful sufficient condition for a distribution to be determined by its moments is

(3.11) Theorem. If $\limsup_{k \to \infty} \mu_{2k}^{1/2k}/2k = r < \infty$ then there is at most one d.f. F with $\mu_k = \int x^k dF(x)$ for all positive integers k.

Remark. This is slightly stronger than **Carleman's condition**

$$\sum_{k=1}^{\infty} 1/\mu_{2k}^{1/2k} = \infty$$

which is also sufficient for the conclusion of (3.11).

Proof   Let F be any d.f. with the moments μ_k and let $\nu_k = \int |x|^k dF(x)$. The Cauchy-Schwarz inequality implies $\nu_{2k+1}^2 \le \mu_{2k}\mu_{2k+2}$ so

$$\limsup_{k \to \infty}(\nu_k^{1/k})/k = r < \infty$$

Multiplying (3.6) by $e^{i\theta x}$ and taking $x = X$ we have

$$\left| e^{i\theta X}\left(e^{itX} - \sum_{m=0}^{n-1} \frac{(itX)^m}{m!} \right) \right| \le \frac{|tX|^n}{n!}$$

Taking expected values and using Exercise 3.14 gives

$$\left| \varphi(\theta + t) - \varphi(\theta) - t\varphi'(\theta) \ldots - \frac{t^{n-1}}{(n-1)!}\varphi^{(n-1)}(\theta) \right| \le \frac{|t|^n}{n!}\nu_n$$

Using the last result, the fact that $\nu_k \le (r + \epsilon)^k k^k$ for large k, and the trivial bound $e^k \ge k^k/k!$ (expand the left-hand side in its power series), we see that for any θ

$$(*) \qquad \varphi(\theta + t) = \varphi(\theta) + \sum_{m=1}^{\infty} \frac{t^m}{m!}\varphi^{(m)}(\theta) \quad \text{for } |t| < 1/er$$

Let G be another distribution with the given moments and ψ its ch.f. Since $\varphi(0) = \psi(0) = 1$, it follows from $(*)$ and induction that $\varphi(t) = \psi(t)$ for $|t| \le k/3r$ for all k, so the two ch.f.'s coincide and the distributions must be equal. $\square$

Combining (3.11) with the discussion that began our consideration of the moment problem.

(3.12) Theorem. Suppose $\int x^k dF_n(x)$ has a limit μ_k for each k and

$$\limsup_{k\to\infty} \mu_{2k}^{1/2k}/2k < \infty$$

then F_n converges weakly to the unique distribution with these moments.

EXERCISE 3.27. Let $G(x) = P(|X| < x)$, $\lambda = \sup\{x : G(x) < 1\}$, and $\nu_k = E|X|^k$. Show that $\nu_k^{1/k} \to \lambda$, so (3.12) holds if $\lambda < \infty$.

EXERCISE 3.28. Suppose $|X|$ has density $Cx^\alpha \exp(-x^\lambda)$ on $(0,\infty)$. Changing variables $y = x^\lambda$, $dy = \lambda x^{\lambda-1} dx$

$$E|X|^n = \int_0^\infty C\lambda y^{(n+\alpha)/\lambda} \exp(-y) y^{1/\lambda-1} dy = C\lambda\Gamma((n+\alpha+1)/\lambda)$$

Use the identity $\Gamma(x+1) = x\Gamma(x)$ for $x \geq 0$ to conclude that (3.12) is satisfied for $\lambda \geq 1$ but not for $\lambda < 1$. This shows the normal ($\lambda = 2$) and gamma ($\lambda = 1$) distributions are determined by their moments.

Our results so far have been for the so-called **Hamburger moment problem**. If we assume *a priori* that the distribution is concentrated on $[0,\infty)$ we have the **Stieltjes moment problem**. There is a 1-1 correspondence between $X \geq 0$ and symmetric distributions on **R** given by $X \to \xi X^2$ where $\xi \in \{-1,1\}$ is independent of X and takes its two values with equal probability. From this we see that

$$\limsup_{k\to\infty} \nu_k^{1/2k}/2k < \infty$$

is sufficient for there to be a unique distribution on $[0,\infty)$ with the given moments. The next example shows that for nonnegative random variables, the last result is close to the best possible.

Counterexample 3. Let $\lambda \in (0,1/2)$, $\beta = \tan(\lambda\pi)$, $-1 \leq a \leq 1$ and

$$f_a(x) = c_\lambda \exp(-x^\lambda)(1 + a\sin(\beta x^\lambda)) \quad \text{for } x \geq 0$$

where $1/c_\lambda = \int_0^\infty \exp(-x^\lambda)\,dx$.

By imitating the calculations in Counterexample 2, it is easy to see that the f_a are probability densities that have the same moments. This example seems to be due to Stoyanov (1987) p. 92-93. The special case $\lambda = 1/4$ is widely known.

2.4. Central Limit Theorems

We are now ready for the main business of the chapter. We will first prove the central limit theorem for

a. i.i.d. sequences

(4.1) **Theorem.** Let $X_1, X_2, \ldots$ be i.i.d. with $EX_i = \mu$, $\mathrm{var}(X_i) = \sigma^2 \in (0, \infty)$. If $S_n = X_1 + \cdots + X_n$ then

$$(S_n - n\mu)/\sigma n^{1/2} \Rightarrow \chi$$

where χ has the standard normal distribution.

Proof By considering $X_i' = X_i - \mu$, it suffices to prove the result when $\mu = 0$. From (3.8)

$$\varphi(t) = E\exp(itX_1) = 1 - \frac{\sigma^2 t^2}{2} + o(t^2)$$

so

$$E\exp(itS_n/\sigma n^{1/2}) = \left(1 - \frac{t^2}{2n} + o(n^{-1})\right)^n$$

From (1.3) it should be clear that the last quantity $\to \exp(-t^2/2)$ as $n \to \infty$, which with (3.4) completes the proof. However (1.3) is a fact about real numbers so we need to extend it to the complex case to complete the proof.

(4.2) **Theorem.** If $c_n \to c \in \mathbf{C}$ then $(1 + c_n/n)^n \to e^c$.

The proof is based on two simple facts:

(4.3) **Lemma.** Let $z_1, \ldots, z_n$ and $w_1, \ldots, w_n$ be complex numbers of modulus $\leq \theta$. Then

$$\left|\prod_{m=1}^n z_m - \prod_{m=1}^n w_m\right| \leq \theta^{n-1} \sum_{m=1}^n |z_m - w_m|$$

Proof The result is true for $n = 1$. To prove it for $n > 1$ observe that

$$\left|\prod_{m=1}^n z_m - \prod_{m=1}^n w_m\right| \leq \left|z_1 \prod_{m=2}^n z_m - z_1 \prod_{m=2}^n w_m\right| + \left|z_1 \prod_{m=2}^n w_m - w_1 \prod_{m=2}^n w_m\right|$$

$$\leq \theta \left|\prod_{m=2}^n z_m - \prod_{m=2}^n w_m\right| + \theta^{n-1}|z_1 - w_1|$$

and use induction.   □

(4.4) Lemma. If b is a complex number with $|b| \leq 1$ then $|e^b - (1+b)| \leq |b|^2$.

Proof $e^b - (1+b) = b^2/2! + b^3/3! + b^4/4! + \ldots$ so if $|b| \leq 1$ then

$$|e^b - (1+b)| \leq \frac{|b|^2}{2}(1 + 1/2 + 1/2^2 + \ldots) = |b|^2 \qquad □$$

Proof of (4.2)   Let $z_m = (1 + c_n/n)$, $w_m = \exp(c_n/n)$, and $\gamma > |c|$. For large n, $|c_n| < \gamma$ and $|c_n|/n \leq 1$, so it follows from (4.3) and (4.4) that as $n \to \infty$

$$|(1 + c_n/n)^n - e^{c_n}| \leq \left(1 + \frac{\gamma}{n}\right)^{n-1} n \left|\frac{c_n}{n}\right|^2 \leq e^\gamma \frac{\gamma^2}{n} \to 0 \qquad □$$

To get a feel for what the central limit theorem says, we will look at some concrete cases.

Example 4.1. Roulette. A roulette wheel has slots numbered 1-36 (18 red and 18 black) and two slots numbered 0 and 00 that are painted green. Players can bet \$1 that the ball will land in a red (or black) slot and win \$1 if it does. If we let X_i be the winnings on the ith play then $X_1, X_2, \ldots$ are i.i.d. with $P(X_i = 1) = 18/38$ and $P(X_i = -1) = 20/38$.

$$EX_i = -1/19 \quad \text{and} \quad \text{var}(X) = EX^2 - (EX)^2 = 1 - (1/19)^2 = .9972$$

We are interested in

$$P(S_n \geq 0) = P\left(\frac{S_n - n\mu}{\sigma\sqrt{n}} \geq \frac{-n\mu}{\sigma\sqrt{n}}\right)$$

Taking $n = 361 = 19^2$ and replacing σ by 1 to keep computations simple,

$$\frac{-n\mu}{\sigma\sqrt{n}} = \frac{361 \cdot (1/19)}{\sqrt{361}} = 1$$

So the central limit theorem and our table of the normal distribution in the back of the book tells us that

$$P(S_n \geq 0) \approx P(\chi \geq 1) = 1 - .8413 = .1587$$

In words, after 361 spins of the roulette wheel the casino will have won $19 of your money on the average but there is a probability of about .16 that you will be ahead.

Example 4.2. Coin flips. Let $X_1, X_2, \ldots$ be i.i.d. with $P(X_i = 0) = P(X_i = 1) = 1/2$. If $X_i = 1$ indicates that a heads occured on the ith toss then $S_n = X_1 + \cdots + X_n$ is the total number of heads at time n.

$$EX_i = 1/2 \quad \text{and} \quad \text{var}(X) = EX^2 - (EX)^2 = 1/2 - 1/4 = 1/4$$

So the central limit theorem tells us $(S_n - n/2)/\sqrt{n/4} \Rightarrow \chi$. Our table of the normal distribution tells us that

$$P(\chi > 2) = 1 - .9773 = .0227$$

so $P(|\chi| \le 2) = 1 - 2(.0227) = .9546$, or plugging into the central limit theorem

$$.95 \approx P((S_n - n/2)/\sqrt{n/4} \in [-2, 2]) = P(S_n - n/2 \in [-\sqrt{n}, \sqrt{n}])$$

Taking $n = 10,000$ this says that 95% of the time the number of heads will be between 4900 and 5100.

Example 4.3. Normal approximation to the binomial. Let $X_1, X_2, \ldots$ and S_n be as in the previous example. To estimate $P(S_{16} = 8)$ using the central limit theorem we regard 8 as the interval $[7.5, 8.5]$. Since $\mu = 1/2$, and $\sigma\sqrt{n} = 2$ for $n = 16$

$$P(|S_{16} - 8| \le .5) = P\left(\frac{|S_n - n\mu|}{\sigma\sqrt{n}} \le .25\right)$$
$$\approx P(|\chi| \le .25) = 2(.5987 - .5) = .1974$$

Even though n is small this agrees well with the exact probability

$$\binom{16}{8} 2^{-16} = \frac{13 \cdot 11 \cdot 10 \cdot 9}{65,536} = .1964.$$

The computations above motivate the **histogram correction**, which is important in using the normal approximation for small n. For example, if we are going to approximate $P(S_{16} \le 11)$ then we regard this probability as $P(S_{16} \le 11.5)$. One obvious reason for doing this is to get the same answer if we regard $P(S_{16} \le 11) = 1 - P(S_{16} \ge 12)$.

EXERCISE 4.1. Suppose you roll a die 180 times. Use the normal approximation (with the histogram correction) to estimate the probability you will get less than 25 sixes.

Example 4.4. Normal approximation to the Poisson. Let Z_λ have a Poisson distribution with mean λ. If $X_1, X_2, \ldots$ are independent and have Poisson distributions with mean 1, then $S_n = X_1 + \cdots + X_n$ has a Poisson distribution with mean n. Since $\mathrm{var}(X_i) = 1$, the central limit theorem implies:

$$(S_n - n)/n^{1/2} \Rightarrow \chi \quad \text{as } n \to \infty$$

To deal with values of λ that are not integers, let N_1, N_2, N_3 be independent Poisson with means $[\lambda]$, $\lambda - [\lambda]$, and $[\lambda] + 1 - \lambda$. If we let $S_{[\lambda]} = N_1$, $Z_\lambda = N_1 + N_2$ and $S_{[\lambda]+1} = N_1 + N_2 + N_3$ then $S_{[\lambda]} \leq Z_\lambda \leq S_{[\lambda]+1}$ and using the limit theorem for the S_n it follows that

$$(Z_\lambda - \lambda)/\lambda^{1/2} \Rightarrow \chi \quad \text{as } \lambda \to \infty$$

Example 4.5. Pairwise independence is good enough for the strong law of large numbers (see (7.1) in Chapter 1). It is not good enough for the central limit theorem. Let $\xi_1, \xi_2, \ldots$ be i.i.d. with $P(\xi_i = 1) = P(\xi_i = -1) = 1/2$. We will arrange things so that for $n \geq 1$

$$S_{2^n} = \xi_1(1 + \xi_2) \cdots (1 + \xi_{n+1}) = \begin{cases} \pm 2^n & \text{with prob } 2^{-n-1} \\ 0 & \text{with prob } 1 - 2^{-n} \end{cases}$$

To do this we let $X_1 = \xi_1$, $X_2 = \xi_1\xi_2$, and for $m = 2^{n-1} + j$, $0 < j \leq 2^{n-1}$, $n \geq 2$ let $X_m = X_j\xi_{n+1}$. Each X_m is a product of a different set of ξ_j's so they are pairwise independent.

EXERCISES

4.2. Let $X_1, X_2, \ldots$ be i.i.d. with $EX_i = 0$, and let $S_n = X_1 + \cdots + X_n$. (a) Use the central limit theorem and Kolmogorov's zero-one law to conclude that $\limsup S_n/\sqrt{n} = \infty$ a.s. (b) Use an argument by contradiction to show that $S_n/\sqrt{n}$ does not converge in probability. Hint: consider $n = m!$.

4.3. Let $X_1, X_2, \ldots$ be i.i.d. and let $S_n = X_1 + \cdots + X_n$. Assume that $S_n/\sqrt{n} \Rightarrow$ a limit and conclude that $EX_i^2 < \infty$. Sketch: Suppose $EX_i^2 = \infty$. Let $X_1', X_2', \ldots$ be an independent copy of the original sequence. Let $Y_i = X_i - X_i'$, $U_i = Y_i 1_{(|Y_i| \leq A)}$, $V_i = Y_i 1_{(|Y_i| > A)}$, and observe that for any K

$$P\left(\sum_{m=1}^{n} Y_m \geq K\sqrt{n}\right) \geq P\left(\sum_{m=1}^{n} U_m \geq K\sqrt{n}, \sum_{m=1}^{n} V_m \geq 0\right)$$

$$\geq \frac{1}{2} P\left(\sum_{m=1}^{n} U_m \geq K\sqrt{n}\right) \geq \frac{1}{5}$$

for large n if A is large enough. Since K is arbitrary this is a contradiction.

4.4. Let $X_1, X_2, \ldots$ be i.i.d. with $X_i \geq 0$, $EX_i = 1$, and $\text{var}(X_i) = \sigma^2 \in (0, \infty)$. Show that $2(\sqrt{S_n} - \sqrt{n}) \Rightarrow \sigma\chi$.

4.5. Self-normalized sums. Let $X_1, X_2, \ldots$ be i.i.d. with $EX_i = 0$ and $EX_i^2 = \sigma^2 \in (0, \infty)$. Then

$$\sum_{m=1}^{n} X_m \Big/ \left(\sum_{m=1}^{n} X_m^2 \right)^{1/2} \Rightarrow \chi$$

4.6. Random index central limit theorem. Let $X_1, X_2, \ldots$ be i.i.d. with $EX_i = 0$ and $EX_i^2 = \sigma^2 \in (0, \infty)$, and let $S_n = X_1 + \cdots + X_n$. Let N_n be a sequence of nonnegative integer valued random variables and a_n a sequence of integers with $N_n/a_n \to 1$ in probability. Show that

$$S_{N_n}/\sigma\sqrt{a_n} \Rightarrow \chi$$

Hint: Use Kolmogorov's inequality ((7.2) in Chapter 1) to conclude that if $X_n = S_{N_n}/\sigma\sqrt{a_n}$ and $Y_n = S_{a_n}/\sigma\sqrt{a_n}$ then $X_n - Y_n \to 0$ in probability.

4.7. A central limit theorem in renewal theory. Let $Y_1, Y_2, \ldots$ be i.i.d. positive random variables with $EY_i = \mu$ and $\text{var}(Y_i) = \sigma^2$. Let $S_n = Y_1 + \cdots + Y_n$ and $N_t = \sup\{m : S_m \leq t\}$. Apply the previous exercise to $X_i = Y_i - \mu$ to prove that as $t \to \infty$

$$(\mu N_t - t)/(\sigma^2 t/\mu)^{1/2} \Rightarrow \chi$$

4.8. A second proof of the renewal CLT. Let $Y_1, Y_2, \ldots, S_n$, and N_t be as in the last exercise. Let $u = [t/\mu]$, $D_t = S_u - t$. Use Kolmogorov's inequality to show

$$P(|S_{u+m} - (S_u + m\mu)| > t^{2/5} \text{ for some } m \in [-t^{3/5}, t^{3/5}]) \to 0 \quad \text{as } t \to \infty$$

Conclude $|N_t - (t - D_t)/\mu|/ t^{1/2} \to 0$ in probability and then obtain the result in the previous exercise.

Our next step is to generalize the central limit theorem to

b. Triangular arrays

(4.5) The Lindeberg-Feller theorem. For each n, let $X_{n,m}$, $1 \leq m \leq n$, be independent random variables with $EX_{n,m} = 0$. Suppose

(i) $\sum_{m=1}^{n} EX_{n,m}^2 \to \sigma^2 > 0$

(ii) For all $\epsilon > 0$, $\lim_{n\to\infty} \sum_{m=1}^n E(|X_{n,m}|^2; |X_{n,m}| > \epsilon) = 0$.

Then $S_n = X_{n,1} + \cdots + X_{n,n} \Rightarrow \sigma\chi$ as $n \to \infty$

Remarks. In words, the theorem says that a sum of a large number of small independent errors has approximately a normal distribution. To see that (4.5) contains our first central limit theorem, let $Y_1, Y_2 \ldots$ be i.i.d. with $EY_i = 0$ and $EY_i^2 = \sigma^2 \in (0, \infty)$, and let $X_{n,m} = Y_m/n^{1/2}$. Then $\sum_{m=1}^n EX_{n,m}^2 = \sigma^2$ and if $\epsilon > 0$

$$\sum_{m=1}^n E(|X_{n,m}|^2; |X_{n,m}| > \epsilon) = nE(|Y_1/n^{1/2}|^2; |Y_1/n^{1/2}| > \epsilon)$$

$$= E(|Y_1|^2; |Y_1| > \epsilon n^{1/2}) \to 0$$

by the dominated convergence theorem since $EY_1^2 < \infty$.

Proof   Let $\varphi_{n,m}(t) = E\exp(itX_{n,m})$, $\sigma_{n,m}^2 = EX_{n,m}^2$. By (3.4) it suffices to show that

$$\prod_{m=1}^n \varphi_{n,m}(t) \to \exp(-t^2\sigma^2/2)$$

Let $z_{n,m} = \varphi_{n,m}(t)$ and $w_{n,m} = (1 - t^2\sigma_{n,m}^2/2)$. By (3.7)

$$\begin{aligned}
|z_{n,m} - w_{n,m}| &\leq E(|tX_{n,m}|^3/3! \wedge 2|tX_{n,m}|^2/2!) \\
&\leq E(|tX_{n,m}|^3/6; |X_{n,m}| \leq \epsilon) + E(|tX_{n,m}|^2; |X_{n,m}| > \epsilon) \\
&\leq \frac{\epsilon t^3}{6} E(|X_{n,m}|^2; |X_{n,m}| \leq \epsilon) + t^2 E(|X_{n,m}|^2; |X_{n,m}| > \epsilon)
\end{aligned}$$

Summing $m = 1$ to n, letting $n \to \infty$, and using (i) and (ii) gives

$$\limsup_{n\to\infty} \sum_{m=1}^n |z_{n,m} - w_{n,m}| \leq \frac{\epsilon t^3}{6}\sigma^2$$

Since $\epsilon > 0$ is arbitrary, it follows that the sequence converges to 0. Our next step is to use (4.3) with $\theta = 1$ to get

$$\left| \prod_{m=1}^n \varphi_{n,m}(t) - \prod_{m=1}^n (1 - t^2\sigma_{n,m}^2/2) \right| \to 0$$

To check the hypotheses of (4.3) note that since $\varphi_{n,m}$ is a ch.f. $|\varphi_{n,m}(t)| \leq 1$ for all n, m. For the terms in the second product we note that

$$\sigma_{n,m}^2 \leq \epsilon^2 + E(|X_{n,m}|^2 ; |X_{n,m}| > \epsilon)$$

and ϵ is arbitrary so (ii) implies $\sup_m \sigma_{n,m}^2 \to 0$ and thus if n is large $1 \geq 1 - t^2 \sigma_{n,m}^2/2 > -1$ for all m.

To complete the proof now we apply Exercise 1.1 with $a_{n,m} = t^2 \sigma_{n,m}^2/2$. We have just shown $\sup_m \sigma_{n,m}^2 \to 0$. (i) implies

$$\sum_{m=1}^{n} a_{n,m} \to \sigma^2 t^2/2$$

so $\prod_{m=1}^{n}(1 - t^2 \sigma_{n,m}^2/2) \to \exp(-t^2 \sigma^2/2)$ and the proof is complete. $\square$

Example 4.6.   Cycles in a random permutation and record values. Continuing the analysis of Examples 5.4 and 6.2 in Chapter 1, let $Y_1, Y_2, \ldots$ be independent with $P(Y_m = 1) = 1/m$, and $P(Y_m = 0) = 1 - 1/m$. $EY_m = 1/m$ and $\text{var}(Y_m) = 1/m - 1/m^2$. So if $S_n = Y_1 + \cdots + Y_n$ then $ES_n \sim \log n$ and $\text{var}(S_n) \sim \log n$. Let

$$X_{n,m} = (Y_m - 1/m)/(\log\ n)^{1/2}$$

$EX_{n,m} = 0$, $\sum_{m=1}^{n} EX_{n,m}^2 \to 1$, and for any $\epsilon > 0$

$$\sum_{m=1}^{n} E(|X_{n,m}|^2; |X_{n,m}| > \epsilon) \to 0$$

since the sum is 0 as soon as $(\log n)^{-1/2} < \epsilon$. Applying (4.5) now gives

$$(\log n)^{-1/2} \left(S_n - \sum_{m=1}^{n} \frac{1}{m} \right) \Rightarrow \chi$$

Observing that
$$\sum_{m=1}^{n-1} \frac{1}{m} \geq \int_{1}^{n} x^{-1}\, dx = \log n \geq \sum_{m=2}^{n} \frac{1}{m}$$

shows $|\log n - \sum_{m=1}^{n} 1/m| \leq 1$ and the conclusion can be written as

$$(S_n - \log n)/(\log n)^{1/2} \Rightarrow \chi$$

Example 4.7.   The converse of the three series theorem. Recall the set-up of (8.4) in Chapter 1. Let $X_1, X_2, \ldots$ be independent, let $A > 0$, and let

$Y_m = X_m 1_{(|X_m| \le A)}$. In order that $\sum_{n=1}^{\infty} X_n$ converges (i.e., $\lim_{N \to \infty} \sum_{n=1}^{N} X_n$ exists) it is necessary that:

(i) $\displaystyle\sum_{n=1}^{\infty} P(|X_n| > A) < \infty$, (ii) $\displaystyle\sum_{n=1}^{\infty} EY_n$ converges, and (iii) $\displaystyle\sum_{n=1}^{\infty} \text{var}(Y_n) < \infty$

Proof The necessity of the first condition is clear. For if that sum is infinite, $P(|X_n| > A \text{ i.o.}) > 0$ and $\lim_{n \to \infty} \sum_{m=1}^{n} X_m$ cannot exist. Suppose next that the sum in (i) is finite but the sum in (iii) is infinite. Let

$$c_n = \sum_{m=1}^{n} \text{var}(Y_m) \quad \text{and} \quad X_{n,m} = (Y_m - EY_m)/c_n^{1/2}$$

$EX_{n,m} = 0$, $\sum_{m=1}^{n} EX_{n,m}^2 = 1$, and for any $\epsilon > 0$

$$\sum_{m=1}^{n} E(|X_{n,m}|^2; |X_{n,m}| > \epsilon) \to 0$$

since the sum is 0 as soon as $2A/c_n^{1/2} < \epsilon$. Applying (4.5) now gives that if $S_n = X_{n,1} + \cdots + X_{n,n}$ then $S_n \Rightarrow \chi$. Now

(i) if $\lim_{n \to \infty} \sum_{m=1}^{n} X_m$ exists, $\lim_{n \to \infty} \sum_{m=1}^{n} Y_m$ exists.

(ii) if we let $T_n = (\sum_{m \le n} Y_m)/c_n^{1/2}$ then $T_n \Rightarrow 0$.

The last two results and Exercise 2.10 imply $(S_n - T_n) \Rightarrow \chi$. Since

$$S_n - T_n = - \left(\sum_{m \le n} EY_m \right) /c_n^{1/2}$$

is not random this is absurd.

Finally assume the series in (i) and (iii) are finite. (7.3) in Chapter 1 implies that $\lim_{n \to \infty} \sum_{m=1}^{n} (Y_m - EY_m)$ exists, so if $\lim_{n \to \infty} \sum_{m=1}^{n} X_m$ and hence $\lim_{n \to \infty} \sum_{m=1}^{n} Y_m$ does, taking differences shows that (ii) holds. $\square$

Example 4.8. Infinite variance. Suppose $X_1, X_2, \ldots$ are i.i.d. and have $P(X_1 > x) = P(X_1 < -x)$ and $P(|X_1| > x) = x^{-2}$ for $x \ge 1$.

$$E|X_1|^2 = \int_0^{\infty} 2x P(|X_1| > x)\, dx = \infty$$

but it turns out that when $S_n = X_1 + \cdots + X_n$ is suitably normalized it converges to a normal distribution. Let

$$Y_{n,m} = X_m 1_{(|X_m| \leq n^{1/2} \log \log n)}$$

The truncation level $c_n = n^{1/2} \log \log n$ is chosen large enough to make

$$\sum_{m=1}^{n} P(Y_{n,m} \neq X_m) \leq n P(|X_1| > c_n) \to 0$$

However, we want the variance of $Y_{n,m}$ to be as small as possible, so we keep the truncation close to the lowest possible level.

Our next step is to show $EY_{n,m}^2 \sim \log n$. For this we need upper and lower bounds. Since $P(|Y_{n,m}| > x) \leq P(|X_1| > x)$ and is 0 for $x > c_n$ we have

$$EY_{n,m}^2 \leq \int_0^{c_n} 2y P(|X_1| > y)\, dy = 1 + \int_1^{c_n} 2/y\, dy$$
$$= 1 + 2 \log c_n = 1 + \log n + 2 \log \log \log n \sim \log n$$

In the other direction we observe $P(|Y_{n,m}| > x) = P(|X_1| > x) - P(|X_1| > c_n)$ and the right-hand side is $\geq (1 - (\log \log n)^{-2}) P(|X_1| > x)$ when $x \leq \sqrt{n}$ so

$$EY_{n,m}^2 \geq (1 - (\log \log n)^{-2}) \int_1^{\sqrt{n}} 2/y\, dy \sim \log n$$

If $S_n' = Y_{n,1} + \cdots + Y_{n,n}$ then $\mathrm{var}(S_n') \sim n \log n$, so we apply (4.5) to $X_{n,m} = Y_{n,m}/(n \log n)^{1/2}$. Things have been arranged so that (i) is satisfied. Since $|Y_{n,m}| \leq n^{1/2} \log \log n$, the sum in (ii) is 0 for large n, and it follows that $S_n'/(n \log n)^{1/2} \Rightarrow \chi$. Since the choice of c_n guarantees $P(S_n \neq S_n') \to 0$, the same result holds for S_n.

Remark. In Section 2.6 we will see that if we replace $P(|X_1| > x) = x^{-2}$ in Example 4.9 by $P(|X_1| > x) = x^{-\alpha}$ where $0 < \alpha < 2$, then $S_n/n^{1/\alpha} \Rightarrow$ to a limit which is not χ. The last word on convergence to the normal distribution is the next result due to Lévy.

(4.6) Theorem. Let $X_1, X_2, \ldots$ be i.i.d. and $S_n = X_1 + \cdots + X_n$. In order that there exist constants a_n and $b_n > 0$ so that $(S_n - a_n)/b_n \Rightarrow \chi$, it is necessary and sufficient that

$$y^2 P(|X_1| > y)/E(|X_1|^2; |X_1| \leq y) \to 0.$$

A proof can be found in Gnedenko and Kolmogorov (1954), a reference that contains the last word on many results about sums of independent random variables.

EXERCISES

In the next five problems $X_1, X_2, \ldots$ are independent and $S_n = X_1 + \cdots + X_n$.

4.9. Suppose $P(X_m = m) = P(X_m = -m) = m^{-2}/2$,

$$P(X_m = 1) = P(X_m = -1) = (1 - m^{-2})/2$$

Show that $\operatorname{var}(S_n)/n \to 2$ but $S_n/\sqrt{n} \Rightarrow \chi$. The trouble here is that $X_{n,m} = X_m/\sqrt{n}$ does not satisfy (ii).

4.10. Show that if $|X_i| \le M$ and $\sum_n \operatorname{var}(X_n) = \infty$ then

$$(S_n - ES_n)/\sqrt{\operatorname{var}(S_n)} \Rightarrow \chi$$

4.11. Suppose $EX_i = 0$, $EX_i^2 = 1$ and $E|X_i|^{2+\delta} \le C$ for some $0 < \delta, C < \infty$. Show that $S_n/\sqrt{n} \Rightarrow \chi$.

4.12. Prove **Lyapunov's Theorem.** Let $\alpha_n = \{\operatorname{var}(S_n)\}^{1/2}$. If there is a $\delta > 0$ so that

$$\lim_{n \to \infty} \alpha_n^{-(2+\delta)} \sum_{m=1}^{n} E(|X_m - EX_m|^{2+\delta}) = 0$$

then $(S_n - ES_n)/\alpha_n \Rightarrow \chi$. Note that the previous exercise is a special case of this result.

4.13. Suppose $P(X_j = j) = P(X_j = -j) = 1/2j^\beta$ and $P(X_j = 0) = 1 - j^{-\beta}$. Show that (i) If $\beta > 1$ then $S_n \to S_\infty$ a.s. (ii) if $\beta < 1$ then $S_n/n^{(3-\beta)/2} \Rightarrow c\chi$. (iii) if $\beta = 1$ then $S_n/n \Rightarrow \aleph$ where

$$E \exp(it\aleph) = \exp\left(-\int_0^1 x^{-1}(1 - \cos xt)\, dx\right)$$

*c. Prime divisors (Erdös-Kac)

Our aim here is to prove that an integer picked at random from $\{1, 2, \ldots, n\}$ has about

$$\log \log n + \chi(\log \log n)^{1/2}$$

prime divisors. Since $\exp(e^4) = 5.15 \times 10^{23}$, this result does not apply to most numbers we encounter in "everyday life." The first step in derving this result is to give a

Second proof of (4.5) The first step is to let

$$h_n(\epsilon) = \sum_{m=1}^{n} E(X_{n,m}^2; |X_{n,m}| > \epsilon)$$

and observe

(4.7) Lemma. $h_n(\epsilon) \to 0$ for each fixed $\epsilon > 0$ so we can pick $\epsilon_n \to 0$ so that $h_n(\epsilon_n) \to 0$.

Proof Let N_m be chosen so that $h_n(1/m) \le 1/m$ for $n \ge N_m$ and $m \to N_m$ is increasing. Let $\epsilon_n = 1/m$ for $N_m \le n < N_{m+1}$, and $= 1$ for $n < N_1$. When $N_m \le n < N_{m+1}$, $\epsilon_n = 1/m$, so $|h_n(\epsilon_n)| = |h_n(1/m)| \le 1/m$ and the desired result follows. $\square$

Let $X'_{n,m} = X_{n,m} 1_{(|X_{n,m}| > \epsilon_n)}$, $Y_{n,m} = X_{n,m} 1_{(|X_{n,m}| \le \epsilon_n)}$, and $Z_{n,m} = Y_{n,m} - EY_{n,m}$. Clearly $|Z_{n,m}| \le 2\epsilon_n$. Using $X_{n,m} = X'_{n,m} + Y_{n,m}$, $Z_{n,m} = Y_{n,m} - EY_{n,m}$, $EY_{n,m} = -EX'_{n,m}$, the variance of the sum is the sum of the variances, and $\text{var}(W) \le EW^2$, we have

$$E\left(\sum_{m=1}^{n} X_{n,m} - \sum_{m=1}^{n} Z_{n,m}\right)^2 = E\left(\sum_{m=1}^{n} X'_{n,m} - EX'_{n,m}\right)^2$$

$$= \sum_{m=1}^{n} E(X'_{n,m} - EX'_{n,m})^2 \le \sum_{m=1}^{n} E(X'_{n,m})^2 \to 0$$

as $n \to \infty$, by the choice of ϵ_n.

Let $S_n = \sum_{m=1}^{n} X_{n,m}$ and $T_n = \sum_{m=1}^{n} Z_{n,m}$. The last computation shows $S_n - T_n \to 0$ in L^2 and hence in probability by (5.3) in Chapter 1. Thus by Exercise 2.10 it suffices to show $T_n \Rightarrow \sigma\chi$. (i) implies $ES_n^2 \to \sigma^2$. We have just shown that $E(S_n - T_n)^2 \to 0$ so the triangle inequality for the L^2 norm implies $ET_n^2 \to \sigma^2$. To compute higher moments we observe

$$T_n^r = \sum_{k=1}^{r} \sum_{r_i} \frac{r!}{r_1! \cdots r_k!} \frac{1}{k!} \sum_{i_j} Z_{n,i_1}^{r_1} \cdots Z_{n,i_k}^{r_k}$$

where $\sum_{r_i}$ extends over all k-tuples of positive integers with $r_1 + \cdots + r_k = r$ and $\sum_{i_j}$ extends over all k-tuples of distinct integers with $1 \le i \le n$. If we let

$$A_n(r_1, ..., r_k) = \sum_{i_j} EZ_{n,i_1}^{r_1} \cdots EZ_{n,i_k}^{r_k}$$

then

$$ET_n^r = \sum_{k=1}^{r} \sum_{r_i} \frac{r!}{r_1! \cdots r_k!} \frac{1}{k!} A_n(r_1, ... r_k)$$

To evaluate the limit of ET_n^r we observe:

(a) If some $r_j = 1$, then $A_n(r_1, ... r_k) = 0$ since $EZ_{n,i_j} = 0$.

(b) If all $r_j = 2$ then

$$\sum_{i_j} EZ_{n,i_1}^2 \cdots EZ_{n,i_k}^2 \le \left(\sum_{m=1}^{n} EZ_{n,m}^2 \right)^k \to \sigma^{2k}$$

To argue the other inequality we note that for any $1 \le a < b \le k$ we can estimate the sum over all the $i_1, \ldots, i_k$ with $i_a = i_b$ by replacing EZ_{n,i_a}^2 by $(2\epsilon_n)^2$ to get (the factor $\binom{k}{2}$ giving the number of ways to pick $1 \le a < b \le k$)

$$\left(\sum_{m=1}^{n} EZ_{n,m}^2 \right)^k - \sum_{i_j} EZ_{n,i_1}^2 \cdots EZ_{n,i_k}^2 \le \binom{k}{2} (2\epsilon_n)^2 \left(\sum_{m=1}^{n} EZ_{n,m}^2 \right)^{k-1} \to 0$$

(c) If all the $r_i \ge 2$ but some $r_j > 2$ then using

$$E|Z_{n,i_j}|^{r_j} \le (2\epsilon_n)^{r_j - 2} EZ_{n,i_j}^2$$

we have

$$|A_n(r_1, ... r_k)| \le \sum_{i_j} E|Z_{n,i_1}|^{r_1} \cdots E|Z_{n,i_k}|^{r_k}$$

$$\le (2\epsilon_n)^{r-2k} A_n(2, ... 2) \to 0$$

When r is odd some r_j must be $= 1$ or ≥ 3 so $ET_n^r \to 0$ by (a) and (c). If $r = 2k$ is even, (a)–(c) imply

$$ET_n^r \to \frac{\sigma^{2k}(2k)!}{2^k k!} = E(\sigma\chi)^r$$

and the result follows from (3.12). $\square$

Turning to the result for prime divisors, let P_n denote the uniform distribution on $\{1, \ldots, n\}$. If $P_\infty(A) \equiv \lim P_n(A)$ exists the limit is called the density of $A \subset \mathbf{Z}$. Let A_p be the set of integers divisible by p. Clearly if p is a prime $P_\infty(A_p) = 1/p$ and $q \ne p$ is another prime

$$P_\infty(A_p \cap A_q) = 1/pq = P_\infty(A_p)P_\infty(A_q)$$

Even though P_∞ is not a probability measure (since $P(\{i\}) = 0$ for all i), we can interpret this as saying that the events of being divisible by p and q are independent. Let $\delta_p(n) = 1$ if n is divisible by p, and $= 0$ otherwise, and

$$g(n) = \sum_{p \leq n} \delta_p(n) \quad \text{be the number of prime divisors of } n$$

this and future sums on p being over the primes. Intuitively the $\delta_p(n)$ behave like X_p that are i.i.d. with

$$P(X_p = 1) = 1/p \quad \text{and} \quad P(X_p = 0) = 1 - 1/p$$

The mean and variance of $\sum_{p \leq n} X_p$ are

$$\sum_{p \leq n} 1/p \quad \text{and} \quad \sum_{p \leq n} 1/p(1 - 1/p)$$

respectively. It is known that

$$(*) \qquad \qquad \sum_{p \leq n} 1/p = \log\log n + O(1)$$

(see Hardy and Wright (1959) Chapter XXII), while anyone can see $\sum_p 1/p^2 < \infty$, so applying (4.5) to X_p and making a small leap of faith gives us:

(4.8) Erdös-Kac central limit theorem. As $n \to \infty$

$$P_n \left(m \leq n : g(m) - \log\log n \leq x(\log\log n)^{1/2} \right) \to P(\chi \leq x)$$

Proof   We begin by showing that we can ignore the primes "near" n. Let

$$\alpha_n = n^{1/\log\log n}$$
$$\log \alpha_n = \log n / \log\log n$$
$$\log\log \alpha_n = \log\log n - \log\log\log n$$

The sequence α_n has two nice properties:

(a) $\left(\sum_{\alpha_n < p \leq n} 1/p \right) / (\log\log n)^{1/2} \to 0$ by $(*)$

Proof of (a)   By $(*)$

$$\sum_{\alpha_n < p \leq n} 1/p = \sum_{p \leq n} 1/p - \sum_{p \leq \alpha_n} 1/p$$
$$= \log\log n - \log\log \alpha_n + O(1)$$
$$= \log\log\log n + O(1) \qquad \qquad \square$$

(b) If $\epsilon > 0$ then $\alpha_n \leq n^\epsilon$ for large n and hence $\alpha_n^r/n \to 0$ for all $r < \infty$.

Proof of (b) $1/\log\log n \to 0$ as $n \to \infty$. $\qquad\qquad\qquad\qquad\qquad$ $\square$

Let $g_n(m) = \sum_{p \leq \alpha_n} \delta_p(m)$ and let E_n denote expected value w.r.t. P_n.

$$E_n\left(\sum_{\alpha_n < p \leq n} \delta_p\right) = \sum_{\alpha_n < p \leq n} P_n(m : \delta_p(m) = 1) \leq \sum_{\alpha_n < p \leq n} 1/p$$

so by (a) it is enough to prove the result for g_n. Let

$$S_n = \sum_{p \leq \alpha_n} X_p$$

where the X_p are the independent random variables introduced above. Let $b_n = ES_n$ and $a_n^2 = \text{var}(S_n)$. (a) tells us that b_n and a_n^2 are both

$$\log\log n + o((\log\log n)^{1/2})$$

so it suffices to show

$$P_n(m : g_n(m) - b_n \leq x a_n) \to P(\chi \leq x)$$

An application of (4.5) shows $(S_n - b_n)/a_n \Rightarrow \chi$, and since $|X_p| \leq 1$ it follows from the second proof of (4.5) that

$$E((S_n - b_n)/a_n)^r \to E\chi^r \quad \text{for all } r$$

Using notation from that proof (and replacing i_j by p_j)

$$ES_n^r = \sum_{k=1}^{r}\sum_{r_i} \frac{r!}{r_1! \cdots r_k!} \frac{1}{k!} \sum_{p_j} E(X_{p_1}^{r_1} \cdots X_{p_k}^{r_k})$$

Since $X_p \in \{0,1\}$ the summand is

$$E(X_{p_1} \cdots X_{p_k}) = 1/(p_1 \cdots p_k)$$

A little thought reveals that

$$E_n(\delta_{p_1} \cdots \delta_{p_k}) \leq \frac{1}{n}[n/(p_1 \cdots p_k)]$$

The two moments differ by $\leq 1/n$ so

$$|E(S_n^r) - E_n(g_n^r)| = \sum_{k=1}^{r} \sum_{r_i} \frac{r!}{r_1! \cdots r_k!} \frac{1}{k!} \sum_{p_j} \frac{1}{n}$$

$$\leq \frac{1}{n} \Big(\sum_{p \leq \alpha_n} 1 \Big)^r \leq \frac{\alpha_n^r}{n} \to 0$$

by (b). Now

$$E(S_n - b_n)^r = \sum_{m=0}^{r} \binom{r}{m} ES_n^m (-b_n)^{r-m}$$

$$E(g_n - b_n)^r = \sum_{m=0}^{r} \binom{r}{m} Eg_n^m (-b_n)^{r-m}$$

so subtracting and using our bound on $|E(S_n^r) - E_n(g_n^r)|$ with $r = m$

$$|E(S_n - b_n)^r - E(g_n - b_n)^r| \leq \sum_{m=0}^{r} \binom{r}{m} \frac{1}{n} \alpha_n^m b_n^{r-m} = (\alpha_n + b_n)^r / n \to 0$$

since $b_n \leq \alpha_n$. This is more than enough to conclude that

$$E\left((g_n - b_n)/a_n \right)^r \to E\chi^r$$

and the desired result follows from (3.12). $\square$

*d. Rates of convergence (Berry-Esseen)

(4.9) Theorem. Let $X_1, X_2, \ldots$ be i.i.d. with $EX_i = 0$, $EX_i^2 = \sigma^2$, and $E|X_i|^3 = \rho < \infty$. If $F_n(x)$ is the distribution of $(X_1 + \cdots + X_n)/\sigma\sqrt{n}$ and $\mathcal{N}(x)$ is the standard normal distribution then

$$|F_n(x) - \mathcal{N}(x)| \leq 3\rho/\sigma^3 \sqrt{n}$$

Remarks. The reader should note that the inequality holds for all n and x, but since $\rho \geq \sigma^3$ it only has nontrivial content for $n \geq 10$. It is easy to see that the rate cannot be faster than $n^{-1/2}$. When $P(X_i = 1) = P(X_i = -1) = 1/2$, symmetry and (1.4) imply

$$F_{2n}(0) = \frac{1}{2}\{1 + P(S_{2n} = 0)\} = \frac{1}{2}(1 + (\pi n)^{-1/2}) + o(n^{-1/2})$$

The constant 3 is not the best known (van Beek (1972) gets 0.8), but as Feller brags "our streamlined method yields a remarkably good bound even though it avoids the usual messy numerical calculations." The hypothesis $E|X|^3$ is needed to get the rate $n^{-1/2}$. Heyde (1967) has shown that for $0 < \delta < 1$

$$\sum_{n=1}^{\infty} n^{-1+\delta/2} \sup_x |F_n(x) - \mathcal{N}(x)| < \infty$$

if and only if $E|X|^{2+\delta} < \infty$. For this and more on rates of convergence see Hall (1982).

Proof Since neither side of the inequality is affected by scaling we can suppose without loss of generality that $\sigma^2 = 1$. The first phase of the argument is to derive an inequality, (4.11), that relates the difference between the two distributions to the distance between their ch.f.'s. Polya's density (see Example 3.8 and use (3.1e))

$$h_L(x) = \frac{1 - \cos Lx}{\pi L x^2}$$

has ch.f. $\omega_L(\theta) = (1 - |\theta/L|)^+$ for $|\theta| \le L$. We will use H_L for its distribution function. We will convolve the distributions under consideration with H_L to get ch.f. that have compact support. The first step is to show that convolution with H_L does not reduce the difference between the distributions too much.

(4.10) Lemma. Let F and G be distribution functions with $G'(x) \le \lambda < \infty$. Let $\Delta(x) = F(x) - G(x)$, $\eta = \sup |\Delta(x)|$, $\Delta_L = \Delta * H_L$, and $\eta_L = \sup |\Delta_L(x)|$. Then

$$\eta_L \ge \frac{\eta}{2} - \frac{12\lambda}{\pi L} \quad \text{or} \quad \eta \le 2\eta_L + \frac{24\lambda}{\pi L}$$

Proof Δ goes to 0 at $\pm\infty$, G is continuous, and F is a d.f., so there is an x_0 with $\Delta(x_0) = \eta$ or $\Delta(x_0-) = -\eta$. By looking at the d.f.'s of (-1) times the r.v.'s in the second case, we can suppose without loss of generality that $\Delta(x_0) = \eta$. Since $G'(x) \le \lambda$ and F is nondecreasing $\Delta(x_0 + s) \ge \eta - \lambda s$. Letting $\delta = \eta/2\lambda$, and $t = x_0 + \delta$, we have

$$\Delta(t - x) \ge \begin{cases} (\eta/2) + \lambda x & \text{for } |x| \le \delta \\ -\eta & \text{otherwise} \end{cases}$$

To estimate the convolution Δ_L we observe

$$2 \int_\delta^\infty h_L(x)\, dx \le 2 \int_\delta^\infty 2/(\pi L x^2) dx = 4/(\pi L \delta)$$

Looking at $(-\delta, \delta)$ and its complement separately, and noticing symmetry implies $\int_{-\delta}^{\delta} x h_L(x)\, dx = 0$ we have

$$\eta_L \ge \Delta_L(t) \ge \frac{\eta}{2}\left(1 - \frac{4}{\pi L \delta}\right) - \eta\frac{4}{\pi L \delta} = \frac{\eta}{2} - \frac{6\eta}{\pi L \delta} = \frac{\eta}{2} - \frac{12\lambda}{\pi L}$$

proving (4.10).   □

(4.11) **Lemma.** Let K_1 and K_2 be d.f. with mean 0 whose ch.f. κ_i are integrable

$$K_1(x) - K_2(x) = (2\pi)^{-1} \int -e^{-itx} \frac{\kappa_1(t) - \kappa_2(t)}{it} \, dt$$

Proof Since the κ_i are integrable, the inversion formula (3.3) implies that the density $k_i(x)$ has

$$k_i(y) = (2\pi)^{-1} \int e^{-ity} \kappa_i(t) \, dt$$

Subtracting the last expression with $i = 2$ from the one with $i = 1$ then integrating from a to x and letting $\Delta K = K_1 - K_2$ gives

$$\Delta K(x) - \Delta K(a) = (2\pi)^{-1} \int_a^x \int e^{-ity} \{\kappa_1(t) - \kappa_2(t)\} \, dt \, dy$$

$$= (2\pi)^{-1} \int \{e^{-ita} - e^{-itx}\} \frac{\kappa_1(t) - \kappa_2(t)}{it} \, dt$$

the application of Fubini's theorem being justified since the κ_i are integrable in t and we are considering a bounded interval in y.

The factor $1/it$ could cause problems near zero but we have supposed that the K_i have mean 0, so $\{1 - \kappa_i(t)\}/t \to 0$ by Exercise 3.14, and hence $(\kappa_1(t) - \kappa_2(t))/it$ is bounded and continuous. The factor $1/it$ improves the integrability for large t so $(\kappa_1(t) - \kappa_2(t))/it$ is integrable. Letting $a \to -\infty$ and using the Riemann-Lebesgue lemma (Exercise 4.5 in the Appendix) gives (4.11).   □

Let φ_F and φ_G be the ch.f.'s of F and G. Applying (4.11) to $F_L = F * H_L$ and $G_L = G * H_L$, gives

$$|F_L(x) - G_L(x)| \le \frac{1}{2\pi} \int |\varphi_F(t)\omega_L(t) - \varphi_G(t)\omega_L(t)| \frac{dt}{|t|}$$

$$\le \frac{1}{2\pi} \int_{-L}^{L} |\varphi_F(t) - \varphi_G(t)| \frac{dt}{|t|}$$

since $|\omega_L(t)| \le 1$. Using (4.10) now we have

$$|F(x) - G(x)| \le \frac{1}{\pi} \int_{-L}^{L} |\varphi_F(\theta) - \varphi_G(\theta)| \frac{d\theta}{|\theta|} + \frac{24\lambda}{\pi L}$$

where $\lambda = \sup_x G'(x)$. Plugging in $F = F_n$ and $G = \mathcal{N}$ gives

$$(4.12) \qquad |F_n(x) - \mathcal{N}(x)| \le \frac{1}{\pi} \int_{-L}^{L} |\varphi^n(\theta/\sqrt{n}) - \psi(\theta)| \frac{d\theta}{|\theta|} + \frac{24\lambda}{\pi L}$$

and it remains to estimate the right hand side. This phase of the argument is fairly routine but there is a fair amount of algebra. To save the reader from trying to improve the inequalities along the way in hopes of getting a better bound, we would like to observe that we have used the fact that $C = 3$ to get rid of the cases $n \leq 9$, and we use $n \geq 10$ in (e).

To estimate the second term in (4.12) we observe that

(a)
$$\sup_x G'(x) = G'(0) = (2\pi)^{-1/2} = .39894 < 2/5$$

For the first we observe that if $|\alpha|, |\beta| \leq \gamma$

(b)
$$|\alpha^n - \beta^n| \leq \sum_{m=0}^{n-1} |\alpha^{n-m}\beta^m - \alpha^{n-m-1}\beta^{m+1}| \leq n|\alpha - \beta|\gamma^{n-1}$$

Using (3.7) now gives (recall we are supposing $\sigma^2 = 1$)

(c) $$|\varphi(t) - 1 + t^2/2| \leq \rho|t|^3/6$$
(d) $$|\varphi(t)| \leq 1 - t^2/2 + \rho|t|^3/6 \quad \text{if } t^2 \leq 2$$

Let $L = 4\sqrt{n}/3\rho$. If $|\theta| \leq L$ then by (d) and the fact $\rho|\theta|/\sqrt{n} \leq 4/3$

$$|\varphi(\theta/\sqrt{n})| \leq 1 - \theta^2/2n + \rho|\theta|^3/6n^{3/2}$$
$$\leq 1 - 5\theta^2/18n \leq \exp(-5\theta^2/18n)$$

since $1 - x \leq e^{-x}$. We will now apply (b) with

$$\alpha = \varphi(\theta/\sqrt{n}) \qquad \beta = \exp(-\theta^2/2n) \qquad \gamma = \exp(-5\theta^2/18n)$$

Since we are supposing $n \geq 10$

(e) $$\gamma^{n-1} \leq \exp(-\theta^2/4)$$

For the other part of (b) we write

$$n|\alpha - \beta| \leq n|\varphi(\theta/\sqrt{n}) - 1 + \theta^2/2n| + n|1 - \theta^2/2n - \exp(-\theta^2/2n)|$$

To bound the first term on the right-hand side, observe (c) implies

$$n|\varphi(\theta/\sqrt{n}) - 1 + \theta^2/2n| \leq \rho|\theta|^3/6n^{1/2}$$

For the second term note that if $0 < x < 1$ then we have an alternating series with decreasing terms so

$$|e^{-x} - (1 - x)| = \left| -\frac{x^2}{2!} + \frac{x^3}{3!} - \cdots \right| \leq \frac{x^2}{2}$$

Taking $x = \theta^2/2n$ it follows that for $|\theta| \leq L \leq \sqrt{2n}$

$$n|1 - \theta^2/2n - \exp(-\theta^2/2n)| \leq \theta^4/8n$$

Combining this with our estimate on the first term gives

(f) $\qquad\qquad\qquad n|\alpha - \beta| \leq \rho|\theta|^3/6n^{1/2} + \theta^4/8n$

Using (f) and (e) in (b), gives

(g)
$$\frac{1}{|\theta|}|\varphi^n(\theta/\sqrt{n}) - \exp(-\theta^2/2)| \leq \exp(-\theta^2/4)\left\{ \frac{\rho\theta^2}{6n^{1/2}} + \frac{|\theta|^3}{8n} \right\}$$
$$\leq \frac{1}{L}\exp(-\theta^2/4)\left\{ \frac{2\theta^2}{9} + \frac{|\theta|^3}{18} \right\}$$

since $\rho/\sqrt{n} = 4/3L$, and $1/n = 1/\sqrt{n} \cdot 1/\sqrt{n} \leq 4/3L \cdot 1/3$ since $\rho \geq 1$ and $n \geq 10$. Using (g) and (a) in (4.12) gives

$$\pi L|F_n(x) - \mathcal{N}(x)| \leq \int \exp(-\theta^2/4)\left\{ \frac{2\theta^2}{9} + \frac{|\theta|^3}{18} \right\} d\theta + 9.6$$

Recalling $L = 4\sqrt{n}/3\rho$, we see that the last result is of the form $|F_n(x) - \mathcal{N}(x)| \leq C\rho/\sqrt{n}$. To evaluate the constant we observe

$$\int (2\pi a)^{-1/2} x^2 \exp(-x^2/2a) dx = a$$

and writing $x^3 = 2x^2 \cdot x/2$ and integrating by parts

$$2\int_0^\infty x^3 \exp(-x^2/4)\, dx = 2\int_0^\infty 4x \exp(-x^2/4)\, dx$$
$$= -16e^{-x^2/4}\Big|_0^\infty = 16$$

This gives us

$$|F_n(x) - \mathcal{N}(x)| \leq \frac{1}{\pi} \cdot \frac{3}{4}\left(\frac{2}{9} \cdot 2 \cdot \sqrt{4\pi} + \frac{16}{18} + 9.6 \right)\frac{\rho}{\sqrt{n}} < 3\frac{\rho}{\sqrt{n}}$$

For the last step you have to get out your calculator or trust Feller.

*2.5. Local Limit Theorems

In Section 2.1 we saw that if $X_1, X_2, \ldots$ are i.i.d. with $P(X_1 = 1) = P(X_1 = -1) = 1/2$ and k_n is a sequence of integers with $2k_n/(2n)^{1/2} \to x$ then

$$P(S_{2n} = 2k_n) \sim (\pi n)^{-1/2} \exp(-x^2/2)$$

In this section we will prove two theorems that generalize the last result. We begin with two definitions. A random variable X has a **lattice distribution** if there are constants b and $h > 0$ so that $P(X \in b + h\mathbf{Z}) = 1$, where $b + h\mathbf{Z} = \{b + hz : z \in \mathbf{Z}\}$. The largest h for which the last statement holds is called the **span** of the distribution.

Example 5.1. If $P(X = 1) = P(X = -1) = 1/2$ then X has a lattice distribution with span 2. When h is 2, one possible choice is $b = -1$.

The next result relates the last definition to the characteristic function. To check (ii) in its statement note that in the last example $E(e^{itX}) = \cos t$ has $|\cos(t)| = 1$ when $t = n\pi$.

(5.1) Theorem. There are only three possibilities.

(i) $|\varphi(t)| < 1$ for all $t \neq 0$.

(ii) There is a $\lambda > 0$ so that $|\varphi(\lambda)| = 1$ and $|\varphi(t)| < 1$ for $0 < t < \lambda$. In this case X has a lattice distribution with span $2\pi/\lambda$.

(iii) $|\varphi(t)| = 1$ for all t. In this case $X = b$ a.s. for some b.

Proof We begin with (ii). It suffices to show that $|\varphi(t)| = 1$ if and only if $P(X \in b + (2\pi/t)\mathbf{Z}) = 1$ for some b. First if $P(X \in b + (2\pi/t)\mathbf{Z}) = 1$ then

$$\varphi(t) = Ee^{itX} = e^{itb} \sum_{n \in \mathbf{Z}} e^{i2\pi n} P(X = b + (2\pi/t)n) = e^{itb}$$

Conversely, if $|\varphi(t)| = 1$, then there is equality in the inequality $|Ee^{itX}| \leq E|e^{itX}|$ so the distribution of e^{itX} must be concentrated at some point e^{itb}, and $P(X \in b + (2\pi/t)\mathbf{Z}) = 1$.

To prove trichotomy now, we suppose that (i) and (ii) do not hold, i.e., there is a sequence $t_n \downarrow 0$ so that $|\varphi(t_n)| = 1$. The first paragraph shows that there is a b_n so that $P(X \in b_n + (2\pi/t_n)\mathbf{Z}) = 1$. Without loss of generality we can pick $b_n \in (-\pi/t_n, \pi/t_n]$. As $n \to \infty$, $P(X \notin (-\pi/t_n, \pi/t_n]) \to 0$ so it follows that $P(X = b_n) \to 1$. This is only possible if $b_n = b$ for $n \geq N$, and $P(X = b) = 1$. $\qquad \square$

We call the three cases in (5.1): (i) **nonlattice**, (ii) **lattice**, and (iii) **degenerate**. The reader should notice that this means that lattice random variables are by definition nondegenerate. Before we turn to the main business of this section, we would like to introduce one more special case. If X is a lattice distribution and we can take $b = 0$, i.e., $P(X \in h\mathbf{Z}) = 1$, then X is said to be **arithmetic**. In this case if $\lambda = 2\pi/h$ then $\varphi(\lambda) = 1$ and φ is periodic: $\varphi(t + \lambda) = \varphi(t)$.

Our first local limit theorem is for the lattice case. Let $X_1, X_2, \ldots$ be i.i.d. with $EX_i = 0$, $EX_i^2 = \sigma^2 \in (0, \infty)$, and having a common lattice distribution with span h. If $S_n = X_1 + \cdots + X_n$ and $P(X_i \in b + h\mathbf{Z}) = 1$ then $P(S_n \in nb + h\mathbf{Z}) = 1$. We put

$$p_n(x) = P(S_n/\sqrt{n} = x) \quad \text{for } x \in \mathcal{L}_n = \{(nb + hz)/\sqrt{n} : z \in \mathbf{Z}\}$$

and

$$n(x) = (2\pi\sigma^2)^{-1/2} \exp(-x^2/2\sigma^2) \quad \text{for } x \in (-\infty, \infty)$$

(5.2) Theorem. Under the hypotheses above, as $n \to \infty$

$$\sup_{x \in \mathcal{L}_n} \left| \frac{n^{1/2}}{h} p_n(x) - n(x) \right| \to 0$$

Remark. To explain the statement, note that if we followed the approach in Example 4.3 then we would conclude that for $x \in \mathcal{L}_n$

$$p_n(x) \approx \int_{x-h/2\sqrt{n}}^{x+h/2\sqrt{n}} n(y)\, dy \approx \frac{h}{\sqrt{n}} n(x)$$

Proof Let Y be a random variable with $P(Y \in a + \theta\mathbf{Z}) = 1$ and $\psi(t) = E\exp(itY)$. It follows from part (iii) of Exercise 3.2 that

$$P(Y = x) = \frac{1}{2\pi/\theta} \int_{-\pi/\theta}^{\pi/\theta} e^{-itx}\psi(t)\, dt$$

Using this formula with $\theta = h/\sqrt{n}$, $\psi(t) = E\exp(itS_n/\sqrt{n}) = \varphi^n(t/\sqrt{n})$, and then multiplying each side by $1/\theta$ gives

$$\frac{n^{1/2}}{h} p_n(x) = \frac{1}{2\pi} \int_{-\pi\sqrt{n}/h}^{\pi\sqrt{n}/h} e^{-itx}\varphi^n(t/\sqrt{n})\, dt$$

Using the inversion formula (3.3) for $n(x)$, which has ch.f. $\exp(-\sigma^2 t^2/2)$, gives

$$n(x) = \frac{1}{2\pi} \int e^{-itx} \exp(-\sigma^2 t^2/2)\, dt$$

Subtracting the last two equations gives (recall $\pi > 1$, $|e^{-itx}| \leq 1$)

$$\left| \frac{n^{1/2}}{h} p_n(x) - n(x) \right| \leq \int_{-\pi\sqrt{n}/h}^{\pi\sqrt{n}/h} |\varphi^n(t/\sqrt{n}) - \exp(-\sigma^2 t^2/2)|\, dt$$
$$+ \int_{\pi\sqrt{n}/h}^{\infty} \exp(-\sigma^2 t^2/2)\, dt$$

The right-hand side is independent of x, so to prove (5.2) it suffices to show that it approaches 0. The second integral clearly $\to 0$. To estimate the first integral, we observe that $\varphi^n(t/\sqrt{n}) \to \exp(-\sigma^2 t^2/2)$ so the integrand goes to 0 and it is now just a question of "applying the dominated convergence theorem."

To do this, we will divide the integral into three pieces. The bounded convergence theorem implies that for any $A < \infty$ the integral over $(-A, A)$ approaches 0. To estimate the integral over $(-A, A)^c$, we observe that since $EX_i = 0$ and $EX_i^2 = \sigma^2$, formula (3.7) and the triangle inequality imply that

$$|\varphi(u)| \leq |1 - \sigma^2 u^2/2| + \frac{u^2}{2} E(\min(|u| \cdot |X|^3, 6|X|^2))$$

The last expected value $\to 0$ as $u \to 0$. This means we can pick $\delta > 0$ so that if $|u| < \delta$, it is $\leq \sigma^2/2$ and hence

$$|\varphi(u)| \leq 1 - \sigma^2 u^2/2 + \sigma^2 u^2/4 = 1 - \sigma^2 u^2/4 \leq \exp(-\sigma^2 u^2/4)$$

since $1 - x \leq e^{-x}$. Applying the last result to $u = t/\sqrt{n}$ we see that for $t \leq \delta\sqrt{n}$

(5.3) $$|\varphi(t/\sqrt{n})^n| \leq \exp(-\sigma^2 t^2/4)$$

So the integral over $(-\delta\sqrt{n}, \delta\sqrt{n}) - (-A, A)$ is smaller than

$$2 \int_{A}^{\delta\sqrt{n}} \exp(-\sigma^2 t^2/4)\, dt$$

which is small if A is large.

To estimate the rest of the integral we observe that since X has span h, (5.1) implies $|\varphi(u)| \neq 1$ for $u \in [\delta, \pi/h]$. φ is continuous so there is an $\eta < 1$ so that $|\varphi(u)| \leq \eta < 1$ for $|u| \in [\delta, \pi/h]$. Letting $u = t/\sqrt{n}$ again, we see that the integral over $[-\pi\sqrt{n}/h, \pi\sqrt{n}/h] - (-\delta\sqrt{n}, \delta\sqrt{n})$ is smaller than

$$2 \int_{\delta\sqrt{n}}^{\pi\sqrt{n}/h} \eta^n + \exp(-\sigma^2 t^2/2)\, dt$$

which $\to 0$ as $n \to \infty$. This completes the proof of (5.2). $\qquad\square$

We turn now to the nonlattice case. Let $X_1, X_2, \ldots$ be i.i.d. with $EX_i = 0$, $EX_i^2 = \sigma^2 \in (0, \infty)$, and having a common characteristic function $\varphi(t)$ that has $|\varphi(t)| < 1$ for all $t \neq 0$. Let $S_n = X_1 + \cdots + X_n$ and $n(x) = (2\pi\sigma^2)^{-1/2} \exp(-x^2/2\sigma^2)$.

(5.4) Theorem. Under the hypotheses above, if $x_n/\sqrt{n} \to x$ and $a < b$

$$\sqrt{n} P(S_n \in (x_n + a, \, x_n + b)) \to (b - a)n(x)$$

Remark. The proof of (5.4) has to be a little devious because the assumption above does not give us much control over the behavior of φ. For a bad example let $q_1, q_2, \ldots$ be an enumeration of the positive rationals which has $q_n \leq n$. Suppose

$$P(X = q_n) = P(X = -q_n) = 1/2^{n+1}$$

In this case $EX = 0$, $EX^2 < \infty$, and the distribution is nonlattice. However the characteristic function has $\limsup_{t\to\infty} |\varphi(t)| = 1$.

Proof To tame bad ch.f.'s we use a trick. Let $\delta > 0$

$$h_0(y) = \frac{1}{\pi} \cdot \frac{1 - \cos \delta y}{\delta y^2}$$

be the density of the Polya's distribution and let $h_\theta(x) = e^{i\theta x} h_0(x)$. If we introduce the Fourier transform

$$\hat{g}(u) = \int e^{iuy} g(y)\, dy$$

then it follows from Example 3.8 that

$$\hat{h}_0(u) = \begin{cases} 1 - |u/\delta| & \text{if } |u| \leq \delta \\ 0 & \text{otherwise} \end{cases}$$

and it is easy to see that $\hat{h}_\theta(u) = \hat{h}_0(u + \theta)$. We will show that for any θ

(*) $$\sqrt{n}\, Eh_\theta(S_n - x_n) \to n(x) \int h_\theta(y)\, dy$$

Before proving (∗) we will show it implies (5.4). Let

$$\mu_n(A) = \sqrt{n}P(S_n - x_n \in A), \quad \text{and} \quad \mu(A) = n(x)|A|$$

where $|A| = $ the Lebesgue measure of A. Let

$$\alpha_n = \sqrt{n}\,Eh_0(S_n - x_n) \quad \text{and} \quad \alpha = n(x)\int h_0(y)\,dy = n(x)$$

Finally, define probability measures by

$$\nu_n(B) = \frac{1}{\alpha_n}\int_B h_0(y)\mu_n(dy), \quad \text{and} \quad \nu(B) = \frac{1}{\alpha}\int_B h_0(y)\mu(dy)$$

Taking $\theta = 0$ in (∗) we see $\alpha_n \to \alpha$ and so (∗) implies

(∗∗)
$$\int e^{i\theta y}\nu_n(dy) \to \int e^{i\theta y}\nu(dy)$$

Since this holds for all θ, it follows from (3.4) that $\nu_n \Rightarrow \nu$. Now if $|a|, |b| < 2\pi/\delta$ then the function

$$k(y) = \frac{1}{h_0(y)}\cdot 1_{(a,b)}(y)$$

is bounded and continuous a.s. with respect to ν so it follows from (2.3) that

$$\int k(y)\nu_n(dy) \to \int k(y)\nu(dy)$$

Since $\alpha_n \to \alpha$ this implies

$$\sqrt{n}P(S_n \in (x_n + a, x_n + b)) \to (b - a)n(x)$$

which is the conclusion of (5.4).

Turning now to the proof of (∗), the inversion formula (3.3) implies

$$h_0(x) = \frac{1}{2\pi}\int e^{-iux}\hat{h}_0(u)\,du$$

Recalling the definition of h_θ, using the last result, and changing variables $u = v + \theta$ we have

$$h_\theta(x) = e^{i\theta x}h_0(x) = \frac{1}{2\pi}\int e^{-i(u-\theta)x}\hat{h}_0(u)\,du$$

$$= \frac{1}{2\pi}\int e^{-ivx}\hat{h}_\theta(v)\,dv$$

since $\hat{h}_\theta(v) = \hat{h}_0(v+\theta)$. Letting F_n be the distribution of $S_n - x_n$ and integrating gives

$$E h_\theta(S_n - x_n) = \frac{1}{2\pi} \int \int e^{-iux} \hat{h}_\theta(u) \, du \, dF_n(x)$$

$$= \frac{1}{2\pi} \int \int e^{-iux} \, dF_n(x) \hat{h}_\theta(u) \, du$$

by Fubini's theorem. (Recall $\hat{h}_\theta(u)$ has compact support and F_n is a distribution function.) Using (3.1e) we see that the last expression

$$= \frac{1}{2\pi} \int \varphi(-u)^n e^{iux_n} \hat{h}_\theta(u) \, du$$

To take the limit as $n \to \infty$ of this integral, let $[-M, M]$ be an interval with $\hat{h}_\theta(u) = 0$ for $u \notin [-M, M]$. By (5.3) above we can pick δ so that for $|u| < \delta$

(5.5)
$$|\varphi(u)| \le \exp(-\sigma^2 u^2/4)$$

Let $I = [-\delta, \delta]$ and $J = [-M, M] - I$. Since $|\varphi(u)| < 1$ for $u \ne 0$ and φ is continuous, there is a constant $\eta < 1$ so that $|\varphi(u)| \le \eta < 1$ for $u \in J$. Since $|\hat{h}_\theta(u)| \le 1$, this implies that

$$\left| \frac{\sqrt{n}}{2\pi} \int_J \varphi(-u)^n e^{iux_n} \hat{h}_\theta(u) \, du \right| \le \frac{\sqrt{n}}{2\pi} \cdot 2M\eta^n \to 0$$

as $n \to \infty$. For the integral over I, change variables $u = t/\sqrt{n}$ to get

$$\frac{1}{2\pi} \int_{-\delta\sqrt{n}}^{\delta\sqrt{n}} \varphi(-t/\sqrt{n})^n e^{itx_n/\sqrt{n}} \hat{h}_\theta(t/\sqrt{n}) \, dt$$

The central limit theorem implies $\varphi(-t/\sqrt{n})^n \to \exp(-\sigma^2 t^2/2)$. Using (5.5) now and the dominated convergence theorem gives (recall $x_n/\sqrt{n} \to x$)

$$\frac{\sqrt{n}}{2\pi} \int_I \varphi(-u)^n e^{iux_n} \hat{h}_\theta(u) \, du \to \frac{1}{2\pi} \int \exp(-\sigma^2 t^2/2) e^{itx} \hat{h}_\theta(0) \, dt$$

$$= n(x)\hat{h}_\theta(0) = n(x) \int h_\theta(y) \, dy$$

by the inversion formula (3.3) and the definition of $\hat{h}_\theta(0)$. This proves (*) and completes the proof of (5.4).   □

2.6. Poisson Convergence

We begin with the

a. Basic limit theorem

which is sometimes facetiously called the "weak law of small numbers" or the "law of rare events." These names derive from the fact that the Poisson appears as the limit of a sum of indicators of events that have small probablilities.

(6.1) Theorem. For each n let $X_{n,m}$, $1 \leq m \leq n$ be independent random variables with $P(X_{n,m} = 1) = p_{n,m}$, $P(X_{n,m} = 0) = 1 - p_{n,m}$. Suppose

(i) $\sum_{m=1}^{n} p_{n,m} \to \lambda \in (0, \infty)$,

and (ii) $\max_{1 \leq m \leq n} p_{n,m} \to 0$.

If $S_n = X_{n,1} + \cdots + X_{n,n}$ then $S_n \Rightarrow Z$ where Z is Poisson(λ).

Here Poisson(λ) is shorthand for Poisson distribution with mean λ, that is,

$$P(Z = k) = e^{-\lambda} \lambda^k / k!$$

Note that in the spirit of the Lindeberg-Feller theorem, no single term contributes very much to the sum. In contrast to that theorem, the contributions, when positive, are not small.

First proof   Let $\varphi_{n,m}(t) = E(\exp(itX_{n,m})) = (1 - p_{n,m}) + p_{n,m}e^{it}$ and let $S_n = X_{n,1} + \cdots + X_{n,n}$. Then

$$E \exp(itS_n) = \prod (1 + p_{n,m}(e^{it} - 1))$$

Let $0 \leq p \leq 1$. $|\exp(p(e^{it} - 1))| = \exp(p\mathrm{Re}\,(e^{it} - 1)) \leq 1$ and $|1 + p(e^{it} - 1)| \leq 1$ since it is on the line segment connecting 1 to e^{it}. Using (4.3) with $\theta = 1$ and then (4.4) which is valid when $\max_m p_{n,m} \leq 1/2$ since $|e^{it} - 1| \leq 2$

$$\left| \exp\left(\sum_{m=1}^{n} p_{n,m}(e^{it} - 1) \right) - \prod_{m=1}^{n} \{1 + p_{n,m}(e^{it} - 1)\} \right|$$

$$\leq \sum_{m=1}^{n} \left| \exp(p_{n,m}(e^{it} - 1)) - \{1 + p_{n,m}(e^{it} - 1)\} \right|$$

$$\leq \sum_{m=1}^{n} p_{n,m}^2 |e^{it} - 1|^2$$

Using $|e^{it} - 1| \leq 2$ again, it follows that the last expression

$$\leq 4 \left(\max_{1 \leq m \leq n} p_{n,m} \right) \sum_{m=1}^{n} p_{n,m} \to 0$$

by assumptions (i) and (ii). The last conclusion and $\sum_{m=1}^{n} p_{n,m} \to \lambda$ imply

$$E \exp(itS_n) \to \exp(\lambda(e^{it} - 1))$$

To complete the proof now we consult Example 3.2 for the ch.f. of the Poisson distribution and apply (3.4).   □

We will now consider some concrete situations in which (6.1) can be applied. In each case we are considering a situation in which $p_{n,m} = c/n$ so we approximate the distribution of the sum by a Poisson with mean c.

Example 6.1. In a calculus class with 400 students, the number of students who have their birthday on the day of the final exam has approximately a Poisson distribution with mean $400/365 = 1.096$. This means that the probability no one was born on that date is about $e^{-1.096} = .334$. Similar reasoning shows that the number of babies born on a given day or the number of people who arrive at a bank between 1:15 and 1:30 should have a Poisson distribution.

Example 6.2. Suppose we roll two dice 36 times. The probability of "double ones" (one on each die) is $1/36$ so the number of times this occurs should have approximately a Poisson distribution with mean 1. Comparing the Poisson approximation with exact probabilities shows that the agreement is good even though the number of trials is small.

k	0	1	2	3
Poisson	.3678	.3678	.1839	.0613
exact	.3627	.3730	.1865	.0604

After we give the second proof of (6.1), (see (6.5)) we will discuss rates of convergence. Those results will show that for large n the largest discrepancy occurs for $k = 1$ and is about $1/2en$ ($= .0051$ in this case).

Example 6.3. Let $\xi_{n,1}, \ldots, \xi_{n,n}$ be independent and uniformly distributed over $[-n, n]$. Let $X_{n,1} = 1$ if $\xi_n \in (a, b)$, $= 0$ otherwise. S_n is the number of points that land in (a, b). $p_{n,m} = (b - a)/2n$ so $\sum_m p_{n,m} = (b - a)/2$. This shows (i) and (ii) in (6.1) hold, and we conclude that $S_n \Rightarrow Z$, a Poisson r.v. with mean $(b - a)/2$. A two dimensional version of the last theorem might explain why the statistics of flying bomb hits in the South of London during World War II fit a Poisson distribution. As Feller, Vol. I (1968) p.160–161 reports, the area was divided into 576 areas of 1/4 square kilometers each. The total number of hits was 537 for an average of .9323 per cell. The table below compares N_k the number of cells with k hits with the predictions of the Poisson approximation.

k	0	1	2	3	4	≥ 5
N_k	229	211	93	35	7	1
Poisson	226.74	211.39	98.54	30.62	7.14	1.57

For other observations fitting a Poisson distribution, see Feller, Vol. I (1968) Section VI.7.

Our second proof of (6.1) requires a little more work but provides information about the rate of convergence. (See (6.5) below.) We begin by defining the **total variation distance** between two measures on a countable set S.

$$\|\mu - \nu\| \equiv \sum_z |\mu(z) - \nu(z)| = 2 \sup_{A \subset S} |\mu(A) - \nu(A)|$$

The first equality is a definition. To prove the second, note that for any A

$$\sum_z |\mu(z) - \nu(z)| \geq |\mu(A) - \nu(A)| + |\mu(A^c) - \nu(A^c)| = 2|\mu(A) - \nu(A)|$$

and there is equality when $A = \{z : \mu(z) \geq \nu(z)\}$.

EXERCISE 6.1. Show that (i) $d(\mu, \nu) = \|\mu - \nu\|$ defines a metric on probability measures on $\mathbf{Z}$ and (ii) $\|\mu_n - \mu\| \to 0$ if and only if $\mu_n(x) \to \mu(x)$ for each $x \in \mathbf{Z}$, which by Exercise 2.8 is equivalent to $\mu_n \Rightarrow \mu$.

EXERCISE 6.2. Show that $\|\mu - \nu\| \leq 2\delta$ if and only if there are random variables X and Y with distributions μ and ν so that $P(X \neq Y) \leq \delta$.

The next three lemmas are the keys to our second proof.

(6.2) **Lemma.** If $\mu_1 \times \mu_2$ denotes the product measure on $\mathbf{Z} \times \mathbf{Z}$ that has $(\mu_1 \times \mu_2)(x, y) = \mu_1(x)\mu_2(y)$ then

$$\|\mu_1 \times \mu_2 - \nu_1 \times \nu_2\| \leq \|\mu_1 - \nu_1\| + \|\mu_2 - \nu_2\|$$

Proof $\|\mu_1 \times \mu_2 - \nu_1 \times \nu_2\| = \sum_{x,y} |\mu_1(x)\mu_2(y) - \nu_1(x)\nu_2(y)|$

$$\leq \sum_{x,y} |\mu_1(x)\mu_2(y) - \nu_1(x)\mu_2(y)| + \sum_{x,y} |\nu_1(x)\mu_2(y) - \nu_1(x)\nu_2(y)|$$

$$= \sum_y \mu_2(y) \sum_x |\mu_1(x) - \nu_1(x)| + \sum_x \nu_1(x) \sum_y |\mu_2(y) - \nu_2(y)|$$

$$= \|\mu_1 - \nu_1\| + \|\mu_2 - \nu_2\| \qquad\qquad \square$$

(6.3) Lemma. If $\mu_1 * \mu_2$ denotes the convolution of μ_1 and μ_2, that is,

$$\mu_1 * \mu_2(x) = \sum_y \mu_1(x - y)\mu_2(y)$$

then $\|\mu_1 * \mu_2 - \nu_1 * \nu_2\| \le \|\mu_1 \times \mu_2 - \nu_1 \times \nu_2\|$

Proof $\|\mu_1 * \mu_2 - \nu_1 * \nu_2\| = \sum_x \left| \sum_y \mu_1(x - y)\mu_2(y) - \sum_y \nu_1(x - y)\nu_2(y) \right|$

$$\le \sum_x \sum_y |\mu_1(x - y)\mu_2(y) - \nu_1(x - y)\nu_2(y)|$$

$$= \|\mu_1 \times \mu_2 - \nu_1 \times \nu_2\| \qquad \square$$

(6.4) Lemma. Let μ be the measure with $\mu(1) = p$ and $\mu(0) = 1 - p$. Let ν be a Poisson distribution with mean p. Then $\|\mu - \nu\| \le 2p^2$.

Proof $\|\mu - \nu\| = |\mu(0) - \nu(0)| + |\mu(1) - \nu(1)| + \sum_{n \ge 2} \nu(n)$

$$= |1 - p - e^{-p}| + |p - p\, e^{-p}| + 1 - e^{-p}(1 + p)$$

Since $1 - x \le e^{-x} \le 1$ for $x \ge 0$, the above

$$= e^{-p} - 1 + p + p(1 - e^{-p}) + 1 - e^{-p} - pe^{-p}$$
$$= 2p(1 - e^{-p}) \le 2p^2 \qquad \square$$

Second proof of (6.1)   Let $\mu_{n,m}$ be the distribution of $X_{n,m}$. Let μ_n be the distribution of S_n. Let $\nu_{n,m}$, ν_n, and ν be Poisson distributions with means $p_{n,m}$, $\lambda_n = \sum_{m \le n} p_{n,m}$, and λ respectively. Since $\mu_n = \mu_{n,1} * \cdots * \mu_{n,n}$ and $\nu_n = \nu_{n,1} * \cdots * \nu_{n,n}$, (6.3), (6.2), and (6.4) imply

$$(6.5) \qquad \|\mu_n - \nu_n\| \le \sum_{m=1}^{n} \|\mu_{n,m} - \nu_{n,m}\| \le 2 \sum_{m=1}^{n} p_{n,m}^2$$

Using the definition of total variation distance now gives

$$\sup_A |\mu_n(A) - \nu_n(A)| \le \sum_{m=1}^{n} p_{n,m}^2$$

(i) and (ii) in (6.1) imply that the right-hand side $\to 0$. Since $\nu_n \Rightarrow \nu$ as $n \to \infty$, the result follows. $\qquad \square$

Remark. The proof above is due to Hodges and Le Cam (1960). By different methods C. Stein (1987) (see (43) on p. 89) has proved

$$\sup_A |\mu_n(A) - \nu_n(A)| \leq (\lambda \vee 1)^{-1} \sum_{m=1}^{n} p_{n,m}^2$$

Rates of convergence. When $p_{n,m} = 1/n$, (6.5) becomes

$$\sup_A |\mu_n(A) - \nu_n(A)| \leq 1/n$$

To assess the quality of this bound we will compare the Poisson and binomial probabilities for k successes.

k	Poisson	Binomial
0	e^{-1}	$\left(1 - \frac{1}{n}\right)^n$
1	e^{-1}	$n \cdot n^{-1}\left(1 - \frac{1}{n}\right)^{n-1} = \left(1 - \frac{1}{n}\right)^{n-1}$
2	$e^{-1}/2!$	$\binom{n}{2}n^{-2}\left(1 - \frac{1}{n}\right)^{n-2} = \left(1 - \frac{1}{n}\right)^{n-1}/2!$
3	$e^{-1}/3!$	$\binom{n}{3}n^{-3}\left(1 - \frac{1}{n}\right)^{n-3} = \left(1 - \frac{2}{n}\right)\left(1 - \frac{1}{n}\right)^{n-2} \Big/ 3!$

Since $(1-x) \leq e^{-x}$, we have $\mu_n(0) - \nu_n(0) \leq 0$. Expanding

$$\log(1 + x) = x - \frac{x^2}{2} + \frac{x^3}{3} - \cdots$$

gives

$$(n-1)\log\left(1 - \frac{1}{n}\right) = -\frac{n-1}{n} - \frac{n-1}{2n^2} - \cdots = -1 + \frac{1}{2n} + O(n^{-2})$$

So

$$n\left(\left(1 - \frac{1}{n}\right)^{n-1} - e^{-1}\right) = ne^{-1}\left(\exp\{1/2n + O(n^{-2})\} - 1\right) \to e^{-1}/2$$

and it follows that

$$n(\mu_n(1) - \nu_n(1)) \to e^{-1}/2$$
$$n(\mu_n(2) - \nu_n(2)) \to e^{-1}/4$$

For $k \geq 3$, using $(1-2/n) \leq (1-1/n)^2$ and $(1-x) \leq e^{-x}$ shows $\mu_n(k) - \nu_n(k) \leq 0$, so

$$\sup_{A \subset \mathbf{Z}} |\mu_n(A) - \nu_n(A)| \approx 3/4en$$

There is a large literature on Poisson approximations for dependent events. Here we consider

b. Two examples with dependence

that can be treated by exact calculations.

Example 6.4. Matching. Let π be a random permutation of $\{1, 2, \ldots, n\}$, let $X_{n,m} = 1$ if m is a fixed point (0 otherwise), and let $S_n = X_{n,1} + \cdots + X_{n,n}$ be the number of fixed points. We want to compute $P(S_n = 0)$. (For a more exciting story consider men checking hats or wives swapping husbands.) Let $A_{n,m} = \{X_{n,m} = 1\}$. The inclusion-exclusion formula implies

$$
\begin{aligned}
P\left(\cup_{m=1}^{n} A_m\right) = {} & \sum_{m} P(A_m) - \sum_{\ell < m} P(A_\ell \cap A_m) \\
& + \sum_{k < \ell < m} P(A_k \cap A_\ell \cap A_m) - \cdots \\
= {} & n \cdot \frac{1}{n} - \binom{n}{2} \frac{(n-2)!}{n!} + \binom{n}{3} \frac{(n-3)!}{n!} - \cdots
\end{aligned}
$$

since the number of permutations with k specified fixed points is $(n - k)!$ Cancelling some factorials gives

$$
P(S_n > 0) = \sum_{m=1}^{n} \frac{(-1)^{m-1}}{m!} \quad \text{so} \quad P(S_n = 0) = \sum_{m=0}^{n} \frac{(-1)^m}{m!}
$$

Recognizing the second sum as the first $n + 1$ terms in the expansion of e^{-1} gives

$$
|P(S_n = 0) - e^{-1}| = \left| \sum_{m=n+1}^{\infty} \frac{(-1)^m}{m!} \right|
$$

$$
\leq \frac{1}{(n+1)!} \left| \sum_{k=0}^{\infty} (n+2)^{-k} \right| = \frac{1}{(n+1)!} \cdot \left(1 - \frac{1}{n+2} \right)^{-1}
$$

a much better rate of convergence than $1/n$. To compute the other probabilities we observe that by considering the locations of the fixed points

$$
\begin{aligned}
P(S_n = k) &= \binom{n}{k} \frac{1}{n(n-1) \cdots (n-k+1)} P(S_{n-k} = 0) \\
&= \frac{1}{k!} P(S_{n-k} = 0) \to e^{-1}/k!
\end{aligned}
$$

Example 6.5.   Occupancy problem. Suppose that r balls are placed at random into n boxes. It follows from the Poisson approximation to the Binomial that if $n \to \infty$ and $r/n \to c$, then the number of balls in a given box will approach a Poisson distribution with mean c. The last observation should explain why the fraction of occupied boxes approached e^{-c} in Example 5.5 of Chapter 1. Here we will show:

(6.6) Theorem. If $ne^{-r/n} \to \lambda \in [0, \infty)$ the number of empty boxes approaches a Poisson distribution with mean λ.

Proof   To see where the answer comes from, notice that in the Poisson approximation the probability that a given box is empty is $e^{-r/n} \approx \lambda/n$, so if the occupancy of the various boxes were independent, the result would follow from (6.1). To prove the result we begin by observing

$$P(\text{ boxes } i_1, i_2, \ldots, i_k \text{ are empty }) = \left(1 - \frac{k}{n}\right)^r$$

If we let $p_m(r, n) = $ the probability exactly m boxes are empty when r balls are put in n boxes then $P(\text{ no empty box }) = 1 - P(\text{ at least one empty box })$. So by inclusion-exclusion

(a)
$$p_0(r, n) = \sum_{k=0}^{n} (-1)^k \binom{n}{k} \left(1 - \frac{k}{n}\right)^r$$

By considering the locations of the empty boxes

(b)
$$p_m(r, n) = \binom{n}{m} \left(1 - \frac{m}{n}\right)^r p_0(r, n - m)$$

To evaluate the limit of $p_m(r, n)$ we begin by showing that if $ne^{-r/n} \to \lambda$ then

(c)
$$\binom{n}{m} \left(1 - \frac{m}{n}\right)^r \to \lambda^m / m!$$

One half of this is easy. Since $(1 - x) \le e^{-x}$ and $ne^{-r/n} \to \lambda$

(d)
$$\binom{n}{m} \left(1 - \frac{m}{n}\right)^r \le \frac{n^m}{m!} e^{-mr/n} \to \lambda^m / m!$$

For the other direction observe $\binom{n}{m} \ge (n - m)^m / m!$ so

$$\binom{n}{m} \left(1 - \frac{m}{n}\right)^r \ge \left(1 - \frac{m}{n}\right)^{m+r} n^m / m!$$

Now $(1 - m/n)^m \to 1$ as $n \to \infty$ and $1/m!$ is a constant. To deal with the rest we note that if $0 \le t \le 1/2$ then

$$\log(1 - t) = -t - t^2/2 - t^3/3 \cdots$$

$$\ge -t - \frac{t^2}{2}\left(1 + 2^{-1} + 2^{-2} + \cdots\right) = -t - t^2$$

so we have

$$\log\left(n^m\left(1 - \frac{m}{n}\right)^r\right) \ge m\log n - rm/n - r(m/n)^2$$

Our assumption $ne^{-r/n} \to \lambda$ means

$$r = n\log n - n\log\lambda + o(n)$$

so $r(m/n)^2 \to 0$. Multiplying the last display by m/n and rearranging gives $m\log n - rm/n \to m\log\lambda$. Combining the last two results shows

$$\liminf_{n \to \infty} n^m\left(1 - \frac{m}{n}\right)^r \ge \lambda^m$$

and (c) follows. From (a), (c), and the dominated convergence theorem (using (d) to get the domination) we get

(e) if $e^{-r/n} \to \lambda$ then $p_0(r, n) \to \sum_{k=0}^{\infty}(-1)^k\frac{\lambda^k}{k!} = e^{-\lambda}$

For fixed m, $(n - m)e^{-r/(n-m)} \to \lambda$, so it follows from (e) that $p_0(r, n - m) \to e^{-\lambda}$. Combining this with (b) and (c) completes the proof of (6.6).   □

Example 6.6.   Coupon collector's problem.   Let $X_1, X_2, \ldots$ be i.i.d. uniform on $\{1, 2, \ldots, n\}$ and $T_n = \inf\{m : \{X_1, \ldots X_m\} = \{1, 2, \ldots, n\}\}$. Since $T_n \le m$ if and only if m balls fill up all n boxes, it follows from (6.6) that

$$P(T_n - n\log n \le nx) \to \exp(-e^{-x})$$

Proof   If $r = n\log n + nx$ then $ne^{-r/n} \to e^{-x}$.   □

For a concrete example consider: What is the probability that in a village of 2190 $(= 6 \cdot 365)$ people all birthdays are represented? Do you think the answer is much different for 1825 $(= 5 \cdot 365)$ people?

Solution   Here $n = 365$ so $365\log 365 = 2153$ and

$$P(T_{365} \le 2190) = P((T_{365} - 2153)/365 \le 37/365)$$

$$\approx \exp(-e^{-0.1014}) = \exp(-0.9036) = 0.4051$$

$$P(T_{365} \le 1825) = P((T_{365} - 2153)/365 \le -328/365)$$

$$\approx \exp(-e^{0.8986}) = \exp(-2.4562) = 0.085$$

As we observed in Example 5.3 of Chapter 1, if we let $\tau_k^n = \inf\{m : |\{X_1, \ldots, X_m\}| = k\}$, then $\tau_1^n = 1$ and for $2 \le k \le n$, $\tau_k^n - \tau_{k-1}^n$ are independent and have a geometric distribution with parameter $1 - (k-1)/n$.

EXERCISE 6.3. Suppose $k/n^{1/2} \to \lambda \in [0, \infty)$ and show that $\tau_k^n - k \Rightarrow$ Poisson($\lambda^2/2$). Hint: this is easy if you use (6.7) below.

EXERCISE 6.4. Let $\mu_{n,k} = E\tau_k^n$ and $\sigma_{n,k}^2 = \text{var}(\tau_k^n)$. Suppose $k/n \to a \in (0,1)$, and use the Lindeberg-Feller theorem to show $(\tau_k^n - \mu_{n,k})/\sqrt{n} \Rightarrow \sigma\chi$.

The last result is true when $k/n^{1/2} \to \infty$ and $n - k \to \infty$, see Baum and Billingsley (1966). Results for $k = n - j$ can be obtained from (6.6), so we have examined all the possibilities.

c. Poisson processes

(6.1) generalizes trivially to give the following result.

(6.7) **Theorem.** Let $X_{n,m}$, $1 \le m \le n$ be independent nonnegative integer valued random variables with $P(X_{n,m} = 1) = p_{n,m}$, $P(X_{n,m} \ge 2) = \epsilon_{n,m}$.

(i) $\sum_{m=1}^n p_{n,m} \to \lambda \in (0, \infty)$,

(ii) $\max_{1 \le m \le n} p_{n,m} \to 0$,

and (iii) $\sum_{m=1}^n \epsilon_{n,m} \to 0$.

If $S_n = X_{n,1} + \cdots + X_{n,n}$ then $S_n \Rightarrow Z$ where Z is Poisson(λ).

Proof   Let $X'_{n,m} = 1$ if $X_{n,m} = 1$, and 0 otherwise. Let $S'_n = X'_{n,1} + \cdots + X'_{n,n}$. (i)-(ii) and (6.1) imply $S'_n \Rightarrow Z$, (iii) tells us $P(S_n \ne S'_n) \to 0$ and the result follows from the converging together lemma, Exercise 2.10. $\qquad\square$

The next result, which uses (6.7), explains why the Poisson distribution comes up so frequently in applications. Let $N(s,t)$ be the number of arrivals at a bank or an ice cream parlor in the time interval $(s,t]$. Suppose

(i) the numbers of arrivals in disjoint intervals are independent,

(ii) the distribution of $N(s,t)$ only depends on $t - s$,

(iii) $P(N(0,h) = 1) = \lambda h + o(h)$,

and (iv) $P(N(0,h) \ge 2) = o(h)$.

Here, the two $o(h)$ stand for functions $g_1(h)$ and $g_2(h)$ with $g_i(h)/h \to 0$ as $h \to 0$.

(6.8) Theorem. If (i)-(iv) hold then $N(0,t)$ has a Poisson distribution with mean λt.

Proof Let $X_{n,m} = N((m-1)t/n, mt/n)$ for $1 \leq m \leq n$ and apply (6.7). □

A family of random variables N_t, $t \geq 0$ satisfying:

(i) if $0 = t_0 < t_1 < \ldots < t_n$, $N(t_k) - N(t_{k-1})$, $1 \leq k \leq n$ are independent,

(ii) $N(t) - N(s)$ is Poisson$(\lambda(t-s))$,

is called a **Poisson process with rate** λ. To understand how N_t behaves, it is useful to have another method to construct it. Let $\xi_1, \xi_2, \ldots$ be independent random variables with $P(\xi_i > t) = e^{-\lambda t}$ for $t \geq 0$. Let $T_n = \xi_1 + \cdots + \xi_n$ and $N_t = \sup\{n : T_n \leq t\}$ where $T_0 = 0$. In the language of renewal theory (see (7.3) in Chapter 1), T_n is the time of the nth arrival and N_t is the number of arrivals by time t. To check that N_t is a Poisson process we begin by recalling (see Exercise 4.8 in Chapter 1)

$$P(T_n = s) = \frac{\lambda^n s^{n-1}}{(n-1)!} e^{-\lambda s} \quad \text{for} \quad s \geq 0$$

i.e., the distribution of T_n has a density given by the right-hand side. Now

$$P(N_t = 0) = P(T_1 > t) = e^{-\lambda t}$$

and for $n \geq 1$

$$P(N_t = n) = P(T_n \leq t < T_{n+1}) = \int_0^t P(T_n = s) P(\xi_{n+1} > t - s)\, ds$$

$$= \int_0^t \frac{\lambda^n s^{n-1}}{(n-1)!} e^{-\lambda s} e^{-\lambda(t-s)}\, ds = e^{-\lambda t} \frac{(\lambda t)^n}{n!}$$

The last two formulas show that N_t has a Poisson distribution with mean λt. To check that the number of arrivals in disjoint intervals is independent we observe

$$P(T_{n+1} \geq u | N_t = n) = P(T_{n+1} \geq u, T_n \leq t)/P(N_t = n)$$

To compute the numerator we observe

$$P(T_{n+1} \geq u, T_n \leq t) = \int_0^t P(T_n = s) P(\xi_{n+1} \geq u - s)\, ds$$

$$= \int_0^t \frac{\lambda^n s^{n-1}}{(n-1)!} e^{-\lambda s} e^{-\lambda(u-s)} ds = e^{-\lambda u} \frac{(\lambda t)^n}{n!}$$

The denominator is $P(N_t = n) = e^{-\lambda t}(\lambda t)^n/n!$, so

$$P(T_{n+1} \geq u | N_t = n) = e^{-\lambda u}/e^{-\lambda t} = e^{-\lambda(u-t)}$$

or rewriting things $P(T_{n+1} - t \geq s | N_t = n) = e^{-\lambda s}$. Let $T_1' = T_{N(t)+1} - t$, and $T_2' = T_{N(t)+k} - T_{N(t)+k-1}$ for $k \geq 2$. The last computation shows that T_1' is independent of N_t. If we observe that

$$P(T_n \leq t, T_{n+1} \geq u, T_{n+k} - T_{n+k-1} \geq v_k, k = 2, \dots, K)$$

$$= P(T_n \leq t, T_{n+1} \geq u) \prod_{k=2}^{K} P(\xi_{n+k} \geq v_k)$$

then it follows that

(a) $T_1', T_2', \dots$ are i.i.d. and independent of N_t.

The last observation shows that the arrivals after time t are independent of N_t and have the same distribution as the original sequence. From this it follows easily that:

(b) If $0 = t_0 < t_1 \dots < t_n$ then $N(t_i) - N(t_{i-1})$, $i = 1, \dots, n$ are independent.

To see this, observe that the vector $(N(t_2) - N(t_1), \dots, N(t_n) - N(t_{n-1}))$ is $\sigma(T_k', k \geq 1)$ measurable and hence is independent of $N(t_1)$. Then use induction to conclude

$$P(N(t_i) - N(t_{i-1}) = k_i, i = 1, \dots, n) = \prod_{i=1}^{n} \exp(-\lambda(t_i - t_{i-1})) \frac{\lambda(t_i - t_{i-1}))^{k_i}}{k_i!}$$

Remark. The key to the proof of (a) is the lack of memory property of the exponential distribution:

$$(*) \qquad\qquad P(T > t + s | T > t) = P(T > s)$$

which implies that the location of the first arrival after t is independent of what occurred before time t and has an exponential distribution.

EXERCISE 6.5. Show that if $P(T > 0) = 1$ and $(*)$ holds then there is a $\lambda > 0$ so that $P(T > t) = e^{-\lambda t}$ for $t \geq 0$. Hint: First show that this holds for $t = m2^{-n}$.

EXERCISE 6.6. Show that (iii) and (iv) in (6.8) can be replaced by

(v) If $N_{s-} = \lim_{r \uparrow s} N_r$ then $P(N_s - N_{s-} \geq 2$ for some $s) = 0$.

That is, if (i), (ii), and (v) hold then there is a $\lambda \geq 0$ so that $N(0, t)$ has a Poisson distribution with mean λt. Prove this by showing: (a) If $u(s) = P(N_s = 0)$ then (i) and (ii) imply $u(r)u(s) = u(r+s)$. It follows that $u(s) = e^{-\lambda s}$ for some $\lambda \geq 0$, so (iii) holds. (b) if $v(s) = P(N_s \geq 2)$ and $A_n = \{N_{k/n} - N_{(k-1)/n} \geq 2$ for some $k \leq n\}$ then (v) implies $P(A_n) \to 0$ as $n \to \infty$ and (iv) holds.

EXERCISE 6.7. Let T_n be the time of the nth arrival in a rate λ Poisson process. Let $U_1, U_2, \ldots, U_n$ be uniform on (0,1) and let V_k^n be the kth smallest number in $\{U_1, \ldots, U_n\}$. Show that $(V_1^n, \ldots, V_n^n)$ and $(T_1/T_{n+1}, \ldots, T_n/T_{n+1})$ have the same distribution.

Spacings. The last result can be used to study the spacings between the order statistics of i.i.d. uniforms. We use notation of Exercise 6.7 in the next four exercises, taking $\lambda = 1$ and letting $V_0^n = 0$, and $V_{n+1}^n = 1$.

EXERCISE 6.8. Smirnov (1949) $nV_k^n \Rightarrow T_k$.

EXERCISE 6.9. Weiss (1955) $n^{-1} \sum_{m=1}^n 1_{(n(V_i^n - V_{i-1}^n) > x)} \to e^{-x}$ in probability.

EXERCISE 6.10. $(n/\log n) \max_{1 \leq m \leq n+1} V_m^n - V_{m-1}^n \to 1$ in probability.

EXERCISE 6.11. $P(n^2 \min_{1 \leq m \leq n} V_m^n - V_{m-1}^n > x) \to e^{-x}$.

For the rest of the section we concentrate on the Poisson process itself.

EXERCISE 6.12. **Thinning.** Let N have a Poisson distribution with mean λ and let $X_1, X_2, \ldots$ be an independent i.i.d. sequence with $P(X_i = j) = p_j$ for $j = 0, 1, \ldots, k$. Let $N_j = |\{m \leq N : X_m = j\}|$. Show that $N_0, N_1, \ldots, N_k$ are independent and N_j has a Poisson distribution with mean λp_j.

In the important special case $X_i \in \{0, 1\}$ the result says that if we thin a Poisson process by flipping a coin with probability p of heads to see if we keep the arrival then the result is a Poisson process with rate λp.

EXERCISE 6.13. **Poissonization and the occupancy problem.** If we put a Poisson number of balls with mean r in n boxes and let N_i be the number of balls in box i then the last exercise implies $N_1, \ldots, N_n$ are independent and have a Poisson distribution with mean r/n. Use this observation to prove (6.6).

Hint: if $r = n \log n - (\log \lambda)n + o(n)$ and $s_i = n \log n - (\log \mu_i)n$ with $\mu_2 < \lambda < \mu_1$ then the normal approximation to the Poisson tells us $P(\text{Poisson}(s_1) < r < \text{Poisson}(s_2)) \to 1$ as $n \to \infty$.

Compound Poisson process. At the arrival times $T_1, T_2, \ldots$ of a Poisson process with rate λ, groups of customers of size $\xi_1, \xi_2, \ldots$ arrive at an ice cream parlor. Suppose the ξ_i are i.i.d. and independent of the $T'_j s$. This is a **compound Poisson process.** The result of Exercise 6.12 shows that $N_t^k =$ the number of groups of size k to arrive in $[0, t]$ are independent Poisson's with mean $p_k \lambda t$.

A Poisson process on a measure space $(S, \mathcal{S}, \mu)$ is a random map $m : \mathcal{S} \to \{0, 1, \ldots\}$ with the following property: if $A_1, \ldots, A_n$ are disjoint sets with $\mu(A_i) < \infty$ then $m(A_1), \ldots, m(A_n)$ are independent and have Poisson distributions with means $\mu(A_i)$. μ is called the **mean measure** of the process. Exercise 6.12 implies that if $\mu(S) < \infty$ we can construct m by the following recipe: let $X_1, X_2, \ldots$ be i.i.d. elements of S with distribution $\nu(\cdot) = \mu(\cdot)/\mu(S)$, let N be an independent Poisson random variable with mean $\mu(S)$, and let $m(A) = |A \cap \{X_1, \ldots, X_N\}|$. To extend the construction to infinite measure spaces, e.g., $S = \mathbf{R}^d$, $\mathcal{S} =$ Borel sets, $\mu =$ Lebesgue measure, divide the space up into disjoint sets of finite measure and put independent Poisson processes on each set.

*2.7. Stable Laws

Let $X_1, X_2, \ldots$ be i.i.d. and $S_n = X_1 + \cdots + X_n$. In Section 2.4 we saw that if $EX_i = \mu$ and $\mathrm{var}(X_i) = \sigma^2 \in (0, \infty)$ then

$$(S_n - n\mu)/\sigma n^{1/2} \Rightarrow \chi$$

In this section we will investigate the case $EX_1^2 = \infty$ and give necessary and sufficient conditions for the existence of constants a_n and b_n so that

$$(S_n - b_n)/a_n \Rightarrow Y \quad \text{where } Y \text{ is nondegenerate}$$

We begin with an example. Suppose the distribution of X_i has

(7.1a) $\qquad\qquad P(X_1 > x) = P(X_1 < -x)$

(7.1b) $\qquad\qquad P(|X_1| > x) = x^{-\alpha} \quad \text{for } x \geq 1$

where $0 < \alpha < 2$. If $\varphi(t) = E \exp(itX_1)$ then

$$1 - \varphi(t) = \int_1^\infty (1 - e^{itx}) \frac{\alpha}{2|x|^{\alpha+1}} \, dx + \int_{-\infty}^{-1} (1 - e^{itx}) \frac{\alpha}{2|x|^{\alpha+1}} \, dx$$

$$= \alpha \int_1^\infty \frac{1 - \cos(tx)}{x^{\alpha+1}} \, dx$$

Changing variables $tx = u$, $dx = du/t$ the last integral becomes

$$= \alpha \int_t^\infty \frac{1 - \cos u}{(u/t)^{\alpha+1}} \frac{du}{t} = t^\alpha \alpha \int_t^\infty \frac{1 - \cos u}{u^{\alpha+1}} du$$

As $u \to 0$, $1 - \cos u \sim u^2/2$. So $(1 - \cos u)/u^{\alpha+1} \sim u^{-\alpha+1}/2$ which is integrable, since $\alpha < 2$ implies $-\alpha + 1 > -1$. If we let

$$C = \alpha \int_0^\infty \frac{1 - \cos u}{u^{\alpha+1}} du < \infty$$

and observe (7.1a) implies $\varphi(t) = \varphi(-t)$, then the results above show

(7.2) $$1 - \varphi(t) \sim C|t|^\alpha \text{ as } t \to 0$$

Let $X_1, X_2, \ldots$ be i.i.d. with the distribution given in (7.1) and let $S_n = X_1 + \cdots + X_n$.

$$E \exp(itS_n/n^{1/\alpha}) = \varphi(t/n^{1/\alpha})^n = (1 - \{1 - \varphi(t/n^{1/\alpha})\})^n$$

As $n \to \infty$, $n(1 - \varphi(t/n^{1/\alpha})) \to C|t|^\alpha$, so it follows from (4.2) that

$$E \exp(itS_n/n^{1/\alpha}) \to \exp(-C|t|^\alpha)$$

From part (ii) of (3.4) it follows that the expression on the right is the characteristic function of some Y and

(7.3) $$S_n/n^{1/\alpha} \Rightarrow Y$$

To prepare for our general result, we will now give another proof of (7.3). If $0 < a < b$ and $an^{1/\alpha} > 1$ then

$$P(an^{1/\alpha} < X_1 < bn^{1/\alpha}) = \frac{1}{2}(a^{-\alpha} - b^{-\alpha})n^{-1}$$

so it follows from (6.1) that

$$N_n(a, b) \equiv |\{m \le n : X_m/n^{1/\alpha} \in (a, b)\}| \Rightarrow N(a, b)$$

where $N(a, b)$ has a Poisson distribution with mean $(a^{-\alpha} - b^{-\alpha})/2$. An easy extension of the last result shows that if $A \subset \mathbf{R} - (-\delta, \delta)$ and $\delta n^{1/\alpha} > 1$ then

$$P(X_1/n^{1/\alpha} \in A) = n^{-1} \int_A \frac{\alpha}{2|x|^{\alpha+1}} dx$$

so $N_n(A) \equiv |\{m \leq n : X_m/n^{1/\alpha} \in A\}| \Rightarrow N(A)$, where $N(A)$ has a Poisson distribution with mean

$$\mu(A) = \int_A \frac{\alpha}{2|x|^{\alpha+1}} \, dx < \infty$$

The limiting family of random variables $N(A)$ is called a **Poisson process on** $(-\infty, \infty)$ **with mean measure** μ. (See the end of Section 2.6 for more on this process.) Notice that for any $\epsilon > 0$, $\mu(\epsilon, \infty) = \epsilon^{-\alpha}/2 < \infty$, so $N(\epsilon, \infty) < \infty$.

The last paragraph describes the limiting behavior of the random set

$$\mathcal{X}_n = \{X_m/n^{1/\alpha} : 1 \leq m \leq n\}$$

To describe the limit of $S_n/n^{1/\alpha}$, we will "sum up the points". Let $\epsilon > 0$ and

$$I_n(\epsilon) = \{m \leq n : |X_m| > \epsilon n^{1/\alpha}\}$$
$$\hat{S}_n(\epsilon) = \sum_{m \in I_n(\epsilon)} X_m \qquad \bar{S}_n(\epsilon) = S_n - \hat{S}_n(\epsilon)$$

$I_n(\epsilon) =$ the indices of the "big terms," i.e., those $> \epsilon n^{1/\alpha}$ in magnitude. $\hat{S}_n(\epsilon)$ is the sum of the big terms, and $\bar{S}_n(\epsilon)$ is the rest of the sum. The first thing we will do is show that the contribution of $\bar{S}_n(\epsilon)$ is small if ϵ is. Let

$$\bar{X}_m(\epsilon) = X_m 1_{(|X_m| \leq \epsilon n^{1/\alpha})}$$

Symmetry implies $E\bar{X}_m(\epsilon) = 0$, so $E(\bar{S}_n(\epsilon)^2) = nE\bar{X}_1(\epsilon)^2$.

$$E\bar{X}_1(\epsilon)^2 = \int_0^\infty 2y P(\bar{X}_1(\epsilon) > y) \, dy \leq \int_0^1 2y \, dy + \int_1^{\epsilon n^{1/\alpha}} 2y \, y^{-\alpha} \, dy$$
$$= 1 + \frac{2}{2-\alpha} \epsilon^{2-\alpha} n^{2/\alpha-1} - \frac{2}{2-\alpha} \leq \frac{2\epsilon^{2-\alpha}}{2-\alpha} n^{2/\alpha-1}$$

where we have used $\alpha < 2$ in computing the integral and $\alpha > 0$ in the final inequality. From this it follows that

(7.4) $$E(\bar{S}_n(\epsilon)/n^{1/\alpha})^2 \leq \frac{2\epsilon^{2-\alpha}}{2-\alpha}$$

To compute the limit of $\hat{S}_n(\epsilon)/n^{1/\alpha}$, we observe that $|I_n(\epsilon)|$ has a binomial distribution with success probability $p = \epsilon^{-\alpha}/n$. Given $|I_n(\epsilon)| = m$, $\hat{S}_n(\epsilon)/n^{1/\alpha}$ is the sum of m independent random variables with a distribution F_n^ϵ that is symmetric about 0 and has

$$1 - F_n^\epsilon(x) = P(X_1/n^{1/\alpha} > x \,|\, |X_1|/n^{1/\alpha} > \epsilon) = x^{-\alpha}/2\epsilon^{-\alpha} \quad \text{for } x \geq \epsilon$$

The last distribution is the same as that of ϵX_1, so if $\varphi(t) = E \exp(itX_1)$, the distribution F_n^ϵ has characteristic function $\varphi(\epsilon t)$. Combining the observations in this paragraph gives

$$E \exp(it\hat{S}_n(\epsilon)/n^{1/\alpha}) = \sum_{m=0}^{n} \binom{n}{m} (\epsilon^{-\alpha}/n)^m (1 - \epsilon^{-\alpha}/n)^{n-m} \varphi(\epsilon t)^m$$

Writing

$$\binom{n}{m} \frac{1}{n^m} = \frac{1}{m!} \frac{n(n-1)\cdots(n-m+1)}{n^m} \le \frac{1}{m!}$$

noting $(1-\epsilon^{-\alpha}/n)^n \le \exp(-\epsilon^{-\alpha})$ and using the dominated convergence theorem

(7.5)
$$E \exp(it\hat{S}_n(\epsilon)/n^{1/\alpha}) \to \sum_{m=0}^{\infty} \exp(-\epsilon^{-\alpha})(\epsilon^{-\alpha})^m \varphi(\epsilon t)^m/m!$$
$$= \exp(-\epsilon^{-\alpha}\{1 - \varphi(\epsilon t)\})$$

To get (7.3) now, we use the following generalization of (4.7).

(7.6) Lemma. If $h_n(\epsilon) \to g(\epsilon)$ for each $\epsilon > 0$ and $g(\epsilon) \to g(0)$ as $\epsilon \to 0$ then we can pick $\epsilon_n \to 0$ so that $h_n(\epsilon_n) \to g(0)$.

Proof Let N_m be chosen so that $|h_n(1/m) - g(1/m)| \le 1/m$ for $n \ge N_m$ and $m \to N_m$ is increasing. Let $\epsilon_n = 1/m$ for $N_m \le n < N_{m+1}$ and $= 1$ for $n < N_1$. When $N_m \le n < N_{m+1}$, $\epsilon_n = 1/m$ so it follows from the triangle inequality and the definition of ϵ_n that

$$|h_n(\epsilon_n) - g(0)| \le |h_n(1/m) - g(1/m)| + |g(1/m) - g(0)|$$
$$\le 1/m + |g(1/m) - g(0)|$$

When $n \to \infty$, we have $m \to \infty$ and the result follows. $\square$

Let $h_n(\epsilon) = E \exp(it\hat{S}_n(\epsilon)/n^{1/\alpha})$ and $g(\epsilon) = \exp(-\epsilon^{-\alpha}\{1 - \varphi(\epsilon t)\})$. (7.2) implies $1 - \varphi(t) \sim C|t|^\alpha$ as $t \to 0$ so

$$g(\epsilon) \to \exp(-C|t|^\alpha) \quad \text{as } \epsilon \to 0$$

and (7.6) implies we can pick $\epsilon_n \to 0$ with $h_n(\epsilon_n) \to \exp(-C|t|^\alpha)$. Introducing Y with $E \exp(itY) = \exp(-C|t|^\alpha)$, it follows that $\hat{S}_n(\epsilon_n)/n^{1/\alpha} \Rightarrow Y$. If $\epsilon_n \to 0$ then (7.4) implies

$$\bar{S}_n(\epsilon_n)/n^{1/\alpha} \Rightarrow 0$$

and (7.3) follows from the converging together lemma, Exercise 2.10. $\square$

Once we give one final definition, we will state and prove the general result alluded to above. L is said to be **slowly varying**, if

$$\lim_{x \to \infty} L(tx)/L(x) = 1 \quad \text{for all } t > 0$$

EXERCISE 7.1. Show that $L(t) = \log t$ is slowly varying but t^ϵ is not if $\epsilon \neq 0$.

(7.7) Theorem. Suppose $X_1, X_2, \ldots$ are i.i.d. with a distribution that satisfies

(i) $\lim_{x \to \infty} P(X_1 > x)/P(|X_1| > x) = \theta \in [0, 1]$

(ii) $P(|X_1| > x) = x^{-\alpha} L(x)$

where $\alpha < 2$ and L is slowly varying. Let $S_n = X_1 + \cdots + X_n$

$$a_n = \inf\{x : P(|X_1| > x) \leq n^{-1}\} \quad \text{and} \quad b_n = nE(X_1 1_{(|X_1| \leq a_n)})$$

As $n \to \infty$, $(S_n - b_n)/a_n \Rightarrow Y$ where Y has a nondegenerate distribution.

Remark. This is not much of a generalization of the example but the conditions are necessary for the existence of constants a_n and b_n so that $(S_n - b_n)/a_n \Rightarrow Y$, where Y is nondegenerate. Proofs of necessity can be found in Chapter 9 of Breiman (1968) or in Gnedenko and Kolmogorov (1954). (7.13) gives the ch.f. of Y. The reader can skip to that point without much loss.

Proof It is not hard to see that (ii) implies

(7.8)
$$nP(|X_1| > a_n) \to 1$$

To prove this note that $nP(|X_1| > a_n) \leq 1$ and let $\epsilon > 0$. Taking $x = a_n/(1+\epsilon)$ and $t = 1 + 2\epsilon$, (ii) implies

$$(1 + 2\epsilon)^{-\alpha} = \lim_{n \to \infty} \frac{P(|X_1| > (1 + 2\epsilon)a_n/(1 + \epsilon))}{P(|X_1| > a_n/(1 + \epsilon))} \leq \liminf_{n \to \infty} \frac{P(|X_1| > a_n)}{1/n}$$

proving (7.8) since ϵ is arbitrary. Combining (7.8) with (i) and (ii) gives

(7.9)
$$nP(X_1 > xa_n) \to \theta x^{-\alpha} \quad \text{for } x > 0$$

so $|\{m \leq n : X_m > xa_n\}| \Rightarrow \text{Poisson}(\theta x^{-\alpha})$. The last result leads, as before, to the conclusion that $\mathcal{X}_n = \{X_m/a_n : 1 \leq m \leq n\}$ converges to a Poisson process on $(-\infty, \infty)$ with mean measure

$$\mu(A) = \int_{A \cap (0,\infty)} \theta \alpha |x|^{-(\alpha+1)} \, dx + \int_{A \cap (-\infty,0)} (1 - \theta)\alpha |x|^{-(\alpha+1)} \, dx$$

To sum up the points, let $I_n(\epsilon) = \{m \leq n : |X_m| > \epsilon n^{1/\alpha}\}$

$$\hat{\mu}(\epsilon) = EX_m 1_{(\epsilon a_n < |X_m| \leq a_n)} \quad \hat{S}_n(\epsilon) = \sum_{m \in I_n(\epsilon)} X_m$$

$$\bar{\mu}(\epsilon) = EX_m 1_{(|X_m| \leq \epsilon a_n)}$$

$$\bar{S}_n(\epsilon) = (S_n - b_n) - (\hat{S}_n(\epsilon) - n\hat{\mu}(\epsilon)) = \sum_{m=1}^{n} \{X_m 1_{(|X_m| \leq \epsilon a_n)} - \bar{\mu}(\epsilon)\}$$

If we let $\bar{X}_m(\epsilon) = X_m 1_{(|X_m| \leq \epsilon a_n)}$ then

$$E(\bar{S}_n(\epsilon)/a_n)^2 = n \operatorname{var}(\bar{X}_1(\epsilon)/a_n) \leq n E(\bar{X}_1(\epsilon)/a_n)^2$$

$$E(\bar{X}_1(\epsilon)/a_n)^2 \leq \int_0^\epsilon 2y P(|X_1| > y a_n)\, dy$$

$$= P(|X_1| > a_n) \int_0^\epsilon 2y \frac{P(|X_1| > y a_n)}{P(|X_1| > a_n)}\, dy$$

We would like to use (7.8), and (ii) to conclude

$$n E(\bar{X}_1(\epsilon)/a_n)^2 \to \int_0^\epsilon 2y\, y^{-\alpha}\, dy = \frac{2}{2-\alpha} \epsilon^{2-\alpha}$$

and hence

(7.10) $$\limsup_{n \to \infty} E(\bar{S}_n(\epsilon)/a_n)^2 \leq \frac{2\epsilon^{2-\alpha}}{2-\alpha}$$

To justify interchanging the limit and the integral and complete the proof of (7.10), we show the following (take $\delta < 2 - \alpha$):

Lemma. For any $\delta > 0$ there is C so that for all $t \geq t_0$ and $y \leq 1$

$$P(|X_1| > yt)/P(|X_1| > t) \leq Cy^{-\alpha-\delta}$$

Proof   (ii) implies that as $t \to \infty$

$$P(|X_1| > t/2)/P(|X_1| > t) \to 2^\alpha$$

so for $t \geq t_0$ we have

$$P(|X_1| > t/2)/P(|X_1| > t) \leq 2^{\alpha+\delta}$$

Iterating and stopping the first time $t/2^m < t_0$ we have for all $n \geq 1$

$$P(|X_1| > t/2^n)/P(|X_1| > t) \leq C2^{(\alpha+\delta)n}$$

where $C = 1/P(|X_1| > t_0)$. Applying the last result to the first n with $1/2^n < y$ and noticing $y \leq 1/2^{n-1}$ we have

$$P(|X_1| > yt)/P(|X_1| > t) \leq C2^{\alpha+\delta}y^{-\alpha-\delta}$$

$\square$

To compute the limit of $\hat{S}_n(\epsilon)$, we observe that $|I_n(\epsilon)| \Rightarrow \text{Poisson}(\epsilon^{-\alpha})$. Given $|I_n(\epsilon)| = m$, $\hat{S}_n(\epsilon)/n^{1/\alpha}$ is the sum of m independent random variables with distribution F_n^ϵ that has

$$1 - F_n^\epsilon(x) = P(X_1/a_n > x \,|\, |X_1|/a_n > \epsilon) \to \theta x^{-\alpha}/\epsilon^{-\alpha}$$
$$F_n^\epsilon(-x) = P(X_1/a_n < -x \,|\, |X_1|/a_n > \epsilon) \to (1-\theta)|x|^{-\alpha}/\epsilon^{-\alpha}$$

for $x \geq \epsilon$. If we let $\psi_n^\epsilon(t)$ denote the characteristic function of F_n^ϵ, then (3.4) implies

$$\psi_n^\epsilon(t) \to \psi^\epsilon(t) = \int_\epsilon^\infty e^{itx}\theta\epsilon^\alpha\alpha x^{-(\alpha+1)}dx + \int_{-\infty}^{-\epsilon} e^{itx}(1-\theta)\epsilon^\alpha\alpha|x|^{-(\alpha+1)}\,dx$$

as $n \to \infty$. So repeating the proof of (7.5) gives

$$E\exp(it\hat{S}_n(\epsilon)/a_n) \to \exp(-\epsilon^{-\alpha}\{1 - \psi^\epsilon(t)\})$$
$$= \exp\left(\int_\epsilon^\infty (e^{itx} - 1)\theta\alpha x^{-(\alpha+1)}\,dx \right.$$
$$\left. + \int_{-\infty}^{-\epsilon} (e^{itx} - 1)(1-\theta)\alpha|x|^{-(\alpha+1)}dx\right)$$

where we have used $\epsilon^{-\alpha} = \int_\epsilon^\infty \alpha x^{-(\alpha+1)}\,dx$. To bring in

$$\hat{\mu}(\epsilon) = EX_m 1_{(\epsilon a_n < |X_m| \leq a_n)}$$

we observe that (7.9) implies $nP(xa_n < X_m \leq ya_n) \to \theta(x^{-\alpha} - y^{-\alpha})$. So

$$n\hat{\mu}(\epsilon)/a_n \to \int_\epsilon^1 x\theta\alpha x^{-(\alpha+1)}\,dx + \int_{-1}^{-\epsilon} x(1-\theta)\alpha|x|^{-(\alpha+1)}\,dx$$

From this it follows that $E \exp(it\{\hat{S}_n(\epsilon) - n\hat{\mu}(\epsilon)\}/a_n) \to$

$$\exp\left(\int_1^\infty (e^{itx} - 1)\theta\alpha x^{-(\alpha+1)}\,dx\right.$$

$$+ \int_\epsilon^1 (e^{itx} - 1 - itx)\theta\alpha x^{-(\alpha+1)}\,dx$$

(7.11)

$$+ \int_{-1}^{-\epsilon} (e^{itx} - 1 - itx)(1-\theta)\alpha|x|^{-(\alpha+1)}\,dx$$

$$\left.+ \int_{-\infty}^{-1} (e^{itx} - 1)(1-\theta)\alpha|x|^{-(\alpha+1)}\,dx\right)$$

The last expression is messy but $e^{itx} - 1 - itx \sim -t^2 x^2/2$ as $t \to 0$, so we need to subtract the itx to make

$$\int_0^1 (e^{itx} - 1 - itx)x^{-(\alpha+1)}dx \quad \text{converge when } \alpha \geq 1$$

To reduce the number of integrals from four to two, we can write the limit as $\epsilon \to 0$ of the left-hand side of (7.11) as

$$\exp\left(itc + \int_0^\infty \left(e^{itx} - 1 - \frac{itx}{1+x^2}\right)\theta\alpha x^{-(\alpha+1)}\,dx\right.$$

(7.12)

$$\left.+ \int_{-\infty}^0 \left(e^{itx} - 1 - \frac{itx}{1+x^2}\right)(1-\theta)\alpha|x|^{-(\alpha+1)}\,dx\right)$$

where c is a constant. Combining (7.10) and (7.11) using (7.6) it follows easily that $(S_n - b_n)/a_n \Rightarrow Y$ where Ee^{itY} is given in (7.12).   □

EXERCISE 7.2. Show that when $\alpha < 1$, centering is unnecessary, i.e., we can let $b_n = 0$.

By doing some calculus (see Breiman (1968), p. 204-206) one can rewrite (7.12) as

(7.13) $$\exp(itc - b|t|^\alpha\{1 + i\kappa\,\text{sgn}\,(t)w_\alpha(t)\})$$

where $-1 \leq \kappa \leq 1$, $(\kappa = 2\theta - 1)$ and

$$w_\alpha(t) = \begin{cases} \tan(\pi\alpha/2) & \text{if } \alpha \neq 1 \\ (2/\pi)\log|t| & \text{if } \alpha = 1 \end{cases}$$

The reader should note that while we have assumed $0 < \alpha < 2$ throughout the developments above, if we set $\alpha = 2$ then the term with κ vanishes and (7.13) reduces to the characteristic function of the normal distribution with mean c and variance $2b$.

The distributions whose characteristic functions are given in (7.13) are called **stable laws**. α is commonly called the **index**. When $\alpha = 1$ and $\kappa = 0$, we have the Cauchy distribution. Apart from the Cauchy and the normal, there is only one other case in which the density is known: when $\alpha = 1/2$, $\kappa = 1$ the density is

(7.14)
$$(2\pi y^3)^{-1/2} \exp(-1/2y) \quad \text{for } y \geq 0$$

One can calculate the ch.f. and verify our claim. However, later (see Section 7.4) we will be able to check the claim without effort, so we leave the somewhat tedious calculation to the reader.

We are now finally ready to treat some examples

Example 7.1. Let $X_1, X_2, \ldots$ be i.i.d. with a density that is symmetric about 0, and continuous and positive at 0. We claim that

$$\frac{1}{n}\left(\frac{1}{X_1} + \cdots + \frac{1}{X_n}\right) \Rightarrow \text{a Cauchy distribution } (\alpha = 1, \kappa = 0)$$

To verify this note that

$$P(1/X_i > x) = P(0 < X_i < x^{-1})$$
$$= \int_0^{x^{-1}} f(y)\, dy \sim f(0)/x$$

as $x \to \infty$. A similar calculation shows $P(1/X_i < -x) \sim f(0)/x$ so in (7.7) (i) holds with $\theta = 1/2$ and (ii) holds with $\alpha = 1$. The scaling constant $a_n \sim 2f(0)n$ while the centering constant vanishes since we have supposed the distribution of X is symmetric about 0.

Remark. Readers who want a challenge should try to drop the symmetry assumption, assuming for simplicity that f is differentiable at 0.

Example 7.2. Let $X_1, X_2, \ldots$ be i.i.d. with $P(X_i = 1) = P(X_i = -1) = 1/2$, let $S_n = X_1 + \cdots + X_n$, and let $\tau = \inf\{n \geq 1 : S_n = 1\}$. In Chapter 3 (see the discussion after (3.4)) we will show

$$P(\tau > 2n) \sim \pi^{-1/2} n^{-1/2} \quad \text{as } n \to \infty$$

Let $\tau_1, \tau_2, \ldots$ be independent with the same distribution as τ, and let $T_n = \tau_1 + \cdots + \tau_n$. Results in Section 3.1 imply that T_n has the same distribution as the nth time S_m hits 0. We claim that T_n/n^2 converges to the stable law with $\alpha = 1/2$, $\kappa = 1$ and note that this is the key to the derivation of (7.14). To prove the claim note that in (7.7) (i) holds with $\theta = 1$ and (ii) holds with $\alpha = 1/2$. The scaling constant $a_n \sim Cn^2$. Since $\alpha < 1$, Exercise 7.2 implies the centering constant is unnecessary.

Example 7.3. Assume n objects $X_{n,1}, \ldots, X_{n,n}$ are placed independently and at random in $[-n, n]$. Let

$$F_n = \sum_{m=1}^{n} \operatorname{sgn}(X_{n,m})/|X_{n,m}|^p$$

be the net force exerted on 0. We will now show that if $p > 1/2$ then

$$\lim_{n \to \infty} E \exp(it F_n) = \exp(-c|t|^{1/p})$$

To do this it is convenient to let $X_{n,m} = nY_m$ where the Y_i are i.i.d. on $[-1, 1]$. Then

$$F_n = n^{-p} \sum_{m=1}^{n} \operatorname{sgn}(Y_m)/|Y_m|^p$$

Letting $Z_m = \operatorname{sgn}(Y_m)/|Y_m|^p$, Z_m is symmetric about 0 with $P(|Z_m| > x) = P(|Y_m| < x^{-1/p})$ so in (7.7) (i) holds with with $\theta = 1/2$ and (ii) holds with $\alpha = 1/p$. The scaling constant $a_n \sim Cn^p$ and the centering constant is 0 by symmetry.

EXERCISE 7.3. Show that (i) If $p < 1/2$ then $F_n/n^{1/2-p} \Rightarrow c\chi$.
(ii) If $p = 1/2$ then $F_n/(\log n)^{1/2} \Rightarrow c\chi$.

Example 7.4. In the examples above we have had $b_n = 0$. To get a feel for the centering constants consider $X_1, X_2, \ldots$ i.i.d. with

$$P(X_i > x) = \theta x^{-\alpha} \qquad P(X_i < -x) = (1-\theta)x^{-\alpha}$$

where $0 < \alpha < 2$. In this case $a_n = n^{1/\alpha}$ and

$$b_n = n \int_1^{n^{1/\alpha}} (2\theta - 1)\alpha x^{-\alpha} \, dx \sim \begin{cases} cn & \alpha > 1 \\ cn \log n & \alpha = 1 \\ cn^{1/\alpha} & \alpha < 1 \end{cases}$$

When $\alpha < 1$ the centering is the same size as the scaling and can be ignored. When $\alpha > 1$, $b_n \sim n\mu$ where $\mu = EX_i$.

Our next result explains the name **stable laws**. A random variable Y is said to have a **stable law** if for every integer $k > 0$ there are constants a_k and b_k so that if $Y_1, \ldots, Y_k$ are i.i.d. and have the same distribution as Y, then $(Y_1 + \ldots + Y_k - b_k)/a_k =_d Y$. The last definition makes half of the next result obvious.

(7.15) Theorem. Y is the limit of $(X_1 + \cdots + X_k - b_k)/a_k$ for some i.i.d. sequence X_i if and only if Y has a stable law.

Proof If Y has a stable law we can take $X_1, X_2, \ldots$ i.i.d. with distribution Y. To go the other way let

$$Z_n = (X_1 + \cdots + X_n - b_n)/a_n$$

and $S_n^j = X_{(j-1)n+1} + \cdots + X_{jn}$. A little arithmetic shows

$$Z_{nk} = (S_n^1 + \cdots + S_n^k - b_{nk})/a_{nk}$$
$$a_{nk} Z_{nk} = (S_n^1 - b_n) + \cdots + (S_n^k - b_n) + (kb_n - b_{nk})$$
$$a_{nk} Z_{nk}/a_n = (S_n^1 - b_n)/a_n + \cdots + (S_n^k - b_n)/a_n + (kb_n - b_{nk})/a_n$$

The first k terms on the right-hand side $\Rightarrow Y_1 + \cdots + Y_k$ as $n \to \infty$ where $Y_1, \ldots, Y_k$ are independent and have the same distribution as Y, and $Z_{nk} \Rightarrow Y$. So the desired result follows from the following. Take $W_n = Z_{nk}$ and

$$W_n' = \frac{a_{kn}}{a_n} Z_{nk} - \frac{kb_n - b_{nk}}{a_n}$$

(7.16) The convergence of types theorem. If $W_n \Rightarrow W$ and there are constants $\alpha_n > 0$, β_n so that $W_n' = \alpha_n W_n + \beta_n \Rightarrow W'$ where W and W' are nondegenerate, then there are constants α and β so that $\alpha_n \to \alpha$ and $\beta_n \to \beta$.

Proof Let $\varphi_n(t) = E \exp(itW_n)$.

$$\psi_n(t) = E \exp(it(\alpha_n W_n + \beta_n)) = \exp(it\beta_n)\varphi_n(\alpha_n t)$$

If φ and ψ are the characteristic functions of W and W' then

(a) $\qquad \varphi_n(t) \to \varphi(t) \qquad \psi_n(t) = \exp(it\beta_n)\varphi_n(\alpha_n t) \to \psi(t)$

Take a subsequence $\alpha_{n(m)}$ that converges to a limit $\alpha \in [0, \infty]$. Our first step is to observe $\alpha = 0$ is impossible. If this happens then using the uniform convergence proved in Exercise 3.16

(b) $\qquad |\psi_n(t)| = |\varphi_n(\alpha_n t)| \to 1$

$|\psi(t)| \equiv 1$, and the limit is degenerate by (5.1). Letting $t = u/\alpha_n$ and interchanging the roles of φ and ψ shows $\alpha = \infty$ is impossible. If α is a subsequential limit, then arguing as in (b) gives $|\psi(t)| = |\varphi(\alpha t)|$. If there are two subsequential limits $\alpha' < \alpha$, using the last equation for both limits implies $|\varphi(u)| = |\varphi(u\alpha'/\alpha)|$. Iterating gives $|\varphi(u)| = |\varphi(u(\alpha'/\alpha)^k)| \to 1$ as $k \to \infty$, contradicting our assumption that W' is nondegenerate, so $\alpha_n \to \alpha \in [0, \infty)$.

To conclude that $\beta_n \to \beta$ now, we observe that (ii) of Exercise 3.16 implies $\varphi_n \to \varphi$ uniformly on compact sets so $\varphi_n(\alpha_n t) \to \varphi(\alpha t)$. If δ is small enough so that $|\varphi(\alpha t)| > 0$ for $|t| \leq \delta$, it follows from (a) and another use of Exercise 3.16 that

$$\exp(it\beta_n) = \frac{\psi_n(t)}{\varphi_n(\alpha t)} \to \frac{\psi(t)}{\varphi(\alpha t)}$$

uniformly on $[-\delta, \delta]$. $\exp(it\beta_n)$ is the ch.f. of a pointmass at β_n. Using (3.5) now as in the proof of (3.4), it follows that the sequence of distributions that are pointmasses at β_n is tight, i.e., β_n is bounded. If $\beta_{n_m} \to \beta$ then $\exp(it\beta) = \psi(t)/\varphi(\alpha t)$ for $|t| \leq \delta$, so there can only be one subsequential limit. □

(7.15) justifies calling the distributions with characteristic functions given by (7.13) or (7.12) stable laws. To complete the story we should mention that these are the only stable laws. Again, see Chapter 9 of Breiman (1968) or Gnedenko and Kolmogorov (1954). The next example shows that it is sometimes useful to know what all the possible limits are.

Example 7.4. The Holtsmark distribution. ($\alpha = 3/2$, $\kappa = 0$). Suppose stars are distributed in space according to a Poisson process with density t and their masses are i.i.d. Let X_t be the x-component of the gravitational force at 0 when the density is t. A change of density $1 \to t$ corresponds to a change of length $1 \to t^{-1/3}$, and gravitational attraction follows an inverse square law so

(7.17) $$X_t \overset{d}{=} t^{3/2} X_1$$

If we imagine thinning the Poisson process by rolling an n sided die then Exercise 6.12 implies

$$X_t \overset{d}{=} X^1_{t/n} + \cdots + X^n_{t/n}$$

where the random variables on the right-hand side are independent and have the same distribution as $X_{t/n}$. It follows from (7.15) that X_t has a stable law. The scaling propety (7.17) implies $\alpha = 3/2$. Since $X_t =_d -X_t$, $\kappa = 0$.

EXERCISES

EXERCISE 7.4. Let Y be a stable law with $\kappa = 1$. Use the limit theorem (7.7) to conclude that $Y \geq 0$ if $\alpha < 1$.

EXERCISE 7.5. Let X be symmetric stable with index α. (i) Use (3.5) to show that $E|X|^p < \infty$ for $p < \alpha$. (ii) Use the second proof of (7.3) to show that $P(|X| \geq x) \geq Cx^{-\alpha}$ so $E|X|^\alpha = \infty$.

EXERCISE 7.6. Let $Y, Y_1, Y_2, \ldots$ be independent and have a stable law with index α. (7.15) implies there are constants α_k and β_k so that $Y_1 + \cdots + Y_k$ and $\alpha_k Y + \beta_k$ have the same distribution. Use the proof of (7.15), (7.7) and Exercise 7.2 to conclude that (i) $\alpha_k = k^{1/\alpha}$, (ii) if $\alpha < 1$ then $\beta_k = 0$.

EXERCISE 7.7. Let Y be a stable law with index $\alpha < 1$ and $\kappa = 1$. Exercise 7.4 implies that $Y \geq 0$ so we can define its Laplace transform $\psi(\lambda) = E\exp(-\lambda Y)$. The previous exercise implies that for any integer $n \geq 1$ we have $\psi(\lambda)^n = \psi(n^{1/\alpha}\lambda)$. Use this to conclude $E\exp(-\lambda Y) = \exp(-c\lambda^\alpha)$.

EXERCISE 7.8. (i) Show that if X is symmetric stable with index α and $Y \geq 0$ is an independent stable with index $\beta < 1$ then $XY^{1/\alpha}$ is symmetric stable with index $\alpha\beta$. (ii) Let W_1 and W_2 be independent standard normals. Check that $1/W_2^2$ has the density given in (7.14) and use this to conclude that W_1/W_2 has a Cauchy distribution.

*2.8. Infinitely Divisible Distributions

In the last section we identified the distributions that can appear as the limit of normalized sums of i.i.d.r.v.'s. In this section we will describe those that are limits of sums

$$(*) \qquad\qquad S_n = X_{n,1} + \cdots + X_{n,n}$$

where the $X_{n,m}$ are i.i.d. Note the verb "describe." We will prove almost nothing in this section, just state some of the most important facts to bring the reader up to cocktail party literacy.

A sufficient condition for Z to be a limit of sums of the form $(*)$ is that Z has an **infinitely divisible distribution**, i.e., for each n there is an i.i.d. sequence $Y_{n,1}, \ldots, Y_{n,n}$ so that

$$Z \overset{d}{=} Y_{n,1} + \cdots + Y_{n,n}$$

Our first result shows that this condition is also necessary.

(8.1) Theorem. Z is a limit of sums of type $(*)$ if and only if Z has an infinitely divisible distribution.

Proof   As remarked above we only have to prove necessity. Write

$$S_{2n} = (X_{2n,1} + \cdots + X_{2n,n}) + (X_{2n,n+1} + \cdots + X_{2n,2n}) \equiv Y_n + Y_n'$$

The random variables Y_n and Y_n' are independent and have the same distribution. If $S_n \Rightarrow Z$ then the distributions of Y_n are a tight sequence since

$$P(Y_n > y)^2 = P(Y_n > y)P(Y_n' > y) \le P(S_{2n} > 2y)$$

and similarly $P(Y_n < -y)^2 \le P(S_{2n} < -2y)$. If we take a subsequence n_k so that $Y_{n_k} \Rightarrow Y$ (and hence $Y_{n_k}' \Rightarrow Y'$) then $Z =_d Y + Y'$. A similar argument shows that Z can be divided into $n > 2$ pieces and the proof is complete.   □

With (8.1) established, we turn now to examples. In the first three cases the distribution is infinitely divisible because it is a limit of sums of the form $(*)$. The number gives the relevant limit theorem.

Example 8.1. Normal distribution. (4.1)

Example 8.2. Stable Laws. (7.7)

Example 8.3. Poisson distribution. (6.1)

Example 8.4. Compound Poisson distribution. Let $\xi_1, \xi_2, \ldots$ be i.i.d. and $N(\lambda)$ be an independent Poisson r.v. with mean λ. Then $Z = \xi_1 + \cdots + \xi_{N(\lambda)}$ has an infinitely divisible distribution. (Let $X_{n,j} =_d \xi_1 + \cdots + \xi_{N(\lambda/n)}$.) For developments below we would like to observe that if $\varphi(t) = E \exp(it\xi_i)$ then

$$(8.2) \qquad E \exp(itZ) = \sum_{n=0}^{\infty} e^{-\lambda} \frac{\lambda^n}{n!} \varphi(t)^n = \exp(-\lambda(1 - \varphi(t)))$$

Exercise 8.1. Show that the gamma distribution is infinitely divisible.

The next two exercises give examples of distributions that are not infinitely divisible.

Exercise 8.2. Show that the distribution of a bounded r.v. Z is infinitely divisible if and only if Z is constant. Hint: Show $\mathrm{var}(Z) = 0$.

EXERCISE 8.3. Show that if μ is infinitely divisible, its ch.f. φ never vanishes. Hint: Look at $\psi = |\varphi|^2$, which is also infinitely divisible, to avoid taking nth roots of complex numbers then use Exercise 3.20.

Example 8.4 is a son of 8.3 but a father of 8.1 and 8.2. To explain this remark we observe that if $\xi = \epsilon$ and $-\epsilon$ with probability $1/2$ each then $\varphi(t) = (e^{i\epsilon t} + e^{-i\epsilon t})/2 = \cos(\epsilon t)$. So if $\lambda = \epsilon^{-2}$ then (8.2) implies

$$E \ \exp(itZ) = \exp(-\epsilon^2(1 - \cos(\epsilon t))) \to \exp(-t^2/2)$$

as $\epsilon \to 0$. In words, the normal distribution is a limit of compound Poisson distributions. To see that Example 8.2 is also a special case (using the notation from the proof of (7.7)), let

$$I_n(\epsilon) = \{m \le n : |X_m| > \epsilon a_n\}$$
$$\hat{S}_n(\epsilon) = \sum_{m \in I_n(\epsilon)} X_m$$
$$\bar{S}_n(\epsilon) = S_n - \hat{S}_n(\epsilon)$$

If $\epsilon_n \to 0$ then $\bar{S}_n(\epsilon_n)/a_n \Rightarrow 0$. If ϵ is fixed then as $n \to \infty$ we have $|I_n(\epsilon)| \Rightarrow$ Poisson($\epsilon^{-\alpha}$) and $\hat{S}_n(\epsilon)/a_n \Rightarrow$ a compound Poisson distribution:

$$E \exp(it\hat{S}_n(\epsilon)/a_n) \to \exp(-\epsilon^{-\alpha}\{1 - \psi^\epsilon(t)\})$$

Combining the last two observations and using the proof of (7.7) shows that stable laws are limits of compound Poisson distributions. The formula (7.12) for the limiting ch.f.

(8.3)
$$\exp\left(itc + \int_0^\infty \left(e^{itx} - 1 - \frac{itx}{1 + x^2}\right) \theta \alpha x^{-(\alpha+1)} \, dx \right.$$
$$\left. + \int_{-\infty}^0 \left(e^{itx} - 1 - \frac{itx}{1 + x^2}\right)(1 - \theta)\alpha|x|^{-(\alpha+1)} \, dx\right)$$

helps explain:

(8.4) **Lévy-Khintchine Theorem.** Z has an infinitely divisible distribution if and only if its characteristic function has

$$\log \varphi(t) = ict - \frac{\sigma^2 t^2}{2} + \int \left(e^{itx} - 1 - \frac{itx}{1 + x^2}\right) \mu(dx)$$

where μ is a measure with $\mu(\{0\}) = 0$ and $\int \frac{x^2}{1+x^2}\mu(dx) < \infty$.

For a proof see Breiman, (1968) Section 9.5., or Feller II, (1971) Section XVII.2. μ is called the **Lévy measure** of the distribution. Comparing with (8.3) and recalling the proof of (7.7) suggests the following interpretation of μ: if $\sigma^2 = 0$ then Z can be built up by making a Poisson process on **R** with mean measure μ and then summing up the points. As in the case of stable laws we have to sum the points in $[-\epsilon, \epsilon]^c$ subtract an appropriate constant and let $\epsilon \to 0$.

EXERCISE 8.4. What is the Lévy measure for the limit $\aleph$ in part (iii) of Exercise 4.13?

The theory of infinitely divisible distributions is simpler in the case of finite variance. In this case we have

(8.5) **Kolmogorov's Theorem.** Z has an infinitely divisible distribution with mean 0 and finite variance if and only if its ch.f. has

$$\log \varphi(t) = \int (e^{itx} - 1 - itx)x^{-2}\, \nu(dx)$$

Here the integrand is $-t^2/2$ at 0, ν is called the **canonical measure** and $\text{var}(Z) = \nu(\mathbf{R})$.

To explain the formula, note that if Z_λ has a Poisson distribution with mean λ

$$E \exp(itx(Z_\lambda - \lambda)) = \exp(\lambda(e^{itx} - 1 - itx))$$

so the measure for $Z = x(Z_\lambda - \lambda)$ has $\nu(\{x\}) = \lambda x^2$.

*2.9. Limit Theorems in $\mathbf{R}^d$

Let $X = (X_1, \ldots, X_d)$ be a random vector. We define its **distribution function** by $F(x) = P(X \le x)$. Here $x \in \mathbf{R}^d$, and $X \le x$ means $X_i \le x_i$ for $i = 1, \ldots, d$. As in one dimension, F has three obvious properties:

(i) It is nondecreasing, i.e., if $x \le y$ then $F(x) \le F(y)$.

(ii) $\lim_{x \to \infty} F(x) = 1$, $\lim_{x_i \to -\infty} F(x) = 0$.

(iii) F is right continuous, i.e., $\lim_{y \downarrow x} F(y) = F(x)$.

Here $x \to \infty$ means each coordinate x_i goes to ∞, $x_i \to -\infty$ means we let $x_i \to -\infty$ keeping the other coordinates fixed, and $y \downarrow x$ means each coordinate $y_i \downarrow x_i$.

In one dimension any function with properties (i)-(iii) is the distribution of some random variable. See (1.2) in Chapter 1. In $d \geq 2$ this is not the case. Suppose $d = 2$ and let $a_1 < b_1$, $a_2 < b_2$.

$$P(X \in (a_1, b_1] \times (a_2, b_2]) = F(b_1, b_2) - F(a_1, b_2) - F(b_1, a_2) + F(a_1, a_2)$$

so if F is going to be a distribution function the last quantity has to be ≥ 0. The next example shows that this is not guaranteed by (i)-(iii).

$$F(x_1, x_2) = \begin{cases} 1 & \text{if } x_1, x_2 \geq 1 \\ 2/3 & \text{if } x_1 \geq 1 \text{ and } 0 \leq x_2 < 1 \\ 2/3 & \text{if } x_2 \geq 1 \text{ and } 0 \leq x_1 < 1 \\ 0 & \text{otherwise} \end{cases}$$

If $0 < a_1, a_2 < 1 \leq b_1, b_2 < \infty$ then

$$F(b_1, b_2) - F(a_1, b_2) - F(b_1, a_2) + F(a_1, a_2) = 1 - 2/3 - 2/3 + 0 = -1/3$$

A little thought reveals that F is the distribution function of the measure with

$$\mu(\{(0, 1)\}) = \mu(\{(1, 0)\}) = 2/3 \qquad \mu(\{(1, 1)\}) = -1/3$$

To formulate the additional condition we need to guarantee that F is the distribution function of a probability measure, let

$$A = (a_1, b_1] \times \cdots \times (a_d, b_d]$$
$$V = \{a_1, b_1\} \times \cdots \times \{a_d, b_d\}$$

$V =$ the vertices of the rectangle A. If $v \in V$ let

$$\text{sgn}(v) = (-1)^{\# \text{ of } a\text{'s in } v}$$

The inclusion-exclusion formula implies

$$P(X \in A) = \sum_{v \in V} \text{sgn}(v) F(v)$$

So if we use $\Delta_A F$ to denote the right-hand side we need

(iv) $\Delta_A F \geq 0$ for all rectangles A.

The last condition guarantees that the measure assigned to each rectangle is ≥ 0. A standard result from measure theory (see (1.6) in the Appendix) now implies there is a unique probability measure with distribution F.

EXERCISE 9.1. If F is the distribution of $(X_1, \ldots, X_d)$ then $F_i(x) = P(X_i \leq x)$ are its **marginal distributions**. How can they be obtained from F?

EXERCISE 9.2. Let $F_1, \ldots, F_d$ be distributions on **R**. Show that for any $\alpha \in [-1, 1]$

$$F(x_1, \ldots, x_d) = \left\{ 1 + \alpha \prod_{i=1}^{d}(1 - F_i(x_i)) \right\} \prod_{j=1}^{d} F_j(x_j)$$

is a d.f. with the given marginals. The case $\alpha = 0$ corresponds to independent r.v.'s.

EXERCISE 9.3. A distribution F is said to have a **density** f if

$$F(x_1, \ldots, x_k) = \int_{-\infty}^{x_1} \cdots \int_{-\infty}^{x_k} f(y)\, dy_k \ldots dy_1$$

Show that if f is continuous, $\partial^k F / \partial x_1 \ldots \partial x_k = f$.

If F_n and F are distribution functions on $\mathbf{R}^d$, we say that F_n **converges weakly** to F, and write $F_n \Rightarrow F$, if $F_n(x) \to F(x)$ at all continuity points of F. Our first task is to show that there are enough continuity points for this to be a sensible definition. For a concrete example consider

$$F(x, y) = \begin{cases} 1 & \text{if } x \geq 0,\, y \geq 1 \\ y & \text{if } x \geq 0,\, 0 \leq y < 1 \\ 0 & \text{otherwise} \end{cases}$$

F is the distribution function of $(0, Y)$ where Y is uniform on $(0,1)$. Notice that this distribution has no atoms, but F is discontinuous at $(0, y)$ when $y > 0$.

Keeping the last example in mind, observe that if $x_n < x$, i.e., $x_{n,i} < x_i$ for all coordinates i, and $x_n \uparrow x$ as $n \to \infty$ then

$$F(x) - F(x_n) = P(X \leq x) - P(X \leq x_n) \downarrow P(X \leq x) - P(X < x)$$

In $d = 2$ the last expression is the probability X lies in

$$\{(a, x_2) : a \leq x_1\} \cup \{(x_1, b) : b \leq x_2\}$$

Let $H_c^i = \{x : x_i = c\}$ be the hyperplane where the ith coordinate is c. For each i, the H_c^i are disjoint so $D^i = \{c : P(X \in H_c^i) > 0\}$ is at most countable. It is easy to see that if x has $x_i \notin D^i$ for all i then F is continuous at x. This gives us more than enough points to reconstruct F.

As in Section 2.2, it will be useful to have several equivalent definitions of weak convergence. In Chapter 7 we will need to know that this is valid for an arbitrary metric space (S, ρ) so we will prove the result in that generality and insert another equivalence that will be useful there. f is said to be **Lipschitz continuous** if there is a constant C so that $|f(x) - f(y)| \le C\rho(x, y)$.

(9.1) Theorem. The following statements are equivalent to $X_n \Rightarrow X_\infty$.

(i) $Ef(X_n) \to Ef(X_\infty)$ for all bounded continuous f.

(ii) $Ef(X_n) \to Ef(X_\infty)$ for all bounded Lipschitz continuous f.

(iii) For all closed sets K, $\limsup_{n \to \infty} P(X_n \in K) \le P(X_\infty \in K)$.

(iv) For all open sets G, $\liminf_{n \to \infty} P(X_n \in G) \ge P(X_\infty \in G)$.

(v) For all sets A with $P(X_\infty \in \partial A) = 0$, $\lim_{n \to \infty} P(X_n \in A) = P(X_\infty \in A)$.

(vi) Let $D_f =$ the set of discontinuities of f. For all bounded functions f with $P(X_\infty \in D_f) = 0$, we have $Ef(X_n) \to Ef(X_\infty)$.

Proof We will begin by showing that (i)–(vi) are equivalent.

(i) implies (ii): Trivial.

(ii) implies (iii): Let $\rho(x, K) = \inf\{\rho(x, y) : y \in K\}$, $\varphi_j(r) = (1 - jr)^+$, and $f_j(x) = \varphi_j(\rho(x, K))$. f_j is Lipschitz continuous, has values in $[0, 1]$, and $\downarrow 1_K(x)$ as $j \uparrow \infty$. So

$$\limsup_{n \to \infty} P(X_n \in K) \le \lim_{n \to \infty} Ef_j(X_n) = Ef_j(X_\infty) \downarrow P(X_\infty \in K) \text{ as } j \uparrow \infty$$

(iii) is equivalent to (iv): As in the proof of (2.3) this follows easily from two facts: A is open if and only if A^c is closed; $P(A) + P(A^c) = 1$.

(iii) and (iv) imply (v): Let $K = \bar{A}$, $G = A^o$, and reason as in the proof of (2.3).

(v) implies (vi): Suppose $|f(x)| \le K$ and pick $\alpha_0 < \alpha_1 < \ldots < \alpha_\ell$ so that $P(f(X_\infty) = \alpha_i) = 0$ for $0 \le i \le \ell$, $\alpha_0 < -K < K < \alpha_\ell$, and $\alpha_i - \alpha_{i-1} < \epsilon$. This is always possible since $\{\alpha : P(f(X_\infty) = \alpha) > 0\}$ is a countable set. Let $A_i = \{x : \alpha_{i-1} < f(x) \le \alpha_i\}$. $\partial A_i \subset \{x : f(x) \in \{\alpha_{i-1}, \alpha_i\}\} \cup D_f$, so $P(X_\infty \in \partial A_i) = 0$, and it follows from (v) that

$$\sum_{i=1}^{\ell} \alpha_i P(X_n \in A_i) \to \sum_{i=1}^{\ell} \alpha_i P(X_\infty \in A_i)$$

The definition of the α_i implies

$$0 \le \sum_{i=1}^{\ell} \alpha_i P(X_n \in A_i) - Ef(X_n) \le \epsilon \quad \text{for } 1 \le n \le \infty$$

Since ϵ is arbitrary, it follows that $Ef(X_n) \to Ef(X_\infty)$.

(vi) implies (i): Trivial.

It remains to show that the five conditions are equivalent to weak convergence ($\Rightarrow$).

(v) implies ($\Rightarrow$) : If F is continuous at x, then $A = (-\infty, x_1] \times \ldots \times (-\infty, x_d]$ has $\mu(\partial A) = 0$, so $F_n(x) = P(X_n \in A) \to P(X_\infty \in A) = F(x)$.

($\Rightarrow$) implies (iv): Let $D^i = \{c : P(X_\infty \in H_c^i) > 0\}$ where $H_c^i = \{x : x^i = c\}$. We say a rectangle $A = (a_1, b_1] \times \ldots \times (a_d, b_d]$ is good if a_i, $b_i \notin D^i$ for all i. ($\Rightarrow$) implies that for all good rectangles $P(X_n \in A) \to P(X_\infty \in A)$. This is also true for B that are a finite disjoint union of good rectangles. Now any open set G is an increasing limit of B_k's that are a finite disjoint union of good rectangles, so

$$\liminf_{n \to \infty} P(X_n \in G) \ge \liminf_{n \to \infty} P(X_n \in B_k) = P(X_\infty \in B_k) \uparrow P(X_\infty \in G)$$

as $k \to \infty$. The proof of (9.1) is complete. $\qquad\square$

Remark. In Section 2.2 we proved that (i)-(v) are consequences of weak convergence by constructing r.v.'s with the given distributions so that $X_n \to X_\infty$ a.s. This can be done in $\mathbf{R}^d$ (or any complete separable metric space) but the construction is rather messy. See Billingsley (1979) p. 337-340 for a proof in $\mathbf{R}^d$.

EXERCISE 9.4. Let X_n be random vectors. Show that if $X_n \Rightarrow X$ then the coordinates $X_{n,i} \Rightarrow X_i$.

A sequence of probability measures μ_n is said to be **tight** if for any $\epsilon > 0$ there is an M so that $\liminf_{n \to \infty} \mu_n([-M, M]^d) \ge 1 - \epsilon$.

(9.2) Theorem. If μ_n is tight, then there is a weakly convergent subsequence.

Proof   Let F_n be the associated distribution functions, and let $q_1, q_2, \ldots$ be an enumeration of $\mathbf{Q}^d$ = the points in $\mathbf{R}^d$ with rational coordinates. By a diagonal argument like the one in the proof of (2.5), we can pick a subsequence so that $F_{n(k)}(q) \to G(q)$ for all $q \in \mathbf{Q}^d$. Let

$$F(x) = \inf\{G(q) : q \in \mathbf{Q}^d, q > x\}$$

where $q > x$ means $q_i > x_i$ for all i. It is easy to see that F is right continuous. To check that it is a distribution function, we observe that if A is a rectangle with vertices in $\mathbf{Q}^d$ then $\Delta_A F_n \ge 0$ for all n, so $\Delta_A G \ge 0$, and taking limits we see that the last conclusion holds for F for all rectangles A. Tightness implies that F has properties (i) and (ii) of a distribution F. We leave it to the reader to check that $F_n \Rightarrow F$. The proof of (2.5) works if you read inequalities such as $r_1 < r_2 < x < s$ as the corresponding relations between vectors. $\square$

The **characteristic function** of $(X_1, \ldots, X_d)$ is $\varphi(t) = E \exp(it \cdot X)$ where $t \cdot X = t_1 X_1 + \cdots + t_d X_d$ is the usual dot product of two vectors.

(9.3) Inversion formula. If $A = [a_1, b_1] \times \ldots \times [a_d, b_d]$ with $\mu(\partial A) = 0$ then

$$\mu(A) = \lim_{T \to \infty} (2\pi)^{-d} \int_{[-T,T]^d} \prod_{j=1}^{d} \psi_j(t_j) \varphi(t)\, dt$$

where $\psi_j(s) = (\exp(-isa_j) - \exp(-isb_j))/is$.

Proof   Fubini's theorem implies

$$\int_{[-T,T]^d} \int \prod_{j=1}^{d} \psi_j(t_j) \exp(it_j x_j)\, \mu(dx)\, dt$$

$$= \int \prod_{j=1}^{d} \int_{-T}^{T} \psi_j(t_j) \exp(it_j x_j)\, dt_j\, \mu(dx)$$

It follows from the proof of (3.2) that

$$\int_{-T}^{T} \psi_j(t_j) \exp(it_j x_j)\, dt_j \to \pi \left(1_{(a_j, b_j)}(x) + 1_{[a_j, b_j]}(x) \right)$$

so the desired conclusion follows from the bounded convergence theorem. $\square$

EXERCISE 9.5. Let φ be the ch.f. of a distribution F on $\mathbf{R}$. What is the distribution on $\mathbf{R}^d$ that corresponds to the ch.f. $\psi(t_1, \ldots, t_d) = \varphi(t_1 + \cdots + t_d)$?

EXERCISE 9.6. Show that random variables $X_1, \ldots, X_k$ are independent if and only if

$$\varphi_{X_1, \ldots X_k}(t) = \prod_{j=1}^{k} \varphi_{X_j}(t_j)$$

(9.4) Continuity theorem. Let X_n, $1 \leq n \leq \infty$ be random vectors with ch.f. φ_n. A necessary and sufficient condition for $X_n \Rightarrow X_\infty$ is that $\varphi_n(t) \rightarrow \varphi_\infty(t)$.

Proof $\exp(it \cdot x)$ is bounded and continuous, so if $X_n \Rightarrow X_\infty$ then $\varphi_n(t) \rightarrow \varphi_\infty(t)$. To prove the other direction it suffices, as in the proof of (3.4), to prove that the sequence is tight. To do this, we observe that if we fix $\theta \in \mathbf{R}^d$ then for all $s \in \mathbf{R}$, $\varphi_n(s\theta) \rightarrow \varphi_\infty(s\theta)$ so it follows from (3.4) that the distributions of $\theta \cdot X_n$ are tight. Applying the last observation to the d unit vectors $e_1, \ldots, e_d$, shows that the distributions of X_n are tight and completes the proof. $\square$

Remark. As before, if $\varphi_n(t) \rightarrow \varphi_\infty(t)$ with $\varphi_\infty(t)$ continuous at 0, then $\varphi_\infty(t)$ is the ch.f. of some X_∞ and $X_n \Rightarrow X_\infty$.

(9.4) has an important corollary.

(9.5) Cramér-Wold device. A sufficient condition for $X_n \Rightarrow X_\infty$ is that $\theta \cdot X_n \Rightarrow \theta \cdot X_\infty$ for all $\theta \in \mathbf{R}^d$.

Proof The indicated condition implies $E \exp(i\theta \cdot X_n) \rightarrow E \exp(i\theta \cdot X_\infty)$ for all $\theta \in \mathbf{R}^d$. $\square$

(9.5) leads immediately to

(9.6) The central limit theorem in $\mathbf{R}^d$. Let $X_1, X_2, \ldots$ be i.i.d. random vectors with $EX_n = \mu$, and finite covariances

$$\Gamma_{ij} = E((X_{n,i} - \mu_i)(X_{n,j} - \mu_j))$$

If $S_n = X_1 + \cdots + X_n$ then $(S_n - n\mu)/n^{1/2} \Rightarrow \chi$, where χ has a multivariate normal distribution with mean 0 and covariance Γ, i.e.,

$$E \exp(i\theta \cdot \chi) = \exp\left(-\sum_i \sum_j \theta_i \theta_j \Gamma_{ij}/2\right)$$

Proof By considering $X'_n = X_n - \mu$ we can suppose without loss of generality that $\mu = 0$. Let $\theta \in \mathbf{R}^d$. $\theta \cdot X_n$ is a random variable with mean 0 and variance

$$E\left(\sum_i \theta_i X_{n,i}\right)^2 = \sum_i \sum_j E\left(\theta_i \theta_j X_{n,i} X_{n,j}\right) = \sum_i \sum_j \theta_i \theta_j \Gamma_{ij}$$

so it follows from the one-dimensional central limit theorem and (9.5) that $S_n/n^{1/2} \Rightarrow \chi$ where

$$E \exp(i\theta \cdot \chi) = \exp\left(-\sum_i \sum_j \theta_i \theta_j \Gamma_{ij}/2\right) \qquad \square$$

To illustrate the use of (9.6), we consider two examples. In each $e_1, \ldots, e_d$ are the d unit vectors.

Example 9.1. Simple random walk on $\mathbf{Z}^d$. Let $X_1, X_2, \ldots$ be i.i.d. with

$$P(X_n = +e_i) = P(X_n = -e_i) = 1/2d \quad \text{for } i = 1, \ldots, d$$

$EX_n^i = 0$ and if $i \neq j$ then $EX_n^i X_n^j = 0$ since both components cannot be nonzero simultaneously. So the covariance matrix is $\Gamma_{ij} = (1/2d)I$.

Example 9.2. Let $X_1, X_2, \ldots$ be i.i.d. with $P(X_n = e_i) = 1/6$ for $i = 1, 2, \ldots, 6$. In words, we are rolling a die and keeping track of the numbers which come up. $EX_{n,i} = 1/6$ and $EX_{n,i}X_{n,j} = 0$ for $i \neq j$, so $\Gamma_{ij} = (1/6)(5/6)$ when $i = j$ and $= -(1/6)^2$ when $i \neq j$. In this case the limiting distribution is concentrated on $\{x : \sum_i x_i = 0\}$.

Our treatment of the central limit theorem would not be complete without some discussion of the **multivariate normal distribution**. We begin by observing that $\Gamma_{ij} = \Gamma_{ji}$ and if $EX_i = 0$ and $EX_i X_j = \Gamma_{i,j}$

$$\sum_i \sum_j \theta_i \theta_j \Gamma_{ij} = E\left(\sum_i \theta_i X_i\right)^2 \geq 0$$

so Γ is symmetric and nonnegative definite. A well known result implies that there is an orthogonal matrix U (i.e., one with $U^t U = I$, the identity matrix) so that $\Gamma = U^t V U$, where $V \geq 0$ is a diagonal matrix. Let W be the nonnegative diagonal matrix with $W^2 = V$. If we let $A = WU$ then $\Gamma = A^t A$. Let Y be a d-dimensional vector whose components are independent and have normal distributions with mean 0 and variance 1. If we view vectors as $1 \times d$ matrices and let $\chi = YA$, then χ has the desired normal distribution. To check this observe that

$$\theta \cdot YA = \sum_i \theta_i \sum_j Y_j A_{ji}$$

has a normal distribution with mean 0 and variance

$$\sum_j \left(\sum_i A_{ji} \theta_i \right)^2 = \sum_j \left(\sum_i \theta_i A_{ij}^t \right) \left(\sum_k A_{jk} \theta_k \right) = \theta A^t A \theta^t = \theta \Gamma \theta^t$$

so $E(\exp(i\theta \cdot \chi)) = \exp(-(\theta \Gamma \theta^t)/2)$.

If the covariance matrix has rank d we say that the normal distribution is **nondegenerate**. In this case its density function is given by

$$(2\pi)^{-d/2} (\det \Gamma)^{-1/2} \exp \left(-\sum_{i,j} y_i \Gamma_{ij}^{-1} y_j /2 \right)$$

The joint distribution in degenerate cases can be computed by using a linear transformation to reduce to the nondegenerate case. For instance in Example 9.2 we can look at the distribution of $(X_1, \ldots, X_5)$.

EXERCISE 9.7. Suppose $(X_1, \ldots, X_d)$ has a multivariate normal distribution with mean vector θ and covariance Γ. Show $X_1, \ldots, X_d$ are independent if and only if $\Gamma_{ij} = 0$ for $i \neq j$. In words, uncorrelated random variables with a joint normal distribution are independent.

EXERCISE 9.8. Show that $(X_1, \ldots, X_d)$ has a multivariate normal distribution with mean vector θ and covariance Γ if and only if every linear combination $c_1 X_1 + \cdots + c_d X_d$ has a normal distribution with mean $c\theta^t$ and variance $c\Gamma c^t$.

3 Random Walks

Let $X_1, X_2, \ldots$ be i.i.d. taking values in $\mathbf{R}^d$ and let $S_n = X_1 + \ldots + X_n$. S_n is a **random walk**. In the last chapter we were primarily concerned with the distribution of S_n. In this one we will look at properties of the sequence $S_1(\omega), S_2(\omega), \ldots$ For example, does the last sequence return to (or near) 0 infinitely often? The first section introduces stopping times, a concept that will be very important in this and the next two chapters. After the first section is completed, the remaining three can be read in any order or skipped without much loss. The second section is not starred since it contains some basic facts about random walks.

3.1. Stopping Times

Most of the results in this section are valid for i.i.d. X's taking values in some nice measurable space $(S, \mathcal{S})$ and will be proved in that generality. For several reasons it is convenient to use the special probability space from the proof of Kolmogorov's extension theorem:

$$\Omega = \{(\omega_1, \omega_2, \ldots) : \omega_i \in S\}$$
$$\mathcal{F} = \mathcal{S} \times \mathcal{S} \times \ldots$$
$$P = \mu \times \mu \times \ldots \qquad \mu \text{ is the distribution of } X_i$$
$$X_n(\omega) = \omega_n$$

So throughout this section, we will suppose (without loss of generality) that our random variables are constructed on this special space.

Before taking up our main topic, we will prove a 0-1 law that, in the i.i.d. case, generalizes Kolmogorov's. To state the new 0-1 law we need two definitions. A **finite permutation** of $\mathbf{N} = \{1, 2, \ldots\}$ is a map π from $\mathbf{N}$ onto $\mathbf{N}$ so that $\pi(i) \neq i$ for only finitely many i. If π is a finite permutation of $\mathbf{N}$ and $\omega \in S^{\mathbf{N}}$ we define $(\pi\omega)_i = \omega_{\pi(i)}$. In words, the coordinates of ω are rearranged according to π. Since $X_i(\omega) = \omega_i$ this is the same as rearranging the random variables. An event A is **permutable** if $\pi^{-1}A \equiv \{\omega : \pi\omega \in A\}$ is equal to A

for any finite permutation π, or in other words, if its occurrence is not affected by rearranging the random variables. The collection of permutable events is a σ-field. It is called the **exchangeable** σ-field and denoted by $\mathcal{E}$.

To see the reason for interest in permutable events, suppose $S = \mathbf{R}$ and let $S_n(\omega) = X_1(\omega) + \cdots + X_n(\omega)$. Two examples of permutable events are

(i) $\{\omega : S_n(\omega) \in A \text{ i.o.}\}$

(ii) $\{\omega : \limsup_{n \to \infty} S_n(\omega)/c_n \geq 1\}$

In each case the event is permutable because $S_n(\omega) = S_n(\pi\omega)$ for large n. The list of examples can be enlarged considerably by observing:

(iii) All events in the tail σ-field $\mathcal{T}$ are permutable.

To see this observe that if $B \in \sigma(X_{n+1}, X_{n+2}, \ldots)$ then the occurrence of B is unaffected by a permutation of $X_1, \ldots, X_n$. (i) shows that the converse of (iii) is false. The next result shows that for an i.i.d. sequence there is no difference between $\mathcal{E}$ and $\mathcal{T}$. They are both trivial.

(1.1) Hewitt-Savage 0-1 law. If $X_1, X_2, \ldots$ are i.i.d. and $A \in \mathcal{E}$ then $P(A) \in \{0, 1\}$.

Proof   Let $A \in \mathcal{E}$. As in the proof of Kolmogorov's 0-1 law, we will show A is independent of itself, i.e., $P(A) = P(A \cap A) = P(A)P(A)$ so $P(A) \in \{0, 1\}$. Let $A_n \in \sigma(X_1, \ldots, X_n)$ so that

(a) $$P(A_n \Delta A) \to 0$$

Here $A \Delta B = (A - B) \cup (B - A)$ is the symmetric difference. The existence of the A_n's is proved in Exercise 3.1 in the Appendix. A_n can be written as $\{\omega : (\omega_1, \ldots, \omega_n) \in B_n\}$ with $B_n \in \mathcal{S}^n$. Let

$$\pi(j) = \begin{cases} j+n & \text{if } 1 \leq j \leq n \\ j-n & \text{if } n+1 \leq j \leq 2n \\ j & \text{if } j \geq 2n+1 \end{cases}$$

Observing that π^2 is the identity (so we don't have to worry about whether to write π or π^{-1}) and the coordinates are i.i.d. (so the permuted coordinates are) gives

(b) $$P(\omega : \omega \in A_n \Delta A) = P(\omega : \pi\omega \in A_n \Delta A)$$

Now $\{\omega : \pi\omega \in A\} = \{\omega : \omega \in A\}$, since A is permutable, and

$$\{\omega : \pi\omega \in A_n\} = \{\omega : (\omega_{n+1}, \ldots, \omega_{2n}) \in B_n\}$$

If we use A'_n to denote the last event then we have

(c) $$\{\omega : \pi\omega \in A_n \Delta A\} = \{\omega : \omega \in A'_n \Delta A\}$$

Combining (b) and (c) gives

(d) $$P(A_n \Delta A) = P(A'_n \Delta A)$$

Now $A - C \subset (A - B) \cup (B - C)$ and with a similar inequality for $C - A$ implies $A \Delta C \subset (A \Delta B) \cup (B \Delta C)$. The last inequality, (d), and (a) imply

$$P(A_n \Delta A'_n) \leq P(A_n \Delta A) + P(A \Delta A'_n) \to 0$$

The last result implies

$$0 \leq P(A_n) - P(A_n \cap A'_n)$$
$$\leq P(A_n \cup A'_n) - P(A_n \cap A'_n) = P(A_n \Delta A'_n) \to 0$$

so $P(A_n \cap A'_n) \to P(A)$. But A_n and A'_n are independent, so

$$P(A_n \cap A'_n) = P(A_n)P(A'_n) \to P(A)^2$$

(Recall $P(A'_n) = P(A_n)$.) This shows $P(A) = P(A)^2$, and proves (1.1).   □

A typical application of (1.1) is

(1.2) Theorem. For a random walk on **R**, there are only four possibilities, one of which has probability one.
  (i) $S_n = 0$ for all n.
  (ii) $S_n \to \infty$.
  (iii) $S_n \to -\infty$.
  (iv) $-\infty = \liminf S_n < \limsup S_n = \infty$.

Proof   (1.1) implies $\limsup S_n$ is a constant $c \in [-\infty, \infty]$. Let $S'_n = S_{n+1} - X_1$. Since S'_n has the same distribution as S_n it follows that $c = c - X_1$. If c is finite, subtracting c from both sides we conclude $X_1 \equiv 0$ and (i) occurs. Turning the last statement around we see that if $X_1 \not\equiv 0$ then $c = -\infty$ or ∞. The same analysis applies to the liminf. Discarding the impossible combination $\limsup S_n = -\infty$ and $\liminf S_n = +\infty$ we have proved the result.   □

EXERCISE 1.1. **Symmetric random walk.** Let $X_1, X_2, \ldots \in \mathbf{R}$ be i.i.d. with a distribution that is symmetric about 0 and nondegenerate (i.e., $P(X_i = 0) < 1$). Show that we are in case (iv) of (1.2).

EXERCISE 1.2. Let $X_1, X_2, \ldots$ be i.i.d. with $EX_i = 0$ and $EX_i^2 = \sigma^2 \in (0, \infty)$. Use the central limit theorem to conclude that we are in case (iv) of (1.2). Later in Exercise 1.10 you will show that $EX_i = 0$ and $P(X_i = 0) < 1$ is sufficient.

The special case in which $P(X_i = 1) = P(X_i = -1) = 1/2$ is called **simple random walk**. Since a simple random walk cannot skip over any integers, it follows from either exercise above that with probability one it visits every integer infinitely many times.

Let $\mathcal{F}_n = \sigma(X_1, \ldots, X_n) = $ the information known at time n. A random variable N taking values in $\{1, 2, \ldots\} \cup \{\infty\}$ is said to be a **stopping time** or an **optional random variable** if for every $n < \infty$, $\{N = n\} \in \mathcal{F}_n$. If we think of S_n as giving the (logarithm of the) price of a stock at time n, and N as the time we sell it, then the last definition says that the decision to sell at time n must be based on the information known at that time. The last interpretation gives one explanation for the second name. N is a time at which we can exercise an option to buy a stock. Chung prefers the second name because N is "usually rather a momentary pause after which the process proceeds again: time marches on!"

The canonical example of a stopping time is $N = \inf\{n : S_n \in A\}$, the **hitting time of A**. To check that this is a stopping time, we observe that

$$\{N = n\} = \{S_1 \in A^c, \ldots, S_{n-1} \in A^c, S_n \in A\} \in \mathcal{F}_n$$

Two concrete examples of hitting times that have appeared above are

Example 1.1. $N = \min\{n : S_n > \lambda\}$ from the proof of (8.2) in Chapter 1.

Example 1.2. If the $X_i \geq 0$ and $N_t = \sup\{n : S_n \leq t\}$ is the random variable that first appeared in Example 7.1 in Chapter 1, then $N_t + 1 = \inf\{n : S_n > t\}$ is a stopping time.

The next result allows us to construct new examples from the old ones.

EXERCISE 1.3. If S and T are stopping times then $S \wedge T$ and $S \vee T$ are stopping times. Since constant times are stopping times, it follows that $S \wedge n$ and $S \vee n$ are stopping times.

EXERCISE 1.4. Suppose S and T are stopping times. Is $S + T$ a stopping time? Give a proof or a counterexample.

Associated with each stopping time N is a σ-field $\mathcal{F}_N = $ the information known at time N. Formally $\mathcal{F}_N$ is the collection of sets A that have $A \cap \{N =$

$n\} \in \mathcal{F}_n$ for all $n < \infty$, i.e., when $N = n$, A must be measurable with respect to the information known at time n. Trivial but important examples of sets in $\mathcal{F}_N$ are $\{N \leq n\}$, i.e., N is measurable with respect to $\mathcal{F}_N$.

EXERCISE 1.5. Show that if $Y_n \in \mathcal{F}_n$ and N is a stopping time, $Y_N \in \mathcal{F}_N$. As a corollary of this result we see that if $f : S \rightarrow \mathbf{R}$ is measurable, $T_n = \sum_{m \leq n} f(X_m)$, and $M_n = \max_{m \leq n} T_m$ then T_N and $M_N \in \mathcal{F}_N$. An important special case is $S = \mathbf{R}^d$, $f(x) = x$.

EXERCISE 1.6. Show that if $M \leq N$ are stopping times then $\mathcal{F}_M \subset \mathcal{F}_N$.

EXERCISE 1.7. Show that if $L \leq M$ and $A \in \mathcal{F}_L$ then

$$N = \begin{cases} L & \text{on } A \\ M & \text{on } A^c \end{cases} \quad \text{is a stopping time}$$

Our first result about $\mathcal{F}_N$ is

(1.3) Theorem. Let $X_1, X_2, \ldots$ be i.i.d., $\mathcal{F}_n = \sigma(X_1, \ldots, X_n)$ and N be a stopping time. Conditional on $\{N < \infty\}$, $\{X_{N+n}, n \geq 1\}$ is independent of $\mathcal{F}_N$ and has the same distribution as the original sequence.

Proof By (2.2) in the appendix it is enough to show that if $A \in \mathcal{F}_N$ and $B_j \in \mathcal{S}$ for $1 \leq j \leq k$ then

$$P(A, N < \infty, X_{N+j} \in B_j, 1 \leq j \leq k) = P(A \cap \{N < \infty\}) \prod_{j=1}^{k} \mu(B_j)$$

where $\mu(B) = P(X_i \in B)$. The method ("divide and conquer") is one that we will see many times below. We break things down according to the value of N in order to replace N by n and reduce to the case of a fixed time.

$$P(A, N = n, X_{N+j} \in B_j, 1 \leq j \leq k) = P(A, N = n, X_{n+j} \in B_j, 1 \leq j \leq k)$$

$$= P(A \cap \{N = n\}) \prod_{j=1}^{k} \mu(B_j)$$

since $A \cap \{N = n\} \in \mathcal{F}_n$ and that σ-field is independent of $X_{n+1}, \ldots, X_{n+k}$. Summing over n now gives the desired result. $\square$

To delve further into properties of stopping times we recall we have supposed $\Omega = S^N$, and define the **shift** $\theta : \Omega \rightarrow \Omega$ by

$$(\theta\omega)(n) = \omega(n + 1) \qquad n = 1, 2, \ldots$$

In words we drop the first coordinate, and shift the others one place to the left. The iterates of θ are defined by composition. Let $\theta^1 = \theta$ and for $k \geq 2$ let $\theta^k = \theta \circ \theta^{k-1}$. Clearly $(\theta^k \omega)(n) = \omega(n+k)$, $n = 1, 2, \ldots$ To extend the last definition to stopping times we let

$$\theta^N \omega = \begin{cases} \theta^n \omega & \text{on } \{N = n\} \\ \Delta & \text{on } \{N = \infty\} \end{cases}$$

Here Δ is an extra point which we add to Ω. According to the only joke in Blumenthal and Getoor (1968), Δ is a "cemetary or heaven depending upon your point of view". Seriously, Δ is a convenience in making definitions, like the next one.

Example 1.3. Returns to 0. For a concrete example of the use of θ, suppose $S = \mathbf{R}^d$ and let

$$\tau(\omega) = \inf\{n : \omega_1 + \cdots + \omega_n = 0\}$$

where $\inf \emptyset = \infty$, and we set $\tau(\Delta) = \infty$. If we let $\tau_2(\omega) = \tau(\omega) + \tau(\theta^\tau \omega)$ then on $\{\tau < \infty\}$,

$$\tau(\theta^\tau \omega) = \inf\{n : (\theta^\tau \omega)_1 + \cdots + (\theta^\tau \omega)_n = 0\}$$
$$= \inf\{n : \omega_{\tau+1} + \cdots + \omega_{\tau+n} = 0\}$$
$$\tau(\omega) + \tau(\theta^\tau \omega) = \inf\{m > \tau : \omega_1 + \cdots + \omega_m = 0\}$$

So τ_2 is the time of the second visit to 0 (and thanks to the conventions $\theta^\infty \omega = \Delta$ and $\tau(\Delta) = \infty$, this is true for all ω). The last computation generalizes easily to show that if we let

$$\tau_n = \tau_{n-1} + \tau(\theta^{\tau_{n-1}} \omega)$$

then τ_n is the time of the nth visit to 0.

If we have any stopping time T we can define its iterates by $T_0 = 0$ and

$$T_n = T_{n-1} + T(\theta^{T_{n-1}} \omega) \quad \text{for } n \geq 1$$

If we assume $P = \mu \times \mu \times \ldots$ then

(1.4) $$P(T_n < \infty) = P(T < \infty)^n$$

Proof We will prove this by induction. The result is trivial when $n = 1$. Suppose now that it is valid for $n-1$. Applying (1.3) to $N = T_{n-1}$ we see that $T(\theta^{T_{n-1}} \omega) < \infty$ is independent of $T^{n-1} < \infty$, and has the same probability as $T < \infty$ so

$$P(T_n < \infty) = P(T_{n-1} < \infty, T(\theta^{T_{n-1}} \omega) < \infty)$$
$$= P(T_{n-1} < \infty)P(T < \infty) = P(T < \infty)^n$$

by the induction hypothesis. $\qquad\qquad\qquad\qquad\qquad\qquad\qquad$ $\square$

Letting $t_n = T(\theta^{T_{n-1}})$ we can extend (1.3) to

(1.5) Theorem. Suppose $P(T < \infty) = 1$. Then the "random vectors"

$$V_n = (t_n, X_{T_{n-1}+1}, \ldots, X_{T_n})$$

are independent and identically distributed.

Proof It is clear from (1.3) that V_n and V_1 have the same distribution. The independence follows from (1.3) and induction since $V_1, \ldots, V_{n-1} \in \mathcal{F}(T_{n-1})$.
$\square$

Example 1.4. Ladder variables. Let $\alpha(\omega) = \inf\{n : \omega_1 + \cdots + \omega_n > 0\}$ where $\inf \emptyset = \infty$, and set $\alpha(\Delta) = \infty$. Let $\alpha_0 = 0$ and let $\alpha_k = \alpha_{k-1} + \alpha(\theta^{\alpha_{k-1}}\omega)$ for $k \geq 1$. At time α_k the random walk is at a record high value. The next three exercises investigate these times.

EXERCISE 1.8. (i) If $P(\alpha < \infty) < 1$ then $P(\sup S_n < \infty) = 1$.
(ii) If $P(\alpha < \infty) = 1$ then $P(\sup S_n = \infty) = 1$.

EXERCISE 1.9. Let $\beta(\omega) = \inf\{n : S_n < 0\}$. Prove that the four possibilities in (1.2) correspond to the four combinations of $P(\alpha < \infty) < 1$ or $= 1$, and $P(\beta < \infty) < 1$ or $= 1$.

EXERCISE 1.10. Let $S_0 = 0$, $\bar{\beta}(\omega) = \inf\{n \geq 1 : S_n \leq 0\}$ and

$$A_m^n = \{0 \geq S_m, S_1 \geq S_m, \ldots, S_{m-1} \geq S_m, S_m < S_{m+1}, \ldots, S_m < S_n\}$$

(i) Show $1 = \sum_{m=0}^n P(A_m^n) = \sum_{m=0}^n P(\alpha > m)P(\bar{\beta} > n - m)$.
(ii) Let $n \to \infty$ and conclude $E\alpha = 1/P(\bar{\beta} = \infty)$.

EXERCISE 1.11. (i) Combine the last exercise with the proof of (ii) in Exercise 1.8 to conclude that if $EX_i = 0$ then $P(\bar{\beta} = \infty) = 0$. (ii) Show that if we assume in addition that $P(X_i = 0) < 1$ then $P(\beta = \infty) = 0$ and Exercise 1.9 implies we are in case (iv) of (1.2).

Our final result about stopping times is:

(1.6) Wald's equation. Let $X_1, X_2, \ldots$ be i.i.d. with $E|X_i| < \infty$. If N is a stopping time with $EN < \infty$ then $ES_N = EX_1 EN$.

Proof First suppose the $X_i \geq 0$.

$$ES_N = \int S_N dP = \sum_{n=1}^{\infty} \int S_n 1_{\{N=n\}} dP = \sum_{n=1}^{\infty} \sum_{m=1}^{n} \int X_m 1_{\{N=n\}} dP$$

Since the $X_i \geq 0$, we can interchange the order of summation (i.e., use Fubini's theorem) to conclude that the last expression

$$= \sum_{m=1}^{\infty} \sum_{n=m}^{\infty} \int X_m 1_{\{N=n\}} dP = \sum_{m=1}^{\infty} \int X_m 1_{\{N \geq m\}} dP$$

Now $\{N \geq m\} = \{N \leq m-1\}^c \in \mathcal{F}_{m-1}$ and is independent of X_m so the last expression

$$= \sum_{m=1}^{\infty} EX_m P(N \geq m) = EX_1 EN$$

To prove the result in general we run the last argument backwards. If we have $EN < \infty$ then

$$\infty > \sum_{m=1}^{\infty} E|X_m| P(N \geq m) = \sum_{m=1}^{\infty} \sum_{n=m}^{\infty} \int |X_m| 1_{\{N=n\}} dP$$

The last formula shows that the double sum converges absolutely in one order, so Fubini's theorem gives

$$\sum_{m=1}^{\infty} \sum_{n=m}^{\infty} \int X_m 1_{\{N=n\}} dP = \sum_{n=1}^{\infty} \sum_{m=1}^{n} \int X_m 1_{\{N=n\}} dP$$

Using the independence of $\{N \geq m\} \in \mathcal{F}_{m-1}$ and X_m, and rewriting the last identity it follows that

$$\sum_{m=1}^{\infty} EX_m P(N \geq m) = ES_N$$

Since the left-hand side is $EN\, EX_1$ the proof is complete. $\qquad \square$

EXERCISE 1.12. Let $X_1, X_2, \ldots$ be i.i.d. uniform on (0,1), let $S_n = X_1 + \cdots + X_n$, and let $T = \inf\{n : S_n > 1\}$. Show that $P(T > n) = 1/n!$, so $ET = e$ and $ES_T = e/2$.

Example 1.5. Simple random walk. Let $X_1, X_2, \ldots$ be i.i.d. with $P(X_i = 1) = 1/2$ and $P(X_i = -1) = 1/2$. Let $a < 0 < b$ be integers and let $N = \inf\{n :$

$S_n \notin (a,b)\}$. To apply (1.6) we have to check that $EN < \infty$. To do this we observe that if $x \in (a, b)$ then

$$P(x + S_{b-a} \notin (a,b)) \geq 2^{-(b-a)}$$

since $b - a$ steps of size $+1$ in a row will take us out of the interval. Iterating the last inequality it follows that

$$P(N > n(b - a)) \leq \left(1 - 2^{-(b-a)}\right)^n$$

so $EN < \infty$. Applying (1.6) now gives $ES_N = 0$ or

$$bP(S_N = b) + aP(S_N = a) = 0$$

Since $P(S_N = b) + P(S_N = a) = 1$, it follows that $(b - a)P(S_N = b) = -a$, so

$$P(S_N = b) = \frac{-a}{b - a} \qquad P(S_N = a) = \frac{b}{b - a}$$

Letting $T_a = \inf\{n : S_n = a\}$, we can write the last conclusion as

(1.7) $$P(T_a < T_b) = \frac{b}{b - a} \quad \text{for } a < 0 < b$$

Setting $b = M$ and letting $M \to \infty$ gives

$$P(T_a < \infty) \geq P(T_a < T_M) \to 1$$

for all $a < 0$. From symmetry (and the fact that $T_0 \equiv 0$) it follows that

(1.8) $$P(T_x < \infty) = 1 \quad \text{for all } x \in \mathbf{Z}$$

Our final fact about T_x is that $ET_x = \infty$ for $x \neq 0$. To see that note that if $ET_x < \infty$ then (1.6) would imply

$$x = ES_{T_x} = EX_1 ET_x = 0$$

In Section 3.3 we will compute the distribution of T_1 and show that

$$P(T_1 > t) \sim C \, t^{-1/2}$$

EXERCISE 1.13. **Asymmetric simple random walk.** Let $X_1, X_2, \ldots$ be i.i.d. with $P(X_1 = 1) = p > 1/2$ and $P(X_1 = -1) = 1 - p$, and let $S_n = X_1 + \cdots + X_n$. Let $\alpha = \inf\{m : S_m > 0\}$ and $\beta = \inf\{n : S_n < 0\}$.

(i) Use Exercise 1.9 to conclude that $P(\alpha < \infty) = 1$ and $P(\beta < \infty) < 1$.
(ii) If $Y = \inf S_n$ then $P(Y \leq -k) = P(\beta < \infty)^k$.
(iii) Apply Wald's equation to $\alpha \wedge n$ and let $n \to \infty$ to get $E\alpha = 1/EX_1 = 1/(2p - 1)$. Comparing with Exercise 1.10 shows $P(\bar{\beta} = \infty) = 2p - 1$.

EXERCISE 1.14. **An optimal stopping problem.** Let X_n, $n \geq 1$ be i.i.d. with $EX_1^+ < \infty$ and let

$$Y_n = \max_{1 \leq m \leq n} X_m - cn$$

That is, we are looking for a large value of X but we have to pay $c > 0$ for each observation. (i) Let $T = \inf\{n : X_n > a\}$, $p = P(X_n > a)$, and compute EY_T. (ii) Let α (possibly < 0) be the unique solution of $E(X_1 - \alpha)^+ = c$. Show that $EY_T = \alpha$ in this case and use the inequality

$$Y_n \leq \alpha + \sum_{m=1}^{n}((X_m - \alpha)^+ - c)$$

for $n \geq 1$ to conclude that if $\tau \geq 1$ is a stopping time with $E\tau < \infty$ then $EY_\tau \leq \alpha$. The analysis above assumes that you have to play at least once. If the optimal $\alpha < 0$ then you shouldn't play at all.

EXERCISE 1.15. **Wald's second equation.** Let $X_1, X_2, \ldots$ be i.i.d. with $EX_n = 0$ and $EX_n^2 = \sigma^2 < \infty$. If T is a stopping time with $ET < \infty$ then $ES_T^2 = \sigma^2 ET$. Hint: Compute $ES_{T \wedge n}^2$ by induction and show that $S_{T \wedge n}$ is a Cauchy sequence in L^2.

An amusing consequence of the last result is

(1.7) **Theorem.** Let $X_1, X_2, \ldots$ be i.i.d. with $EX_n = 0$ and $EX_n^2 = 1$, and let $T_c = \inf\{n \geq 1 : |S_n| > cn^{1/2}\}$.

$$ET_c \quad \begin{cases} < \infty & \text{for } c < 1 \\ = \infty & \text{for } c \geq 1 \end{cases}$$

Proof   One half of this is easy. If $ET_c < \infty$ then the previous exercise implies $ET_c = E(S_{T_c}^2) > c^2 ET_c$ a contradiction if $c \geq 1$. To prove the other direction we let $\tau = T_c \wedge n$ and observe $S_{\tau-1}^2 \leq c^2(\tau - 1)$, so using the Cauchy-Schwarz inequality

$$E\tau = ES_\tau^2 = ES_{\tau-1}^2 + 2E(S_{\tau-1}X_\tau) + EX_\tau^2$$
$$\leq c^2 E\tau + 2c(E\tau\, EX_\tau^2)^{1/2} + EX_\tau^2$$

To complete the proof now we will show

Lemma. If T is a stopping time with $ET = \infty$ then

$$EX^2_{T \wedge n}/E(T \wedge n) \to 0$$

(1.7) follows for if $\epsilon < 1 - c^2$ and n is large we will have $E\tau \leq (c^2 + \epsilon)E\tau$ a contradiction.

Proof We begin by writing

$$E(X^2_{T \wedge n}) = E(X^2_{T \wedge n}; X^2_{T \wedge n} \leq \epsilon(T \wedge n)) + \sum_{j=1}^n E(X^2_j; T \wedge n = j, X^2_j > \epsilon j)$$

The first term is $\leq \epsilon E(T \wedge n)$. To bound the second, choose $N \geq 1$ so that for $n \geq N$

$$\sum_{j=1}^n E(X^2_j; X^2_j > \epsilon j) < n\epsilon$$

This is possible since the dominated convergence theorem implies $E(X^2_j; X^2_j > \epsilon j) \to 0$ as $j \to \infty$. For the first part of the sum we use a trivial bound

$$\sum_{j=1}^N E(X^2_j; T \wedge n = j, X^2_j > \epsilon j) \leq N E X^2_1$$

To bound the remainder of the sum we note (i) $X^2_j \geq 0$; (ii) $\{T \wedge n \geq j\}$ is $\in \mathcal{F}_{j-1}$ and hence is independent of $X^2_j 1_{(X^2_j > \epsilon j)}$, (iii) use some trivial arithmetic, (iv) use Fubini's theorem and enlarge the range of j, (v) use the choice of N and a trivial inequality

$$\sum_{j=N}^n E(X^2_j; T \wedge n = j, X^2_j > \epsilon j) \leq \sum_{j=N}^n E(X^2_j; T \wedge n \geq j, X^2_j > \epsilon j)$$

$$= \sum_{j=N}^n P(T \wedge n \geq j) E(X^2_j; X^2_j > \epsilon j)$$

$$= \sum_{j=N}^n \sum_{k=j}^\infty P(T \wedge n = k) E(X^2_j; X^2_j > \epsilon j)$$

$$\leq \sum_{k=N}^\infty \sum_{j=1}^k P(T \wedge n = k) E(X^2_j; X^2_j > \epsilon j)$$

$$\leq \sum_{k=N}^\infty \epsilon k P(T \wedge n = k) \leq \epsilon E(T \wedge n)$$

Combining our estimates shows

$$EX_{T \wedge n}^2 \leq 2\epsilon E(T \wedge n) + NEX_1^2$$

Letting $n \to \infty$ and noting $E(T \wedge n) \to \infty$ we have

$$\limsup_{n \to \infty} EX_{T \wedge n}^2 / E(T \wedge n) \leq 2\epsilon$$

where ϵ is arbitrary. $\square$

3.2. Recurrence

Throughout this section S_n will be random walk, i.e., $S_n = X_1 + \cdots + X_n$ where $X_1, X_2, \ldots$ are i.i.d., and we will investigate the question mentioned at the beginning of the chapter. Does the sequence $S_1(\omega), S_2(\omega), \ldots$ return to (or near) 0 infinitely often? The answer to the last question is either Yes or No, and the random walk is called recurrent or transient accordingly. We begin with some definitions that formulate the question precisely and a result that establishes a dichotomy between the two cases.

The number $x \in \mathbf{R}^d$ is said to be a **recurrent value** for the random walk S_n if for every $\epsilon > 0$, $P(\|S_n - x\| < \epsilon \text{ i.o.}) = 1$. Here $\|x\| = \sup |x_i|$. The reader will see the reason for this choice of norm in the proof of (2.5). The Hewitt-Savage 0-1 law, (1.1), implies that if the last probability is < 1 it is 0. Our first result shows that to know the set of recurrent values it is enough to check $x = 0$. A number x is said to be a **possible value** of the random walk if for any $\epsilon > 0$ there is an n so that $P(\|S_n - x\| < \epsilon) > 0$.

(2.1) Theorem. The set $\mathcal{V}$ of recurrent values is either $\emptyset$ or a closed subgroup of $\mathbf{R}$. In the second case $\mathcal{V} = \mathcal{U}$, the set of possible values.

Proof   Suppose $\mathcal{V} \neq \emptyset$. It is clear that $\mathcal{V}^c$ is open, so $\mathcal{V}$ is closed. To prove that $\mathcal{V}$ is a group we will first show that

(*) if $x \in \mathcal{U}$ and $y \in \mathcal{V}$ then $y - x \in \mathcal{V}$.

This statement has been formulated so that once it is established, (2.1) follows easily. Let

$$p_{\delta,m} = P(\|S_n - z\| \geq \delta \text{ for all } n \geq m)$$

If $y - x \notin \mathcal{V}$, there is an $\epsilon > 0$ and $m \geq 1$ so that $p_{2\epsilon,m}(y - x) > 0$. Since $x \in \mathcal{U}$, there is a k so that $P(\|S_k - x\| < \epsilon) > 0$. Since

$$P(\|S_n - S_k - (y - x)\| \geq 2\epsilon \text{ for all } n \geq k + m) = p_{2\epsilon,m}(y - x)$$

and is independent of $\{\|S_k - x\| < \epsilon\}$ it follows that

$$p_{\epsilon,m+k}(y) \geq P(\|S_k - x\| < \epsilon)p_{2\epsilon,m}(y - x) > 0$$

contradicting $y \in V$, so $y - x \in V$.

To conclude V is a group when $V \neq \emptyset$, let $q, r \in V$, and observe: (i) taking $x = y = r$ in (*) shows $0 \in V$, (ii) taking $x = r$, $y = 0$ shows $-r \in V$, and (iii) taking $x = -r$, $y = q$ shows $q + r \in V$. To prove that $V = U$ now, observe that if $u \in U$ taking $x = u$, $y = 0$ shows $-u \in V$ and since V is a group, it follows that $u \in V$. $\qquad\square$

If $V = \emptyset$, the random walk is said to be **transient**, otherwise it is called **recurrent**. Before plunging into the technicalities needed to treat a general random walk we begin by analyzing the special case Polya considered in 1921. Legend has it that Polya thought of this problem while wandering around in a park near Zürich when he noticed that he kept encountering the same young couple. History does not record what the young couple thought.

Example 2.1. Simple random walk on $\mathbf{Z}^d$.

$$P(X_i = e_j) = P(X_i = -e_j) = 1/2d$$

for each of the d unit vectors e_j. To analyze this case we begin with a result that is valid for any random walk. Let $\tau_0 = 0$ and $\tau_n = \inf\{m > \tau_{n-1} : S_m = 0\}$ be the time of the nth return to 0. From (1.4) it follows that

$$P(\tau_n < \infty) = P(\tau_1 < \infty)^n$$

a fact that leads easily to:

(2.2) Theorem. For any random walk, the following are equivalent:
(i) $P(\tau_1 < \infty) = 1$, (ii) $P(S_m = 0 \text{ i.o.}) = 1$, and (iii) $\sum_{m=0}^{\infty} P(S_m = 0) = \infty$.

Proof If $P(\tau_1 < \infty) = 1$, then $P(\tau_n < \infty) = 1$ for all n and $P(S_m = 0 \text{ i.o.}) = 1$. Let

$$V = \sum_{m=0}^{\infty} 1_{(S_m=0)} = \sum_{n=0}^{\infty} 1_{(\tau_n < \infty)}$$

be the number of visits to 0, counting the visit at time 0. Taking expected value and using Fubini's theorem to put the expected value inside the sum:

$$EV = \sum_{m=0}^{\infty} P(S_m = 0) = \sum_{n=0}^{\infty} P(\tau_n < \infty)$$

$$= \sum_{n=0}^{\infty} P(\tau_1 < \infty)^n = \frac{1}{1 - P(\tau_1 < \infty)}$$

The second equality shows (ii) implies (iii), and in combination with the last two shows that if (i) is false then (iii) is false (i.e., (iii) implies (i)).   □

(2.3) Theorem. Simple random walk is recurrent in $d \leq 2$ and transient in $d \geq 3$.

To steal a joke from Kakutani (U.C.L.A. colloquium talk): "A drunk man will find his way home but a drunk bird may get lost forever."

Proof of (2.3)   Let $\rho_d(m) = P(S_m = 0)$. $\rho_d(m)$ is 0 if m is odd. From (1.4) in Chapter 2 we get $\rho_1(2n) \sim (\pi n)^{-1/2}$ as $n \to \infty$. This and (2.2) gives the result in one dimension. Our next step is

Simple random walk is recurrent in two dimensions. Note that in order for $S_{2n} = 0$ we must for some $0 \leq m \leq n$ have m up steps, m down steps, $n - m$ to the left and $n - m$ to the right so

$$\rho_2(2n) = 4^{-2n} \sum_{m=0}^{n} \frac{2n!}{m!\,m!\,(n-m)!\,(n-m)!}$$

$$= 4^{-2n} \binom{2n}{n} \sum_{m=0}^{n} \binom{n}{m}\binom{n}{n-m} = 4^{-2n}\binom{2n}{n}^2 = \rho_1(2n)^2$$

To see the next to last equality consider choosing n students from a class with n boys and n girls and observe that for some $0 \leq m \leq n$ you must choose m boys and $n - m$ girls. Using the asymptotic formula $\rho_1(2n) \sim (\pi n)^{-1/2}$, we get $\rho_2(2n) \sim (\pi n)^{-1}$. Since $\sum n^{-1} = \infty$ the result follows from (2.2).

Remark. For a direct proof of $\rho_2(2n) = \rho_1(2n)^2$ note that if T_n^1 and T_n^2 are independent one dimensional random walks then T_n jumps from x to $x + (1, 1)$, $x + (1, -1)$, $x + (-1, 1)$, and $x + (-1, -1)$ with equal probability, so rotating T_n by 45 degrees and dividing by $\sqrt{2}$ gives S_n.

Simple random walk is transient in three dimensions. Again since the number of steps in the directions $\pm e_i$ must be equal for $i = 1, 2, 3$

$$\rho_3(2n) = 6^{-2n} \sum_{j,k} \frac{(2n)!}{(j!k!(n-j-k)!)^2}$$

$$= 2^{-2n} \binom{2n}{n} \sum_{j,k} \left(3^{-n} \frac{n!}{j!k!(n-j-k)!} \right)^2$$

$$\leq 2^{-2n} \binom{2n}{n} \max_{j,k} 3^{-n} \frac{n!}{j!k!(n-j-k)!}$$

where in the last inequality we have used the fact that if $a_{j,k}$ are ≥ 0 and sum to 1 then $\sum_{j,k} a_{j,k}^2 \leq \max_{j,k} a_{j,k}$. Our last step is to show

$$\max_{j,k} 3^{-n} \frac{n!}{j!k!(n-j-k)!} \leq Cn^{-1}$$

To do this we note that (a) if any of the numbers j, k or $n-j-k$ is $< [n/3]$ increasing the smallest number and decreasing the largest number decreases the denominator (since $x(1-x)$ is maximized at $1/2$) so the maximum occurs when all three numbers are as close as possible to $n/3$; (b) Stirlings' formula implies

$$\frac{n!}{j!k!(n-j-k)!} \sim \frac{n^n}{j^j k^k (n-j-k)^{n-j-k}} \cdot \sqrt{\frac{n}{jk(n-j-k)}} \cdot \frac{1}{2\pi}$$

Taking j and k within 1 of $n/3$ the first term on the right is $\leq C3^n$ and the desired result follows.

Simple random walk is transient in $d > 3$. Let $T_n = (S_n^1, S_n^2, S_n^3)$, $N(0) = 0$ and $N(n) = \inf\{m > N(n-1) : T_m \neq T_{N(n-1)}\}$. It is easy to see that $T_{N(n)}$ is a three dimensional simple random walk. Since $T_{N(n)}$ returns infinitely often to 0 with probability 0 and the first three coordinates are constant in between the $N(n)$, S_n is transient. $\qquad\square$

Remark. Let $\pi_d = P(S_n = 0$ for some $n \geq 1)$ be the probability simple random walk on $\mathbf{Z}^d$ returns to 0. The last display in the proof of (2.2) implies

$$\sum_{n=0}^{\infty} P(S_{2n} = 0) = \frac{1}{1 - \pi_d}$$

In $d = 3$, $P(S_{2n} = 0) \sim Cn^{-3/2}$ so $\sum_{n=N}^{\infty} P(S_{2n} = 0) \sim C'N^{-1/2}$ and the series converges rather slowly. For example if we want to compute the return probability to 5 decimal places we would need 10^{10} terms. At the end of the section we will give another formula that leads very easily to accurate results.

The rest of this section is devoted to proving the following facts about random walks:

(2.7) S_n is recurrent in $d = 1$ if $S_n/n \to 0$ in probability.

(2.8) S_n is recurrent in $d = 2$ if $S_n/n^{1/2} \Rightarrow$ a nondegenerate normal distribution.

(2.12) S_n is transient in $d \geq 3$ if it is "truly three dimensional".

To prove (2.12), we will give a necessary and sufficient condition for recurrence, (2.9), that shows that the conditions in (2.7) and (2.8) are close to the best possible. The first step in deriving these results is to generalize (2.2):

(2.4) Lemma. If $\sum_{n=1}^{\infty} P(\|S_n\| < \epsilon) < \infty$ then $P(\|S_n\| < \epsilon$ i.o.$) = 0$.
If $\sum_{n=1}^{\infty} P(\|S_n\| < \epsilon) = \infty$ then $P(\|S_n\| < 2\epsilon$ i.o.$) = 1$.

Proof   The first conclusion follows from the Borel-Cantelli lemma. To prove the second let $F = \{\|S_n\| < \epsilon$ i.o.$\}^c$. Breaking things down according to the last time $\|S_n\| < \epsilon$,

$$P(F) = \sum_{m=0}^{\infty} P(\|S_m\| < \epsilon, \|S_n\| \geq \epsilon \text{ for all } n \geq m + 1)$$

$$\geq \sum_{m=0}^{\infty} P(\|S_m\| < \epsilon, \|S_n - S_m\| \geq 2\epsilon \text{ for all } n \geq m + 1)$$

$$= \sum_{m=0}^{\infty} P(\|S_m\| < \epsilon)\rho_{2\epsilon,1}$$

where $\rho_{\delta,k} = P(\|S_n\| \geq \delta$ for all $n \geq k)$. Since $P(F) \leq 1$, and

$$\sum_{m=0}^{\infty} P(\|S_m\| < \epsilon) = \infty$$

it follows that $\rho_{2\epsilon,1} = 0$. To extend this conclusion to $\rho_{2\epsilon,k}$ with $k \geq 2$, let

$$A_m = \{\|S_m\| < \epsilon, \|S_n\| \geq \epsilon \text{ for all } n \geq m + k\}$$

Since any ω can be in at most k of the A_m, repeating the argument above gives

$$k \geq \sum_{m=0}^{\infty} P(A_m) \geq \sum_{m=0}^{\infty} P(\|S_m\| < \epsilon)\rho_{2\epsilon,k}$$

So $\rho_{2\epsilon,k} = P(\|S_n\| \geq 2\epsilon$ for all $j \geq k) = 0$, and since k is arbitrary, the desired conclusion follows.   □

Our second step is to show that the convergence or divergence of the sums in (2.4) is independent of ϵ. The previous proof is valid for any norm. For the next one we need $\|x\| = \sup_i |x_i|$.

(2.5) Lemma. Let m be an integer ≥ 2.

$$\sum_{n=0}^{\infty} P(\|S_n\| < m\epsilon) \leq (2m)^d \sum_{n=0}^{\infty} P(\|S_n\| < \epsilon)$$

Proof We begin by observing

$$\sum_{n=0}^{\infty} P(\|S_n\| < m\epsilon) \leq \sum_{n=0}^{\infty} \sum_{k} P(S_n \in k\epsilon + [0,\epsilon)^d)$$

where the inner sum is over $k \in \{-m, \ldots, m-1\}^d$. If we let

$$T_k = \inf\{m \geq 0 : S_m \in k\epsilon + [0,\epsilon)^d\}$$

then breaking things down according to the value of T_k, and using Fubini's theorem gives

$$\sum_{n=0}^{\infty} P(S_n \in k\epsilon + [0,\epsilon)^d) = \sum_{m=0}^{\infty} \sum_{n=0}^{\infty} P(S_n \in k\epsilon + [0,\epsilon)^d, T_k = m)$$

$$\leq \sum_{m=0}^{\infty} \sum_{n=m}^{\infty} P(\|S_n - S_m\| < \epsilon, T_k = m)$$

Since $\{T_k = m\}$ and $\{\|S_n - S_m\| < \epsilon\}$ are independent, the last sum

$$= \sum_{m=0}^{\infty} P(T_k = m) \sum_{\ell=0}^{\infty} P(\|S_\ell\| < \epsilon) \leq \sum_{\ell=0}^{\infty} P(\|S_\ell\| < \epsilon)$$

Since there are $(2m)^d$ values of k in $\{-m, \ldots, m-1\}^d$, the proof of (2.5) is complete. $\square$

Combining (2.4) and (2.5) gives

(2.6) Corollary. The convergence (resp. divergence) of $\sum_n P(\|S_n\| < \epsilon)$ for a single value of $\epsilon > 0$ is sufficient for transience (resp. recurrence).

In $d = 1$ if $EX_i = \mu \neq 0$ then the strong law of large numbers implies $S_n/n \to \mu$ so $|S_n| \to \infty$ and S_n is transient. As a converse we have

(2.7) Chung-Fuchs theorem. Suppose $d = 1$. If the weak law of large numbers holds in the form $S_n/n \to 0$ in probability then S_n is recurrent.

Proof Let $u_n(x) = P(|S_n| < x)$ for $x > 0$. (2.5) implies

$$\sum_{n=0}^{\infty} u_n(1) \geq \frac{1}{2m} \sum_{n=0}^{\infty} u_n(m) \geq \frac{1}{2m} \sum_{n=0}^{Am} u_n(n/A)$$

for any $A < \infty$ since $u_n(x) \geq 0$ and is increasing in x. By hypothesis $u_n(n/A) \to 1$, so letting $m \to \infty$ and noticing the right-hand side is $A/2$ times the average of the first Am terms

$$\sum_{n=0}^{\infty} u_n(1) \geq A/2$$

Since A is arbitrary the sum must be ∞ and the desired conclusion follows from (2.6). $\qquad\qquad\square$

(2.8) Theorem. If S_n is a random walk in $\mathbf{R}^2$ and $S_n/n^{1/2} \Rightarrow$ a nondegenerate normal distribution then S_n is recurrent.

Remark. The conclusion is also true if the limit is degenerate but in that case the random walk is essentially one (or zero) dimensional, and the result follows from the Chung-Fuchs theorem.

Proof Let $u(n,m) = P(\|S_n\| < m)$. (2.5) implies

$$\sum_{n=0}^{\infty} u(n,1) \geq (4m^2)^{-1} \sum_{n=0}^{\infty} u(n,m)$$

If $m/\sqrt{n} \to c$ then

$$u(n,m) \to \int_{[-c,c]^2} n(x)\, dx$$

where $n(x)$ is the density of the limiting normal distribution. If we use $\rho(c)$ to denote the right-hand side, and let $n = [\theta m^2]$, it follows that $u([\theta m^2], m) \to \rho(\theta^{-1/2})$. If we write

$$m^{-2} \sum_{n=0}^{\infty} u(n,m) = \int_0^\infty u([\theta m^2], m)\, d\theta$$

let $m \to \infty$, and use Fatou's lemma we get

$$\liminf_{m\to\infty} (4m^2)^{-1} \sum_{n=0}^{\infty} u(n,m) \geq 4^{-1} \int_0^\infty \rho(\theta^{-1/2})\, d\theta$$

Since the normal density is positive and continuous at 0

$$\rho(c) = \int_{[-c,c]^2} n(x)\, dx \sim n(0)(2c)^2$$

as $c \to 0$. So $\rho(\theta^{-1/2}) \sim 4n(0)/\theta$ as $\theta \to \infty$, the integral diverges and back-tracking to the first inequality in the proof it follows that $\sum_{n=0}^{\infty} u(n, 1) = \infty$, proving (2.8). $\qquad \square$

We come now to the promised necessary and sufficient condition for recurrence. Here $\varphi = E \exp(itX_j)$ is the ch.f. of one step of the random walk.

(2.9) Theorem. Let $\delta > 0$. S_n is recurrent if and only if

$$\int_{(-\delta,\delta)^d} \text{Re} \, \frac{1}{1 - \varphi(y)} \, dy = \infty$$

We will prove a weaker result:

(2.9') Theorem. Let $\delta > 0$. S_n is recurrent if and only if

$$\sup_{r<1} \int_{(-\delta,\delta)^d} \text{Re} \, \frac{1}{1 - r\varphi(y)} \, dy = \infty$$

Remark. Half of the work needed to get (2.9) from (2.9') is trivial.

$$0 \le \text{Re} \, \frac{1}{1 - r\varphi(y)} \to \text{Re} \, \frac{1}{1 - \varphi(y)} \quad \text{as } r \to 1$$

so Fatou's lemma shows that if the integral is infinite, the walk is recurrent. The other direction is rather difficult: (2.9') is in Chung and Fuchs (1951) but a proof of (2.9) had to wait for Ornstein (1969) and Stone (1969) to solve the problem independently. Their proofs use a trick to reduce to the case where the increments have a density and then a second trick to deal with that case, so we will not give the details here. The reader can consult either of the sources cited, or Port and Stone (1969) where the result is demonstrated for random walks on Abelian groups.

Proof of (2.9') The first ingredient in the solution is the

(2.10) Parseval relation. Let μ and ν be probability measures on $\mathbf{R}^d$ with ch.f.'s φ and ψ.

$$\int \psi(t) \, \mu(dt) = \int \varphi(x) \, \nu(dx)$$

Proof Since e^{itx} is bounded, Fubini's theorem implies

$$\int \psi(t)\mu(dt) = \int \int e^{itx}\nu(dx)\mu(dt) = \int \int e^{itx}\mu(dt)\nu(dx) = \int \varphi(x)\nu(dx) \quad \square$$

Our second ingredient is a little calculus.

(2.11) Lemma. If $|x| \leq \pi/3$ then $1 - \cos x \geq x^2/4$.

Proof   It suffices to prove the result for $x > 0$. If $z \leq \pi/3$ then $\cos z \geq 1/2$,

$$\sin y = \int_0^y \cos z \, dz \geq \frac{y}{2}$$

$$1 - \cos x = \int_0^x \sin y \, dy \geq \int_0^x \frac{y}{2} \, dy = \frac{x^2}{4} \qquad \square$$

From Example 3.5 in Chapter 2, we see that the density

$$\frac{\delta - |x|}{\delta^2} \quad \text{when} \quad |x| \leq \delta, \qquad 0 \quad \text{otherwise}$$

has ch.f. $2(1 - \cos \delta t)/(\delta t)^2$. Let μ_n denote the distribution of S_n. Using (2.11) (note $\pi/3 \geq 1$) and then (2.10), we have

$$P(\|S_n\| < 1/\delta) \leq 4^d \int \prod_{i=1}^{d} \frac{1 - \cos(\delta t_i)}{(\delta t_i)^2} \mu_n(dt)$$

$$= 2^d \int_{(-\delta, \delta)^d} \prod_{i=1}^{d} \frac{\delta - |x_i|}{\delta^2} \varphi^n(x) \, dx$$

Our next step is to sum from 0 to ∞. To be able to interchange the sum and the integral we first multiply by r^n where $r < 1$.

$$\sum_{n=0}^{\infty} r^n P(\|S_n\| < 1/\delta) \leq 2^d \int_{(-\delta, \delta)^d} \prod_{i=1}^{d} \frac{\delta - |x_i|}{\delta^2} \frac{1}{1 - r\varphi(x)} \, dx$$

Symmetry dictates that the integral on the right is real so we can take the real part without affecting its value. Letting $r \uparrow 1$ and using $(\delta - |x|)/\delta \leq 1$

$$\sum_{n=0}^{\infty} P(\|S_n\| < 1/\delta) \leq \left(\frac{2}{\delta}\right)^d \sup_{r<1} \int_{(-\delta, \delta)^d} \text{Re} \, \frac{1}{1 - r\varphi(x)} \, dx$$

and using (2.6) gives half of (2.9′).

To prove the other direction, we begin by noting that Example 3.8 from Chapter 2 shows that the density

$$\frac{1 - \cos(x/\delta)}{\pi x^2/\delta}$$

has ch.f. $1 - |\delta t|$ when $|t| \leq 1/\delta$, 0 otherwise. Using $1 \geq \prod_{i=1}^{d}(1 - |\delta x_i|)$ and then (2.10)

$$P(\|S_n\| < 1/\delta) \geq \int_{(-1/\delta, 1/\delta)^d} \prod_{i=1}^{d}(1 - |\delta x_i|)\, \mu_n(dx)$$

$$= \int \prod_{i=1}^{d} \frac{1 - \cos(t_i/\delta)}{\pi t_i^2/\delta}\, \varphi^n(t)\, dt$$

Multiplying by r^n and summing gives

$$\sum_{n=0}^{\infty} r^n P(\|S_n\| < 1/\delta) \geq \int \prod_{i=1}^{d} \frac{1 - \cos(t_i/\delta)}{\pi t_i^2/\delta} \frac{1}{1 - r\varphi(t)}\, dt$$

The last integral is real, so its value is unaffected if we integrate only the real part of the integrand. If we do this and apply (2.11) we get

$$\sum_{n=0}^{\infty} r^n P(\|S_n\| < 1/\delta) \geq (4\pi\delta)^{-d} \int_{(-\delta,\delta)^d} \text{Re}\, \frac{1}{1 - r\varphi(t)}\, dt$$

Letting $r \uparrow 1$ and using (2.6) now, completes the proof of (2.9'). $\qquad\square$

We will now consider some examples. Our goal in $d = 1$ and $d = 2$ is to convince you that the conditions in (2.7) and (2.8) are close to the best possible.

d = 1. Consider the symmetric stable laws that have ch.f. $\varphi(t) = \exp(-|t|^\alpha)$. To avoid using facts that we have not proved, we will obtain our conclusions from (2.9'). It is not hard to use that form of the criterion in this case since

$$1 - r\varphi(t) \downarrow 1 - \exp(-|t|^\alpha) \quad \text{as } r \uparrow 1$$
$$1 - \exp(-|t|^\alpha) \sim |t|^\alpha \quad \text{as } t \to 0$$

From this it follows that the corresponding random walk is transient for $\alpha < 1$ and recurrent for $\alpha \geq 1$. The case $\alpha > 1$ is covered by (2.7) since these random walks have mean 0. The result for $\alpha = 1$ is new because the Cauchy distribution does not satisfy $S_n/n \to 0$ in probability. The random walks with $\alpha < 1$ are interesting because (1.2) implies (see Exercise 1.1)

$$-\infty = \liminf S_n < \limsup S_n = \infty$$

but $P(|S_n| < M \text{ i.o.}) = 0$ for any $M < \infty$.

Remark. The stable law examples are misleading in one respect. Shepp (1964) has proved that recurrent random walks may have arbitrarily large tails. To be precise, given a function $\epsilon(x) \downarrow 0$ as $x \uparrow \infty$, there is a recurrent random walk with $P(|X_1| \geq x) \geq \epsilon(x)$ for large x.

d = 2. Let $\alpha < 2$, and let $\varphi(t) = \exp(-|t|^\alpha)$ where $|t| = (t_1^2 + t_2^2)^{1/2}$. φ is the characteristic function of a random vector (X_1, X_2) that has two nice properties:

(i) the distribution of (X_1, X_2) is invariant under rotations,

(ii) X_1 and X_2 have symmetric stable laws with index α.

Again $1 - r\varphi(t) \downarrow 1 - \exp(-|t|^\alpha)$ as $r \uparrow 1$ and $1 - \exp(-|t|^\alpha) \sim |t|^\alpha$ as $t \to 0$. Changing to polar coordinates and noticing

$$2\pi \int_0^\delta dx\, x\, x^{-\alpha} < \infty$$

when $1 - \alpha > -1$ shows the random walks with ch.f. $\exp(-|t|^\alpha)$, $\alpha < 2$ are transient. When $p < \alpha$ we have $E|X_1|^p < \infty$ by Exercise 7.5 in Chapter 2, so these examples show that (2.8) is reasonably sharp.

d $\geq$ 3. The integral $2\pi \int_0^\delta dx\, x^{d-1}\, x^{-2} < \infty$ so if a random walk is recurrent in $d \geq 3$, its ch.f. must $\to 1$ faster than t^2. In Exercise 3.19 of Chapter 2 we observed that (in one dimension) if $\varphi(r) = 1 + o(r^2)$ then $\varphi(r) \equiv 1$. By considering $\varphi(r\theta)$ where r is real and θ is a fixed vector, the last conclusion generalizes easily to $\mathbf{R}^d$, $d > 1$ and suggests that once we exclude walks that stay on a plane through 0 no three dimensional random walks are recurrent.

   A random walk in $\mathbf{R}^3$ is **truly three dimensional** if the distribution of X_1 has $P(X_1 \cdot \theta \neq 0) > 0$ for all $\theta \neq 0$.

(2.12) Theorem. No truly three dimensional random walk is recurrent.

Proof   We will deduce the result from (2.9′). We begin with some arithmetic. If z is complex, the conjugate of $1 - z$ is $1 - \bar{z}$, so

$$\frac{1}{1-z} = \frac{1-\bar{z}}{|1-z|^2} \quad \text{and} \quad \mathrm{Re}\,\frac{1}{1-z} = \frac{\mathrm{Re}\,(1-z)}{|1-z|^2}$$

If $z = a + bi$ with $a \leq 1$ then using the previous formula and dropping the b^2 from the denominator

$$\mathrm{Re}\,\frac{1}{1-z} = \frac{1-a}{(1-a)^2 + b^2} \leq \frac{1}{1-a}$$

Taking $z = r\varphi(t)$ and supposing for the second inequality that $0 \le \operatorname{Re} \varphi(t) \le 1$ we have

(a) $$\operatorname{Re} \frac{1}{1 - r\varphi(t)} \le \frac{1}{\operatorname{Re}(1 - r\varphi(t))} \le \frac{1}{\operatorname{Re}(1 - \varphi(t))}$$

The last calculation shows that it is enough to estimate

$$\operatorname{Re}(1 - \varphi(t)) = \int \{1 - \cos(x \cdot t)\}\mu(dx) \ge \int_{|x \cdot t| < \pi/3} \frac{|x \cdot t|^2}{4} \mu(dx)$$

by (2.11). Writing $t = \rho\theta$ where $\theta \in S = \{x : |x| = 1\}$ gives

(b) $$\operatorname{Re}(1 - \varphi(\rho\theta)) \ge \frac{\rho^2}{4} \int_{|x \cdot \theta| < \pi/3\rho} |x \cdot \theta|^2 \mu(dx)$$

Fatou's lemma implies that if we let $\rho \to 0$ and $\theta(\rho) \to \theta$ then

(c) $$\liminf_{\rho \to 0} \int_{|x \cdot \theta(\rho)| < \pi/3\rho} |x \cdot \theta(\rho)|^2 \mu(dx) \ge \int |x \cdot \theta|^2 \mu(dx) > 0$$

I claim this implies that for $\rho < \rho_0$

(d) $$\inf_{\theta \in S} \int_{|x \cdot \theta| < \pi/3\rho} |x \cdot \theta|^2 \mu(dx) = C > 0$$

To get the last conclusion, observe that if it is false then for $\rho = 1/n$ there is a θ_n so that

$$\int_{|x \cdot \theta_n| < n\pi/3} |x \cdot \theta_n|^2 \mu(dx) \le 1/n$$

All the θ_n lie in S, a compact set, so if we pick a convergent subsequence we contradict (c). Combining (b) and (d) gives

$$\operatorname{Re}(1 - \varphi(\rho\theta)) \ge C\rho^2/4$$

Using the last result and (a) then changing to polar coordinates we see that if δ is small (so $\operatorname{Re} \varphi(y) \ge 0$ on $(-\delta, \delta)^d$)

$$\int_{(-\delta,\delta)^d} \operatorname{Re} \frac{1}{1 - r\varphi(y)} \, dy \le \int_0^{\delta\sqrt{d}} d\rho\, \rho^{d-1} \int d\theta \frac{1}{\operatorname{Re}(1 - \varphi(\rho\theta))}$$

$$\le C' \int_0^1 d\rho\, \rho^{d-3} < \infty$$

when $d > 2$ so the desired result follows from (2.9'). $\qquad\qquad\qquad\square$

Remark. The analysis becomes much simpler when we consider random walks on $\mathbf{Z}^d$. The inversion formula given in Exericse 3.1 of Chapter 2 implies

$$P(S_n = 0) = (2\pi)^{-d} \int_{(-\pi,\pi)^d} \varphi^n(t)\, dt$$

Multiplying by r^n and summing gives

$$\sum_{n=0}^{\infty} r^n P(S_n = 0) = (2\pi)^{-d} \int_{(-\pi,\pi)^d} \frac{1}{1 - r\varphi(t)}\, dt$$

In the case of simple random walk in $d = 3$, $\varphi(t) = \frac{1}{3}\sum_{j=1}^{3} \cos t_j$ is real.

$$\frac{1}{1 - r\varphi(t)} \uparrow \frac{1}{1 - \varphi(t)} \quad \text{when } \varphi(t) > 0$$

$$0 \le \frac{1}{1 - r\varphi(t)} \le 1 \quad \text{when } \varphi(t) \le 0$$

So using the monotone and bounded convergence theorems

$$\sum_{n=0}^{\infty} P(S_n = 0) = (2\pi)^{-3} \int_{(-\pi,\pi)^3} \left(1 - \frac{1}{3}\sum_{m=1}^{3} \cos x_i \right)^{-1} dx$$

This integral was first evaluated by Watson in 1939 in terms of elliptic integrals, which could be found in tables. Glasser and Zucker (1977) showed that it was

$$(\sqrt{6}/32\pi^3)\Gamma(1/24)\Gamma(5/24)\Gamma(7/24)\Gamma(11/24) = 1.516386059137\ldots$$

so it follows from the remark after the proof of (2.3) that

$$\pi_3 = .340537329544\ldots$$

For numerical results in $4 \le d \le 9$, see Kondo and Hara (1987).

*3.3. Visits to 0, Arcsine Laws

In the last section we took a broad look at the recurrence of random walks. In this section we will take a deep look at one example: simple random walk (on $\mathbf{Z}$). To steal a line from Chung, "We shall treat this by combinatorial methods

as an antidote to the analytic skulduggery above". The developments here
follow Chapter III of Feller, vol. I. To facilitate discussion we will think of the
sequence $S_1, S_2, \ldots, S_n$ as being represented by a polygonal line with segments
$(k-1, S_{k-1}) \to (k, S_k)$. A **path** is a polygonal line that is a possible outcome
of simple random walk. To count the number of paths from (0,0) to (n, x) it
is convenient to introduce a and b defined by: $a = (n+x)/2$ is the number of
positive steps in the path and $b = (n-x)/2$ is the number of negative steps.
Notice that $n = a + b$ and $x = a - b$. If $-n \le x \le n$ and $n - x$ is even, the
a and b defined above are nonnegative integers, and the number of paths from
(0,0) to (n, x) is

$$(*) \qquad\qquad N_{n,x} = \binom{n}{a}$$

Otherwise the number of paths is 0.

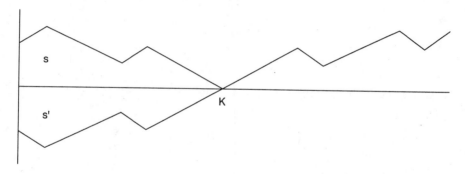

Figure 3.3.1

(3.1) Reflection principle. If $x, y > 0$ then the number of paths from $(0, x)$
to (n, y) that touch the x-axis = the number of paths from $(0, -x)$ to (n, y).

Proof Suppose $(0, s_0), (1, s_1), \ldots, (n, s_n)$ is a path from $(0, x)$ to (n, y). Let
$K = \inf\{k : s_k = 0\}$. Let $s'_k = -s_k$ for $k \le K$, $s'_k = s_k$ for $K \le k \le n$.
Then (k, s'_k), $0 \le k \le n$, is a path from $(0, -x)$ to (n, y). Conversely, if
$(0, t_0), (1, t_1), \ldots, (n, t_n)$ is a path from $(0, -x)$ to (n, y) then it must cross 0.
Let $K = \inf\{k : t_k = 0\}$. Let $t'_k = -t_k$ for $k \le K$, $t'_k = t_k$ for $K \le k \le n$.
Then (k, t'_k), $0 \le k \le n$, is a path from $(0, -x)$ to (n, y) that touches the x-
axis at K. (See Figure 3.3.1.) The last two observations set up a one-to-one
correspondence between the two classes of paths, so their numbers must be
equal. $\qquad\square$

From (3.1) we get a result first proved in 1878.

(3.2) The Ballot Theorem. Suppose that in an election candidate A gets α votes, and candidate B gets β votes where $\beta < \alpha$. The probability that throughout the counting A always leads B is $(\alpha - \beta)/(\alpha + \beta)$.

Proof Let $x = \alpha - \beta$, $n = \alpha + \beta$. Clearly there are as many such outcomes as there are paths from $(1,1)$ to (n, x) that do not touch the axis. The reflection principle implies that the number of paths from $(1,1)$ to (n, x) that touch the axis = the number of paths from $(1,-1)$ to (n, x), so by $(*)$ the number of paths from $(1,1)$ to (n, x) that do not touch the axis is

$$N_{n-1,x-1} - N_{n-1,x+1} = \binom{n-1}{\alpha-1} - \binom{n-1}{\alpha}$$

$$= \frac{(n-1)!}{(\alpha-1)!(n-\alpha)!} - \frac{(n-1)!}{\alpha!(n-\alpha-1)!}$$

$$= \frac{\alpha - (n-\alpha)}{n} \cdot \frac{n!}{\alpha!(n-\alpha)!} = \frac{\alpha-\beta}{\alpha+\beta} N_{n,x}$$

since $n = \alpha + \beta$, proving (3.2). $\square$

Using the ballot theorem we can compute the distribution of the time to hit 0.

(3.3) Lemma. $P(S_1 \neq 0, \ldots, S_{2n} \neq 0) = P(S_{2n} = 0)$.

Proof $P(S_1 > 0, \ldots, S_{2n} > 0) = \sum_{r=1}^{\infty} P(S_1 > 0, \ldots, S_{2n-1} > 0, S_{2n} = 2r)$. From the proof of (3.2) we see that the number of paths from $(0,0)$ to $(2n, 2r)$ that do not touch the axis at positive times (= the number of paths from $(1,1)$ to $(2n, 2r)$ that do not touch the axis) is

$$N_{2n-1,2r-1} - N_{2n-1,2r+1}$$

If we let $p_{n,x} = P(S_n = x)$ then this implies

$$P(S_1 > 0, \ldots, S_{2n-1} > 0, S_{2n} = 2r) = \frac{1}{2}(p_{2n-1,2r-1} - p_{2n-1,2r+1})$$

Summing from $r = 1$ to ∞ gives

$$P(S_1 > 0, \ldots, S_{2n} > 0) = \frac{1}{2}p_{2n-1,1} = \frac{1}{2}P(S_{2n} = 0)$$

Symmetry implies $P(S_1 < 0, \ldots, S_{2n} < 0) = (1/2)P(S_{2n} = 0)$ and the proof is complete. $\square$

Let $R = \inf\{m \geq 1 : S_m = 0\}$. Combining (3.3) with (1.4) from Chapter 2 gives

$$(3.4) \qquad P(R > 2n) = P(S_{2n} = 0) \sim \pi^{-1/2} n^{-1/2}$$

Since $P(R > x) / P(|R| > x) = 1$, it follows from (7.7) in Chapter 2 that R is in the domain of attraction of the stable law with $\alpha = 1/2$ and $\kappa = 1$. This implies that if R_n is the time of the nth return to 0 then $R_n/n^2 \Rightarrow Y$, the indicated stable law. In Example 7.2 in Chapter 2, we considered $\tau = T_1$ where $T_x = \inf\{n : S_n = x\}$. Since $S_1 \in \{-1, 1\}$ and $T_1 =_d T_{-1}$, $R =_d 1 + T_1$, and it follows that $T_n/n^2 \Rightarrow Y$, the same stable law. In Example 6.6 of Chapter 7, we will use this observation to find the density of the limit.

This completes our discussion of visits to 0. We turn now to the arcsine laws. The first one concerns

$$L_{2n} = \sup\{m \leq 2n : S_m = 0\}$$

It is remarkably easy to compute the distribution of L_{2n}.

(3.5) Lemma. Let $u_{2m} = P(S_{2m} = 0)$. Then $P(L_{2n} = 2k) = u_{2k} u_{2n-2k}$.

Proof $P(L_{2n} = 2k) = P(S_{2k} = 0, S_{2k+1} \neq 0, \ldots, S_{2n} \neq 0)$, so the desired result follows from (3.3). $\qquad \square$

(3.6) Arcsine law for the last visit to 0. For $0 < a < b < 1$

$$P(a \leq L_{2n}/2n \leq b) \rightarrow \int_a^b \pi^{-1}(x(1-x))^{-1/2}\, dx$$

To see the reason for the name substitute $y = x^{1/2}$, $dy = (1/2)x^{-1/2}\, dx$ in the integral to obtain

$$\int_{\sqrt{a}}^{\sqrt{b}} \frac{2}{\pi}(1 - y^2)^{-1/2}\, dy = \frac{2}{\pi}\{\arcsin(\sqrt{b}) - \arcsin(\sqrt{a})\}$$

The symmetry of the limit distribution implies

$$P(L_{2n}/2n \leq 1/2) \rightarrow 1/2$$

In gambling terms, if two people were to bet \$1 on a coin flip every day of the year then with probability 1/2 one of the players will be ahead from July 1 to the end of the year, an event that would undoubtedly cause the other player to complain about his bad luck. The next result deals directly with the amount of time one player is ahead.

Proof of (3.6) From the asymptotic formula for u_{2n} it follows that if $k/n \to x$ then

$$nP(L_{2n} = 2k) \to \pi^{-1}(x(1-x))^{-1/2}$$

To get from this to (3.6) we let $2na_n =$ the smallest even integer $\geq 2na$, let $2nb_n =$ the largest even integer $\leq 2nb$, and let $f_n(x) = nP(L_{2n} = k)$ for $2k/2n \leq x < 2(k+1)/2n$ so we can write

$$P(a \leq L_{2n}/2n \leq b) = \sum_{k=na_n}^{nb_n} P(L_{2n} = 2k) = \int_{a_n}^{b_n+1/n} f_n(x)\,dx$$

Our first result implies that uniformly on compact sets

$$f_n(x) \to f(x) = \pi^{-1}(x(1-x))^{-1/2}$$

The uniformity of the convergence implies

$$\sup_{a_n \leq x \leq b_n+1/n} f_n(x) \to \sup_{a \leq x \leq b} f(x) < \infty$$

if $0 < a \leq b < 1$ so the bounded convergence theorem gives

$$\int_{a_n}^{b_n+1/n} f_n(x)\,dx \to \int_a^b f(x)\,dx \qquad \square$$

(3.7) Theorem. Let π_{2n} be the number of segments $(k-1, S_{k-1}) \to (k, S_k)$ that lie above the axis (i.e., in $\{(x,y) : x \geq 0\}$), and let $u_m = P(S_m = 0)$.

$$P(\pi_{2n} = 2k) = u_{2k} u_{2n-2k}$$

and consequently, if $0 < a < b < 1$

$$P(a \leq \pi_{2n}/2n \leq b) \to \int_a^b \pi^{-1}(x(1-x))^{-1/2}\,dx$$

Remark. Since $\pi_{2n} =_d L_{2n}$, the second conclusion follows from the proof of (3.6). The reader should note that the limiting density $\pi^{-1}(x(1-x))^{-1/2}$ has a minimum at $x = 1/2$, and $\to \infty$ as $x \to 0$ or 1. An equal division of steps between the positive and negative side is therefore the least likely possibility, and completely one sided divisions have the highest probability.

Proof   Let $\beta_{2k,2n}$ denote the probability of interest. We will prove $\beta_{2k,2n} = u_{2k}u_{2n-2k}$ by induction. When $n = 1$ it is clear that

$$\beta_{0,2} = \beta_{2,2} = 1/2 = u_0 u_2$$

For a general n, first suppose $k = n$. From the proof of (3.3) we have

$$\frac{1}{2}u_{2n} = P(S_1 > 0, \ldots, S_{2n} > 0)$$
$$= P(S_1 = 1, S_2 - S_1 \geq 0, \ldots, S_{2n} - S_1 \geq 0)$$
$$= \frac{1}{2}P(S_1 \geq 0, \ldots, S_{2n-1} \geq 0)$$
$$= \frac{1}{2}P(S_1 \geq 0, \ldots, S_{2n} \geq 0) = \frac{1}{2}\beta_{2n,2n}$$

The next to last equality follows from the observation that if $S_{2n-1} \geq 0$ then $S_{2n-1} \geq 1$ and hence $S_{2n} \geq 0$.

The last computation proves the result for $k = n$. Since $\beta_{0,2n} = \beta_{2n,2n}$, the result is also true when $k = 0$. Suppose now that $1 \leq k \leq n - 1$. In this case if R is the time of the first return to 0, then $R = 2m$ with $0 < m < n$. Letting $f_{2m} = P(R = 2m)$ and breaking things up according to whether the first excursion was on the positive or negative side gives

$$\beta_{2k,2n} = \frac{1}{2}\sum_{m=1}^{k} f_{2m}\beta_{2k-2m,2n-2m} + \frac{1}{2}\sum_{m=1}^{n-k} f_{2m}\beta_{2k,2n-2m}$$

Using the induction hypothesis, it follows that

$$\beta_{2k,2n} = \frac{1}{2}u_{2n-2k}\sum_{m=1}^{k} f_{2m}u_{2k-2m} + \frac{1}{2}u_{2k}\sum_{m=1}^{n-k} f_{2m}u_{2n-2k-2m}$$

By considering the time of the first return to 0, we see

$$u_{2k} = \sum_{m=1}^{k} f_{2m} \qquad u_{2n-2k} = \sum_{m=1}^{n-k} f_{2m}u_{2n-2k-2m}$$

and the desired result follows.   □

Our derivation of (3.7) relied heavily on special properties of simple random walk. There is a closely related result due to E. Sparre-Andersen that is valid for very general random walks. However, notice that the hypothesis (ii) in (3.8) below excludes simple random walk.

(3.8) Theorem. Let $\nu_n = |\{k : 1 \le k \le n, S_k > 0\}|$. Then

(i) $P(\nu_n = k) = P(\nu_k = k)P(\nu_{n-k} = 0)$

(ii) If the distribution of X_1 is symmetric and $P(S_m = 0) = 0$ for all $m \ge 1$ then

$$P(\nu_n = k) = u_{2k}u_{2n-2k}$$

where $u_{2m} = 2^{-2m}\binom{2m}{m}$ is the probability simple random walk is 0 at time $2m$.

(iii) Under the hypotheses of (ii),

$$P(a \le \nu_n/n \le b) \to \int_a^b \pi^{-1}(x(1-x))^{-1/2}\,dx \quad \text{for } 0 < a < b < 1$$

Proof Taking things in reverse order, (iii) is an immediate consequence of (ii) and the proof of (3.6). Our next step is to show (ii) follows from (i) by induction. When $n = 1$, our assumptions imply $P(\nu_1 = 0) = 1/2 = u_0u_2$. If $n > 1$ and $1 < k < n$ then (i) and the induction hypothesis imply

$$P(\nu_n = k) = u_{2k}u_0 \cdot u_0 u_{2n-2k} = u_{2k}u_{2n-2k}$$

since $u_0 = 1$. To handle the cases $k = 0$ and $k = n$, we note that (3.5) implies

$$\sum_{k=0}^{n} u_{2k}u_{2n-2k} = 1$$

We have $\sum_{k=0}^{n} P(\nu_n = k) = 1$ and our assumptions imply $P(\nu_n = 0) = P(\nu_n = n)$, so these probabilities must be equal to u_0u_{2n}.

The proof of (i) is tricky and requires careful definitions since we are not supposing X_1 is symmetric or that $P(S_m = 0) = 0$. Let $\nu'_n = |\{k : 1 \le k \le n, S_k \le 0\}| = n - \nu_n$.

$$M_n = \max_{0 \le j \le n} S_j \qquad \ell_n = \min\{j : 0 \le j \le n, S_j = M_n\}$$

$$M'_n = \min_{0 \le j \le n} S_j \qquad \ell'_n = \max\{j : 0 \le j \le n, S_j = M'_n\}$$

The first symmetry is straightforward.

(3.9) Lemma. (ℓ_n, S_n) and $(n - \ell'_n, S_n)$ have the same distribution.

Proof If we let $T_k = S_n - S_{n-k} = X_n + \cdots + X_{n-k+1}$ then T_k $0 \le k \le n$ has the same distribution as S_k, $0 \le k \le n$. Clearly

$$\max_{0 \le k \le n} T_k = S_n - \min_{0 \le k \le n} S_{n-k}$$

and the set of k for which the extrema are attained are the same.

The second symmetry is much less obvious.

(3.10) Lemma. (ℓ_n, S_n) and (ν_n, S_n) have the same distribution. (ℓ'_n, S_n) and (ν'_n, S_n) have the same distribution.

Remark. (i) follows from (3.10) and the trivial observation

$$P(\ell_n = k) = P(\ell_k = k)P(\ell_{n-k} = 0)$$

so once (3.10) is established, the proof of (3.8) will be complete.

Proof of (3.10) When $n = 1$, $\{\ell_1 = 0\} = \{S_1 \le 0\} = \{\nu_1 = 0\}$, and $\{\ell'_1 = 0\} = \{S_1 > 0\} = \{\nu'_1 = 0\}$. We shall prove the general case by induction, supposing that both statements have been proved when n is replaced by $n - 1$. Let

$$G(y) = P(\ell_{n-1} = k, S_{n-1} \le y)$$
$$H(y) = P(\nu_{n-1} = k, S_{n-1} \le y)$$

On $\{S_n \le 0\}$, we have $\ell_{n-1} = \ell_n$, and $\nu_{n-1} = \nu_n$ so if $F(y) = P(X_1 \le y)$ then for $x \le 0$

(*) $\qquad P(\ell_n = k, S_n \le x) = \displaystyle\int F(x - y)\, dG(y)$

$$= \int F(x - y)\, dH(y) = P(\nu_n = k, S_n \le x)$$

On $\{S_n > 0\}$, we have $\ell'_{n-1} = \ell'_n$, and $\nu'_{n-1} = \nu'_n$ so repeating the last computation shows that for $x \ge 0$

$$P(\ell'_n = n - k, S_n > x) = P(\nu'_n = n - k, S_n > x)$$

Since (ℓ_n, S_n) has the same distribution as $(n - \ell'_n, S_n)$ and $\nu'_n = n - \nu_n$, it follows that for $x \ge 0$

$$P(\ell_n = k, S_n > x) = P(\nu_n = k, S_n > x)$$

Setting $x = 0$ in the last result and $(*)$, and adding gives

$$P(\ell_n = k) = P(\nu_n = k)$$

Subtracting the last two equations, and combining the result with $(*)$ gives

$$P(\ell_n = k, S_n \leq x) = P(\nu_n = k, S_n \leq x)$$

for all x. Since (ℓ_n, S_n) has the same distribution as $(n - \ell'_n, S_n)$ and $\nu'_n = n - \nu_n$, it follows that

$$P(\ell'_n = n - k, S_n > x) = P(\nu'_n = n - k, S_n > x)$$

for all x. This completes the proof of (3.10) and hence of (3.8). $\square$

*3.4. Renewal Theory

Let $\xi_1, \xi_2, \ldots$ be i.i.d. positive random variables with distribution F and define a sequence of times by $T_0 = 0$, and $T_k = T_{k-1} + \xi_k$ for $k \geq 1$. As explained in Section 1.7, we think of ξ_i as the lifetime of the ith lightbulb, and T_k is the time the kth bulb burns out. A second interpretation from Section 2.6 is that T_k is the time of arrival of the kth customer. To have a neutral terminology we will refer to the T_k as renewals. The term renewal refers to the fact that the process "starts afresh" at T_k, i.e., $\{T_{k+j} - T_k, j \geq 1\}$ has the same distribution as $\{T_j, j \geq 1\}$.

Departing slightly from the notation in Sections 1.7 and 2.6, we let $N_t = \inf\{k : T_k > t\}$. N_t is the number of renewals in $[0, t]$, counting the renewal at time 0. In Chapter 1 we showed (see (7.5)) that

(4.1) Theorem. As $t \to \infty$, $N_t/t \to 1/\mu$ a.s. where $\mu = E\xi_i \in (0, \infty]$ and $1/\infty = 0$.

Our first result concerns the asymptotic behavior of $U(t) = EN_t$.

(4.2) Theorem. As $t \to \infty$, $U(t)/t \to 1/\mu$.

Proof   We will apply Wald's equation to the stopping time N_t. The first step is to show that $P(\xi_i > 0) > 0$ implies $EN_t < \infty$. To do this pick $\delta > 0$ so that $P(\xi_i > \delta) = \epsilon > 0$ and pick K so that $K\delta \geq t$. Since K consecutive $\xi'_i s$ that are $> \delta$ will make $S_n > t$, we have

$$P(N_t > mK) \leq (1 - \epsilon^K)^m$$

and $EN_t < \infty$. If $\mu < \infty$, applying Wald's equation now gives

$$\mu EN_t = ES_{N_t} \geq t$$

so $U(t) \geq t/\mu$. The last inequality is trivial when $\mu = \infty$ so it holds in general.

Turning to the upper bound, we observe that if $P(\xi_i \leq c) = 1$, then repeating the last argument shows $\mu EN_t = ES_{N_t} \leq t + c$ and the result holds for bounded distributions. If we let $\bar{\xi}_i = \xi_i \wedge c$ and define $\bar{T}_n$ and $\bar{N}_t$ in the obvious way then

$$EN_t \leq E\bar{N}_t \leq (t + c)/E(\bar{\xi}_i)$$

Letting $t \to \infty$ and then $c \to \infty$ gives $\limsup_{t \to \infty} EN_t/t \leq 1/\mu$ and the proof is complete. $\qquad\square$

EXERCISE 4.1. Show that $t/E(X_i \wedge t) \leq U(t) \leq 2t/E(X_i \wedge t)$.

EXERCISE 4.2. Deduce (4.2) from (4.1) by showing

$$\limsup_{t \to \infty} E(N_t/t)^2 < \infty.$$

Hint: Use a comparison like the one in the proof of (4.1).

EXERCISE 4.3. Customers arrive at times of a Poisson process with rate 1. If the server is occupied they leave. (Think of a public telephone or prostitute.) If not they enter service and require a service time with a distribution F that has mean μ. Show that the times at which customers enter service are a renewal process with mean $\mu + 1$, and use (4.1) to conclude that the asymptotic fraction of customers served is $1/(\mu + 1)$.

To take a closer look at when the renewals occur we let

$$U(A) = \sum_{n=0}^{\infty} P(T_n \in A)$$

U is called the **renewal measure**. We absorb the old definition, $U(t) = EN_t$, into the new one by regarding $U(t)$ as shorthand for $U([0, t])$. This should not cause problems since $U(t)$ is the distribution function for the renewal measure. The asymptotic behavior of $U(t)$ depends upon whether the distribution F is **arithmetic**, i.e., concentrated on $\{\delta, 2\delta, 3\delta, \ldots\}$ for some $\delta > 0$, or **nonarithmetic**, i.e., not arithmetic. We will treat the first case in Chapter 5 as an application of Markov chains, so we will restrict our attention to the second case here.

(4.3) Blackwell's renewal theorem. If F is nonarithmetic then $U([t, t+h]) \rightarrow h/\mu$ as $t \rightarrow \infty$.

We will prove the result in the case $\mu < \infty$ by "coupling" following Lindvall (1977) (and Athreya, McDonald, and Ney (1978)). To set the stage for the proof, we need a definition and some preliminary computations. If $T_0 \geq 0$ is independent of $\xi_1, \xi_2, \ldots$ and has distribution G, then $T_k = T_{k-1} + \xi_k$, $k \geq 1$ defines a **delayed renewal process**, and G is the **delay distribution**. If we let $N_t = \inf\{k : T_k > t\}$ as before and set $V(t) = EN_t$, then breaking things down according to the value of T_0 gives

$$(4.4) \qquad V(t) = \int_0^t U(t-s)\, dG(s)$$

The last integral, and all similar expressions below, is intended to include the contribution of any mass G has at 0. If we let $U(r) = 0$ for $r < 0$ then the last equation can be written as $V = U * G$ where $*$ denotes convolution.

Applying similar reasoning to U gives

$$(4.5) \qquad U(t) = 1 + \int_0^t U(t-s)\, dF(s)$$

or introducing convolution notation

$$U = 1_{[0,\infty)}(t) + U * F.$$

Convolving each side with G (and recalling $G * U = U * G$) gives

$$(4.6) \qquad V = G * U = G + V * F$$

We know $U(t) \sim t/\mu$. Our next step is to find a G so that $V(t) = t/\mu$. Plugging what we want into (4.6) gives

$$t/\mu = G(t) + \int_0^t \frac{t-y}{\mu}\, dF(y)$$

or

$$G(t) = t/\mu - \int_0^t \frac{t-y}{\mu}\, dF(y)$$

The integration by parts formula is

$$\int_0^t K(y)\, dH(y) = H(t)K(t) - H(0)K(0) - \int_0^t H(y)\, dK(y)$$

If we let $H(y) = (y - t)/\mu$ and $K(y) = 1 - F(y)$, then

$$\frac{1}{\mu} \int_0^t 1 - F(y) \, dy = \frac{t}{\mu} - \int_0^t \frac{t - y}{\mu} \, dF(y)$$

so we have

(4.7)
$$G(t) = \frac{1}{\mu} \int_0^t 1 - F(y) \, dy$$

It is comforting to note that $\mu = \int_{[0,\infty)} 1 - F(y) \, dy$, so the last formula defines a probability distribution. When the delay distribution G is the one given in (4.7), we call the result the **stationary renewal process**. Something very special happens when $F(t) = 1 - \exp(-\lambda t)$, $t \geq 0$ where $\lambda > 0$ (i.e., the renewal process is a rate λ Poisson process). In this case $\mu = 1/\lambda$ so $G(t) = F(t)$.

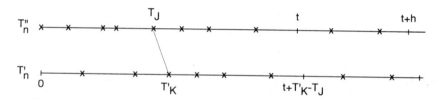

Figure 3.4.1

Proof of (4.3) for $\mu < \infty$ Let T_n be a renewal process (with $T_0 = 0$) and T_n' be an independent stationary renewal process. Our first goal is to find J and K so that $|T_J - T_K'| < \epsilon$ and the increments $\{T_{J+i} - T_J, i \geq 1\}$ and $\{T_{K+i}' - T_K', i \geq 1\}$ are i.i.d. sequences independent of what has come before.

Let $\eta_1, \eta_2, \ldots$ and $\eta_1', \eta_2', \ldots$ be i.i.d. independent of T_n and T_n' and take the values 0 and 1 with probability $1/2$ each. Let $\nu_n = \eta_1 + \cdots + \eta_n$ and $\nu_n' = 1 + \eta_1' + \cdots + \eta_n'$, $S_n = T_{\nu_n}$ and $S_n' = T_{\nu_n'}'$. The increments of $S_n - S_n'$ are 0 with probability at least $1/4$, and the support of their distribution is symmetric and contains the support of the ξ_k so if the distribution of the ξ_k is nonarithmetic the random walk $S_n - S_n'$ is irreducible. Since the increments of $S_n - S_n'$ have mean 0, $N = \inf\{n : |S_n - S_n'| < \epsilon\}$ has $P(N < \infty) = 1$ and we can let $J = \nu_N$ and $K = \nu_N'$. Let

$$T_n'' = \begin{cases} T_n & \text{if } J \geq n \\ T_J + T_{K+(n-J)}' - T_K' & \text{if } J < n \end{cases}$$

In other words, the increments $T_{J+i}'' - T_J''$ are the same as $T_{K+i}' - T_K'$ for $i \geq 1$. See Figure 3.4.1 for a picture.

It is easy to see from the construction that T_n and T_n'' have the same distribution. If we let

$$N'[s,t] = |\{n : T_n' \in [s,t]\}| \quad \text{and} \quad N''[s,t] = |\{n : T_n'' \in [s,t]\}|$$

be the number of renewals in $[s,t]$ in the two processes, then on $\{T_J \le t\}$

$$N''[t,t+h] = N'[t+T_K'-T_J, t+h+T_K'-T_J] \begin{cases} \ge N'[t+\epsilon, t+h-\epsilon] \\ \le N'[t-\epsilon, t+h+\epsilon] \end{cases}$$

To relate the expected number of renewals in the two processes, we observe that even if we condition on the location of all the renewals in $[0,s]$, the expected number of renewals in $[s, s+t]$ is at most $U(t)$, since the worst thing that could happen is to have a renewal at time s. Combining the last two observations, we see that if $\epsilon < h/2$ (so $[t+\epsilon, t+h-\epsilon]$ has positive length)

$$U([t,t+h]) = EN''[t,t+h] \ge E(N'[t+\epsilon, t+h-\epsilon]; T_J \le t)$$
$$\ge \frac{h-2\epsilon}{\mu} - P(T_J > t)U(h)$$

since $EN'[t+\epsilon, t+h-\epsilon] = (h-2\epsilon)/\mu$ and $\{T_J > t\}$ is determined by the renewals of T in $[0,t]$ and the renewals of T' in $[0, t+\epsilon]$. For the other direction we observe

$$U([t,t+h]) \le E(N'[t-\epsilon, t+h+\epsilon]; T_J \le t) + E(N''[t,t+h]; T_J > t)$$
$$\le \frac{h+2\epsilon}{\mu} + P(T_J > t)U(h)$$

The desired result now follows from the fact that $P(T_J > t) \to 0$ and $\epsilon < h/2$ is arbitrary. □

Remark. In the first edition, we followed Athreya, McDonald, and Ney too closely and repeated their mistaken claim that the coupling can be done taking $J = K = \inf\{k : |T_k - T_k'| < \epsilon\}$. To see this is not correct, suppose that the interrenewal times have $P(\xi_j = 1) = P(\xi_j = 1 + \pi) = 1/2$. This distribution is nonarithmetic but $\xi_j - \xi_j'$ is concentrated on $\{-\pi, 0, \pi\}$ so we cannot couple unless $|T_0' - \pi| < \epsilon$. This problem was pointed out to us by Torgny Lindvall. The remedy used above is different from the one in his (1977) proof.

Proof of (4.3) for $\mu = \infty$ In this case there is no stationary renewal process, so we have to resort to other methods. Let

$$\beta = \limsup_{t \to \infty} U(t, t+1] = \lim_{k \to \infty} U(t_k, t_k + 1]$$

for some sequence $t_k \to \infty$. We want to prove that $\beta = 0$ for then by addition the previous conclusion holds with 1 replaced by any integer n and by monotonicity with n replaced by any $h < n$ and this gives us the result in (4.3). Fix i and let

$$a_{k,j} = \int_{(j-1,j]} U(t_k - y, t_k + 1 - y] \, dF^{i*}(y)$$

By considering the location of T_i we get

(a) $$\lim_{k \to \infty} \sum_{j=1}^{\infty} a_{k,j} = \lim_{k \to \infty} \int U(t_k - y, t_k + 1 - y] \, dF^{i*}(y) = \beta$$

Since β is the lim sup we must have

(b) $$\limsup_{k \to \infty} a_{k,j} \leq \beta \cdot P(T_i \in (j-1, j])$$

We want to conclude from (a) and (b) that

(c) $$\liminf_{k \to \infty} a_{k,j} \geq \beta \cdot P(T_i \in (j-1, j])$$

To do this we observe that by considering the location of the first renewal in $(j-1, j]$

(d) $$0 \leq a_{k,j} \leq U(1) P(T_i \in (j-1, j])$$

(c) is trivial when $\beta = 0$ so we can suppose $\beta > 0$. To argue by contradiction suppose there exist j_0 and $\epsilon > 0$ so that

$$\liminf_{k \to \infty} a_{k,j_0} \leq \beta \cdot \{P(T_i \in (j_0 - 1, j_0]) - \epsilon\}$$

Pick $k_n \to \infty$ so that

$$a_{k_n, j_0} \to \beta \cdot \{P(T_i \in (j_0 - 1, j_0]) - \epsilon\}$$

Using (d) we can pick $J \geq j_0$ so that

$$\limsup_{n \to \infty} \sum_{j=J+1}^{\infty} a_{k_n, j} \leq U(1) \sum_{j=J+1}^{\infty} P(T_i \in (j-1, j]) \leq \beta \epsilon/2$$

Now an easy argument shows

$$\limsup_{n \to \infty} \sum_{j=1}^{J} a_{k_n, j} \leq \sum_{j=1}^{J} \limsup_{n \to \infty} a_{k_n, j} \leq \beta \left(\sum_{j=1}^{J} P(T_i \in (j-1, j]) - \epsilon \right)$$

by (b) and our assumption. Adding the last two results shows

$$\limsup_{n \to \infty} \sum_{j=1}^{\infty} a_{k_n, j} \le \beta(1 - \epsilon/2)$$

which contradicts (a), and proves (c).

Now if $j - 1 < y \le j$, we have

$$U(t_k - y, t_k + 1 - y] \le U(t_k - j, t_k + 2 - j]$$

so using (c) it follows that for j with $P(T_i \in (j - 1, j]) > 0$, we must have

$$\liminf_{k \to \infty} U(t_k - j, t_k + 2 - j] \ge \beta$$

Summing over i, we see that the last conclusion is true when $U(j - 1, j] > 0$.

The support of U is closed under addition. (If x is in the support of F^{m*} and y is in the support of F^{n*} then $x + y$ is in the support of $F^{(m+n)*}$.) We have assumed F is nonarithmetic, so $U(j - 1, j] > 0$ for $j \ge j_0$. Letting $r_k = t_k - j_0$ and considering the location of the last renewal in $[0, r_k]$ and the index of the T_i gives

$$1 = \sum_{i=0}^{\infty} \int_0^{r_k} (1 - F(r_k - y)) \, dF^{i*}(y) = \int_0^{r_k} (1 - F(r_k - y)) \, dU(y)$$

$$\ge \sum_{n=1}^{\infty} (1 - F(2n)) \, U(r_k - 2n, r_k + 2 - 2n]$$

Since $\liminf_{k \to \infty} U(r_k - 2n, r_k + 2 - 2n] \ge \beta$, and

$$\sum_{n=0}^{\infty} (1 - F(2n)) \ge \mu/2 = \infty$$

β must be 0 and the proof is complete.   □

Remark. Following Lindvall (1977) we have based the proof for $\mu = \infty$ on part of Feller's (1961) proof of the discrete renewal theorem (i.e., for arithmetic distributions). See Freedman (1971b) p. 22–25 for an account of Feller's proof. Purists can find a proof that does everything by coupling in Thorisson (1987).

Our next topic is the **renewal equation**: $H = h + H * F$. Two cases we have seen in (4.5) and (4.6) are

Example 4.1. $h \equiv 1$: $U(t) = 1 + \int_0^t U(t-s) \, dF(s)$

Example 4.2. $h(t) = G(t)$: $V(t) = G(t) + \int_0^t V(t-s) \, dF(s)$

The last equation is valid for an arbitrary delay distribution. If we let G be the distribution in (4.7) and subtract the last two equations, we get

Example 4.3. $H(t) = U(t) - t/\mu$ satifies the renewal equation with $h(t) = \frac{1}{\mu} \int_t^\infty 1 - F(s) \, ds$.

Last but not least, we have an example that is a typical application of the renewal equation.

Example 4.4. Let $x > 0$ be fixed and let $H(t) = P(T_{N(t)} - t > x)$. By considering the value of T_1 we get

$$H(t) = (1 - F(t+x)) + \int_0^t H(t-s) \, dF(s)$$

The examples above should provide motivation for

(4.8) Theorem. If h is bounded then the function

$$H(t) = \int_0^t h(t-s) \, dU(s)$$

is the unique solution of the renewal equation that is bounded on bounded intervals.

Proof   Let $U_n(A) = \sum_{m=0}^n P(T_m \in A)$ and

$$H_n(t) = \int_0^t h(t-s) \, dU_n(s) = \sum_{m=0}^n (h * F^{m*})(t)$$

Here, F^{m*} is the distribution of T_m, and we have extended the definition of h by setting $h(r) = 0$ for $r < 0$. From the last expression it should be clear that

$$H_{n+1} = h + H_n * F$$

The fact that $U(t) < \infty$ implies $U(t) - U_n(t) \to 0$. Since h is bounded,

$$|H_n(t) - H(t)| \le \|h\|_\infty |U(t) - U_n(t)|$$

and $H_n(t) \to H(t)$ uniformly on bounded intervals. To estimate the convolution we note that

$$|H_n * F(t) - H * F(t)| \leq \sup_{s \leq t} |H_n(s) - H(s)|$$

$$\leq \|h\|_\infty |U(t) - U_n(t)|$$

since $U - U_n = \sum_{m=n+1}^\infty F^{m*}$ is increasing in t. Letting $n \to \infty$ in $H_{n+1} = h + H_n * F$, we see that H is a solution of the renewal equation which is bounded on bounded intervals.

To prove uniqueness, we observe that if H_1 and H_2 are two solutions then $K = H_1 - H_2$ satisfies $K = K * F$. If K is bounded on bounded intervals, iterating gives $K = K * F^{n*} \to 0$ as $n \to \infty$, so $H_1 = H_2$. $\square$

The proof of (4.8) is valid when $F(\infty) = P(\xi_i < \infty) < 1$. In this case we have a **terminating renewal process**. After a geometric number of trials with mean $1/(1 - F(\infty))$, $T_n = \infty$. This "trivial case" has some interesting applications.

Example 4.5. Pedestrian delay. A chicken wants to cross a road (we won't ask why) on which the traffic is a Poisson process with rate λ. She needs one unit of time with no arrival to safely cross the road. Let $M = \inf\{t \geq 0 :$ there are no arrivals in $(t, t+1]\}$ be the waiting time until she starts to cross the street. By considering the time of the first arrival we see that $H(t) = P(M \leq t)$ satisfies

$$H(t) = e^{-\lambda} + \int_0^1 H(t - y)\,\lambda e^{-\lambda y}\,dy$$

Comparing with Example 4.1 and using (4.8) we see that

$$H(t) = e^{-\lambda} \sum_{n=0}^\infty F^{n*}(t)$$

We could have gotten this answer without renewal theory by noting

$$P(M \leq t) = \sum_{n=0}^\infty P(T_n \leq t, T_{n+1} = \infty)$$

The last representation allows us to compute the mean of M. Let μ be the mean of the interarrival time given that it is < 1, and note that the lack of memory property of the exponential distribution implies

$$\mu = \int_0^1 x\lambda e^{-\lambda x}\,dx = \int_0^\infty - \int_1^\infty = \frac{1}{\lambda} - \left(1 + \frac{1}{\lambda}\right)e^{-\lambda}$$

Then by considering the number of renewals in our terminating renewal process

$$EM = \sum_{n=0}^{\infty} e^{-\lambda}(1 - e^{-\lambda})^n n\mu = (e^{\lambda} - 1)\mu$$

since if X is a geometric with success probability $e^{-\lambda}$ then $EM = \mu E(X - 1)$.

Example 4.6. Cramér's estimates of ruin. Consider an insurance company that collects money at rate c and experiences i.i.d. claims at the arrival times of a Poisson process N_t with rate 1. If its inital capital is x, its wealth at time t is

$$W_x(t) = x + ct - \sum_{m=1}^{Nt} Y_i$$

Here $Y_1, Y_2, \ldots$ are i.i.d. with distribution G and mean μ. Let

$$R(x) = P(W_x(t) \geq 0 \text{ for all } t)$$

be the probability of never going bankrupt starting with capital x. By considering the time and size of the first claim:

(a) $$R(x) = \int_0^{\infty} e^{-s} \int_0^{x+cs} R(x + cs - y)\, dG(y)\, ds$$

This does not look much like a renewal equation, but with some ingenuity it can be transformed into one. Changing variables $t = x + cs$

$$R(x)e^{-x/c} = \int_x^{\infty} e^{-t/c} \int_0^t R(t - y)\, dG(y)\, \frac{dt}{c}$$

Differentiating w.r.t. x and then mutliplying by $e^{x/c}$

$$R'(x) = \frac{1}{c}R(x) - \int_0^x R(x - y)\, dG(y) \cdot \frac{1}{c}$$

Integrating x from 0 to w

(b) $$R(w) - R(0) = \frac{1}{c}\int_0^w R(x)\, dx - \frac{1}{c}\int_0^w \int_0^x R(x - y)\, dG(y)\, dx$$

Interchanging the order of integration in the double integral, letting

$$S(w) = \int_0^w R(x)\, dx$$

using $dG = -d(1 - G)$ and then integrating by parts

$$-\frac{1}{c}\int_0^w \int_y^w R(x - y)\,dx\,dG(y) = -\frac{1}{c}\int_0^w S(w - y)\,dG(y)$$

$$= \frac{1}{c}\int_0^w S(w - y)\,d(1 - G)(y)$$

$$= \frac{1}{c}\left\{-S(w) + \int_0^w (1 - G(y))R(w - y)\,dy\right\}$$

Plugging this into (b), we finally have a renewal equation

(c) $$\qquad R(w) = R(0) + \int_0^w R(w - y)\frac{1 - G(y)}{c}\,dy$$

It took some cleverness to arrive at the last equation but it is straightforward to analyze. First we dismiss a trivial case. If $\mu > c$,

$$\frac{1}{t}\left(ct - \sum_{m=1}^{Nt} Y_i\right) \to c - \mu < 0 \quad \text{a.s.}$$

so $R(x) \equiv 0$. When $\mu < c$

$$F(x) = \int_0^x \frac{1 - G(y)}{c}\,dy$$

is a defective probability distribution with $F(\infty) = \mu/c$. Our renewal equation can be written as

(d) $$\qquad R = R(0) + R * F$$

so comparing with Example 4.1 and using (4.8) tells us $R(w) = R(0)U(w)$. To complete the solution we have to compute the constant $R(0)$. Letting $w \to \infty$ and noticing $R(w) \to 1$, $U(w) \to (1 - F(\infty))^{-1} = (1 - \mu/c)^{-1}$ we have $R(0) = 1 - \mu/c$.

The basic fact about solutions of the renewal equation (in the nonterminating case) is:

(4.9) **The renewal theorem.** If F is nonarithmetic and h is directly Riemann integrable then as $t \to \infty$

$$H(t) \to \frac{1}{\mu}\int_0^\infty h(s)\,ds$$

Intuitively this holds since (4.8) implies

$$H(t) = \int_0^t h(t-s)\, dU(s)$$

and (4.3) implies $dU(s) \to ds/\mu$ as $s \to \infty$. We will define directly Riemann integrable in a minute. We will start doing the proof and then figure out what we need to assume.

Proof Suppose

$$h(s) = \sum_{k=0}^{\infty} a_k 1_{[k\delta, (k+1)\delta)}(s)$$

where $\sum_{k=0}^{\infty} |a_k| < \infty$. Since $U([t, t+\delta]) \le U([0,\delta]) < \infty$, it follows easily from (4.3) that

$$\int_0^t h(t-s)dU(s) = \sum_{k=0}^{\infty} a_k U((t-(k+1)\delta, t-k\delta]) \to \frac{1}{\mu} \sum_{k=0}^{\infty} a_k \delta$$

(Pick K so that $\sum_{k \ge K} |a_k| \le \epsilon/2U([0,\delta])$ and then T so that

$$|a_k| \cdot |U((t-(k+1)\delta, t-k\delta]) - \delta/\mu| \le \frac{\epsilon}{2K}$$

for $t \ge T$ and $0 \le k < K$.) If h is an arbitrary function on $[0,\infty)$ we let

$$I^\delta = \sum_{k=0}^{\infty} \delta \sup\{h(x) : x \in [k\delta, (k+1)\delta)\}$$

and

$$I_\delta = \sum_{k=0}^{\infty} \delta \inf\{h(x) : x \in [k\delta, (k+1)\delta)\}$$

be upper and lower Riemann sums approximating the integral of h over $[0,\infty)$. Comparing h with the obvious upper and lower bounds that are constant on $[k\delta, (k+1)\delta)$ and using the result for the special case

$$\frac{I_\delta}{\mu} \le \liminf_{t \to \infty} \int_0^t h(t-s)\, dU(s) \le \limsup_{t \to \infty} \int_0^t h(t-s)\, dU(s) \le \frac{I^\delta}{\mu}$$

If I^δ and I_δ both approach the same finite limit I as $\delta \to 0$, then h is said to be **directly Riemann integrable**, and it follows that

$$\int_0^t h(t-s)\, dU(y) \to I/\mu \qquad \qquad \square$$

Remark. The word "direct" in the name refers to the fact that while the Riemann integral over $[0, \infty)$ is usually defined as the limit of integrals over $[0, a]$, we are approximating the integral over $[0, \infty)$ directly.

In checking the new hypothesis in (4.9), the following result is useful.

(4.10) Lemma. If $h(x) \geq 0$ is decreasing with $h(0) < \infty$ and $\int_0^\infty h(x)\, dx < \infty$ then h is directly Riemann integrable.

Proof Because h is decreasing, $I^\delta = \sum_{k=0}^\infty \delta h(k\delta)$ and $I_\delta = \sum_{k=0}^\infty \delta h((k+1)\delta)$. So

$$I^\delta \geq \int_0^\infty h(x)\, dx \geq I_\delta = I^\delta - h(0)\delta$$

proving the desired result. $\qquad\qquad\qquad\qquad\qquad\qquad\qquad\qquad\qquad\qquad\quad \square$

The last result suffices for all our applications, so we leave it to the reader to do

EXERCISE 4.4. If $h \geq 0$ is continuous then h is directly Riemann integrable if and only if $I^\delta < \infty$ for some $\delta > 0$ (and hence for all $\delta > 0$).

Returning now to our examples, we skip the first two because in those cases $h(t) \to 1$ as $t \to \infty$, so h is not integrable in any sense.

Example 4.3, part II. $h(t) = \frac{1}{\mu} \int_{[t,\infty)} 1 - F(s)\, ds$. h is decreasing, $h(0) = 1$, and

$$\mu \int_0^\infty h(t)\, dt = \int_0^\infty \int_t^\infty 1 - F(s)\, ds\, dt$$
$$= \int_0^\infty \int_0^s 1 - F(s)\, dt\, ds = \int_0^\infty s(1 - F(s))\, ds = E(\xi_i^2/2)$$

So if $\nu \equiv E(\xi_i^2) < \infty$, it follows from (4.10), (4.9), and the formula in Example 4.3 that

$$0 \leq U(t) - t/\mu \to \nu/2\mu^2 \quad \text{as } t \to \infty$$

When the renewal process is a rate λ Poisson process, i.e., $P(\xi_i > t) = e^{-\lambda t}$, $N(t) - 1$ has a Poisson distribution with mean λt, so $U(t) = 1 + \lambda t$. Acccording to Feller, Vol. II (1971) p. 385, if the ξ_i are uniform on (0,1) then

$$U(t) = \sum_{k=0}^n (-1)^k e^{t-k} (t-k)^k / k! \quad \text{for } n \leq t \leq n+1$$

As he says, the exact expression "reveals little about the nature of U. The asymptotic formula $0 \leq U(t) - 2t \to 2/3$ is much more interesting."

Example 4.4, part II. $h(t) = 1 - F(t + x)$. Again h is decreasing, but this time $h(0) \leq 1$ and the integral of h is finite when $\mu = E(\xi_i) < \infty$. Applying (4.10) and (4.9) now gives

$$P(T_{N(t)} - t > x) \to \frac{1}{\mu} \int_0^\infty h(s) \, ds = \frac{1}{\mu} \int_x^\infty 1 - F(t) \, dt$$

so (when $\mu < \infty$) the distribution of the **residual waiting time** $T_{N(t)} - t$ converges to the delay distribution that produces the stationary renewal process. This fact also follows from our proof of (4.3).

Using the method employed to study Example 4.4, one can analyze various other aspects of the asymptotic behavior of renewal processes. To avoid repeating ourselves

We assume throughout that F is nonarithmetic, and in problems where the mean appears we assume it is finite.

EXERCISE 4.5. Let $A_t = t - T_{N(t)-1}$ be the "age" at time t, i.e., the amount of time since the last renewal. If we fix $x > 0$ then $H(t) = P(A_t > x)$ satisfies the renewal equation

$$H(t) = (1 - F(t)) \cdot 1_{(x,\infty)}(t) + \int_0^t H(t - s) \, dF(s)$$

so $P(A_t > x) \to \frac{1}{\mu} \int_{(x,\infty)}(1 - F(t)) dt$, which is the limit distribution for the residual lifetime $B_t = T_{N(t)} - t$.

Remark. The last result can be derived from Example 4.4 by noting that if $t > x$ then $P(A_t \geq x) = P(B_{t-x} > x) = P($ no renewal in $(t - x, t])$. To check the placement of the strict inequality, recall $N_t = \inf\{k : T_k > t\}$ so we always have $A_s \geq 0$ and $B_s > 0$.

EXERCISE 4.6. Use the renewal equation in the last problem and (4.8) to conclude that if T is a rate λ Poisson process A_t has the same distribution as $\xi_i \wedge t$.

EXERCISE 4.7. Let $A_t = t - T_{N(t)-1}$ and $B_t = T_{N(t)} - t$. Show that

$$P(A_t > x, B_t > y) \to \frac{1}{\mu} \int_{x+y}^\infty (1 - F(t)) \, dt$$

EXERCISE 4.8. **Alternating renewal process.** Let $\xi_1, \xi_2, \ldots > 0$ be i.i.d. with dist F_1 and let $\eta_1, \eta_2, \ldots > 0$ be i.i.d. with distribution F_2. Let $T_0 = 0$ and for $k \geq 1$ let $S_k = T_{k-1} + \xi_k$ and $T_k = S_k + \eta_k$. In words, we have a machine that works for an amount of time ξ_k, breaks down, and then requires η_k units of time to be repaired. Let $F = F_1 * F_2$ and let $H(t)$ be the probability the machine is working at time t. Show that if F is nonarithmetic then as $t \to \infty$

$$H(t) \to \mu_1/(\mu_1 + \mu_2)$$

where μ_i is the mean of F_i.

EXERCISE 4.9. Write a renewal equation for $H(t) = P($ number of renewals in $[0, t]$ is odd) and use the renewal theorem to show that $H(t) \to 1/2$. Note: this is a special case of the previous exercise.

EXERCISE 4.10. **Renewal densities.** Show that if $F(t)$ has a directly Riemann integrable density function $f(t)$ then the $V = U - 1_{[0,\infty)}$ has a density v that satisfies

$$v(t) = f(t) + \int_0^t v(t - s)\, dF(s)$$

Use the renewal theorem to conclude that if f is directly Riemann integrable then $v(t) \to 1/\mu$ as $t \to \infty$.

Finally we have an example that would have been given right after (4.1), but was delayed because we had not yet defined a delayed renewal process.

Example 4.7. Patterns in coin tossing. Let X_n, $n \geq 1$ take values H and T with probability $1/2$ each. Let $T_0 = 0$ and $T_m = \inf\{n > T_{m-1} : (X_n, \ldots, X_{n+k-1}) = (i_1, \ldots, i_k)\}$ where $(i_1, \ldots, i_k)$ is some pattern of heads and tails. It is easy to see that the T_j form a delayed renewal process, i.e., $t_j = T_j - T_{j-1}$ are independent for $j \geq 1$ and identically distributed for $j \geq 2$. To see that the distribution of t_1 may be different let $(i_1, i_2, i_3) = (H, H, H)$. In this case $P(t_1 = 1) = 1/8$, $P(t_2 = 1) = 1/2$.

EXERCISE 4.11. (i) Show that for any pattern of length k, $Et_j = 2^k$ for $j \geq 2$. (ii) Compute Et_1 when the pattern is HH, and when it is HT. Hint: For HH observe

$$Et_1 = P(HH) + P(HT)E(t_1 + 2) + P(T)E(t_1 + 1)$$

4 Martingales

A martingale X_n can be thought of as the fortune at time n of a player who is betting on a fair game; submartingales (supermartingales) as the outcome of betting on a favorable (unfavorable) game. There are two basic facts about martingales. The first is that you cannot make money betting on them (see (2.7)), and in particular if you choose to stop playing at some bounded time N then your expected winnings EX_N are equal to your initial fortune X_0. (We are supposing for the moment that this is not random.) Our second fact, (2.10), concerns submartingales. To use a heuristic we learned from Mike Brennan, "they are the stochastic analogues of nondecreasing sequences and so if they are bounded above (to be precise, $\sup_n EX_n^+ < \infty$) they converge almost surely." As the material in Section 4.3 shows, this result has diverse applications. Later sections give sufficient conditions for martingales to converge in L^p, $p > 1$ (Section 4.4) and in L^1 (Section 4.5); consider martingales indexed by $n \leq 0$ (Section 4.6); and give sufficient conditions for $EX_N = EX_0$ to hold for unbounded stopping times (Section 4.7). The last result is quite useful for studying the behavior of random walks and other systems.

4.1. Conditional Expectation

We begin with a definition that is important for this chapter and the next one. After giving the definition, we will consider several examples to explain it. Given are a probability space $(\Omega, \mathcal{F}_o, P)$, a σ-field $\mathcal{F} \subset \mathcal{F}_o$, and a random variable $X \in \mathcal{F}_o$ with $E|X| < \infty$. We define the **conditional expectation of** X **given** $\mathcal{F}$, $E(X|\mathcal{F})$, to be any random variable Y that has

(i) $Y \in \mathcal{F}$, i.e., is $\mathcal{F}$ measurable

(ii) for all $A \in \mathcal{F}$, $\int_A X \, dP = \int_A Y \, dP$

Any Y satisfying (i) and (ii) is said to be a **version of** $E(X|\mathcal{F})$. The first thing to be settled is that the conditional expectation exists and is unique. We tackle the second claim first but start with a technical point.

If Y satisfies (i) and (ii) then it is integrable.

Proof   Letting $A = \{Y > 0\} \in \mathcal{F}$, using (ii) twice, and then adding

$$\int_A Y \, dP = \int_A X \, dP \le \int_A |X| \, dP$$

$$\int_{A^c} -Y \, dP = \int_{A^c} -X \, dP \le \int_{A^c} |X| \, dP$$

So we have $E|Y| \le E|X|$. $\square$

Uniqueness. If Y' also satisfies (i) and (ii) then

$$\int_A Y \, dP = \int_A Y' \, dP \quad \text{for all } A \in \mathcal{F}$$

Taking $A = \{Y - Y' \ge \epsilon > 0\}$ we see

$$0 = \int_A Y - Y' \, dP \ge \epsilon P(A)$$

so $P(A) = 0$. Since this holds for all ϵ we have $Y \le Y'$ a.s. and interchanging the roles of Y and Y' we have $Y = Y'$ a.s. Technically, all equalities such as $Y = E(X|\mathcal{F})$ should be written as $Y = E(X|\mathcal{F})$ a.s. but we have ignored this point in previous chapters and will continue to do so.

EXERCISE 1.1. Generalize the last argument to show that if $X_1 = X_2$ on $B \in \mathcal{F}$ then $E(X_1|\mathcal{F}) = E(X_2|\mathcal{F})$ a.s. on B.

Existence. To start, we recall ν is said to be **absolutely continuous with respect to** μ (abbreviated $\nu \ll \mu$) if $\mu(A) = 0$ implies $\nu(A) = 0$, and we use (8.6) from the Appendix:

Radon-Nikodym Theorem. Let μ and ν be σ-finite measures on $(\Omega, \mathcal{F})$. If $\nu \ll \mu$, there is a function $f \in \mathcal{F}$ so that for all $A \in \mathcal{F}$

$$\int_A f \, d\mu = \nu(A)$$

f is usually denoted $d\nu/d\mu$ and called the **Radon-Nikodym derivative**.

The last theorem easily gives the existence of conditional expectation. Suppose first that $X \ge 0$. Let $\mu = P$ and

$$\nu(A) = \int_A X \, dP \quad \text{for } A \in \mathcal{F}$$

The dominated convergence theorem implies ν is a measure (see Exercise 5.8 in the Appendix) and the definition of the integral implies $\nu \ll \mu$. The Radon-Nikodym derivative $d\nu/d\mu \in \mathcal{F}$ and for any $A \in \mathcal{F}$ has

$$\int_A X \, dP = \nu(A) = \int_A \frac{d\nu}{d\mu} \, dP$$

Taking $A = \Omega$ we see that $d\nu/d\mu \geq 0$ is integrable and we have shown that $d\nu/d\mu$ is a version of $E(X|\mathcal{F})$.

To treat the general case now write $X = X^+ - X^-$, let $Y_1 = E(X^+|\mathcal{F})$ and $Y_2 = E(X^-|\mathcal{F})$. Now $Y_1 - Y_2 \in \mathcal{F}$ is integrable and for all $A \in \mathcal{F}$ we have

$$\int_A X \, dP = \int_A X^+ \, dP - \int_A X^- \, dP$$
$$= \int_A Y_1 \, dP - \int_A Y_2 \, dP = \int_A (Y_1 - Y_2) \, dP$$

This shows $Y_1 - Y_2$ is a version of $E(X|\mathcal{F})$ and completes the proof. □

a. Examples

Intuitively, we think of $\mathcal{F}$ as describing the information we have at our disposal – for each $A \in \mathcal{F}$, we know whether or not A has occurred. $E(X|\mathcal{F})$ is then our "best guess" of the value of X given the information we have. Some examples should help to clarify this and connect $E(X|\mathcal{F})$ with other definitions of conditional expectation.

Example 1.1. If $X \in \mathcal{F}$ then $E(X|\mathcal{F}) = X$, i.e., if we know X then our "best guess" is X itself. Since X always satisfies (ii), the only thing that can keep X from being $E(X|\mathcal{F})$ is condition (i). A special case of this example is $X = c$ where c is a constant.

Example 1.2. At the other extreme from perfect information is no information. Suppose X is independent of $\mathcal{F}$, i.e., for all $B \in \mathcal{R}$ and $A \in \mathcal{F}$

$$P(\{X \in B\} \cap A) = P(X \in B)P(A)$$

We claim that in this case $E(X|\mathcal{F}) = EX$, i.e., if you don't know anything about X then the best guess is the mean EX. To check the defintion note that $EX \in \mathcal{F}$ so (i). To verify (ii), we observe that if $A \in \mathcal{F}$ then since X and $1_A \in \mathcal{F}$ are independent (4.8) in Chapter 1 implies

$$\int_A X \, dP = E(X 1_A) = EX \, E1_A = \int_A EX \, dP$$

The reader should note that here and in what follows the game is "guess and verify". We come up with a formula for the conditional expectation, and then check that it satisfies (i) and (ii).

Example 1.3. In this example we relate the new definition of conditional expectation to the first one taught in an undergraduate probability course. Suppose $\Omega_1, \Omega_2, \ldots$ is a finite or infinite partition of Ω into disjoint sets each of which has positive probability, and let $\mathcal{F} = \sigma(\Omega_1, \Omega_2, \ldots)$ be the σ-field generated by these sets. Then

$$E(X|\mathcal{F}) = \frac{E(X;\Omega_i)}{P(\Omega_i)} \quad \text{on } \Omega_i$$

In words, the information in Ω_i tells us which element of the partition our outcome lies in, and given this information the best guess for X is the average value of X over Ω_i. To prove our guess is correct, observe that the proposed formula is constant on each Ω_i, so it is measurable with respect to $\mathcal{F}$. To verify (ii), it is enough to check the equality for $A = \Omega_i$, but this is trivial:

$$\int_{\Omega_i} \frac{E(X;\Omega_i)}{P(\Omega_i)} \, dP = E(X;\Omega_i) = \int_{\Omega_i} X \, dP$$

A degenerate but important special case is $\mathcal{F} = \{\emptyset, \Omega\}$, the trivial σ-field. In this case $E(X|\mathcal{F}) = EX$.

To continue the connection with undergraduate notions let

$$P(A|\mathcal{G}) = E(1_A|\mathcal{G})$$
$$P(A|B) = P(A \cap B)/P(B)$$

and observe that in the last example $P(A|\mathcal{F}) = P(A|\Omega_i)$ on Ω_i.

EXERCISE 1.2. **Bayes' formula.** Let $G \in \mathcal{G}$ and show that

$$P(G|A) = \int_G P(A|\mathcal{G}) \, dP \left/ \int_\Omega P(A|\mathcal{G}) \, dP \right.$$

When $\mathcal{G}$ is the σ-field generated by a partition this reduces to the usual Bayes' formula

$$P(G_i|A) = P(A|G_i)P(G_i) \left/ \sum_j P(A|G_j)P(G_j) \right.$$

The definition of conditional expectation given a σ-field contains conditioning on a random variable as a special case. We define

$$E(X|Y) = E(X|\sigma(Y))$$

where $\sigma(Y)$, the σ-field generated by Y.

Example 1.4. To continue making connection with definitions of conditional expectation from undergraduate probability, suppose X and Y have joint density $f(x, y)$, i.e.,

$$P((X,Y) \in B) = \int_B f(x,y)\,dx\,dy \quad \text{for } B \in \mathcal{R}^2$$

and suppose for simplicity that $\int f(x,y)\,dx > 0$ for all y. We claim that in this case if $E|g(X)| < \infty$ then $E(g(X)|Y) = h(Y)$ where

$$h(y) = \int g(x)f(x,y)\,dx \Big/ \int f(x,y)\,dx$$

To "guess" this formula note that treating the probability densities $P(Y = y)$ as if they were real probabilities

$$P(X = x|Y = y) = \frac{P(X = x, Y = y)}{P(Y = y)} = \frac{f(x,y)}{\int f(x,y)\,dx}$$

so integrating against the conditional probability density we have

$$E(g(X)|Y = y) = \int g(x)P(X = x|Y = y)\,dx$$

To "verify" the proposed formula now, observe $h(Y) \in \sigma(Y)$ so (i) holds. To check (ii), observe that if $A \in \sigma(Y)$ then $A = \{\omega : Y(\omega) \in B\}$ for some $B \in \mathcal{R}$, so

$$E(h(Y); A) = \int_B \int h(y)f(x,y)\,dx\,dy = \int_B \int g(x)f(x,y)\,dx\,dy$$
$$= E(g(X)1_B(Y)) = E(g(X); A)$$

Remark. To drop the assumption that $\int f(x,y)\,dx > 0$ define h by

$$h(y) \int f(x,y)\,dx = \int g(x)f(x,y)\,dx$$

(i.e., h can be anything where $\int f(x,y)\,dx = 0$) and observe this is enough for the proof.

Example 1.5. Suppose X and Y are independent. Let φ be a function with $E|\varphi(X,Y)| < \infty$ and let $g(x) = E(\varphi(x,Y))$. We will now show that

$$E(\varphi(X,Y)|X) = g(X)$$

Proof It is clear that $g(X) \in \sigma(X)$. To check (ii) note that if $A \in \sigma(X)$ then $A = \{X \in C\}$ so using (3.9) and (4.6) in Chapter 1, then the definition of g, and (3.9) in Chapter 1 again

$$\int_A \varphi(X,Y)\, dP = E\{\varphi(X,Y)1_C(X)\}$$

$$= \int \int \varphi(x,y)1_C(x)\, \nu(dy)\, \mu(dx)$$

$$= \int 1_C(x)g(x)\, \mu(dx) = \int_A g(Y)\, dP \qquad \square$$

Example 1.6. Borel's paradox. Let X be a randomly chosen point on the earth, let θ be its longitude and φ be its latitude. It is customary to take $\theta \in [0, 2\pi)$ and $\varphi \in (-\pi/2, \pi/2]$ but we can equally well take $\theta \in [0, \pi)$ and $\varphi \in (-\pi, \pi]$. In words, the new longitude specifies the great circle on which the point lies, and then φ gives the angle.

At first glance it might seem that if X is uniform on the globe then θ and the angle φ on the great circle should both be uniform over their possible values. θ is uniform but φ is not. The paradox completely evaporates once we realize that in the new or in the traditional formulation φ is independent of θ, so the conditional distribution is the unconditional one, which is not uniform since there is more land near the equator than near the North Pole.

b. Properties

Conditional expectation has many of the same properties that ordinary expectation does.

(1.1a) Linearity. $E(aX + Y|\mathcal{F}) = aE(X|\mathcal{F}) + E(Y|\mathcal{F})$

Proof We need to check that the right-hand side is a version of the left. It clearly is $\mathcal{F}$-measurable. To check (ii) we observe that if $A \in \mathcal{F}$ then by linearity of the integral and the defining properties of $E(X|\mathcal{F})$ and $E(Y|\mathcal{F})$

$$\int_A \{aE(X|\mathcal{F}) + E(Y|\mathcal{F})\}\, dP = a\int_A E(X|\mathcal{F})\, dP + \int_A E(Y|\mathcal{F})\, dP$$

$$= a\int_A X\, dP + \int_A Y\, dP = \int_A aX + Y\, dP \qquad \square$$

(1.1b) Monotonicity. If $X \leq Y$ then $E(X|\mathcal{F}) \leq E(Y|\mathcal{F})$.

Proof
$$\int_A E(X|\mathcal{F})\,dP = \int_A X\,dP \leq \int_A Y\,dP = \int_A E(Y|\mathcal{F})\,dP$$

Letting $A = \{E(X|\mathcal{F}) - E(Y|\mathcal{F}) \geq \epsilon > 0\}$ we see that the indicated set has probability 0 for all $\epsilon > 0$. □

EXERCISE 1.3. Prove **Chebyshev's inequality.** If $a > 0$

$$P(|X| \geq a|\mathcal{F}) \leq a^{-2}E(X^2|\mathcal{F})$$

(1.1c) Monotone convergence theorem. If $X_n \geq 0$ and $X_n \uparrow X$ with $EX < \infty$ then $E(X_n|\mathcal{F}) \uparrow E(X|\mathcal{F})$.

Proof Let $Y_n = X - X_n$. It suffices to show that $E(Y_n|\mathcal{F}) \downarrow 0$. Since $Y_n \downarrow$, (1.1b) implies $Z_n \equiv E(Y_n|\mathcal{F}) \downarrow$ a limit Z_∞. If $A \in \mathcal{F}$

$$\int_A Z_n\,dP = \int_A Y_n\,dP$$

Letting $n \to \infty$, noting $Y_n \downarrow 0$, and using the dominated convergence theorem gives that $\int_A Z_\infty\,dP = 0$ for all $A \in \mathcal{F}$, so $Z_\infty \equiv 0$. □

Remark. By applying the last result to $Y_1 - Y_n$ we see that if $Y_n \downarrow Y$ and we have $E|Y_1|, E|Y| < \infty$ then $E(Y_n|\mathcal{F}) \downarrow E(Y|\mathcal{F})$.

EXERCISE 1.4. Suppose $X \geq 0$ and $EX = \infty$. (There is nothing to prove when $EX < \infty$). Show there is a unique $\mathcal{F}$-measurable Y with $0 \leq Y \leq \infty$ so that

$$\int_A X\,dP = \int_A Y\,dP \quad \text{for all } A \in \mathcal{F}$$

Hint: Let $X_M = X \wedge M$, $Y_M = E(X_M|\mathcal{F})$, and let $M \to \infty$.

(1.1d) Jensen's inequality. If φ is convex and $E|X|, E|\varphi(X)| < \infty$ then

$$\varphi(E(X|\mathcal{F})) \leq E(\varphi(X)|\mathcal{F})$$

Proof If φ is linear the result is trivial so we will suppose φ is not linear. We do this so that if we let $S = \{(a,b) : a,b \in \mathbf{Q},\ ax + b \leq \varphi(x) \text{ for all } x\}$ then $\varphi(x) = \sup\{ax + b : (a,b) \in S\}$. (See the proof of (3.2) in Chapter 1 for more details.) If $\varphi(x) \geq ax + b$ then (1.1b) and (1.1a) imply

$$E(\varphi(X)|\mathcal{F}) \geq a\,E(X|\mathcal{F}) + b \quad \text{a.s.}$$

Taking the sup over $(a, b) \in S$ gives

$$E(\varphi(X)|\mathcal{F}) \geq \varphi(E(X|\mathcal{F})) \quad \text{a.s.} \qquad \square$$

Remark. Here we have written a.s. by the inequalities to stress that there is an exceptional set for each a, b so we have to take the sup over a countable set.

EXERCISE 1.5. Imitate the proof in the remark after (5.2) in the appendix to prove the conditional Cauchy-Schwarz inequality.

$$E(XY|\mathcal{G})^2 \leq E(X^2|\mathcal{G})E(Y^2|\mathcal{G})$$

(1.1e) Conditional expectation is a contraction in L^p, $p \geq 1$.

Proof (1.1d) implies $|E(X|\mathcal{F})|^p \leq E(|X|^p|\mathcal{F})$. Taking expected values gives

$$E(|E(X|\mathcal{F})|^p) \leq E(E(|X|^p|\mathcal{F})) = E|X|^p \qquad \square$$

In the last equality we have used an identity that is an immediate consequence of the definition:

(1.1f) $E(E(Y|\mathcal{F})) = E(Y)$.

Proof Use property (ii) with $A = \Omega$.

Conditional expectation also has properties (like (1.1f)) that have no analogue for "ordinary" expectation.

(1.2) **Theorem.** If $\mathcal{F}_1 \subset \mathcal{F}_2$ then (i) $E(E(X|\mathcal{F}_1)|\mathcal{F}_2) = E(X|\mathcal{F}_1)$
(ii) $E(E(X|\mathcal{F}_2)|\mathcal{F}_1) = E(X|\mathcal{F}_1)$.

In words, the smaller σ-field always wins. As the proof will show the first equality is trivial. The second is easy to prove, but in combination with (1.3) is a powerful tool for computing conditional expectations. I have seen it used several times to prove results that are false.

Proof Once we notice that $E(X|\mathcal{F}_1) \in \mathcal{F}_2$, (i) follows from Example 1.1. To prove (ii) notice that $E(X|\mathcal{F}_1) \in \mathcal{F}_1$ and if $A \in \mathcal{F}_1 \subset \mathcal{F}_2$ then

$$\int_A E(X|\mathcal{F}_1) \, dP = \int_A X \, dP = \int_A E(X|\mathcal{F}_2) \, dP \qquad \square$$

EXERCISE 1.6. Give an example on $\Omega = \{a, b, c\}$ in which

$$E(E(X|\mathcal{F}_1)|\mathcal{F}_2) \neq E(E(X|\mathcal{F}_2)|\mathcal{F}_1)$$

The next result shows that for conditional expectation with respect to $\mathcal{F}$, random variables $X \in \mathcal{F}$ are like constants. They can be brought outside the "integral."

(1.3) **Theorem.** If $X \in \mathcal{F}$ and $E|Y|$, $E|XY| < \infty$ then $E(XY|\mathcal{F}) = XE(Y|\mathcal{F})$

Proof The right-hand side $\in \mathcal{F}$, so we have to check (ii). To do this we use the usual four step procedure. First, suppose $X = 1_B$ with $B \in \mathcal{F}$. In this case if $A \in \mathcal{F}$

$$\int_A 1_B E(Y|\mathcal{F}) \, dP = \int_{A \cap B} E(Y|\mathcal{F}) \, dP = \int_{A \cap B} Y \, dP = \int_A 1_B Y \, dP$$

so (ii) holds. The last result extends to simple X by linearity. If $X, Y \geq 0$, let X_n be simple random variables that $\uparrow X$, and use the monotone convergence theorem to conclude that

$$\int_A X E(Y|\mathcal{F}) \, dP = \int_A XY \, dP$$

To prove the result in general, split X and Y into their positive and negative parts. $\qquad\square$

EXERCISE 1.7. Show that when $E|X|$, $E|Y|$, and $E|XY|$ are finite, each statement implies the next one and give examples with $X, Y \in \{-1, 0, 1\}$ a.s. that show the reverse implications are false: (i) X and Y are independent, (ii) $E(Y|X) = EY$, (iii) $E(XY) = EXEY$.

(1.4) **Theorem.** Suppose $EX^2 < \infty$. $E(X|\mathcal{F})$ is the variable $Y \in \mathcal{F}$ that minimizes the "mean square error" $E(X - Y)^2$.

Remark. This result gives a "geometric interpretation" of $E(X|\mathcal{F})$. $L^2(\mathcal{F}_o) = \{Y \in \mathcal{F}_o : EY^2 < \infty\}$ is a Hilbert space, and $L^2(\mathcal{F})$ is a closed subspace. In this case $E(X|\mathcal{F})$ is the projection of X onto $L^2(\mathcal{F})$, i.e., the point in the subspace closest to X. For a picture see Figure 4.1.1.

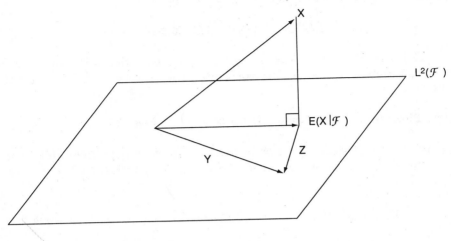

Figure 4.1.1

Proof We begin by observing that if $Z \in L^2(\mathcal{F})$, then (1.3) implies

$$ZE(X|\mathcal{F}) = E(ZX|\mathcal{F})$$

($E|XZ| < \infty$ by the Cauchy-Schwarz inequality.) Taking expected values gives

$$E(ZE(X|\mathcal{F})) = E(E(ZX|\mathcal{F})) = E(ZX)$$

or rearranging,

$$E[Z(X - E(X|\mathcal{F}))] = 0 \quad \text{for } Z \in L^2(\mathcal{F})$$

If $Y \in L^2(\mathcal{F})$ and $Z = Y - E(X|\mathcal{F})$ i.e., $Y = E(X|\mathcal{F}) + Z$ then

$$E(X - Y)^2 = E\{X - E(X|\mathcal{F}) - Z\}^2 = E\{X - E(X|\mathcal{F})\}^2 + EZ^2$$

since the cross product term vanishes. From the last formula it is easy to see $E(X - Y)^2$ is minimized when $Z = 0$. $\qquad \square$

EXERCISE 1.8. Show that if $\mathcal{G} \subset \mathcal{F}$ and $EX^2 < \infty$ then

$$E(\{X - E(X|\mathcal{F})\}^2) + E(\{E(X|\mathcal{F}) - E(X|\mathcal{G})\}^2) = E(\{X - E(X|\mathcal{G})\}^2)$$

Dropping the second term on the left, we get an inequality that says geometrically, the larger the subspace the closer the projection is, or statistically, more information means a smaller mean square error. An important special case occurs when $\mathcal{G} = \{\emptyset, \Omega\}$.

EXERCISE 1.9. Let $\text{var}(X|\mathcal{F}) = E(X^2|\mathcal{F}) - E(X|\mathcal{F})^2$. Show that

$$\text{var}(X) = E(\text{var}(X|\mathcal{F})) + \text{var}(E(X|\mathcal{F}))$$

EXERCISE 1.10. Let $Y_1, Y_2, \ldots$ be i.i.d. with mean μ and variance σ^2, N an independent integer valued r.v. with EN^2 and $X = Y_1 + \cdots + Y_N$. Show that $\text{var}(X) = \sigma^2 EN + \mu^2 \text{var}(N)$. To understand and help remember the formula think about the two special cases in which N or Y is constant.

EXERCISE 1.11. Show that if X and Y are random variables with $E(Y|\mathcal{G}) = X$ and $EY^2 = EX^2$ then $X = Y$ a.s.

EXERCISE 1.12. The result in the last exercise implies that if $EY^2 < \infty$ and $E(Y|\mathcal{G})$ has the same distribution as Y then $E(Y|\mathcal{G}) = Y$ a.s. Prove this under the assumption $E|Y| < \infty$. Hint: The trick is to prove that $\text{sgn}(X) = \text{sgn}(E(X|\mathcal{G}))$ a.s., and then take $X = Y - c$ to get the desired result.

*c. Regular conditional probabilities

Let $(\Omega, \mathcal{F}, P)$ be a probability space, $X : (\Omega, \mathcal{F}) \to (S, \mathcal{S})$ a measurable map, and $\mathcal{G}$ a σ-field $\subset \mathcal{F}$. $\mu : \Omega \times \mathcal{S} \to [0, 1]$ is said to be a **regular conditional distribution** for X given $\mathcal{G}$ if

(i) For each A, $\omega \to \mu(\omega, A)$ is a version of $P(X \in A|\mathcal{G})$.

(ii) For a.e. ω, $A \to \mu(\omega, A)$ is a probability measure on $(S, \mathcal{S})$.

When $S = \Omega$ and X is the identity map, μ is called a **regular conditional probability**.

EXERCISE 1.13. **Continuation of Example 1.4.** Suppose X and Y have a joint density $f(x, y) > 0$. Let

$$\mu(y, A) = \int_A f(x, y)\, dx \bigg/ \int f(x, y)\, dx$$

Show that $\mu(Y(\omega), A)$ is a r.c.d. for X given $\sigma(Y)$.

Regular conditional distributions are useful because they allow us to simultaneously compute the conditional expectation of all functions of X and to generalize properties of ordinary expectation in a more straightforward way.

EXERCISE 1.14. Let $\mu(\omega, A)$ be a r.c.d. for X given $\mathcal{F}$, and let $f : (S, \mathcal{S}) \to (\mathbf{R}, \mathcal{R})$ have $E|f(X)| < \infty$. Start with simple functions and show that

$$E(f(X)|\mathcal{F}) = \int \mu(\omega, dx) f(x) \quad \text{a.s.}$$

EXERCISE 1.15. Use regular conditional probability to get the conditional Hölder inequality from the unconditional one, i.e., show that if p, $q \in (1, \infty)$ with $1/p + 1/q = 1$ then

$$E(|XY| | \mathcal{G}) \le E(|X|^p | \mathcal{G})^{1/p} E(|Y|^q | \mathcal{G})^{1/q}$$

Unfortunately, r.c.d.'s do not always exist. The first example was due to Dieudonné (1948). See Doob (1953), p. 624 or Faden (1985) for more recent developments. Without going into the details of the example, it is easy to see the source of the problem. If $A_1, A_2, \ldots$ are disjoint then (1.1a) and (1.1c) imply

$$P(X \in \cup_n A_n | \mathcal{G}) = \sum_n P(X \in A_n | \mathcal{G}) \quad \text{a.s.}$$

but if $\mathcal{S}$ contains enough countable collections of disjoint sets the exceptional sets may pile up. Fortunately,

(1.6) Theorem. r.c.d.'s exist if $(S, \mathcal{S})$ is nice.

Proof By definition there is a 1-1 map $\varphi : S \to \mathbf{R}$ so that φ and φ^{-1} are measurable. Using monotonicity (1.1b) and throwing away a countable collection of null sets, we find there is a set Ω_o with $P(\Omega_o) = 1$ and a family of random variables $G(q, \omega)$, $q \in \mathbf{Q}$ so that $q \to G(q, \omega)$ is nondecreasing and $\omega \to G(q, \omega)$ is a version of $P(\varphi(X) \le q | \mathcal{G})$. Let $F(x, \omega) = \inf\{G(q, \omega) : q > x\}$. The notation may remind the reader of the proof of (2.5) in Chapter 2. The argument given there shows F is a distribution function. Since $G(q_n, \omega) \downarrow F(x, \omega)$, the remark after (1.1c) implies that $F(x, \omega)$ is a version of $P(\varphi(X) \le x | \mathcal{G})$.

Now for each $\omega \in \Omega_o$, there is a unique measure $\nu(\omega, \cdot)$ on $(\mathbf{R}, \mathcal{R})$ so that $\nu(\omega, (-\infty, x]) = F(x, \omega)$. To check that for each $B \in \mathcal{R}$, $\nu(\omega, B)$ is a version of $P(\varphi(X) \in B | \mathcal{G})$, we observe that the class of B for which this statement is true (this includes the measurability of $\omega \to \nu(\omega, B)$) is a λ-system that contains all sets of the form $(a_1, b_1] \cup \cdots (a_k, b_k]$ where $-\infty \le a_i < b_i \le \infty$, so the desired result follows from the $\pi - \lambda$ theorem. To extract the desired r.c.d. notice that if $A \in \mathcal{S}$, and $B = \varphi(A)$ then $B = (\varphi^{-1})^{-1}(A) \in \mathcal{R}$, and set $\mu(\omega, A) = \nu(\omega, B)$.

The following generalization of (1.6) will be needed in Section 5.1.

EXERCISE 1.16. Suppose X and Y take values in a nice space $(S, \mathcal{S})$ and $\mathcal{G} = \sigma(Y)$. Imitate the proof of (1.6) to show that there is a function $\mu : S \times \mathcal{S} \rightarrow [0, 1]$ so that

(i) for each A, $\mu(Y(\omega), A)$ is a version of $P(X \in A | \mathcal{G})$

(ii) for a.e. ω, $A \rightarrow \mu(Y(\omega), A)$ is a probability measure on $(S, \mathcal{S})$.

4.2. Martingales, Almost Sure Convergence

In this section we will define martingales and their cousins supermartingales and submartingales, and take the first steps in developing their theory. Let $\mathcal{F}_n$ be a **filtration**, i.e., an increasing sequence of σ-fields. A sequence X_n is said to be **adapted** to $\mathcal{F}_n$ if $X_n \in \mathcal{F}_n$ for all n. If a X_n is sequence with

(i) $E|X_n| < \infty$,

(ii) X_n is adapted to $\mathcal{F}_n$,

(iii) $E(X_{n+1} | \mathcal{F}_n) = X_n$ for all n,

then X is said to be a **martingale** (with respect to $\mathcal{F}_n$). If in the last definition $=$ is replaced by $\leq$ or $\geq$ then X is said to be a **supermartingale** or **submartingale** respectively.

Example 2.1. Consider the succesive tosses of a fair coin and let $\xi_n = 1$ if the nth toss is heads and $\xi_n = -1$ if the nth toss is tails. Let $X_n = \xi_1 + \cdots + \xi_n$ and $\mathcal{F}_n = \sigma(\xi_1, \ldots, \xi_n)$ for $n \geq 1$, $X_0 = 0$ and $\mathcal{F}_0 = \{\emptyset, \Omega\}$. I claim that X_n, $n \geq 0$, is a martingale with respect to $\mathcal{F}_n$. To prove this, we observe that $X_n \in \mathcal{F}_n$, $E|X_n| < \infty$, and ξ_{n+1} is independent of $\mathcal{F}_n$, so using the linearity of conditional expectation, (1.1a) and Example 1.2

$$E(X_{n+1} | \mathcal{F}_n) = E(X_n | \mathcal{F}_n) + E(\xi_{n+1} | \mathcal{F}_n) = X_n + E\xi_{n+1} = X_n$$

Note that in this example $\mathcal{F}_n = \sigma(X_1, \ldots, X_n)$ and $\mathcal{F}_n$ is the smallest filtration that X_n is adapted to. In what follows, when the filtration is not mentioned, we will take $\mathcal{F}_n = \sigma(X_1, \ldots, X_n)$.

EXERCISE 2.1. Suppose X_n, $n \geq 1$, is a martingale w.r.t. $\mathcal{G}_n$ and let $\mathcal{F}_n = \sigma(X_1, \ldots, X_n)$. Then $\mathcal{G}_n \supset \mathcal{F}_n$ and X_n is a martingale w.r.t. $\mathcal{F}_n$.

If the coin tosses considered above have $P(\xi_n = 1) \leq 1/2$ then the computation just completed shows $E(X_{n+1} | \mathcal{F}_n) \leq X_n$, i.e., X_n is a supermartingale. In this case X_n corresponds to betting on an unfavorable game so there is nothing "super" about a supermartingale. The name comes from the fact that if f is superharmonic (i.e., f has continuous derivatives of order ≤ 2 and $\partial^2 f / \partial x_1^2 + \cdots + \partial^2 f / \partial x_d^2 \leq 0$) then

$$f(x) \geq \frac{1}{|B(0,r)|} \int_{B(x,r)} f(y)\, dy$$

where $B(x,r) = \{y : |x - y| \leq r\}$ is the ball of radius r, and $|B(0,r)|$ is the volume of the ball of radius r.

EXERCISE 2.2. Suppose f is superharmonic on $\mathbf{R}^d$. Let $\xi_1, \xi_2, \ldots$ be i.i.d. uniform on $B(0,1)$, and define S_n by $S_n = S_{n-1} + \xi_n$ for $n \geq 1$ and $S_0 = x$. Show that $X_n = f(S_n)$ is a supermartingale.

Our first result is an immediate consequence of the definition of a supermartingale. We could take the conclusion of (2.1) as the defintion of supermartingale but then the definition would be harder to check.

(2.1) **Theorem.** If X_n is a supermartingale then for $n > m$, $E(X_n|\mathcal{F}_m) \leq X_m$.

Proof The definition gives the result for $n = m + 1$. Suppose $n = m + k$ with $k \geq 2$. By (1.2),

$$E(X_{m+k}|\mathcal{F}_m) = E(E(X_{m+k}|\mathcal{F}_{m+k-1})|\mathcal{F}_m) \leq E(X_{m+k-1}|\mathcal{F}_m)$$

by the definition and (1.1b). The desired result now follows by induction. $\square$

(2.2) **Corollary.** (i) If X_n is a submartingale then for $n > m$, $E(X_n|\mathcal{F}_m) \geq X_m$.
(ii) If X_n is a martingale then for $n > m$, $E(X_n|\mathcal{F}_m) = X_m$.

Proof To prove (i) note that $-X_n$ is a supermartingale and use (1.1a). For (ii) observe that X_n is a supermartingale and a submartingale. $\square$

Remark. The idea in the proof of (2.2) can be used many times below. To keep from repeating ourselves we will just state the result for either supermartingales or submartingales, and leave it to the reader to translate the result for the other two.

(2.3) **Theorem.** If X_n is a martingale w.r.t. $\mathcal{F}_n$ and φ is a convex function with $E|\varphi(X_n)| < \infty$ for all n then $\varphi(X_n)$ is a submartingale w.r.t. $\mathcal{F}_n$.

Proof By Jensen's inequality and the definition

$$E(\varphi(X_{n+1})|\mathcal{F}_n) \geq \varphi(E(X_{n+1}|\mathcal{F}_n)) = \varphi(X_n)$$ $\square$

(2.4) **Corollary.** Let $p \geq 1$. If X_n is a martingale w.r.t. $\mathcal{F}_n$ and $E|X_n|^p < \infty$ for all n, then $|X_n|^p$ is a submartingale w.r.t. $\mathcal{F}_n$.

(2.5) **Theorem.** If X_n is a submartingale w.r.t. $\mathcal{F}_n$ and φ is an increasing convex function with $E|\varphi(X_n)| < \infty$ for all n, then $\varphi(X_n)$ is a submartingale w.r.t. $\mathcal{F}_n$.

Proof   By Jensen's inequality and the assumptions

$$E(\varphi(X_{n+1})|\mathcal{F}_n) \geq \varphi(E(X_{n+1}|\mathcal{F}_n)) \geq \varphi(X_n) \qquad \square$$

EXERCISE 2.3. Give an example of a submartingale X_n so that X_n^2 is a super-martingale. Hint: X_n does not have to be random.

(2.6) **Corollary.** (i) If X_n is a submartingale then $(X_n - a)^+$ is a submartingale. (ii) If X_n is a supermartingale then $X_n \wedge a$ is a supermartingale.

Let $\mathcal{F}_n$, $n \geq 0$ be a filtration. H_n, $n \geq 1$ is said to be a **predictable sequence**, if $H_n \in \mathcal{F}_{n-1}$ for all $n \geq 1$. In words, the value of H_n may be predicted (with certainty) from the information available at time $n - 1$. In this section we will be thinking of H_n as the amount of money a gambler will bet at time n. This can be based on the outcomes at times $1, \ldots, n - 1$ but not on the outcome at time n!

Once we start thinking of H_n as a gambling system, it is natural to ask how much money we would make if we used it. For concreteness, let us suppose that the game consists of flipping a coin and that for each dollar you bet you win one dollar when the coin comes up heads, and lose your dollar when the coin comes up tails. Let X_n be the net amount of money you would have won at time n if you had bet one dollar each time. If you bet according to a gambling system H then your winnings at time n would be

$$(H \cdot X)_n = \sum_{m=1}^{n} H_m(X_m - X_{m-1})$$

since $X_m - X_{m-1} = +1$ or -1 when the mth toss results in a win or loss respectively.

Let $\xi_m = X_m - X_{m-1}$. A famous gambling system called the "martingale" is defined by $H_1 = 1$ and for $n \geq 2$, $H_n = 2H_{n-1}$ if $\xi_{n-1} = -1$ and $H_n = 1$ if $\xi_{n-1} = 1$. In words, we double our bet when we lose so that if we lose k times and then win our net winnings will be $-1 - 2 \ldots - 2^{k-1} + 2^k = 1$. This system seems to provide us with a "sure thing" as long as $P(\xi_m = 1) > 0$. However, the next result says there is no system for beating an unfavorable game.

(2.7) Theorem. Let X_n, $n \geq 0$, be a supermartingale. If $H_n \geq 0$ is predictable and each H_n is bounded then $(H \cdot X)_n$ is a supermartingale.

Proof Using the fact that conditional expectation is linear, $(H \cdot X)_n \in \mathcal{F}_n$, $H_n \in \mathcal{F}_{n-1}$, and (1.3) we have

$$E((H \cdot X)_{n+1}|\mathcal{F}_n) = (H \cdot X)_n + E(H_{n+1}(X_{n+1} - X_n)|\mathcal{F}_n)$$
$$= (H \cdot X)_n + H_{n+1}E((X_{n+1} - X_n)|\mathcal{F}_n) \leq (H \cdot X)_n$$

since $E((X_{n+1} - X_n)|\mathcal{F}_n) \leq 0$ and $H_{n+1} \geq 0$. $\qquad\qquad\square$

Remark. The same result is obviously true for submartingales and for martingales (in the last case without the restriction $H_n \geq 0$).

The notion of a stopping time, introduced in Section 3.1, is closely related to the concept of a gambling system. Recall that a random variable N is said to be a **stopping time** if $\{N = n\} \in \mathcal{F}_n$ for all $n < \infty$. If you think of N as the time a gambler stops gambling, then the condition above says that the decision to stop at time n must be measurable with respect to the information he has at that time. If we let $H_n = 1_{\{N \geq n\}}$, then $\{N \geq n\} = \{N \leq n - 1\}^c \in \mathcal{F}_{n-1}$, so H_n is predictable, and it follows from (2.7) that $(H \cdot X)_n = X_{N \wedge n} - X_0$ is a supermartingale. Since the constant sequence $Y_n = X_0$ is a supermartingale and the sum of two supermartingales is also, we have:

(2.8) Corollary. If N is a stopping time and X_n is a supermartingale, then $X_{N \wedge n}$ is a supermartingale.

Although you cannot make money with gambling systems, you can prove theorems with them. Suppose X_n, $n \geq 0$, is a submartingale. Let $a < b$, let $N_0 = -1$ and for $k \geq 1$ let

$$N_{2k-1} = \inf\{m > N_{2k-2} : X_m \leq a\}$$
$$N_{2k} = \inf\{m > N_{2k-1} : X_m \geq b\}$$

The N_j are stopping times and $\{N_{2k-1} < m \leq N_{2k}\} = \{N_{2k-1} \leq m - 1\} \cap \{N_{2k} \leq m - 1\}^c \in \mathcal{F}_{m-1}$ so

$$H_m = \begin{cases} 1 & \text{if } N_{2k-1} < m \leq N_{2k} \text{ for some } k \\ 0 & \text{otherwise} \end{cases}$$

defines predictable sequence. $X(N_{2k-1}) \leq a$ and $X(N_{2k}) \geq b$, so between times N_{2k-1} and N_{2k}, X_m crosses from below a to above b. H_m is a gambling system which tries to take advantage of these "upcrossings". In stock market terms, we buy when $X_m \leq a$ and sell when $X_m \geq b$, so every time an upcrossing is completed we make a profit of $\geq (b - a)$. Finally, $U_n = \sup\{k : N_{2k} \leq n\}$ is the number of upcrossings completed by time n. For a picture see Figure 4.2.1. x's mark the locations of the X_m, y's the locations of $Y_m = a + (X_m - a)^+$. The solid lines correspond to increments $X_m - X_{m-1}$ with $H_m = 1$.

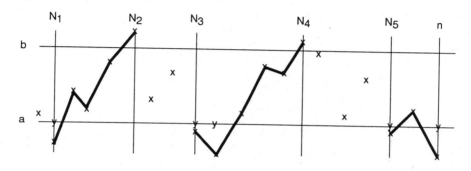

Figure 4.2.1

(2.9) The upcrossing inequality. If X_m, $m \geq 0$, is a submartingale then

$$(b - a)EU_n \leq E(X_n - a)^+ - E(X_0 - a)^+$$

Proof Let $Y_n = a + (X_n - a)^+$. By (2.6), Y_n is a submartingale. Clearly it upcrosses $[a, b]$ the same number of times that X_n does, and we have $(b-a)U_n \leq (H \cdot Y)_n$, since each upcrossing results in a profit $\geq (b-a)$ and a final incomplete upcrossing (if there is one) makes a nonnegative contribution to the right hand side. Let $K_m = 1 - H_m$. Clearly $Y_n - Y_0 = (H \cdot Y)_n + (K \cdot Y)_n$, and it follows from (2.7) that $E(K \cdot Y)_n \geq E(K \cdot Y)_0 = 0$ so $E(H \cdot Y)_n \leq E(Y_n - Y_0)$ proving (2.9). $\qquad \square$

We have proved the result in its classical form, even though this is a little misleading. The key fact is that $E(K \cdot X)_n \geq 0$, i.e., no matter how hard you try you can't lose money betting on a submartingale. From the upcrossing inequality we easily get

(2.10) The martingale convergence theorem. If X_n is a submartingale with $\sup EX_n^+ < \infty$ then as $n \to \infty$, X_n converges a.s. to a limit X with $E|X| < \infty$.

Proof Since $(X - a)^+ \leq X^+ + |a|$, (2.9) implies that

$$EU_n \leq (|a| + EX_n^+)/(b - a)$$

As $n \uparrow \infty$, $U_n \uparrow U$ the number of upcrossings of $[a, b]$ by the whole sequence, so if $\sup EX_n^+ < \infty$ then $EU < \infty$ and hence $U < \infty$ a.s. Since the last conclusion holds for all rational a and b

$$\cup_{a,b \in \mathbf{Q}} \{\liminf X_n < a < b < \limsup X_n\} \quad \text{has probability } 0$$

and hence $\limsup X_n = \liminf X_n$ a.s., i.e., $\lim X_n$ exists a.s. Fatou's lemma guarantees $EX^+ \leq \liminf EX_n^+ < \infty$, so $X < \infty$ a.s. To see $X > -\infty$, we observe that

$$EX_n^- = EX_n^+ - EX_n \leq EX_n^+ - EX_0$$

(since X_n is a submartingale), so another application of Fatou's lemma shows

$$EX^- \leq \liminf_{n \to \infty} EX_n^- \leq \sup_n EX_n^+ - EX_0 < \infty$$

and completes the proof.   □

Remark. To prepare for the proof of (6.1), the reader should note that we have shown that if the number of upcrossings of (a, b) by X_n is finite for all $a, b \in \mathbf{Q}$ then the limit of X_n exists.

An important special case of (2.10) is

(2.11) Corollary. If $X_n \geq 0$ is a supermartingale then as $n \to \infty$, $X_n \to X$ a.s. and $EX \leq EX_0$.

Proof $Y_n = -X_n \leq 0$ is a submartingle with $EY_n^+ = 0$. Since $EX_0 \geq EX_n$, the inequality follows from Fatou's lemma.   □

In the next section we will give several applications of the last two results. We close this one by giving two "counterexamples."

Example 2.2. The first shows that the assumptions of (2.11) (and hence those of (2.10)) do not guarantee convergence in L^1. Let S_n be a symmetric simple random walk with $S_0 = 1$, i.e., $S_n = S_{n-1} + \xi_n$ where $\xi_1, \xi_2, \ldots$ are i.i.d. with $P(X_i = 1) = P(X_i = -1) = 1/2$. Let $N = \inf\{n : S_n = 0\}$ and let $X_n = S_{N \wedge n}$. (2.8) implies that X_n is a nonnegative martingale. (2.11) implies X_n converges to a limit $X_\infty < \infty$ which must be $\equiv 0$, since convergence to $k > 0$ is impossible. (If $X_n = k > 0$ then $X_{n+1} = k \pm 1$.) Since $EX_n = EX_0 = 1$ for all n and $X_\infty = 0$, convergence cannot occur in L^1.

Example 2.2 is an important counterexample to keep in mind as you read the rest of this chapter. The next two are not as important.

Example 2.3. We will now give an example of a martingale with $X_n \to 0$ in probability but not a.s. Let $X_0 = 0$. When $X_{k-1} = 0$, let $X_k = 1$ or -1 with probability $1/2k$ and $= 0$ with probability $1 - 1/k$. When $X_{k-1} \neq 0$ let $X_k = kX_{k-1}$ with probability $1/k$ and $= 0$ with probability $1 - 1/k$. From the construction $P(X_k = 0) = 1 - 1/k$ so $X_k \to 0$ in probability. On the other hand, the second Borel-Cantelli lemma implies $P(X_k = 0 \text{ for } k \geq K) = 0$, and values in $(-1, 1) - \{0\}$ are impossible so X_k does not converge to 0 a.s.

EXERCISE 2.4. Give an example of a martingale X_n with $X_n \to -\infty$ a.s. Hint: Let $X_n = \xi_1 + \cdots + \xi_n$ where the ξ_i are independent (but not identically distributed) with $E\xi_i = 0$.

Our final result is useful in reducing questions about submartingales to questions about martingales.

(2.12) **Doob's decomposition.** Any submartingale X_n, $n \geq 0$, can be written in a unique way as $X_n = M_n + A_n$ where M_n is a martingale and A_n is a predictable sequence with $A_0 = 0$.

Proof We want $X_n = M_n + A_n$, $E(M_n|\mathcal{F}_{n-1}) = M_{n-1}$, and $A_n \in \mathcal{F}_{n-1}$. So we must have

$$E(X_n|\mathcal{F}_{n-1}) = E(M_n|\mathcal{F}_{n-1}) + E(A_n|\mathcal{F}_{n-1})$$
$$= M_{n-1} + A_n = X_{n-1} - A_{n-1} + A_n$$

and it follows that

(a) $A_n - A_{n-1} = E(X_n|\mathcal{F}_{n-1}) - X_{n-1}$

(b) $M_n = X_n - A_n$

Now $A_0 = 0$ and $M_0 = X_0$ by assumption, so we have A_n and M_n defined for all time and we have proved uniqueness. To check that our recipe works, we observe that $A_n - A_{n-1} \geq 0$ since X_n is a submartingale, and induction shows $A_n \in \mathcal{F}_{n-1}$. To see that M_n is a martingale, we use (b), $A_n \in \mathcal{F}_{n-1}$ and (a):

$$E(M_n|\mathcal{F}_{n-1}) = E(X_n - A_n|\mathcal{F}_{n-1})$$
$$= E(X_n|\mathcal{F}_{n-1}) - A_n = X_{n-1} - A_{n-1} = M_{n-1} \qquad \square$$

EXERCISE 2.5. Let $X_n = \sum_{m \leq n} 1_{B_m}$ and suppose $B_n \in \mathcal{F}_n$. What is the Doob decomposition for X_n?

EXERCISES

2.6. Let $\xi_1, \xi_2, \ldots$ be independent with $E\xi_i = 0$ and $\text{var}(\xi_m) = \sigma_m^2$, and let $s_n^2 = \sum_{m=1}^n \sigma_m^2$. Then $S_n^2 - s_n^2$ is a martingale.

2.7. If $\xi_1, \xi_2, \ldots$ are independent and have $E\xi_i = 0$ then

$$X_n^{(k)} = \sum_{1 \leq i_1 < \ldots < i_k \leq n} \xi_{i_1} \cdots \xi_{i_k}$$

is a martingale. When $k = 2$ and $S_n = \xi_1 + \cdots + \xi_n$, $2X_n^{(2)} = S_n^2 - \sum_{m \leq n} \xi_m^2$.

2.8. Generalize (2.6) by showing that if X_n and Y_n are submartingales w.r.t. $\mathcal{F}_n$ then $X_n \vee Y_n$ is also.

2.9. Let $Y_1, Y_2, \ldots$ be nonnegative i.i.d. random variables with $EY_m = 1$ and $P(Y_m = 1) < 1$. (i) Show that $X_n = \prod_{m \leq n} Y_m$ defines a martingale. (ii) Use (2.11) and an argument by contradiction to show $X_n \to 0$ a.s. (iii) Use the strong law of large numbers to conclude $(1/n)\log X_n \to c < 0$.

2.10. Suppose $y_n > -1$ for all n and $\sum |y_n| < \infty$. Show that $\prod_{m=1}^\infty (1 + y_m)$ exists.

2.11. Let X_n and Y_n be positive integrable and adapted to $\mathcal{F}_n$. Suppose

$$E(X_{n+1}|\mathcal{F}_n) \leq (1 + Y_n)X_n$$

with $\sum Y_n < \infty$ a.s. Prove that X_n converges a.s. to a finite limit by finding a closely related supermartingale to which (2.11) can be applied.

2.12. Use the random walks in Exercise 2.2 to conclude that in $d \leq 2$ nonnegative superharmonic functions must be constant. The example $f(x) = |x|^{2-d}$ shows this is false in $d > 2$.

2.13. The switching principle. Suppose X_n^1 and X_n^2 are supermartingales with respect to $\mathcal{F}_n$, and N is a stopping time so that $X_N^1 \geq X_N^2$. Then

$$Y_n = X_n^1 1_{(N>n)} + X_n^2 1_{(N \leq n)} \text{ is a supermartingale.}$$

Since $N + 1$ is a stopping time this implies that

$$Z_n = X_n^1 1_{(N \geq n)} + X_n^2 1_{(N < n)} \text{ is a supermartingale.}$$

2.14. Dubins' inequality. For every positive supermartingale X_n, $n \geq 0$, the number of upcrossings U of $[a, b]$ satisfies

$$P(U \geq k) \leq \left(\frac{a}{b}\right)^k E \min(X_0/a, 1)$$

To prove this we let $N_0 = -1$ and for $j \geq 1$ let

$$N_{2j-1} = \inf\{m > N_{2j-2} : X_m \leq a\}$$
$$N_{2j} = \inf\{m > N_{2j-1} : X_m \geq b\}$$

Let $Y_n = 1$ for $0 \leq n < N_1$ and for $j \geq 1$

$$Y_n = \begin{cases} (b/a)^{j-1}(X_n/a) & \text{for } N_{2j-1} \leq n < N_{2j} \\ (b/a)^j & \text{for } N_{2j} \leq n < N_{2j+1} \end{cases}$$

(i) Use the switching principle in the previous exercise and induction to show that $Z_n^j = Y_{n \wedge N_j}$ is a supermartingale. (ii) Use $EY_{n \wedge N_{2k}} \leq EY_0$ and let $n \to \infty$ to get Dubins' inequality.

4.3. Examples

In this section we will apply the martingale convergence theorem to generalize the second Borel-Cantelli lemma and to study Polya's urn scheme, Radon-Nikodym derivatives, and branching processes. The four topics are independent of each other and are taken up in the order indicated.

a. Bounded increments

Our first result shows that martingales with bounded increments either converge or oscillate between $+\infty$ and $-\infty$.

(3.1) Theorem. Let $X_1, X_2, \ldots$ be a martingale with $|X_{n+1} - X_n| \leq M < \infty$. Let

$$C = \{\lim X_n \text{ exists and is finite}\}$$
$$D = \{\limsup X_n = +\infty \text{ and } \liminf X_n = -\infty\}$$

Then $P(C \cup D) = 1$.

Proof Since $X_n - X_0$ is a martingale, we can without loss of generality suppose that $X_0 = 0$. Let $0 < K < \infty$ and let $N = \inf\{n : X_n \leq -K\}$. $X_{n \wedge N}$ is a martingale with $X_{n \wedge N} \geq -K - M$ a.s. so applying (2.11) to $X_{n \wedge N} + K + M$ shows $\lim X_n$ exists on $\{N = \infty\}$. Letting $K \to \infty$ we see that the limit exists on $\{\liminf X_n > -\infty\}$. Applying the last conclusion to $-X_n$, we see that $\lim X_n$ exists on $\{\limsup X_n < \infty\}$ and the proof is complete. □

EXERCISE 3.1. Let X_n, $n \geq 0$, be a submartingale with $\sup X_n < \infty$. Let $\xi_n = X_n - X_{n-1}$ and suppose $E(\sup \xi_n^+) < \infty$. Show that X_n converges a.s.

EXERCISE 3.2. Give an example of a martingale X_n with $\sup_n |X_n| < \infty$ and $P(X_n = a \text{ i.o.}) = 1$ for $a = -1, 0, 1$. This example shows that it is not enough to have $\sup |X_{n+1} - X_n| < \infty$ in (3.1).

EXERCISE 3.3. (Assumes familiarity with finite state Markov chains.) Fine tune the example for the previous problem so that $P(X_n = 0) \to 1 - 2p$ and $P(X_n = -1)$, $P(X_n = 1) \to p$, where p is your favorite number in $(0,1)$, i.e., you are asked to do this for one value of p that you may choose. This example shows that a martingale can converge in distribution without converging a.s. (or in probability).

EXERCISE 3.4. Let X_n and Y_n be positive integrable and adapted to $\mathcal{F}_n$. Suppose $E(X_{n+1}|\mathcal{F}_n) \le X_n + Y_n$, with $\sum Y_n < \infty$ a.s. Prove that X_n converges a.s. to a finite limit. Hint: Let $N = \inf_k \sum_{m=1}^{k} Y_m > M$, and stop your supermartingale at time N.

(3.2) **Corollary. Second Borel-Cantelli lemma, II.** Let $\mathcal{F}_n$, $n \ge 0$ be a filtration with $\mathcal{F}_0 = \{\emptyset, \Omega\}$ and A_n, $n \ge 1$ a sequence of events with $A_n \in \mathcal{F}_n$. Then

$$\{A_n \text{ i.o.}\} = \left\{ \sum_{n=1}^{\infty} P(A_n|\mathcal{F}_{n-1}) = \infty \right\}$$

Proof If we let $X_0 = 0$ and

$$X_n = \sum_{m=1}^{n} 1_{A_m} - P(A_m|\mathcal{F}_{m-1}) \quad \text{for } n \ge 1$$

then X_n is a martingale with $|X_n - X_{n-1}| \le 1$. Using the notation of (3.1) we have:

$$\text{on } C, \quad \sum_{n=1}^{\infty} 1_{A_n} = \infty \quad \text{if and only if} \quad \sum_{n=1}^{\infty} P(A_n|\mathcal{F}_{n-1}) = \infty$$

$$\text{on } D, \quad \sum_{n=1}^{\infty} 1_{A_n} = \infty \quad \text{and} \quad \sum_{n=1}^{\infty} P(A_n|\mathcal{F}_{n-1}) = \infty$$

Since $P(C \cup D) = 1$, the result follows.   □

EXERCISE 3.5. Let $p_m \in [0,1)$. Use the Borel-Cantelli lemmas to show that $\prod_{m=1}^{\infty}(1 - p_m) = 0$ if and only if $\sum_{m=1}^{\infty} p_m = \infty$.

EXERCISE 3.6. Show $\sum_{n=2}^{\infty} P(A_n| \cap_{m=1}^{n-1} A_m^c) = \infty$ implies $P(\cap_{m=1}^{\infty} A_m^c) = 0$.

b. Polya's urn scheme

An urn contains r red and g green balls. At each time we draw a ball out, then replace it, and add c more balls of the color drawn. Let X_n be the fraction of green balls after the nth draw. To check that X_n is a martingale, note that if there are i red balls and j green balls at time n, then

$$X_{n+1} = \begin{cases} (j+c)/(i+j+c) & \text{with probability } j/(i+j) \\ j/(i+j+c) & \text{with probability } i/(i+j) \end{cases}$$

and we have

$$\frac{j+c}{i+j+c} \cdot \frac{j}{i+j} + \frac{j}{i+j+c} \cdot \frac{i}{i+j} = \frac{(j+c+i)j}{(i+j+c)(i+j)} = \frac{j}{i+j}$$

Since $X_n \geq 0$, (2.11) implies that $X_n \to X_\infty$ a.s. To compute the distribution of the limit we observe (a) the probability of getting green on the first m draws then red on the next $\ell = n - m$ draws is

$$\frac{g}{g+r} \cdot \frac{g+c}{g+r+c} \cdots \frac{g+(m-1)c}{g+r+(m-1)c} \cdot \frac{r}{g+r+mc} \cdots \frac{r+(\ell-1)c}{g+r+(n-1)c}$$

and (b) any other outcome of the first n draws with m green balls drawn and ℓ red balls drawn has the same probability since the denominator remains the same and the numerator is permuted. Consider the special case $c = 1$, $g = 1$, $r = 1$. Let G_n be the number of green balls after the nth draw has been completed and the new ball has been added. It follows from (a) and (b) that

$$P(G_n = m+1) = \binom{n}{m} \frac{m!(n-m)!}{(n+1)!} = \frac{1}{n+1}$$

so X_∞ has a uniform distribution on (0,1). In general, the distribution of X_∞ has density

$$\frac{\Gamma((g+r)/c)}{\Gamma(g/c)\Gamma(r/c)}(1-x)^{\frac{g}{c}-1}x^{\frac{r}{c}-1}$$

This is the **beta distribution** with parameters g/c and r/c. In Example 4.2 we will see that the limit behavior changes drastically if in addition to the c balls of the color chosen we always add one ball of the opposite color.

c. Radon-Nikodym derivatives

Let μ be a finite measure and ν a probability measure on $(\Omega, \mathcal{F})$. Let $\mathcal{F}_n \uparrow \mathcal{F}$ be σ-fields (i.e., $\sigma(\cup \mathcal{F}_n) = \mathcal{F}$). Let μ_n and ν_n be the restrictions of μ and ν to $\mathcal{F}_n$.

(3.3) Theorem. Suppose $\mu_n \ll \nu_n$ for all n. Let $X_n = d\mu_n/d\nu_n$ and let $X = \limsup X_n$. Then

$$\mu(A) = \int_A X d\nu + \mu(A \cap \{X = \infty\})$$

Remark. $\mu_r(A) \equiv \int_A X \, d\nu$ is a measure $\ll \nu$. Since (2.11) implies $\nu(X = \infty) = 0$, $\mu_s(A) \equiv \mu(A \cap \{X = \infty\})$ is singular w.r.t. ν. Thus $\mu = \mu_r + \mu_s$ gives the Lebesgue decomposition of μ (see (8.5) in the Appendix), and $X_\infty = d\mu_r/d\nu$, ν-a.s. Here and in the proof we need to keep track of the measure to which the a.s. refers.

Proof As the reader can probably anticipate:

(3.4) Lemma. X_n (defined on $(\Omega, \mathcal{F}, \nu)$) is a martingale w.r.t. $\mathcal{F}_n$.

Proof We observe that by definition $X_n \in \mathcal{F}_n$. Let $A \in \mathcal{F}_n$. Since $X_n \in \mathcal{F}_n$ and ν_n is the restriction of ν to $\mathcal{F}_n$

$$\int_A X_n \, d\nu = \int_A X_n \, d\nu_n$$

Using the definition of X_n and Exercise 8.7 in the Appendix

$$\int_A X_n \, d\nu_n = \mu_n(A) = \mu(A)$$

the last equality holding since $A \in \mathcal{F}_n$ and μ_n is the restriction of μ to $\mathcal{F}_n$. If $A \in \mathcal{F}_{m-1} \subset \mathcal{F}_m$, using the last result for $n = m$ and $n = m - 1$ gives

$$\int_A X_m d\nu = \mu(A) = \int_A X_{m-1} d\nu$$

so $E(X_m | \mathcal{F}_{m-1}) = X_{m-1}$. $\square$

Since X_n is a nonnegative martingale, (2.11) implies that $X_n \to X$ ν-a.s. We want to check that the equality in the theorem holds. Dividing $\mu(A)$ by $\mu(\Omega)$, we can without loss of generality suppose μ is a probability measure. Let $\rho = (\mu+\nu)/2$, $\rho_n = (\mu_n+\nu_n)/2 = $ the restriction of ρ to $\mathcal{F}_n$. Let $Y_n = d\mu_n/d\rho_n$, $Z_n = d\nu_n/d\rho_n$. $Y_n, Z_n \geq 0$ and $Y_n + Z_n = 2$ (by Exercise 8.6 in the appendix), so Y_n and Z_n are bounded martingales with limits Y and Z. As the reader can probably guess

(*) $$Y = d\mu/d\rho \qquad Z = d\nu/d\rho$$

It suffices to prove the first equality. From the proof of (3.4), if $A \in \mathcal{F}_m \subset \mathcal{F}_n$

$$\mu(A) = \int_A Y_n \, d\rho \to \int_A Y \, d\rho$$

by the bounded convergence theorem. The last computation shows that

$$\mu(A) = \int_A Y \, d\rho \quad \text{for all } A \in \mathcal{G} = \cup_m \mathcal{F}_m$$

$\mathcal{G}$ is a π-system, so the $\pi - \lambda$ theorem implies the equality is valid for all $A \in \mathcal{F} = \sigma(\mathcal{G})$ and ($*$) is proved.

It follows from Exercises 8.8 and 8.9 in the Appendix that $X_n = Y_n/Z_n$. At this point the reader can probably leap to the conclusion that $X = Y/Z$. To get there carefully note $Y + Z = 2$ ρ-a.s. so $\rho(Y = 0, Z = 0) = 0$, and having ruled out $0/0$ we have $X = Y/Z$ ρ-a.s. (Recall $X \equiv \limsup X_n$.) Let $W = (1/Z) \cdot 1_{(Z>0)}$. Using ($*$) then $1 = ZW + 1_{(Z=0)}$, we have

(a) $$\mu(A) = \int_A Y \, d\rho = \int_A YWZ \, d\rho + \int_A 1_{(Z=0)} Y \, d\rho$$

Now ($*$) implies $d\nu = Z \, d\rho$, and it follows from the defintions that

$$YW = X1_{(Z>0)} = X \quad \nu\text{-a.s.}$$

the second equality holding since $\nu(\{Z = 0\}) = 0$. Combining things we have

(b) $$\int_A YWZ \, d\rho = \int_A X \, d\nu$$

To handle the other term we note that ($*$) implies $d\mu = Y \, d\rho$, and it follows from the definitions that $\{X = \infty\} = \{Z = 0\}$ μ-a.s. so

(c) $$\int_A 1_{(Z=0)} Y \, d\rho = \int_A 1_{(X=\infty)} \, d\mu$$

Combining (a), (b), and (c) gives the desired result. $\qquad \square$

Example 3.1. Suppose $\mathcal{F}_n = \sigma(I_{k,n} : 0 \le k < K_n)$ where for each n, $I_{k,n}$ is a partition of Ω, and the $(n + 1)$th partition is a refinement of the nth. In this case the condition $\mu_n \ll \nu_n$ is $\nu(I_{k,n}) = 0$ implies $\mu(I_{k,n}) = 0$, and the martingale $X_n = \mu(I_{k,n})/\nu(I_{k,n})$ on $I_{k,n}$ is an approximation to the Radon-Nikodym derivative. For a concrete example consider $\Omega = [0, 1)$, $I_{k,n} = [k2^{-n}, (k+1)2^{-n})$ for $0 \le k < 2^n$, and $\nu =$ Lebesgue measure.

EXERCISE 3.7. Check by direct computation that the X_n in Example 3.1 is a martingale. Show that if we drop the condition $\mu_n \ll \nu_n$ and set $X_n = 0$ when $\nu(I_{k,n}) = 0$ then $E(X_{n+1}|\mathcal{F}_n) \leq X_n$.

EXERCISE 3.8. Apply (3.3) to Example 3.1 to get a "probabilistic" proof of the Radon-Nikodym theorem. To be precise, suppose $\mathcal{F}$ is **countably generated** (i.e., there is a sequence of sets A_n so that $\mathcal{F} = \sigma(A_n : n \geq 1)$) and show that if μ and ν are σ-finite measures and $\mu \ll \nu$ then there is a function g so that $\mu(A) = \int_A g \, d\nu$.

Remark. Before you object to this as circular reasoning (the Radon-Nikodym theorem was used to define conditional expectation!), observe that the conditional expectations that are needed for Example 3.1 have elementary definitions.

Kakutani dichotomy for infinite product measures. Let μ and ν be measures on sequence space $(\mathbf{R^N}, \mathcal{R^N})$ that make the coordinates $\xi_n(\omega) = \omega_n$ independent. Let $F_n(x) = \mu(\xi_n \leq x)$, $G_n(x) = \nu(\xi_n \leq x)$. Suppose $F_n \ll G_n$ and let $q_n = dF_n/dG_n$. Let $\mathcal{F}_n = \sigma(\xi_m : m \leq n)$, let μ_n and ν_n be the restrictions of μ and ν to $\mathcal{F}_n$, and let $X_n = d\mu_n/d\nu_n$. (3.3) implies that $X_n \to X$ ν-a.s. The convergence of the infinite product is a tail event, so the Kolmogorov 0-1 law implies $\mu(X < \infty) \in \{0,1\}$ and it follows from (3.3) that either $\mu \ll \nu$ or $\mu \perp \nu$. The next result gives a concrete criterion for which of the two alternatives occurs.

(3.5) **Theorem.** $\mu \ll \nu$ or $\mu \perp \nu$ according as $\prod_{m=1}^{\infty} \int \sqrt{q_m} \, dG_m > 0$ or $= 0$.

Proof Jensen's inequality and Exercise 8.7 in the Appendix imply

$$\left(\int \sqrt{q_m} \, dG_m \right)^2 \leq \int q_m \, dG_m = \int dF_m = 1$$

so the infinite product of the integrals is well defined and ≤ 1. Let

$$X_n = \prod_{m \leq n} q_m(\omega_m)$$

as above, and recall that $X_n \to X$ a.s. If the infinite product is 0 then

$$\int X_n^{1/2} \, d\nu = \prod_{m=1}^{n} \int \sqrt{q_m} \, dG_m \to 0$$

Fatou's lemma implies

$$\int X^{1/2} \, d\nu \leq \liminf_{n \to \infty} \int X_n^{1/2} \, d\nu = 0$$

so $X = 0$ ν-a.s., and (3.3) implies $\mu \perp \nu$. To prove the other direction let $Y_n = X_n^{1/2}$. Now $\int q_m \, dG_m = 1$ so if we use E to denote expected value with respect to ν then $EY_m^2 = EX_m = 1$, so

$$E(Y_{n+k} - Y_n)^2 = E(X_{n+k} + X_n - 2X_n^{1/2}X_{n+k}^{1/2}) = 2\left(1 - \prod_{m=n+1}^{n+k} \int \sqrt{q_m} \, dG_m\right)$$

Now $|a - b| = |a^{1/2} - b^{1/2}| \cdot (a^{1/2} + b^{1/2})$, so using Cauchy-Schwarz and the fact $(a + b)^2 \leq 2a^2 + 2b^2$ gives

$$E|X_{n+k} - X_n| = E(|Y_{n+k} - Y_n|(Y_{n+k} + Y_n))$$
$$\leq \left(E(Y_{n+k} - Y_n)^2 E(Y_{n+k} + Y_n)^2\right)^{1/2}$$
$$\leq \left(4E(Y_{n+k} - Y_n)^2\right)^{1/2}$$

From the last two equations it follows that if the infinite product is > 0, then X_n converges to X in L^1, so $P(X = \infty) = 0$ and the desired result follows from (3.3). □

For the next three exercises suppose F_n, G_n are concentrated on $\{0, 1\}$ and have $F_n(0) = 1 - \alpha_n$, $G_n(0) = 1 - \beta_n$.

EXERCISE 3.9. (i) Use (3.5) to find a necessary and sufficient condition for $\mu \ll \nu$. (ii) Suppose that $0 < \epsilon \leq \alpha_n, \beta_n \leq 1 - \epsilon < 1$. Show that in this case the condition is simply $\sum(\alpha_n - \beta_n)^2 < \infty$.

EXERCISE 3.10. Show that if $\sum \alpha_n < \infty$ and $\sum \beta_n = \infty$ in the previous exercise then $\mu \perp \nu$. This shows that the condition $\sum(\alpha_n - \beta_n)^2 < \infty$ is not sufficient for $\mu \ll \nu$ in general.

EXERCISE 3.11. Suppose $0 < \alpha_n, \beta_n < 1$. Show that $\sum|\alpha_n - \beta_n| < \infty$ is sufficient for $\mu \ll \nu$ in general.

d. Branching processes

Let ξ_i^n, $i, n \geq 0$, be i.i.d. nonnegative integer valued random variables. Define a sequence Z_n, $n \geq 0$ by $Z_0 = 1$ and

$$Z_{n+1} = \begin{cases} \xi_1^{n+1} + \cdots + \xi_{Z_n}^{n+1} & \text{if } Z_n > 0 \\ 0 & \text{if } Z_n = 0 \end{cases}$$

Z_n is called a **Galton-Watson process**. The idea behind the definitions is that Z_n is the number of people in the nth generation and each member of the nth generation gives birth independently to an identically distributed number of children. $p_k = P(\xi_i^n = k)$ is called the **offspring distribution**.

(3.6) Lemma. Let $\mathcal{F}_n = \sigma(\xi_i^m : i \geq 1, 1 \leq m \leq n)$ and $\mu = E\xi_i^m$. Then Z_n/μ^n is a martingale w.r.t. $\mathcal{F}_n$.

Proof Clearly $Z_n \in \mathcal{F}_n$.

$$E(Z_{n+1}|\mathcal{F}_n) = \sum_{k=1}^{\infty} E(Z_{n+1} 1_{\{Z_n=k\}}|\mathcal{F}_n)$$

by (1.1a) and (1.1c). On $\{Z_n = k\}$, $Z_{n+1} = \xi_1^{n+1} + \cdots + \xi_k^{n+1}$, so the sum is

$$\sum_{k=1}^{\infty} E((\xi_1^{n+1} + \cdots + \xi_k^{n+1}) 1_{\{Z_n=k\}}|\mathcal{F}_n) = \sum_{k=1}^{\infty} 1_{\{Z_n=k\}} E(\xi_1^{n+1} + \cdots + \xi_k^{n+1}|\mathcal{F}_n)$$

by (1.3). Since each ξ_j^{n+1} is independent of $\mathcal{F}_n$ the last expression

$$= \sum_{k=1}^{\infty} 1_{\{Z_n=k\}} k\mu = \mu Z_n$$

Dividing both sides by μ^{n+1} now gives the desired result. $\qquad\square$

Remark. The reader should notice that in the proof of (3.6) we broke things down according to the value of Z_n to get rid of the random index. A simpler way of doing the last argument (that we will use in the future) is to use Exercise 1.1 to conclude that on $\{Z_n = k\}$

$$E(Z_{n+1}|\mathcal{F}_n) = E(\xi_1^{n+1} + \cdots + \xi_k^{n+1}|\mathcal{F}_n) = k\mu = \mu Z_n$$

Z_n/μ^n is a nonnegative martingale so (2.11) implies $Z_n/\mu^n \to$ a limit a.s. We begin by identifying cases when the limit is trivial.

(3.7) Theorem. If $\mu < 1$ then $Z_n = 0$ for all n sufficiently large, so $Z_n/\mu^n \to 0$.

Proof $E(Z_n/\mu^n) = E(Z_0) = 1$, so $E(Z_n) = \mu^n$. Now $Z_n \geq 1$ on $\{Z_n > 0\}$ so

$$P(Z_n > 0) \leq E(Z_n; Z_n > 0) = E(Z_n) = \mu^n \to 0$$

exponentially fast if $\mu < 1$. $\qquad\square$

The last answer should be intuitive. If each individual on the average gives birth to less than 1 child the species will die out. The next result shows that after we exclude the trivial case in which each individual has exactly one child the same result holds when $\mu = 1$.

(3.8) Theorem. If $\mu = 1$ and $P(\xi_i^m = 1) < 1$ then $Z_n = 0$ for all n sufficiently large.

Proof When $\mu = 1$, Z_n is itself a nonnegative martingle. Since Z_n is integer valued and by (2.11) converges to an a.s. finite limit Z_∞, we must have $Z_n = Z_\infty$ for large n. If $P(\xi_i^m = 1) < 1$ and $k > 0$ then $P(Z_n = k$ for all $n \geq N) = 0$ for any N, so we must have $Z_\infty \equiv 0$. □

When $\mu \leq 1$, the limit of Z_n/μ^n is 0 because the branching process dies out. Our next step is to show.

(3.9) Theorem. If $\mu > 1$ then $P(Z_n > 0$ for all $n) > 0$.

Proof For $s \in [0,1]$, let $\varphi(s) = \sum_{k \geq 0} p_k s^k$ where $p_k = P(\xi_i^m = k)$. φ is the **generating function** for the offspring distribution p_k. Differentiating and referring to (9.2) in the Appendix for the justification gives for $s < 1$

$$\varphi'(s) = \sum_{k=1}^{\infty} k\, p_k s^{k-1} \geq 0$$

$$\varphi''(s) = \sum_{k=2}^{\infty} k(k-1)p_k s^{k-2} \geq 0$$

So φ is increasing and convex, and $\lim_{s \uparrow 1} \varphi'(s) = \sum_{k=1}^{\infty} kp_k = \mu$. Our interest in φ stems from the following facts.

(a) If $\theta_m = P(Z_m = 0)$ then $\theta_m = \sum_{k=0}^{\infty} p_k (\theta_{m-1})^k$.

Proof of (a) If $Z_1 = k$, an event with probability p_k, then $Z_m = 0$ if and only if all k families die out in the remaining $m - 1$ units of time, an independent event with probability θ_{m-1}^k. Summing over the disjoint possibilities for each k gives the desired result. □

(b) If $\varphi'(1) = \mu > 1$ there is a unique $\rho < 1$ so that $\varphi(\rho) = \rho$.

Proof of (b) $\varphi(0) \geq 0$, $\varphi(1) = 1$, and $\varphi'(1) > 1$, so $\varphi(1 - \epsilon) < 1 - \epsilon$ for small ϵ. The last two observations imply the existence of a fixed point. See Figure

4.3.1. To see it is unique observe that $\mu > 1$ implies $p_k > 0$ for some $k > 1$, so $\varphi''(\theta) > 0$ for $\theta > 0$. Since φ is strictly convex, it follows that if $\rho < 1$ is a fixed point then $\varphi(x) < x$ for $x \in (\rho, 1)$.   □

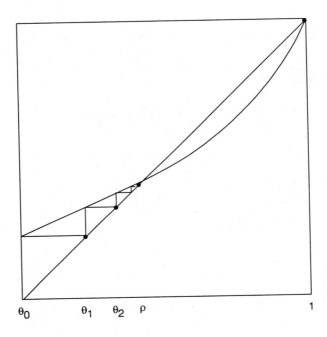

Figure 4.3.1

(c) As $m \uparrow \infty$, $\theta_m \uparrow \rho$.

Proof of (c) $\theta_0 = 0$, $\varphi(\rho) = \rho$, and φ is increasing so induction implies θ_m is increasing and $\theta_n \leq \rho$. Let $\theta_\infty = \lim \theta_m$. Taking limits in $\theta_m = \varphi(\theta_{m-1})$, we see $\theta_\infty = \varphi(\theta_\infty)$. Since $\theta_\infty \leq \rho$, it follows that $\theta_\infty = \rho$.

Combining (a)–(c) shows $P(Z_n = 0$ for some $n) = \lim \theta_n = \rho < 1$ and proves (3.9).   □

The last result shows that when $\mu > 1$ the limit of Z_n/μ^n has a chance of being nonzero. The best result on this question is due to Kesten and Stigum:

(3.10) Theorem. $W = \lim Z_n/\mu^n$ is not $\equiv 0$ if and only if $\sum p_k k \log k < \infty$.

For a proof, see Athreya and Ney (1972) p.24–29. In the next section we will show that $\sum k^2 p_k < \infty$ is sufficient for a nontrivial limit.

EXERCISE 3.12. Show that if $P(\lim Z_n/\mu^n = 0) < 1$ then it is $= \rho$ and hence

$$\{\lim Z_n/\mu^n > 0\} = \{Z_n > 0 \text{ for all } n\} \quad \text{a.s.}$$

EXERCISE 3.13. Galton and Watson who invented the process that bears their names were interested in the survival of family names. Suppose each family has exactly 3 children but coin flips determine their sex. In the 1800s only male children kept the family name so following the male offspring leads to a branching process with $p_0 = 1/8$, $p_1 = 3/8$, $p_2 = 3/8$, $p_3 = 1/8$. Compute the probability ρ that the family name will die out when $Z_0 = 1$.

4.4. Doob's Inequality, Convergence in L^p

We begin by proving a consequence of (2.8).

(4.1) Theorem. If X_n is a submartingale and N is a stopping time with $P(N \le k) = 1$ then

$$EX_0 \le EX_N \le EX_k$$

Remark. Let S_n be a simple random walk with $S_0 = 1$ and let $N = \inf\{n : S_n = 0\}$. (See Example 2.2 for more details.) $ES_0 = 1 > 0 = ES_N$ so the first inequality need not hold for unbounded stopping times. In Section 4.7 we will give conditions that guarantee $EX_0 \le EX_N$ for unbounded N.

Proof (2.8) implies $X_{N \wedge n}$ is a submartingale, so it follows that

$$EX_0 = EX_{N \wedge 0} \le EX_{N \wedge k} = EX_N$$

To prove the other inequality, let $K_n = 1_{\{N < n\}} = 1_{\{N \le n-1\}}$. K_n is predictable so (2.7) implies $(K \cdot X)_n = X_n - X_{N \wedge n}$ is a submartingale and it follows that

$$EX_k - EX_N = E(K \cdot X)_k \ge E(K \cdot X)_0 = 0 \qquad \square$$

EXERCISE 4.1. Show that if $j \leq k$ then $E(X_j; N = j) \leq E(X_k; N = j)$ and sum over j to get a second proof of $EX_N \leq EX_k$.

EXERCISE 4.2. Generalize the proof of (4.1) to show that if X_n is a submartingale and $M \leq N$ are stopping times with $P(N \leq k) = 1$ then $EX_M \leq EX_N$.

EXERCISE 4.3. Use the stopping times from the Exercise 1.7 in Chapter 3 to strengthen the conclusion of the previous exercise to $E(X_N|\mathcal{F}_M) \geq X_M$.

We will see below that (4.1) is very useful. The first indication of this is:

(4.2) **Doob's inequality.** Let X_m be a submartingale, $\bar{X}_n = \max_{0 \leq m \leq n} X_m^+$, $\lambda > 0$, and $A = \{\bar{X}_n \geq \lambda\}$. Then

$$\lambda P(A) \leq EX_n 1_A \leq EX_n^+$$

Proof   Let $N = \inf\{m : X_m \geq \lambda \text{ or } m = n\}$. Since $X_N \geq \lambda$ on A

$$\lambda P(A) \leq EX_N 1_A \leq EX_n 1_A$$

The second inequality follows from the fact that (4.1) implies $EX_N \leq EX_n$ and we have $X_N = X_n$ on A^c. The second inequality in (4.2) is trivial.   □

Example 4.1. If we let $S_n = \xi_1 + \cdots + \xi_n$ where the ξ_m are independent and have $E\xi_m = 0$, $\sigma_m^2 = E\xi_m^2 < \infty$ then (2.3) implies $X_n = S_n^2$ is a submartingale. If we let $\lambda = x^2$ and apply (4.2) to X_n we get Kolmogorov's inequality ((8.2) in Chapter 1):

$$P\left(\max_{1 \leq m \leq n} |S_n| \geq x\right) \leq x^{-2}\text{var}(S_n)$$

Using martingales, one can also prove a lower bound on the maximum that can be used instead of the central limit theorem in our proof of the necessity of the conditions in the three series theorem. (See Example 4.8 in Chapter 2.)

EXERCISE 4.4. Suppose in addition to the conditions introduced above that $|\xi_m| \leq K$, and let $s_n^2 = \sum_{m \leq n} \sigma_m^2$. Exercise 2.6 implies that $S_n^2 - s_n^2$ is a martingale. Use this and (4.1) to conclude

$$P\left(\max_{1 \leq m \leq n} |S_m| \leq x\right) \leq (x + K)^2/\text{var}(S_n)$$

EXERCISE 4.5. Let X_n be a martingale with $X_0 = 0$ and $EX_n^2 < \infty$. Show that

$$P\left(\max_{1 \leq m \leq n} X_m \geq \lambda\right) \leq EX_n^2/(EX_n^2 + \lambda^2)$$

Hint: Use the fact that $(X_n + c)^2$ is a submartingale and optimize over c.

Remark. Some readers may recognize the resemblance to the one sided Chebyshev bound in Exercise 3.6 of Chapter 1. Taking $n = 1$ and choosing an appropriate distribution for X_1 shows that the inequality in Exercise 4.4 is sharp also.

Integrating the inequality in (4.2) gives

(4.3) L^p maximum inequality. If X_n is a submartingale then for $p > 1$,

$$E(\bar{X}_n^p) \le (p/p - 1)^p E(X_n^+)^p$$

Consequently, if Y_n is a martingale and $Y_n^* = \max_{0 \le m \le n} |Y_m|$,

$$E|Y_n^*|^p \le (p/p - 1)^p E(|Y_n|^p)$$

Proof The second inequality follows by applying the first to $X_n = |Y_n|$. To prove the first we will, for reasons that will become clear in a moment, work with $\bar{X}_n \wedge M$ rather than $\bar{X}_n$. Since $\{\bar{X}_n \wedge M \ge \lambda\}$ is always $\{\bar{X}_n \ge \lambda\}$ or $\emptyset$ this does not change the application of (4.2). Using (5.7) in Chapter 1, (4.2), Fubini's theorem, and a little calculus gives

$$E((\bar{X}_n \wedge M)^p) = \int_0^\infty p\lambda^{p-1} P(\bar{X}_n \wedge M \ge \lambda) \, d\lambda$$

$$\le \int_0^\infty p\lambda^{p-1} \left(\lambda^{-1} \int X_n^+ 1_{(\bar{X}_n \wedge M \ge \lambda)} \, dP \right) d\lambda$$

$$= \int X_n^+ \int_0^{\bar{X}_n \wedge M} p\lambda^{p-2} \, d\lambda \, dP$$

$$= \frac{p}{p-1} \int X_n^+ (\bar{X}_n \wedge M)^{p-1} \, dP$$

If we let $q = p/(p - 1)$ be the exponent conjugate to p and apply Hölder's inequality ((3.3) in Chapter 1), we see that the above

$$\le q(E|X_n^+|^p)^{1/p} (E|\bar{X}_n \wedge M|^p)^{1/q}$$

If we divide both sides of the last inequality by $(E|\bar{X}_n \wedge M|^p)^{1/q}$ we get

$$E(|\bar{X}_n \wedge M|^p) \le (p/p - 1)^p E(X_n^+)^p$$

Letting $M \to \infty$ and using the monotone convergence theorem gives the desired result. $\qquad\square$

EXERCISE 4.6. **Another fix for the integrability problem in (4.3).** Note that if $E(X_n^+)^p = \infty$ there is nothing to show and then show $E(X_n^+)^p < \infty$ implies $E\bar{X}_n^p < \infty$.

Example 4.2. (4.3) is false when p = 1. Again the counterexample is provided by Example 2.2. Let S_n be a simple random walk starting from $S_0 = 1$, $N = \inf\{n : S_n = 0\}$, and $X_n = S_{N \wedge n}$. (4.1) implies $EX_n = ES_{N \wedge n} = ES_0 = 1$ for all n. Using (1.7) in Chapter 3 with $a = -1$, $b = M - 1$ we have

$$P\left(\max_m X_m < M\right) = \frac{M-1}{M}$$

so $E(\max_m X_m) = \sum_{M=1}^{\infty} P(\max_m X_m \geq M) = \sum_{M=1}^{\infty} 1/M = \infty$. The monotone convergence theorem implies that $E \max_{m \leq n} X_m \uparrow \infty$ as $n \uparrow \infty$.

The next result gives an extension of (4.3) to $p = 1$. Since this is not one of the most important results, the proof is left to the reader.

(4.4) Theorem. Let X_n be a submartingale and $\log^+ x = \max(\log x, 0)$.

$$E\bar{X}_n \leq (1 - e^{-1})^{-1}\{1 + E(X_n^+ \log^+(X_n^+))\}$$

Remark. The last result is almost the best possible condition for $\sup |X_n| \in L^1$. Gundy has shown that if X_n is a positive martingale that has $X_{n+1} \leq CX_n$ and $EX_0 \log^+ X_0 < \infty$, then $E(\sup X_n) < \infty$ implies $\sup E(X_n \log^+ X_n) < \infty$. For a proof see Neveu (1975) p. 71-73.

EXERCISE 4.7. Prove (4.4) by carrying out the following steps: (i) Imitate the proof of (4.3), but use the trivial bound $P(A) \leq 1$ for $\lambda \leq 1$ to show

$$E(\bar{X}_n \wedge M) \leq 1 + \int X_n^+ \log(\bar{X}_n \wedge M)\, dP$$

(ii) Use calculus to show

$$a \log b \leq a \log a + b/e \leq a \log^+ a + b/e$$

From (4.3) we get the following

(4.5) L^p convergence theorem. If X_n is a martingale with $\sup E|X_n|^p < \infty$ where $p > 1$, then $X_n \to X$ a.s. and in L^p.

Proof $(EX_n^+)^p \leq (E|X_n|)^p \leq E|X_n|^p$ so it follows from the martingale convergence theorem (2.10) that $X_n \to X$ a.s. The second conclusion in (4.3) implies

$$E\left(\sup_{0 \leq m \leq n} |X_m|\right)^p \leq (p/p-1)^p E|X_n|^p$$

Letting $n \to \infty$ and using the monotone convergence theorem implies $\sup |X_n| \in L^p$. Since $|X_n - X|^p \leq (2 \sup |X_n|)^p$, it follows from the dominated convergence theorem, that $E|X_n - X|^p \to 0$. $\square$

The most important special case of the results in this section occurs when $p = 2$. To treat this case, the next two results are useful.

(4.6) Orthogonality of martingale increments. Let X_n be a martingale with $EX_n^2 < \infty$ for all n. If $m \leq n$ and $Y \in \mathcal{F}_m$ has $EY^2 < \infty$ then

$$E((X_n - X_m)Y) = 0$$

Proof The Cauchy-Schwarz inequality implies $E|(X_n - X_m)Y| < \infty$. Using (1.1f), (1.3), and the definition of a martingale,

$$E((X_n - X_m)Y) = E[E((X_n - X_m)Y|\mathcal{F}_m)] = E[Y E((X_n - X_m)|\mathcal{F}_m)] = 0 \ \square$$

(4.7) Conditional variance formula. If X_n is a martingale with $EX_n^2 < \infty$ for all n

$$E((X_n - X_m)^2|\mathcal{F}_m) = E(X_n^2|\mathcal{F}_m) - X_m^2.$$

Remark. This is the conditional analogue of $E(X - EX)^2 = EX^2 - (EX)^2$ and is proved in exactly the same way.

Proof Using the linearity of conditional expectation and then (1.3) we have

$$E(X_n^2 - 2X_n X_m + X_m^2|\mathcal{F}_m) = E(X_n^2|\mathcal{F}_m) - 2X_m E(X_n|\mathcal{F}_m) + X_m^2$$
$$= E(X_n^2|\mathcal{F}_m) - 2X_m^2 + X_m^2 \qquad \square$$

EXERCISE 4.8. Let X_n and Y_n be martingales with $EX_n^2 < \infty$ and $EY_n^2 < \infty$.

$$EX_n Y_n - EX_0 Y_0 = \sum_{m=1}^{n} E(X_m - X_{m-1})(Y_m - Y_{m-1})$$

The next two results generalize (7.3) and (8.2) from Chapter 1. Let X_n, $n \geq 0$, be a martingale and let $\xi_n = X_n - X_{n-1}$ for $n \geq 1$.

EXERCISE 4.9. If $EX_0^2, \sum_{m=1}^{\infty} E\xi_m^2 < \infty$ then $X_n \to X_\infty$ a.s. and in L^2.

EXERCISE 4.10. If $b_m \uparrow \infty$ and $\sum_{m=1}^{\infty} E\xi_m^2/b_m^2 < \infty$ then $X_n/b_n \to 0$ a.s.
In particular if $E\xi_n^2 \leq K < \infty$ and $\sum_{m=1}^{\infty} b_m^{-2} < \infty$ then $X_n/b_n \to 0$ a.s.

Example 4.3.   Branching processes. We continue the study begun at the end of the last section. Using the notation introduced there, we suppose $\mu = E(\xi_i^m) > 1$ and $\text{var}(\xi_i^m) = \sigma^2 < \infty$. Let $X_n = Z_n/\mu^n$. Taking $m = n - 1$ in (4.7) and rearranging we have

$$E(X_n^2|\mathcal{F}_{n-1}) = X_{n-1}^2 + E((X_n - X_{n-1})^2|\mathcal{F}_{n-1})$$

To compute the second term we observe

$$E((X_n - X_{n-1})^2|\mathcal{F}_{n-1}) = E((Z_n/\mu^n - Z_{n-1}/\mu^{n-1})^2|\mathcal{F}_{n-1})$$
$$= \mu^{-2n} E((Z_n - \mu Z_{n-1})^2|\mathcal{F}_{n-1})$$

It follows from Exercise 1.1 that on $\{Z_{n-1} = k\}$,

$$E((Z_n - \mu Z_{n-1})^2|\mathcal{F}_{n-1}) = E\left(\left(\sum_{i=1}^{k} \xi_i^n - \mu k\right)^2 \bigg| \mathcal{F}_{n-1}\right) = k\sigma^2 = Z_{n-1}\sigma^2$$

Combining the last three equations gives

$$EX_n^2 = EX_{n-1}^2 + E(Z_{n-1}\sigma^2/\mu^{2n}) = EX_{n-1}^2 + \sigma^2/\mu^{n+1}$$

since $E(Z_{n-1}/\mu^{n-1}) = EZ_0 = 1$. Now $EX_0^2 = 1$, so $EX_1^2 = 1 + \sigma^2/\mu^2$, and induction gives

$$EX_n^2 = 1 + \sigma^2 \sum_{k=2}^{n+1} \mu^{-k}$$

This shows $\sup_n EX_n^2 < \infty$, so $X_n \to X$ in L^2, and hence $EX_n \to EX$. $EX_n = 1$ for all n, so $EX = 1$ and X is not $\equiv 0$. It follows from Exercise 3.12 that $\{X > 0\} = \{Z_n > 0$ for all n $\}$.

* Square integrable martingales

For the rest of this section we will suppose

$$X_n \text{ is a martingale with } X_0 = 0 \text{ and } EX_n^2 < \infty \text{ for all } n$$

(2.4) imples X_n^2 is a submartingale. It follows from Doob's decomposition (2.12) that we can write $X_n^2 = M_n + A_n$ where M_n is a martingale and from formulas in (2.12) and (4.7) that

$$A_n = \sum_{m=1}^{n} E(X_m^2|\mathcal{F}_{m-1}) - X_{m-1}^2 = \sum_{m=1}^{n} E((X_m - X_{m-1})^2|\mathcal{F}_{m-1})$$

A_n is called the **increasing process** associated with X_n. A_n can be thought of as a path by path measurement of the variance at time n, and $A_\infty = \lim A_n$ as the total variance in the path. (4.9) and (4.10) describe the behavior of the martingale on $\{A_n < \infty\}$ and $\{A_n = \infty\}$ respectively. The key to the proof of the first result is the following

(4.8) Theorem. $E\left(\sup_m |X_m|^2\right) \leq 4EA_\infty$.

Proof   Applying the L^2 maximum inequality (4.3) to X_n gives

$$E\left(\sup_{0\leq m\leq n} |X_m|^2\right) \leq 4EX_n^2 = 4EA_n$$

since $EX_n^2 = EM_n + EA_n$ and $EM_n = EM_0 = EX_0^2 = 0$. Using the monotone convergence theorem now gives the desired result. $\square$

(4.9) Theorem. $\lim_{n\to\infty} X_n$ exists and is finite a.s. on $\{A_\infty < \infty\}$.

Proof   Let $a > 0$. Since $A_{n+1} \in \mathcal{F}_n$, $N = \inf\{n : A_{n+1} > a^2\}$ is a stopping time. Applying (4.8) to $X_{N\wedge n}$ and noticing $A_{N\wedge n} \leq a^2$ gives

$$E\left(\sup_n |X_{N\wedge n}|^2\right) \leq 4a^2$$

so the L^2 convergence theorem (4.5) implies that $\lim X_{N\wedge n}$ exists and is finite a.s. Since a is arbitrary, the desired result follows. $\square$

The next result is a variation on the theme of Exercise 4.10.

(4.10) Theorem. Let $f \geq 1$ be increasing with $\int_0^\infty f(t)^{-2}\, dt < \infty$. Then $X_n/f(A_n) \to 0$ a.s. on $\{A_\infty = \infty\}$.

Proof $H_m = f(A_m)^{-1}$ is bounded and predictable so (2.7) implies

$$Y_n \equiv (H \cdot X)_n = \sum_{m=1}^{n} \frac{X_m - X_{m-1}}{f(A_m)} \quad \text{is a martingale}$$

If B_n is the increasing process associated with Y_n then

$$B_{n+1} - B_n = E((Y_{n+1} - Y_n)^2 | \mathcal{F}_n)$$
$$= E\left(\frac{(X_{n+1} - X_n)^2}{f(A_{n+1})^2} \middle| \mathcal{F}_n \right) = \frac{A_{n+1} - A_n}{f(A_{n+1})^2}$$

since $f(A_{n+1}) \in \mathcal{F}_n$. Our hypotheses on f imply that

$$\sum_{n=0}^{\infty} \frac{A_{n+1} - A_n}{f(A_{n+1})^2} \le \sum_{n=0}^{\infty} \int_{[A_n, A_{n+1})} f(t)^{-2} \, dt < \infty$$

so it follows from (4.9) that $Y_n \to Y_\infty$, and the desired conclusion follows from Kronecker's lemma, (8.1) in Chapter 1. $\qquad\square$

Example 4.4. Let $\epsilon > 0$ and $f(t) = \sqrt{t}(\log t)^{1/2+\epsilon} \vee 1$. Then f satisfies the hypotheses of (4.10). Let $\xi_1, \xi_2, \ldots$ be independent with $E\xi_m = 0$ and $E\xi_m^2 = \sigma_m^2$. In this case $X_n = \xi_1 + \cdots + \xi_n$ is a square integrable martingale with $A_n = \sigma_1^2 + \cdots + \sigma_n^2$ so if $\sum_{i=1}^{\infty} \sigma_i^2 = \infty$ (4.10) implies $X_n/f(A_n) \to 0$ generalizing (8.7) in Chapter 1.

From (4.10) we get a result due to Dubins and Freedman (1965) that extends (6.6) from Chapter 1, and (3.2) above.

(4.11) **Second Borel-Cantelli Lemma, III.** Suppose B_n is adapted to $\mathcal{F}_n$ and let $p_n = P(B_n | \mathcal{F}_{n-1})$. Then

$$\sum_{m=1}^{n} 1_{B(m)} \bigg/ \sum_{m=1}^{n} p_m \to 1 \quad \text{a.s. on} \quad \left\{ \sum_{m=1}^{\infty} p_m = \infty \right\}$$

Proof Define a martingale by $X_0 = 0$ and $X_n - X_{n-1} = 1_{B_n} - P(B_n | \mathcal{F}_{n-1})$ for $n \ge 1$, so that we have

$$\left(\sum_{m=1}^{n} 1_{B(m)} \bigg/ \sum_{m=1}^{n} p_m \right) - 1 = X_n \bigg/ \sum_{m=1}^{n} p_m$$

The increasing process associated with X_n has

$$A_n - A_{n-1} = E((X_n - X_{n-1})^2 | \mathcal{F}_{n-1})$$
$$= E\left((1_{B_n} - p_n)^2 \middle| \mathcal{F}_{n-1} \right) = p_n - p_n^2 \le p_n$$

On $\{A_\infty < \infty\}$, $X_n \to$ a finite limit by (4.9), so on $\{A_\infty < \infty\} \cap \{\sum_m p_m = \infty\}$

$$X_n \bigg/ \sum_{m=1}^{n} p_m \to 0$$

$\{A_\infty = \infty\} = \{\sum_m p_m(1 - p_m) = \infty\} \subset \{\sum_m p_m = \infty\}$ so on $\{A_\infty = \infty\}$ the desired conclusion follows from (4.10) with $f(t) = t \vee 1$. $\qquad\qquad\square$

Remark. The trivial example $B_n = \Omega$ for all n shows we may have $A_\infty < \infty$ and $\sum p_m = \infty$ a.s.

Example 4.5. Bernard Friedman's urn. Consider a variant of Polya's urn (see Section 4.3) in which we add a balls of the color drawn and b balls of the opposite color where $a \geq 0$ and $b > 0$. We will show that if we start with g green balls and r red balls where $g, r > 0$ then the fraction of green balls $g_n \to 1/2$. Let G_n and R_n be the number of green and red balls after the nth draw is completed. Let B_n be the event that the nth ball drawn is green, and let D_n be the number of green balls drawn in the first n draws. It follows from (4.11) that

$$(\star) \qquad D_n \Big/ \sum_{m=1}^{n} g_{m-1} \to 1 \quad \text{a.s. on} \quad \sum_{m=1}^{\infty} g_{m-1} = \infty$$

which always holds since $g_m \geq g/(g + r + (a+b)m)$. At this point the argument breaks into three cases.

Case 1. $a = b = c$. In this case the result is trivial since we always add c balls of each color.

Case 2. $a > b$. We begin with the observation

$$(*) \qquad g_{n+1} = \frac{G_{n+1}}{G_{n+1} + R_{n+1}} = \frac{g + aD_n + b(n - D_n)}{g + r + n(a+b)}$$

If $\limsup_{n\to\infty} g_n \leq x$ then $(*)$ implies $\limsup_{n\to\infty} D_n/n \leq x$ and (since $a > b$)

$$\limsup_{n\to\infty} g_{n+1} \leq \frac{ax + b(1 - x)}{a + b} = \frac{b + (a - b)x}{a + b}$$

The right-hand side is a linear function with slope < 1 and fixed point at $1/2$, so starting with the trivial upper bound $x = 1$ and iterating we conclude that $\limsup g_n \leq 1/2$. Interchanging the roles of red and green shows $\liminf_{n\to\infty} g_n \geq 1/2$ and the result follows.

Case 3. $a < b$. The result is easier to believe in this case since we are adding more balls of the type not drawn, but is a little harder to prove. The trouble is that when $b > a$ and $D_n \leq xn$, the right-hand side of $(*)$ is maximized by taking $D_n = 0$, so we need to also use the fact that if r_n is fraction of red balls then

$$r_{n+1} = \frac{R_{n+1}}{G_{n+1} + R_{n+1}} = \frac{r + bD_n + a(n - D_n)}{g + r + n(a+b)}$$

Combining this with the formula for g_{n+1} it follows that if

$$\limsup_{n \to \infty} g_n \leq x \quad \text{and} \quad \limsup_{n \to \infty} r_n \leq y$$

then

$$\limsup_{n \to \infty} g_n \leq \frac{a(1-y) + by}{a+b} = \frac{a + (b-a)y}{a+b}$$

and

$$\limsup_{n \to \infty} r_n \leq \frac{bx + a(1-x)}{a+b} = \frac{a + (b-a)x}{a+b}$$

Starting with the trivial bounds $x = 1$, $y = 1$ and iterating (observe the two upper bounds are always the same) we conclude as in Case 2 that both limsups are $\leq 1/2$. $\qquad\qquad\qquad\qquad\qquad\qquad\qquad\qquad\qquad\qquad\qquad\quad \square$

Remark. B. Friedman (1949) considered a number of different urn models. The result above is due to Freedman (1965) who proved the result by different methods. The proof above is due to Ornstein and comes from a remark in Freedman's paper.

(4.8) came from using (4.3). If we use (4.2) instead we get a slightly better result.

(4.12) Theorem. $E(\sup_n |X_n|) \leq 3EA_\infty^{1/2}$.

Proof   As in the proof of (4.9) we let $a > 0$ and let $N = \inf\{n : A_{n+1} > a^2\}$. This time however, our starting point is

$$P\left(\sup_m |X_m| > a\right) \leq P(N < \infty) + P\left(\sup_m |X_{N \wedge m}| > a\right)$$

$P(N < \infty) = P(A_\infty > a^2)$. To bound the second term, we apply (4.2) to $X_{N \wedge m}^2$ with $\lambda = a^2$ to get

$$P\left(\sup_{m \leq n} |X_{N \wedge m}| > a\right) \leq a^{-2} E X_{N \wedge n}^2 = a^{-2} E A_{N \wedge n} \leq a^{-2} E(A_\infty \wedge a^2)$$

Letting $n \to \infty$ in the last inequality, substituting the result in the first one, and integrating gives

$$\int_0^\infty P\left(\sup_m |X_m| > a\right) da \leq \int_0^\infty P(A_\infty > a^2)\, da + \int_0^\infty a^{-2} E(A_\infty \wedge a^2)\, da$$

Since $P(A_\infty > a^2) = P(A_\infty^{1/2} > a)$, the first integral is $EA_\infty^{1/2}$. For the second we use (5.7) from Chapter 1 (in the first and fourth steps), Fubini's theorem, and calculus to get

$$\int_0^\infty a^{-2} E(A_\infty \wedge a^2) \, da = \int_0^\infty a^{-2} \int_0^{a^2} P(A_\infty > b) \, db \, da$$

$$= \int_0^\infty P(A_\infty > b) \int_{\sqrt{b}}^\infty a^{-2} \, da \, db$$

$$= \int_0^\infty b^{-1/2} P(A_\infty > b) \, db = 2EA_\infty^{1/2} \qquad \square$$

EXERCISE 4.11. Let $\xi_1, \xi_2, \ldots$ be i.i.d. with $E\xi_i = 0$ and $E\xi_i^2 < \infty$. Let $S_n = \xi_1 + \cdots + \xi_n$. (4.1) implies that for any stopping time N, $ES_{N \wedge n} = 0$. Use (4.12) to conclude that if $EN^{1/2} < \infty$ then $ES_N = 0$.

Remark. Let ξ_i in Exercise 4.11 take the values ± 1 with equal probability, and let $T = \inf\{n : S_n = -1\}$. Since $S_T = -1$ does not have mean 0, it follows that $ET^{1/2} = \infty$. If we recall from (3.4) in Chapter 3 that $P(T > t) \sim Ct^{-1/2}$ we see that the result in Exercise 4.11 is almost the best possible.

4.5. Uniform Integrability, Convergence in L^1

In this section we will give necessary and sufficient conditions for a martingale to converge in L^1. The key to this is the following definition. A collection of random variables X_i, $i \in I$, is said to be **uniformly integrable** if

$$\lim_{M \to \infty} \left(\sup_{i \in I} E(|X_i|; |X_i| > M) \right) = 0$$

If we pick M large enough so that the sup < 1 it follows that

$$\sup_{i \in I} E|X_i| \le M + 1 < \infty$$

This remark will be useful several times below.

A trivial example of a uniformly integrable family is a collection of random variables that are dominated by an integrable random variable, i.e $|X_i| \le Y$ where $EY < \infty$. Our first result gives an interesting example which shows that uniformly integrable families can be very large.

(5.1) Theorem. Given a probability space $(\Omega, \mathcal{F}_o, P)$ and an $X \in L^1$, then $\{E(X|\mathcal{F}) : \mathcal{F} \text{ is a } \sigma\text{-field} \subset \mathcal{F}_o\}$ is uniformly integrable.

Proof If A_n is a sequence of sets with $P(A_n) \to 0$ then the dominated convergence theorem implies $E(|X|; A_n) \to 0$. From the last result, it follows that if $\epsilon > 0$ we can pick $\delta > 0$ so that if $P(A) \le \delta$ then $E(|X|; A) \le \epsilon$. (If not, there are sets A_n with $P(A_n) \le 1/n$ and $E(|X|; A_n) > \epsilon$, a contradiction.)

Pick M large enough so that $E|X|/M \le \delta$. Jensen's inequality and the definition of conditional expectation imply

$$E(\,|E(X|\mathcal{F})|\,;\,|E(X|\mathcal{F})| > M) \le E(\,E(|X|\,|\mathcal{F})\,;\,E(|X|\,|\mathcal{F}) > M)$$
$$= E(\,|X|\,;\,E(|X|\,|\mathcal{F}) > M)$$

since $\{E(|X|\,|\mathcal{F}) > M\} \in \mathcal{F}$. Using Chebyshev's inequality and recalling the definition of M, we have

$$P\{E(|X|\,|\mathcal{F}) > M\} \le E\{E(|X|\,|\mathcal{F})\}/M = E|X|/M \le \delta$$

So by the choice of δ we have

$$E(|E(X|\mathcal{F})|; |E(X|\mathcal{F})| > M) \le \epsilon \quad \text{for all } \mathcal{F}$$

Since ϵ was arbitrary, the collection is uniformly integrable. $\qquad\square$

A common way to check uniform integrability is to use

EXERCISE 5.1. Let $\varphi \ge 0$ be any function with $\varphi(x)/x \to \infty$ as $x \to \infty$, e.g., $\varphi(x) = x^p$ with $p > 1$ or $\varphi(x) = x \log^+ x$. If $E\varphi(|X_i|) \le C$ for all $i \in I$, then $\{X_i : i \in I\}$ is uniformly integrable.

The relevance of uniform integrability to convergence in L^1 is explained by

(5.2) Theorem. If $X_n \to X$ in probability then the following are equivalent:

(i) $\{X_n : n \ge 0\}$ is uniformly integrable.

(ii) $X_n \to X$ in L^1.

(iii) $E|X_n| \to E|X| < \infty$.

Proof **(i) implies (ii)** Let

$$\varphi_M(x) = \begin{cases} M & \text{if } x \ge M \\ x & \text{if } |x| \le M \\ -M & \text{if } x \le -M \end{cases}$$

The triangle inequality implies

$$|X_n - X| \le |X_n - \varphi_M(X_n)| + |\varphi_M(X_n) - \varphi_M(X)| + |\varphi_M(X) - X|$$

Since $|\varphi_M(Y) - Y| = (|Y| - M)^+ \le |Y|1_{(|Y|>M)}$, taking expected value gives

$$E|X_n - X| \le E|\varphi_M(X_n) - \varphi_M(X)| + E(|X_n|; |X_n| > M) + E(|X|; |X| > M)$$

(6.4) in Chapter 1 implies that $\varphi_M(X_n) \to \varphi_M(X)$ in probability, so the first term $\to 0$ by the bounded convergence theorem. (See Exercise 6.3 in Chapter 1.) If $\epsilon > 0$ and M is large, uniform integrability implies that the second term $\le \epsilon$. To bound the third term, we observe that uniform integrability implies $\sup E|X_n| < \infty$, so Fatou's lemma (in the form given in Exercise 6.2 in Chapter 1) implies $E|X| < \infty$ and by making M larger we can make the third term $\le \epsilon$. Combining the last three facts shows $\limsup E|X_n - X| \le 2\epsilon$. Since ϵ is arbitrary this proves (ii).

(ii) implies (iii)   Jensen's inequality implies

$$|E|X_n| - E|X|| \le E||X_n| - |X|| \le E|X_n - X| \to 0$$

(iii) implies (i)   Let $\psi_M(x) = x$ on $[0, M-1]$, $\psi_M = 0$ on $[M, \infty)$, and let ψ_M be linear on $[M-1, M]$. The dominated convergence theorem implies that if M is large $E|X| - E\psi_M(|X|) \le \epsilon/2$. As in the first part of the proof, the bounded convergence theorem implies $E\psi_M(|X_n|) \to E\psi_M(|X|)$, so using (iii) we get that if $n \ge n_0$

$$E(|X_n|; |X_n| > M) \le E|X_n| - E\psi_M(|X_n|)$$
$$\le E|X| - E\psi_M(|X|) + \epsilon/2 < \epsilon$$

By choosing M larger we can make $E(|X_n|; |X_n| > M) \le \epsilon$ for $0 \le n < n_0$, so X_n is uniformly integrable. $\square$

We are now ready to state the main theorems of this section. We have already done all the work, so the proofs are short.

(5.3) Theorem. For a submartingale the following are equivalent:

(i) It is uniformly integrable.

(ii) It converges a.s. and in L^1.

(iii) It converges in L^1.

Proof   (i) implies (ii)   Uniform integrability implies $\sup E|X_n| < \infty$ so the martingale convergence theorem implies $X_n \to X$ a.s., and (5.2) implies $X_n \to X$ in L^1.

(ii) implies (iii)   Trivial

(iii) implies (i) $X_n \to X$ in L^1 implies $X_n \to X$ in probability, (see (5.3) in Chapter 1) so this follows from (5.2).   □

Before proving the analogue of (5.3) for martingales we will isolate two parts of the argument that will be useful later.

(5.4) Lemma. If integrable random variables $X_n \to X$ in L^1 then

$$E(X_n; A) \to E(X; A)$$

Proof $|EX_m 1_A - EX 1_A| \leq E|X_m 1_A - X 1_A| \leq E|X_m - X| \to 0$ □

(5.5) Lemma. If a martingale $X_n \to X$ in L^1 then $X_n = E(X|\mathcal{F}_n)$.

Proof   The martingale property implies that if $m > n$, $E(X_m|\mathcal{F}_n) = X_n$, so if $A \in \mathcal{F}_n$, $E(X_n; A) = E(X_m; A)$. (5.4) implies $E(X_m; A) \to E(X; A)$, so we have $E(X_n; A) = E(X; A)$ for all $A \in \mathcal{F}_n$. Recalling the definition of conditional expectation it follows that $X_n = E(X|\mathcal{F}_n)$.   □

(5.6) Theorem. For a martingale the following are equivalent:

(i) It is uniformly integrable.

(ii) It converges a.s. and in L^1.

(iii) It converges in L^1.

(iv) There is an integrable random variable X so that $X_n = E(X|\mathcal{F}_n)$.

Proof   (i) implies (ii)   Since martingales are also submartingales, this follows from (5.3).

(ii) implies (iii)   Trivial.

(iii) implies (iv)   Follows from (5.5).

(iv) implies (i)   This follows from (5.1).   □

The next result is related to (5.5) but goes in the other direction.

(5.7) Theorem. Suppose $\mathcal{F}_n \uparrow \mathcal{F}_\infty$, i.e., $\mathcal{F}_n$ is an increasing sequence of σ-fields and $\mathcal{F}_\infty = \sigma(\cup_n \mathcal{F}_n)$. As $n \to \infty$,

$$E(X|\mathcal{F}_n) \to E(X|\mathcal{F}_\infty) \quad \text{a.s. and in } L^1$$

Proof The first step is to note that if $m > n$ then (1.2) implies

$$E(E(X|\mathcal{F}_m)|\mathcal{F}_n) = E(X|\mathcal{F}_n)$$

so $Y_n = E(X|\mathcal{F}_n)$ is a martingale. (5.1) implies that Y_n is uniformly integrable, so (5.6) implies that Y_n converges a.s. and in L^1 to a limit Y_∞. The definition of Y_n and (5.5) imply $E(X|\mathcal{F}_n) = Y_n = E(Y_\infty|\mathcal{F}_n)$, and hence

$$\int_A X \, dP = \int_A Y_\infty \, dP \quad \text{for all } A \in \mathcal{F}_n$$

Since X and Y_∞ are integrable, and $\cup_n \mathcal{F}_n$ is a π-system, the $\pi - \lambda$ theorem implies that the last result holds for all $A \in \mathcal{F}_\infty$. Since $Y_\infty \in \mathcal{F}_\infty$, it follows that $Y_\infty = E(X|\mathcal{F}_\infty)$. □

EXERCISE 5.2. Let $Z_1, Z_2, \ldots$ be i.i.d. with $E|Z_i| < \infty$, let θ be an independent r.v. with finite mean, and let $Y_i = Z_i + \theta$. If Z_i is normal$(0,1)$ then in statistical terms we have a sample from a normal population with variance 1 and unknown mean. The distribution of θ is called the **prior distribution**, and $P(\theta \in \cdot|Y_1, \ldots, Y_n)$ is called the **posterior distribution** after n observations. Show that $E(\theta|Y_1, \ldots, Y_n) \to \theta$ a.s.

In the next two exercises $\Omega = [0,1)$, $I_{k,n} = [k2^{-n}, (k+1)2^{-n})$, $\mathcal{F}_n = \sigma(I_{k,n} : 0 \le k < 2^n)$.

EXERCISE 5.3. f is said to be **Lipschitz continuous** if $|f(t) - f(s)| \le K|t - s|$ for $0 \le s, t < 1$. Show that $X_n = (f((k+1)2^{-n}) - f(k2^{-n}))/2^{-n}$ on $I_{k,n}$ defines a martingale, $X_n \to X_\infty$ a.s. and in L^1, and

$$f(b) - f(a) = \int_a^b X_\infty(\omega) \, d\omega$$

EXERCISE 5.4. Suppose f is integrable on $[0,1)$. $E(f|\mathcal{F}_n)$ is a step function and $\to f$ in L^1. From this it follows immediately that if $\epsilon > 0$ there is a step function g on $[0,1]$ with $\int |f - g| \, dx < \epsilon$. This approximation is much simpler than the bare hands approach we used in Exercise 4.3 of the Appendix, but of course we are using a lot of machinery.

An immediate consequence of (5.7) is

(5.8) Lévy's 0-1 law. If $\mathcal{F}_n \uparrow \mathcal{F}_\infty$ and $A \in \mathcal{F}_\infty$ then $E(1_A|\mathcal{F}_n) \to 1_A$ a.s.

To steal a line from Chung: *"The reader is urged to ponder over the meaning of this result and judge for himself whether it is obvious or incredible."* We will now argue for the two points of view.

"It is obvious." $1_A \in \mathcal{F}_\infty$ and $\mathcal{F}_n \uparrow \mathcal{F}_\infty$ so our best guess of 1_A given the information in $\mathcal{F}_n$ should approach 1_A (the best guess given $\mathcal{F}_\infty$).

"It is incredible." Let $X_1, X_2, \ldots$ be independent and suppose $A \in \mathcal{T}$, the tail σ-field. For each n, A is independent of $\mathcal{F}_n$, so $E(1_A|\mathcal{F}_n) = P(A)$. As $n \to \infty$, the left-hand side converges to 1_A a.s., so $P(A) = 1_A$ a.s., and it follows that $P(A) \in \{0, 1\}$, i.e., we have proved Kolmogorov's 0-1 law.

The last argument may not show that (5.8) is "too unusual or improbable to be possible," but this and other applications of (5.8) below show that it is a very useful result.

EXERCISE 5.5. Let X_n be r.v.'s taking values in $[0, \infty)$. Let $D = \{X_n = 0$ for some $n \geq 1\}$ and assume

$$P(D|X_1, \ldots, X_n) \geq \delta(x) > 0 \quad \text{a.s. on } \{X_n \leq x\}$$

Use (5.8) to conclude that $P(D \cup \{\lim_n X_n = \infty\}) = 1$.

EXERCISE 5.6. Let Z_n be a branching process with offspring distribution p_k (see the end of Section 4.3 for definitions). Use the last result to show that if $p_0 > 0$ then $P(\lim_n Z_n = 0 \text{ or } \infty) = 1$.

EXERCISE 5.7. Let $X_n \in [0, 1]$ be adapted to $\mathcal{F}_n$. Let $\alpha, \beta > 0$ with $\alpha + \beta = 1$ and suppose

$$P(X_{n+1} = \alpha + \beta X_n|\mathcal{F}_n) = X_n \qquad P(X_{n+1} = \beta X_n|\mathcal{F}_n) = 1 - X_n$$

Show $P(\lim_n X_n = 0 \text{ or } 1) = 1$ and if $X_0 = \theta$ then $P(\lim_n X_n = 1) = \theta$.

A more technical consequence of (5.8) is

(5.9) Dominated convergence theorem for conditional expectations. Suppose $Y_n \to Y$ a.s. and $|Y_n| \leq Z$ for all n where $EZ < \infty$. If $\mathcal{F}_n \uparrow \mathcal{F}_\infty$ then

$$E(Y_n|\mathcal{F}_n) \to E(Y|\mathcal{F}_\infty) \quad \text{a.s.}$$

Proof   Let $W_N = \sup\{|Y_n - Y_m| : n, m \geq N\}$. $W_N \leq 2Z$ so $EW_N < \infty$. Using monotonicity (1.1b) and applying (5.7) to W_N gives

$$\limsup_{n \to \infty} E(|Y_n - Y||\mathcal{F}_n) \leq \lim_{n \to \infty} E(W_N|\mathcal{F}_n) = E(W_N|\mathcal{F}_\infty)$$

The last result is true for all N and $W_N \downarrow 0$ as $N \uparrow \infty$, so (1.1c) implies $E(W_N|\mathcal{F}_\infty) \downarrow 0$, and Jensen's inequality gives us

$$|E(Y_n|\mathcal{F}_n) - E(Y|\mathcal{F}_n)| \le E(|Y_n - Y||\mathcal{F}_n) \to 0 \quad \text{a.s. as } n \to \infty$$

(5.7) implies $E(Y|\mathcal{F}_n) \to E(Y|\mathcal{F}_\infty)$ a.s. The desired result follows from the last two conclusions and the triangle inequality. $\qquad\square$

EXERCISE 5.8. Show that if $\mathcal{F}_n \uparrow \mathcal{F}_\infty$ and $Y_n \to Y$ in L^1 then $E(Y_n|\mathcal{F}_n) \to E(Y|\mathcal{F}_\infty)$ in L^1.

Example 5.1. Suppose $X_1, X_2, \ldots$ are uniformly integrable and $\to X$ a.s. (5.2) implies $X_n \to X$ in L^1 and combining this with Exercise 5.8 shows $E(X_n|\mathcal{F}) \to E(X|\mathcal{F})$ in L^1. We will now show that $E(X_n|\mathcal{F})$ need not converge a.s. Let $Y_1, Y_2, \ldots$ and $Z_1, Z_2, \ldots$ be independent r.v.'s with

$$P(Y_n = 1) = 1/n \qquad P(Y_n = 0) = 1 - 1/n$$
$$P(Z_n = n) = 1/n \qquad P(Z_n = 0) = 1 - 1/n$$

Let $X_n = Y_n Z_n$. $P(X_n > 0) = 1/n^2$ so the Borel-Cantelli lemma implies $X_n \to 0$ a.s. $E(X_n; |X_n| \ge 1) = n/n^2$, so X_n is uniformly integrable. Let $\mathcal{F} = \sigma(Y_1, Y_2, \ldots)$.

$$E(X_n|\mathcal{F}) = Y_n E(Z_n|\mathcal{F}) = Y_n E Z_n = Y_n$$

Since $Y_n \to 0$ in L^1 but not a.s., the same is true for $E(X_n|\mathcal{F})$.

4.6. Backwards Martingales

A **backwards martingale** (some authors call them reversed) is a martingale indexed by the negative integers, i.e., X_n, $n \le 0$, adapted to an increasing sequence of σ-fields $\mathcal{F}_n$ with

$$E(X_{n+1}|\mathcal{F}_n) = X_n \quad \text{for } n \le -1$$

Because the σ-fields decrease as $n \downarrow -\infty$, the convergence theory for backwards martingales is particularly simple.

(6.1) Theorem. $X_{-\infty} = \lim_{n \to -\infty} X_n$ exists a.s. and in L^1.

Proof Let U_n be the number of upcrossings of $[a, b]$ by $X_{-n}, \ldots, X_0$. The upcrossing inequality (2.9) implies $(b-a)EU_n \le E(X_0 - a)^+$. Letting $n \to \infty$

and using the monotone convergence theorem, we have $EU_\infty < \infty$, so by the remark after the proof of (2.10) the limit exists a.s. The martingale property implies $X_n = E(X_0|\mathcal{F}_n)$, so (5.1) implies X_n is uniformly integrable and (5.2) tells us that the convergence occurs in L^1.   □

EXERCISE 6.1. Show that if $X_0 \in L^p$ the convergence in (6.1) occurs in L^p.

The next result identifies the limit in (6.1).

(6.2) **Theorem.** If $X_{-\infty} = \lim_{n \to -\infty} X_n$ and $\mathcal{F}_{-\infty} = \cap_n \mathcal{F}_n$, then $X_{-\infty} = E(X_0|\mathcal{F}_{-\infty})$.

Proof Clearly $X_{-\infty} \in \mathcal{F}_{-\infty}$. $X_n = E(X_0|\mathcal{F}_n)$, so if $A \in \mathcal{F}_{-\infty} \subset \mathcal{F}_n$ then

$$\int_A X_n \, dP = \int_A X_0 \, dP$$

(6.1) and (5.4) imply $E(X_n; A) \to E(X_{-\infty}; A)$, so

$$\int_A X_{-\infty} \, dP = \int_A X_0 \, dP$$

for all $A \in \mathcal{F}_{-\infty}$, proving the desired conclusion.   □

The next result is (5.7) backwards.

(6.3) **Theorem.** If $\mathcal{F}_n \downarrow \mathcal{F}_{-\infty}$ as $n \downarrow -\infty$ (i.e., $\mathcal{F}_{-\infty} = \cap_n \mathcal{F}_n$), then

$$E(Y|\mathcal{F}_n) \to E(Y|\mathcal{F}_{-\infty}) \quad \text{a.s. and in } L^1$$

Proof $X_n = E(Y|\mathcal{F}_n)$ is a backwards martingale, so (6.1) and (6.2) imply that as $n \downarrow -\infty$, $X_n \to X_{-\infty}$ a.s. and in L^1, where

$$X_{-\infty} = E(X_0|\mathcal{F}_{-\infty}) = E(E(Y|\mathcal{F}_0)|\mathcal{F}_{-\infty}) = E(Y|\mathcal{F}_{-\infty})   □$$

EXERCISE 6.2. Prove the backwards analogue of (5.9). Suppose $Y_n \to Y_{-\infty}$ a.s. as $n \to -\infty$ and $|Y_n| \le Z$ a.s. where $EZ < \infty$. If $\mathcal{F}_n \downarrow \mathcal{F}_{-\infty}$, then $E(Y_n|\mathcal{F}_n) \to E(Y_{-\infty}|\mathcal{F}_{-\infty})$ a.s.

Even though the convergence theory for backwards martingales is easy, there are some nice applications. For the rest of the section, we return to the special space utilized in Section 3.1, so we can utilize definitions given there. That is, we suppose

$$\Omega = \{(\omega_1, \omega_2, \ldots) : \omega_i \in S\}$$
$$\mathcal{F} = S \times S \times \ldots$$
$$X_n(\omega) = \omega_n$$

Let $\mathcal{E}_n$ be the σ-field generated by events that are invariant under permutations that leave $n+1, n+2, \ldots$ fixed. and let $\mathcal{E} = \cap_n \mathcal{E}_n$ be the exchangeable σ-field.

Example 6.1.   The strong law of large numbers.   Let $\xi_1, \xi_2, \ldots$ be i.i.d. with $E|\xi_i| < \infty$. Let $S_n = \xi_1 + \cdots + \xi_n$, let $X_{-n} = S_n/n$, and let

$$\mathcal{F}_{-n} = \sigma(S_n, S_{n+1}, S_{n+2}, \ldots) = \sigma(S_n, \xi_{n+1}, \xi_{n+2}, \ldots)$$

To compute $E(X_{-n}|\mathcal{F}_{-n-1})$, we observe that if $j, k \leq n+1$ symmetry implies $E(\xi_j|\mathcal{F}_{-n-1}) = E(\xi_k|\mathcal{F}_{-n-1})$, so

$$E(\xi_{n+1}|\mathcal{F}_{-n-1}) = \frac{1}{n+1} \sum_{k=1}^{n+1} E(\xi_k|\mathcal{F}_{-n-1})$$

$$= \frac{1}{n+1} E(S_{n+1}|\mathcal{F}_{-n-1}) = \frac{S_{n+1}}{n+1}$$

Since $X_{-n} = (S_{n+1} - \xi_{n+1})/n$, it follows that

$$E(X_{-n}|\mathcal{F}_{-n-1}) = E(S_{n+1}/n|\mathcal{F}_{-n-1}) - E(\xi_{n+1}/n|\mathcal{F}_{-n-1})$$

$$= \frac{S_{n+1}}{n} - \frac{S_{n+1}}{n(n+1)} = \frac{S_{n+1}}{n+1} = X_{-n-1}$$

The last computation shows X_{-n} is a backwards martingale so it follows from (6.1) and (6.2) that $\lim_{n\to\infty} S_n/n = E(X_{-1}|\mathcal{F}_{-\infty})$. Since $\mathcal{F}_{-n} \subset \mathcal{E}_n$, $\mathcal{F}_{-\infty} \subset \mathcal{E}$. The Hewitt-Savage 0-1 law ((1.1) in Chapter 3) says $\mathcal{E}$ is trivial so we have

$$\lim_{n\to\infty} S_n/n = E(X_{-1}) \quad \text{a.s.}$$

Example 6.2. The Ballot Theorem.   Let $\{\xi_j, 1 \leq j \leq n\}$ be i.i.d. nonnegative integer valued r.v.'s, let $S_k = \xi_1 + \cdots + \xi_k$, and let $G = \{S_j < j \text{ for } 1 \leq j \leq n\}$. Then

(6.4) $$P(G|S_n) \geq (1 - S_n/n)^+$$

with equality if $P(\xi_j \leq 2) = 1$.

Remark. To explain the name let $\xi_1, \xi_2, \ldots, \xi_n$ be i.i.d. and take values 0 or 2 with probability 1/2 each. Interpreting 0's and 2's as votes for candidates A and B, we see that

$$G = \{A \text{ leads } B \text{ throughout the counting}\}$$

so $P(G|A \text{ gets } r \text{ votes }) = (1 - 2r/n)^+$, the result in (3.2) in Chapter 3.

Proof The result is trivial when $S_n \geq n$ so suppose $S_n < n$. Computations in Example 6.1 show that $X_{-j} = S_j/j$ is a martingale w.r.t. $\mathcal{F}_{-j} = \sigma(S_j, \ldots, S_n)$. Let $T = \inf\{k \geq -n : X_k \geq 1\}$ and set $T = -1$ if the set is $\emptyset$. Clearly, $X_T \geq 1$ on G^c. Since $G \subset \{T = -1\}$ and $S_1 < 1$ implies $S_1 = 0$, we have $X_T = 0$ on G. Noting $\mathcal{F}_{-n} = \sigma(S_n)$ and using Exercise 4.3

$$P(G^c|S_n) \leq E(X_T|\mathcal{F}_{-n}) = X_{-n} = S_n/n$$

When $P(\xi_j \leq 2) = 1$, $X_T = 1$ on G^c and the $\leq$ is an $=$.   □

Example 6.3. Hewitt-Savage 0-1 law. If $X_1, X_2, \ldots$ are i.i.d. and $A \in \mathcal{E}$ then $P(A) \in \{0, 1\}$.

The key to the new proof is:

(6.5) Lemma. Suppose $X_1, X_2, \ldots$ are i.i.d. and let

$$A_n(\varphi) = \frac{1}{(n)_k} \sum_i \varphi(X_{i_1}, \ldots, X_{i_k})$$

where the sum is over all sequences of distinct integers $1 \leq i_1, \ldots, i_k \leq n$ and

$$(n)_k = n(n-1) \cdots (n-k+1)$$

is the number of such sequences. If φ is bounded

$$A_n(\varphi) \to E\varphi(X_1, \ldots, X_k)   \text{a.s.}$$

Proof $A_n(\varphi) \in \mathcal{E}_n$, so

$$A_n(\varphi) = E(A_n(\varphi)|\mathcal{E}_n) = \frac{1}{(n)_k} \sum_i E(\varphi(X_{i_1}, \ldots, X_{i_k})|\mathcal{E}_n)$$

$$= E(\varphi(X_1, \ldots, X_k)|\mathcal{E}_n)$$

since all the terms in the sum are the same. (6.3) with $\mathcal{F}_{-m} = \mathcal{E}_m$ for $m \geq 1$ implies that

$$E(\varphi(X_1, \ldots, X_k)|\mathcal{E}_n) \to E(\varphi(X_1, \ldots, X_k)|\mathcal{E})$$

We want to show that the limit is $E(\varphi(X_1, \ldots, X_k))$. The first step is to observe that there are $k(n-1)_{k-1}$ terms in $A_n(\varphi)$ involving X_1 and φ is bounded so if we let $1 \in i$ denote the sum over sequences that contain 1.

$$\frac{1}{(n)_k} \sum_{1 \in i} \varphi(X_{i_1}, \ldots, X_{i_k}) \leq \frac{k(n-1)_{k-1}}{(n)_k} \sup \varphi \to 0$$

This shows that

$$E(\varphi(X_1, \ldots, X_k)|\mathcal{E}) \in \sigma(X_2, X_3, \ldots)$$

Repeating the argument for $2, 3, \ldots, k$ shows

$$E(\varphi(X_1, \ldots, X_k)|\mathcal{E}) \in \sigma(X_{k+1}, X_{k+2}, \ldots)$$

Intuitively, if the conditional expectation of a r.v. is independent of the r.v. then

(a) $$E(\varphi(X_1, \ldots, X_k)|\mathcal{E}) = E(\varphi(X_1, \ldots, X_k))$$

To show this, we prove

(b) If $EX^2 < \infty$ and $E(X|\mathcal{G}) \in \mathcal{F}$ with X independent of $\mathcal{F}$ then $E(X|\mathcal{G}) = EX$.

Proof Let $Y = E(X|\mathcal{G})$ and note that (1.1.e) implies $EY^2 \leq EX^2 < \infty$. By independence $EXY = EX\, EY = (EY)^2$ since $EY = EX$. From the geometric interpretation of conditional expectation, (1.4), $E((X - Y)Y) = 0$ so $EY^2 = EXY = (EY)^2$ and $\text{var}(Y) = EY^2 - (EY)^2 = 0$. □

(a) holds for all bounded φ, so $\mathcal{E}$ is independent of $\mathcal{G}_k = \sigma(X_1, \ldots, X_k)$. Since this holds for all k, and $\cup_k \mathcal{G}_k$ is a π-system that contains Ω, (4.1) in Chapter 1 implies $\mathcal{E}$ is independent of $\sigma(\cup_k \mathcal{G}_k) \supset \mathcal{E}$, and we get the usual 0-1 law punch line. If $A \in \mathcal{E}$, it is independent of itself and hence $P(A) = P(A \cap A) = P(A)P(A)$, i.e., $P(A) \in \{0, 1\}$. □

Example 6.4. de Finetti's Theorem. A sequence $X_1, X_2, \ldots$ is said to be **exchangeable** if for each n and permutation π of $\{1, \ldots, n\}$, $(X_1, \ldots, X_n)$ and $(X_{\pi(1)}, \ldots, X_{\pi(n)})$ have the same distribution.

(6.6) **de Finetti's Theorem.** If $X_1, X_2, \ldots$ are exchangeable then conditional on $\mathcal{E}$, $X_1, X_2, \ldots$ are independent and identically distributed.

Proof   Repeating the first calculation in the proof of (6.5) and using the notation introduced there shows that for any exchangeable sequence:

$$A_n(\varphi) = E(A_n(\varphi)|\mathcal{E}_n) = \frac{1}{(n)_k} \sum_i E(\varphi(X_{i_1}, \ldots, X_{i_k})|\mathcal{E}_n)$$

$$= E(\varphi(X_1, \ldots, X_k)|\mathcal{E}_n)$$

since all the terms in the sum are the same. Again (6.3) implies that

(6.7) $$A_n(\varphi) \to E(\varphi(X_1, \ldots, X_k)|\mathcal{E})$$

This time, however, $\mathcal{E}$ may be nontrivial so we cannot hope to show that the limit is $E(\varphi(X_1, \ldots, X_k))$.

Let f and g be bounded functions on $\mathbf{R}^{k-1}$ and $\mathbf{R}$ respectively. If we let $I_{n,k}$ be the set of all sequences of distinct integers $1 \le i_1, \ldots, i_k \le n$ then

$$(n)_{k-1}A_n(f)\, n A_n(g) = \sum_{i \in I_{n,k-1}} f(X_{i_1}, \ldots, X_{i_{k-1}}) \sum_m g(X_m)$$

$$= \sum_{i \in I_{n,k}} f(X_{i_1}, \ldots, X_{i_{k-1}}) g(X_{i_k})$$

$$+ \sum_{i \in I_{n,k-1}} \sum_{j=1}^{k-1} f(X_{i_1}, \ldots, X_{i_{k-1}}) g(X_{i_j})$$

If we let $\varphi(x_1, \ldots, x_k) = f(x_1, \ldots, x_{k-1})g(x_k)$ note that

$$\frac{(n)_{k-1} n}{(n)_k} = \frac{n}{(n-k+1)} \quad \text{and} \quad \frac{(n)_{k-1}}{(n)_k} = \frac{1}{(n-k+1)}$$

then rearrange we have

$$A_n(\varphi) = \frac{n}{n-k+1} A_n(f)A_n(g) - \frac{1}{n-k+1} \sum_{j=1}^{k-1} A_n(\varphi_j)$$

where $\varphi_j(x_1, \ldots, x_{k-1}) = f(x_1, \ldots, x_{k-1})g(x_j)$. Applying (6.7) to φ, f, g, and all the φ_j gives

$$E(f(X_1, \ldots, X_{k-1})g(X_k)|\mathcal{E}) = E(f(X_1, \ldots, X_{k-1})|\mathcal{E})E(g(X_k)|\mathcal{E})$$

It follows by induction that

$$E\left(\prod_{j=1}^k f_j(X_j)\middle|\mathcal{E}\right) = \prod_{j=1}^k E(f_j(X_j)|\mathcal{E}) \qquad \square$$

When the X_i take values in a nice space, there is a regular conditional distribution for $(X_1, X_2, \ldots)$ given $\mathcal{E}$, and the sequence can be represented as a mixture of i.i.d. sequences. Hewitt and Savage (1956) call the sequence **presentable** in this case. For the usual measure theoretic problems the last result is not valid when the X_i take values in an arbitrary measure space. See Dubins and Freedman (1979) and Freedman (1980) for counterexamples.

The simplest special case of (6.7) occurs when the $X_i \in \{0, 1\}$. In this case

(6.8) Theorem. If $X_1, X_2, \ldots$ are exchangeable and take values in $\{0, 1\}$ then there is a probability distribution on $[0, 1]$ so that

$$P(X_1 = 1, \ldots, X_k = 1, X_{k+1} = 0, \ldots, X_n = 0) = \int_0^1 \theta^k (1 - \theta)^{n-k} \, dF(\theta)$$

(6.8) is useful for people concerned about the foundations of statistics (see Section 3.7 of Savage (1972)), since from the palatable assumption of symmetry one gets the powerful conclusion that the sequence is a mixture of i.i.d. sequences. (6.8) has been proved in a variety of different ways. See Feller, Vol. II (1971), p. 228–229 for a proof that is related to the moment problem. Diaconis and Freedman (1980) have a nice proof that starts with the trivial observation that the distribution of a finite exchangeable sequence X_m, $1 \le m \le n$ has the form $p_0 H_{0,n} + \cdots + p_n H_{n,n}$ where $H_{m,n}$ is "drawing without replacement from an urn with m ones and $n - m$ zeros." If $m \to \infty$ and $m/n \to p$ then $H_{m,n}$ approaches product measure with density p. (6.8) follows easily from this and one can get bounds on the rate of convergence.

Exercises

6.3. Prove directly from the definition that if $X_1, X_2, \ldots \in \{0, 1\}$ are exchangeable

$$P(X_1 = 1, \ldots, X_k = 1 | S_n = m) = \binom{n-k}{n-m} \Big/ \binom{n}{m}$$

6.4. If $X_1, X_2, \ldots \in \mathbf{R}$ are exchangeable with $EX_i^2 < \infty$ then $E(X_1 X_2) \ge 0$.

6.5. Use the first few lines of the proof of (6.5) to conclude that if $X_1, X_2, \ldots$ are i.i.d. with $EX_i = \mu$ and $\operatorname{var}(X_i) = \sigma^2 < \infty$ then

$$\binom{n}{2}^{-1} \sum_{1 \le i < j \le n} (X_i - X_j)^2 \to 2\sigma^2$$

4.7. Optional Stopping Theorems

In this section we will prove a number of results that allow us to conclude that if X_n is a submartingale and $M \leq N$ are stopping times then $EX_M \leq EX_N$. Example 2.2 shows that this is not always true, but Exercise 4.2 shows this is true if N is bounded, so our attention will be focused on the case of unbounded N.

(7.1) Theorem. If X_n is a uniformly integrable submartingale then for any stopping time N, $X_{N \wedge n}$ is uniformly integrable.

Proof X_n^+ is a submartingale, so (4.1) implies $EX_{N \wedge n}^+ \leq EX_n^+$. Since X_n^+ is uniformly integrable, it follows from the remark after the definition that

$$\sup_n EX_{N \wedge n}^+ \leq \sup_n EX_n^+ < \infty$$

Using the martingale convergence theorem (2.10) now gives $X_{N \wedge n} \to X_N$ a.s. (here $X_\infty = \lim_n X_n$) and $E|X_N| < \infty$. With this established, the rest is easy since

$$
\begin{aligned}
E(|X_{N \wedge n}|; |X_{N \wedge n}| > K) &= E(|X_N|; |X_N| > K, N \leq n) \\
&\quad + E(|X_n|; |X_n| > K, N > n) \\
&\leq E(|X_N|; |X_N| > K) + E(|X_n|; |X_n| > K)
\end{aligned}
$$

and X_n is uniformly integrable.   □

From the last computation in the proof of (7.1) we get

(7.2) Corollary. If $E|X_N| < \infty$ and $X_n 1_{(N > n)}$ is uniformly integrable, then $X_{N \wedge n}$ is uniformly integrable.

From (7.1) we immediately get

(7.3) Theorem. If X_n is a uniformly integrable submartingale then for any stopping time $N \leq \infty$, we have $EX_0 \leq EX_N \leq EX_\infty$, where $X_\infty = \lim X_n$.

Proof   (4.1) implies $EX_0 \leq EX_{N \wedge n} \leq EX_n$. Letting $n \to \infty$ and observing that (7.1) and (5.3) imply $X_{N \wedge n} \to X_N$ and $X_n \to X_\infty$ in L^1 gives the desired result.   □

From (7.3) we get the following useful corollary.

(7.4) The Optional Stopping Theorem. If $L \leq M$ are stopping times and $Y_{M \wedge n}$ is a uniformly integrable submartingale then $EY_L \leq EY_M$ and

$$Y_L \leq E(Y_M | \mathcal{F}_L)$$

Proof Use the inequality $EX_N \leq EX_\infty$ in (7.3) with $X_n = Y_{M \wedge n}$ and $N = L$. To prove the second result, let $A \in \mathcal{F}_L$ and

$$N = \begin{cases} L & \text{on } A \\ M & \text{on } A^c \end{cases}$$

is a stopping time by Exercise 1.8 in Chapter 3. Using the first result now shows $EY_N \leq EY_M$. Since $N = M$ on A^c it follows from the last inequality and the definition of conditional expectation that

$$E(Y_L; A) \leq E(Y_M; A) = E(E(Y_M | \mathcal{F}_L); A)$$

Since the latter holds for all $A \in \mathcal{F}_L$, the desired result follows. $\square$

The last result is the one we use the most (usually the first inequality with $L = 0$). (7.2) is useful in checking the hypothesis. A typical application is the following generalization of Wald's equation, (1.1) in Chapter 3.

(7.5) Theorem. Suppose X_n is a submartingale and $E(|X_{n+1} - X_n||\mathcal{F}_n) \leq B$ a.s. If $EN < \infty$ then $X_{N \wedge n}$ is uniformly integrable and hence $EX_N \geq EX_0$.

Remark. As usual using the last result twice shows that if X is a martingale then $EX_N = EX_0$. To recover Wald's equation let S_n be a random walk, let $\mu = E(S_n - S_{n-1})$ and apply the martingale result to $X_n = S_n - n\mu$.

Proof We begin by observing that

$$|X_{N \wedge n}| \leq |X_0| + \sum_{m=0}^{\infty} |X_{m+1} - X_m| 1_{(N>m)}$$

To prove uniform integrability, it suffices to show that the right-hand side has finite expectation for then $|X_{N \wedge n}|$ is dominated by an integrable r.v. Now, $\{N > m\} \in \mathcal{F}_m$ so

$$E(|X_{m+1} - X_m|; N > m) = E(E(|X_{m+1} - X_m||\mathcal{F}_m); N > m) \leq BP(N > m)$$

and $E \sum_{m=0}^{\infty} |X_{m+1} - X_m| 1_{(N>m)} \leq B \sum_{m=0}^{\infty} P(N > m) = BEN < \infty.$ $\square$

Before we delve further into applications, we pause to prove one last stopping theorem that does not require uniform integrability.

(7.6) Theorem. If X_n is a nonnegative supermartingale and $N \leq \infty$ is a stopping time, then $EX_0 \geq EX_N$ where $X_\infty = \lim X_n$, (which exists by (2.11)).

Proof   By (4.1) $EX_0 \geq EX_{N \wedge n}$. The monotone convergence theorem implies

$$E(X_N; N < \infty) = \lim_{n \to \infty} E(X_N; N \leq n)$$

and Fatou's lemma implies

$$E(X_N; N = \infty) \leq \liminf_{n \to \infty} E(X_n; N > n)$$

Adding the last two lines and using our first observation

$$EX_N \leq \liminf_{n \to \infty} EX_{N \wedge n} \leq EX_0$$

EXERCISE 7.1. If $X_n \geq 0$ is a supermartingale then $P(\sup X_n > \lambda) \leq EX_0/\lambda$.

Applications to random walks. For the rest of the section, including all the exercises below, $\xi_1, \xi_2, \ldots$ are i.i.d., $S_n = \xi_1 + \cdots + \xi_n$, and $\mathcal{F}_n = \sigma(\xi_1, \ldots, \xi_n)$.

Example 7.1. Asymmetric simple random walk refers to the special case in which $P(\xi_i = 1) = p$ and $P(\xi_i = -1) = q \equiv 1 - p$.

(a) Suppose $p > 1/2$ and let $\varphi(x) = \{(1 - p)/p\}^x$. Then $\varphi(S_n)$ is a martingale.

Proof   Since S_n and ξ_{n+1} are independent, Example 1.5 implies that on $\{S_n = m\}$,

$$E(\varphi(S_{n+1})|\mathcal{F}_n) = p \cdot \left(\frac{1-p}{p}\right)^{m+1} + (1 - p)\left(\frac{1-p}{p}\right)^{m-1}$$

$$= \{1 - p + p\}\left(\frac{1-p}{p}\right)^m = \varphi(S_n) \qquad \square$$

(b) If we let $T_x = \inf\{n : S_n = x\}$ then for $a < 0 < b$

$$P(T_a < T_b) = \frac{\varphi(b) - \varphi(0)}{\varphi(b) - \varphi(a)}$$

Proof Let $N = T_a \wedge T_b$. The first step is to check that $N < \infty$ a.s. This can be done using a slight modification of the argument in Example 1.5 of Chapter 3 or with the following argument. Since $\varphi(S_{N \wedge n})$ is bounded, it is uniformly integrable and (5.6) implies $\lim_n \varphi(S_{N \wedge n})$ exists a.s. and in L^1. Since convergence to an interior point of (a, b) is impossible we must have $N < \infty$ a.s. and

$$\varphi(0) = E\varphi(S_N) = P(T_a < T_b)\varphi(a) + P(T_b < T_a)\varphi(b)$$

Using $P(T_a < T_b) + P(T_b < T_a) = 1$ and solving gives the indicated result. □

(c) If $a < 0$ then $P(\min_n S_n \le a) = P(T_a < \infty) = \{(1-p)/p\}^{-a}$. If $b > 0$ then $P(T_b < \infty) = 1$.

Proof Letting $b \to \infty$ and noting $\varphi(b) \to 0$ gives the first result, since $T_a < \infty$ if and only if $T_a < T_b$ for some b. For the second note that $\varphi(a) \to \infty$ as $a \to -\infty$. □

(d) If $b > 0$ then $ET_b = b/(2p - 1)$.

Proof $X_n = S_n - (p-q)n$ is a martingale. Since $T_b \wedge n$ is a bounded stopping time, (4.1) implies

$$0 = ES_{T_b \wedge n} - (p - q)(T_b \wedge n)$$

Now $b \ge S_{T_b \wedge n} \ge \min_m S_m$ and Exercise 7.3 implies $E(\min_m S_m) > -\infty$, so the dominated convergence theorem implies $ES_{T_b \wedge n} \to ES_{T_b}$ as $n \to \infty$. The monotone convergence theorem implies $E(T_b \wedge n) \uparrow ET_b$ so we have $b = (p - q)ET_b$. □

Remark. The reader should study the technique in this proof of (d) because it is useful in a number of situations (e.g., the exercises below). We apply (4.1) to the bounded stopping time $T_b \wedge n$, then let $n \to \infty$ and use appropriate convergence theorems. Here this is an alternative to showing $ET_b < \infty$ in order to check that $X_{T_b \wedge n}$ is uniformly integrable.

EXERCISE 7.2. Let S_n be an asymmetric simple random walk with $p > 1/2$, and let $\sigma^2 = 1 - (p - q)^2$. Use the fact that $X_n = (S_n - (p-q)n)^2 - \sigma^2 n$ is a martingale to show $\text{var}(T_1) = (1 - (p - q)^2)/(p - q)^3$.

EXERCISE 7.3. Let S_n be a symmetric simple random walk starting at 0, and let $T = \inf\{n : S_n \notin (-a, a)\}$ where a is an integer. (i) Use the fact that $S_n^2 - n$ is a martingale to show that $ET = a^2$. (ii) Find constants b and c so that $Y_n = S_n^4 - 6nS_n^2 + bn^2 + cn$ is a martingale and use this to compute ET^2.

EXERCISE 7.4. Suppose ξ_i is not constant. Let $\varphi(\theta) = E\exp(\theta\xi_1) < \infty$ for $\theta \in (-\delta, \delta)$, and let $\psi(\theta) = \log\varphi(\theta)$. (i) $X_n^\theta = \exp(\theta S_n - n\psi(\theta))$ is a martingale. (ii) ψ is strictly convex. (iii) Show $E\sqrt{X_n^\theta} \to 0$ and conclude that $X_n^\theta \to 0$ a.s.

EXERCISE 7.5. Let S_n be asymmetric simple random walk with $p \geq 1/2$. Let $T_1 = \inf\{n : S_n = 1\}$. Use the martingale of Exercise 7.4 to conclude (i) if $\theta > 0$ then $1 = e^\theta E\varphi(\theta)^{-T_1}$ where $\varphi(\theta) = pe^\theta + qe^{-\theta}$ and $q = 1 - p$. (ii) Set $pe^\theta + qe^{-\theta} = 1/s$ and then solve for $x = e^{-\theta}$ to get

$$Es^{T_1} = (1 - \{1 - 4pqs^2\}^{1/2})/2qs$$

EXERCISE 7.6. Suppose $\varphi(\theta_o) = E\exp(\theta_o\xi_1) = 1$ for some $\theta_o < 0$ and ξ_i is not constant. It follows from the result in Exercise 7.4 that $X_n = \exp(\theta_o S_n)$ is a martingale. Let $T = \inf\{n : S_n \notin (a, b)\}$ and $Y_n = X_{n \wedge T}$. Use (7.4) to conclude that $EX_T = 1$ and $P(S_T \leq a) \leq \exp(-\theta_o a)$.

EXERCISE 7.7. Let S_n be the total assets of an insurance company at the end of year n. In year n, premiums totaling $c > 0$ are received and claims ζ_n are paid where ζ_n is Normal(μ, σ^2) and $\mu < c$. To be precise if $\xi_n = c - \zeta_n$ then $S_n = S_{n-1} + \xi_n$. The company is ruined if its assets drop to 0 or less. Show that if $S_0 > 0$ is nonrandom then

$$P(\text{ ruin }) \leq \exp(-2(c - \mu)S_0/\sigma^2)$$

EXERCISE 7.8. Let Z_n be a branching process with offspring distribution p_k, defined in part d of Section 4.3, and let $\varphi(\theta) = \sum p_k \theta^k$. Suppose $\rho < 1$ has $\varphi(\rho) = \rho$. Show that ρ^{Z_n} is a martingale and use this to conclude $P_x(Z_n = 0$ for some $n \geq 1) = \rho^x$.

5 Markov Chains

The main object of study in this chapter is (temporally homogeneous) Markov chains on a countable state space S. That is, a sequence of r.v.'s X_n, $n \geq 0$, with

$$P(X_{n+1} = j | \mathcal{F}_n) = p(X_n, j)$$

where $\mathcal{F}_n = \sigma(X_0, \ldots, X_n)$, $p(i, j) \geq 0$ and $\sum_j p(i, j) = 1$. The theory focuses on the asymptotic behavior of $p^n(i, j) \equiv P(X_n = j | X_0 = i)$. The basic results are that

(5.2) $$\lim_{n \to \infty} \frac{1}{n} \sum_{m=1}^{n} p^m(i, j) \quad \text{exists always}$$

and under a mild assumption called aperiodicity

(5.5) $$\lim_{n \to \infty} p^n(i, j) \quad \text{exists}$$

In nice situations (i.e., X_n is irreducible and positive recurrent) the limits in (5.2) and (5.5) are a probability distribution that is independent of the starting state i. In words, the chain converges to equilibrium as $n \to \infty$. One of the attractions of Markov chain theory is that these powerful conclusions come out of assumptions that are satisfied in a large number of cases.

5.1. Definitions and Examples

We begin with a very general definition and then gradually specialize to the situation described in the introductory paragraph. Let $(S, \mathcal{S})$ be a measurable space. A sequence of random variables taking values in S is said to be a **Markov chain** with respect to a filtration $\mathcal{F}_n$, if $X_n \in \mathcal{F}_n$ and for all $B \in \mathcal{S}$

$$P(X_{n+1} \in B | \mathcal{F}_n) = P(X_{n+1} \in B | X_n)$$

In words, given the present the rest of the past is irrelevant for predicting the location of X_{n+1}. As usual we turn to an example to illustrate the definition.

Example 1.1. Random walk. Let $X_0, \xi_1, \xi_2, \ldots \in \mathbf{R}^d$ be independent and let $X_n = X_0 + \xi_1 + \cdots + \xi_n$. X_n is a Markov chain with respect to $\mathcal{F}_n = \sigma(X_0, X_1, \ldots, X_n)$. (As in the case of a martingale, this is the smallest filtration we can get away with and the one we will usually use.) To prove our claim we want to show that if μ_j is the distribution of ξ_j then

$$(*) \qquad P(X_{n+1} \in B | \mathcal{F}_n) = \mu_{n+1}(B - X_n) = P(X_{n+1} \in B | X_n)$$

To prove the first equality we note that X_n and ξ_{n+1} are independent and use Example 1.5 in Chapter 4. The second equality follows from

(1.1) Lemma. If $\mathcal{F} \subset \mathcal{G}$ and $E(X|\mathcal{G}) \in \mathcal{F}$ then $E(X|\mathcal{F}) = E(X|\mathcal{G})$.

Proof Since $\mathcal{F} \subset \mathcal{G}$, (1.2) in Chapter 4 implies

$$E(X|\mathcal{F}) = E(E(X|\mathcal{G})|\mathcal{F}) = E(X|\mathcal{G})$$

since $E(X|\mathcal{G}) \in \mathcal{F}$. □

Our next goal is to take the rather abstract object defined at the beginning of the section and, without loss of generality, turn it into something more concrete and easier to work with. We begin with two definitions:

A function $p : S \times S \to \mathbf{R}$ is said to be a **transition probability** if:

(i) For each $x \in S$, $A \to p(x, A)$ is a probability measure on $(S, \mathcal{S})$.

(ii) For each $A \in \mathcal{S}$, $x \to p(x, A)$ is a measurable function.

We say X_n is a Markov chain (w.r.t. $\mathcal{F}_n$) with transition probabilities p_n if

$$P(X_{n+1} \in B | \mathcal{F}_n) = p_n(X_n, B)$$

When $(S, \mathcal{S})$ is nice supposing the existence of a transition probability entails no loss of generality, since the Markov property asserts that the conditional expectation w.r.t. $\mathcal{F}_n$ is the same as w.r.t. $\sigma(X_n)$, and then Exercise 1.16 in Chapter 4 implies the existence of a transition probability. Conversely, if we suppose $(S, \mathcal{S})$ is a nice space then given a sequence of transition probabilities p_n and an **initial distribution** μ on $(S, \mathcal{S})$, we can define a consistent set of finite dimensional distributions by

$$P(X_j \in B_j, 0 \le j \le n) = \int_{B_0} \mu(dx_0) \int_{B_1} p_0(x_0, dx_1)$$

$$(1.2) \qquad \qquad \cdots \int_{B_n} p_{n-1}(x_{n-1}, dx_n)$$

Since we have assumed that $(S, \mathcal{S})$ is nice, Kolmogorov's theorem allows us to construct a probability measure P_μ on **sequence space** $(S^{\{0,1,\ldots\}}, \mathcal{S}^{\{0,1,\ldots\}})$ so that the coordinate maps $X_n(\omega) = \omega_n$ have the desired distributions.

Notation. When $\mu = \delta_x$, a point mass at x, we use P_x as an abbreviation for P_{δ_x}. The measures P_x are the basic objects because once they are defined we can define the P_μ (even for infinite measures μ) by

$$P_\mu(A) = \int \mu(dx)\, P_x(A)$$

Our next step is to check that X_n is a Markov chain (with respect to $\mathcal{F}_n = \sigma(X_0, X_1, \ldots, X_n)$). To prove this, we let $A = \{X_0 \in B_0, X_1 \in B_1, \ldots, X_n \in B_n\}$, $B_{n+1} = B$, and observe that using the definition of the integral, the definition of A and the definition of P_μ

$$\int_A 1_{(X_{n+1} \in B)}\, dP_\mu = P_\mu(A, X_{n+1} \in B)$$

$$= P_\mu(X_0 \in B_0, X_1 \in B_1, \ldots, X_n \in B_n, X_{n+1} \in B)$$

$$= \int_{B_0} \mu(dx_0) \int_{B_1} p_0(x_0, dx_1) \cdots \int_{B_n} p_{n-1}(x_{n-1}, dx_n)\, p_n(x_n, B_{n+1})$$

We would like to assert that the last expression is

$$= \int_A p_n(X_n, B)\, dP_\mu$$

To do this replace $p_n(x_n, B_n)$ by a general function $f(x_n)$. If f is an indicator function, the desired equality is true. Linearity implies that it is valid for simple functions, and the bounded convergence theorem implies that it is valid for bounded measurable f, e.g., $f(x) = p_n(x, B_{n+1})$.

The collection of sets for which

$$\int_A 1_{(X_{n+1} \in B)}\, dP_\mu = \int_A p_n(X_n, B)\, dP_\mu$$

holds is a λ-system, and the collection for which it has been proved is a π-system, so it follows from the $\pi - \lambda$ theorem ((4.2) in Chapter 1) that the equality is true for all $A \in \mathcal{F}_n$. This shows that

$$P(X_{n+1} \in B \mid \mathcal{F}_n) = p_n(X_n, B)$$

and it follows from (1.1) that X_n is a Markov chain with transition probabilities p_n.

At this point we have shown that given a sequence of transition probabilities and an initial distribution we can construct a Markov chain. Conversely,

(1.3) Theorem. If X_n is a Markov chain with transition probabilites p_n, and initial distribution μ then the finite dimensional distributions are given by (1.2).

Proof Our first step is to show that if X_n has transition probability p_n then for any bounded measurable f

$$(1.4) \qquad E(f(X_{n+1})|\mathcal{F}_n) = \int p_n(X_n, dy)f(y)$$

The desired conclusion is a consequence of the next result. Let $\mathcal{H} = $ the collection of bounded functions for which the identity holds.

(1.5) Monotone class theorem. Let $\mathcal{A}$ be a π-system that contains Ω and let $\mathcal{H}$ be a collection of real valued functions that satisfies: (i) If $A \in \mathcal{A}$, then $1_A \in \mathcal{H}$. (ii) If $f, g \in \mathcal{H}$, then $f + g$, and $cf \in \mathcal{H}$ for any real number c. (iii) If $f_n \in \mathcal{H}$ are nonnegative and increase to a bounded function f, then $f \in \mathcal{H}$. Then $\mathcal{H}$ contains all bounded functions measurable with respect to $\sigma(\mathcal{A})$.

Proof The assumption $\Omega \in \mathcal{A}$, (ii), and (iii) imply that $\mathcal{G} = \{A : 1_A \in \mathcal{H}\}$ is a λ-system so by (i) and the $\pi - \lambda$ theorem ((4.2) in Chapter 1) $\mathcal{G} \supset \sigma(\mathcal{A})$. (ii) implies $\mathcal{H}$ contains all simple functions, and (iii) implies that $\mathcal{H}$ contains all bounded measurable functions. $\square$

Returning to our main topic we observe that familiar properties of conditional expectation and (1.4) imply

$$E\left(\prod_{m=0}^{n} f_m(X_m)\right) = E\, E\left(\prod_{m=0}^{n} f_m(X_m)\,\middle|\,\mathcal{F}_{n-1}\right)$$

$$= E\left(\prod_{m=0}^{n-1} f_m(X_m) E(f_n(X_n)|\mathcal{F}_{n-1})\right)$$

$$= E\left(\prod_{m=0}^{n-1} f_m(X_m) \int p_{n-1}(X_{n-1}, dy)f_n(y)\right)$$

The last integral is a bounded measurable function of X_{n-1}, so it follows by induction that if μ is the distribution of X_0 then

$$(1.6) \qquad E\left(\prod_{m=0}^{n} f_m(X_m)\right) = \int \mu(dx_0) f_0(x_0) \int p_0(x_0, dx_1) f_1(x_1)$$

$$\cdots \int p_{n-1}(x_{n-1}, dx_n) f_n(x_n)$$

that is, the finite dimensional distributions coincide with those in (1.2). $\square$

With (1.3) established, it follows that we can describe a Markov chain by giving a sequence of transition probabilities p_n. Having done this, we can and will suppose that the random variables X_n are the coordinate maps ($X_n(\omega) = \omega_n$) on sequence space

$$(\Omega_o, \mathcal{F}) = (S^{\{0,1,\dots\}}, \mathcal{S}^{\{0,1,\dots\}})$$

We choose this representation because it gives us two advantages in investigating the Markov chain: (i) For each initial distribution μ we have a measure P_μ defined by (1.2) that makes X_n a Markov chain with $P_\mu(X_0 \in A) = \mu(A)$. (ii) We have the shift operators θ_n defined in Section 3.1: $(\theta_n \omega)(m) = \omega_{m+n}$.

At this point we have achieved our aim announced earlier in the section of taking the abstract definition and turning it into something easier to work with. Now, we will take one further simplification, this time with loss of generality, and restrict our attention to the **temporally homogeneous** case, in which the transition probability does not depend on n.

Having decided on the framework in which we will investigate things, we can finally give some more examples. In each case S is a countable set and $\mathcal{S}$ = all subsets of S. Let $p(i,j) \geq 0$ and suppose $\sum_j p(i,j) = 1$ for all i. Intuitively, $p(i,j) = P(X_{n+1} = j | X_n = i)$. From $p(i,j)$ we can define a transition probability by

$$p(i, A) = \sum_{j \in A} p(i,j)$$

We will now give five concrete examples that will be our constant companions as the story unfolds. In each case we will just give the transition probability since it is enough to describe the Markov chain.

Example 1.2. Branching processes. $S = \{0, 1, 2, \dots\}$

$$p(i,j) = P\left(\sum_{m=1}^{i} \xi_m = j\right)$$

where $\xi_1, \xi_2, \dots$ are i.i.d. nonnegative integer valued random variables. In words, each of the i individuals at time n (or in generation n) gives birth to an independent and identically distributed number of offspring. Here and in the next four examples, we take the approach that the chain is defined by giving its transition probability. To make the connection with our earlier discussion of branching processes, do

EXERCISE 1.1. Let Z_n be the process defined in part d of Section 4.3. Check that Z_n is a Markov chain with the indicated transition probability.

Example 1.3. Renewal chain. $S = \{0, 1, 2, \ldots\}$, $f_k \geq 0$, and $\sum_{k=1}^{\infty} f_k = 1$.

$$p(0, j) = f_{j+1} \qquad \text{for } j \geq 0$$
$$p(i, i-1) = 1 \qquad \text{for } i \geq 1$$
$$p(i, j) = 0 \qquad \text{otherwise}$$

To explain the definition, let $\xi_1, \xi_2, \ldots$ be i.i.d. with $P(\xi_m = j) = f_j$, let $T_0 = i_0$ and for $k \geq 1$ let $T_k = T_{k-1} + \xi_k$. T_k is the time of the kth arrival in a renewal process that has its first arrival at time i_0. Let

$$Y_m = \begin{cases} 1 & \text{if } m \in \{T_0, T_1, T_2, \ldots\} \\ 0 & \text{otherwise} \end{cases}$$

and let $X_n = \inf\{m - n : m \geq n, Y_m = 1\}$. $Y_m = 1$ if a renewal occurs at time m, and X_n is the amount of time until the first renewal $\geq n$. It is clear that if $X_n = i > 0$ then $X_{n+1} = i - 1$. When $X_n = 0$ we have $T_{N_n} = n$ where $N_n = \inf\{k : T_k \geq n\}$ is a stopping time so (1.3) of Chapter 3 implies ξ_{N_n+1} is independent of $\sigma(X_0, \xi_1, \ldots, \xi_{N_n}) \supset \sigma(X_0, \ldots, X_n)$. We have $p(0, j) = f_{j+1}$ since $\xi_{N_n+1} = j + 1$ implies $X_{n+1} = j$.

Example 1.4. M/G/1 queue. In this model customers arrive according to a Poisson process with rate λ. (M is for Markov and refers to the fact that in a Poisson process the number of arrivals in disjoint time intervals is independent.) Each customer requires an independent amount of service with distribution F. (G is for general service distribution. 1 indicates that there is one server.) Let X_n be the number of customers waiting in the queue at the time the nth customer enters service. To be precise, when $X_0 = x$, the chain starts with x people waiting in line and customer 0 just beginning her service.

The first paragraph is for motivation only. To define our Markov chain X_n, let

$$a_k = \int_0^\infty e^{-\lambda t} \frac{(\lambda t)^k}{k!} \, dF(t)$$

be the probability that k customers arrive during a service time. Let $\xi_1, \xi_2, \ldots$ be i.i.d. with $P(\xi_i = k - 1) = a_k$. We think of ξ_i as the net number of customers to arrive during the ith service time, subtracting one for the customer who completed service, so we define X_n by

$$(*) \qquad\qquad X_{n+1} = (X_n + \xi_{n+1})^+$$

The positive part only takes effect when $X_n = 0$ and $\xi_{n+1} = -1$ and reflects the fact that when the queue has size 0 and no one arrives during the service time the next queue size is 0, since we do not start counting until the next customer arrives and then the queue length will be 0. It is easy to see that the sequence defined in (∗) is a Markov chain with transition probability

$$p(0,0) = a_0 + a_1$$
$$p(j, j-1+k) = a_k \qquad \text{if } j \geq 1 \text{ or } k > 1$$

The formula for a_k is rather complicated and its exact form is not important, so we will simplify things by assuming only that $a_k > 0$ for all $k \geq 0$ and $\sum_{k \geq 0} a_k = 1$.

Example 1.5. Ehrenfest chain. $S = \{0, 1, \ldots, r\}$

$$p(k, k+1) = (r-k)/r$$
$$p(k, k-1) = k/r$$
$$p(i, j) = 0 \qquad \text{otherwise}$$

In words, there is a total of r balls in two urns; k in the first and $r - k$ in the second. We pick one of the r balls at random and move it to the other urn. Ehrenfest used this to model the division of air molecules between two chambers (of equal size and shape) which are connected by a small hole. For an interesting account of this chain see Kac (1947a).

Example 1.6. Birth and death chains. $S = \{0, 1, 2, \ldots\}$ These chains are defined by the restriction $p(i, j) = 0$ when $|i - j| > 1$. The fact that these processes cannot jump over any integers makes it particularly easy to compute things for them.

That should be enough examples for the moment. We conclude this section with some simple calcuations. For a Markov chain on a countable state space, (1.2) says

$$P_\mu(X_k = i_k, 0 \leq k \leq n) = \mu(i_0) \prod_{m=1}^{n} p(i_{m-1}, i_m)$$

When $n = 1$

$$P_\mu(X_1 = j) = \sum_i \mu(i) p(i, j) = \mu p(j)$$

i.e., the product of the row vector μ with the matrix p. When $n = 2$,

$$P_i(X_2 = k) = \sum_j p(i, j) p(j, k) = p^2(i, k)$$

i.e., the second power of the matrix p. Combining the two formulas and generalizing

$$P_\mu(X_n = j) = \sum_i \mu(i) p^n(i, j) = \mu p^n(j)$$

EXERCISES

1.2. Suppose $S = \{1, 2, 3\}$ and

$$p = \begin{pmatrix} .1 & 0 & .9 \\ .7 & .3 & 0 \\ 0 & .4 & .6 \end{pmatrix}$$

Compute $p^2(1, 2)$ and $p^3(2, 3)$ by considering the different ways to get from 1 to 2 in two steps and and from 2 to 3 in three steps.

1.3. Suppose $S = \{0, 1\}$ and

$$p = \begin{pmatrix} 1 - \alpha & \alpha \\ \beta & 1 - \beta \end{pmatrix}$$

Use induction to show that

$$P_\mu(X_n = 0) = \frac{\beta}{\alpha + \beta} + (1 - \alpha - \beta)^n \left\{ \mu(0) - \frac{\beta}{\alpha + \beta} \right\}$$

1.4. Let $\xi_0, \xi_1, \ldots$ be i.i.d. $\in \{H, T\}$, taking each value with probability $1/2$. Show that $X_n = (\xi_n, \xi_{n+1})$ is a Markov chain and compute its transition probability p. What is p^2?

1.5. **Brother sister mating.** In this scheme, two animals are mated and among their direct descendants two individuals of opposite sex are selected at random. These animals are mated and the process continues. Suppose each individual can be one of three genotypes AA, Aa, aa, and suppose that the type of the offspring is determined by selecting a letter from each parent. With these rules the pair of genotypes in the nth generation is a Markov chain with six states:

$$AA, AA \quad AA, Aa \quad AA, aa \quad Aa, Aa \quad Aa, aa \quad aa, aa$$

Compute its transition probability.

1.6. Let $\xi_1, \xi_2, \ldots$ be i.i.d. $\in \{1, 2, \ldots, N\}$ and taking each value with probability $1/N$. Show that $X_n = |\{\xi_1, \ldots, \xi_n\}|$ is a Markov chain and compute its transition probability.

1.7. Let $\xi_1, \xi_2, \ldots$ be i.i.d. $\in \{-1, 1\}$, taking each value with probability $1/2$. Let $S_0 = 0$, $S_n = \xi_1 + \cdots \xi_n$ and $X_n = \max\{S_m : 0 \le m \le n\}$. Show that X_n is not a Markov chain.

1.8. Let $\theta, U_1, U_2, \ldots$ be independent and uniform on $(0, 1)$. Let $X_i = 1$ if $U_i \le \theta$, $= -1$ if $U_i > \theta$, and let $S_n = X_1 + \cdots + X_n$. In words we first pick θ according to the uniform distribution and then flip a coin with probability θ of heads to generate a random walk. (i) Compute $P(X_{n+1} = 1 | X_1, \ldots, X_n)$ and (ii) conclude S_n is a temporally inhomogeneous Markov chain. This is due to the fact that "S_n is a sufficient statistic for estimating θ." The answer to (i) is the estimator of a Bayesian who starts at time n with a uniform prior on $[0, 1]$.

5.2. Extensions of the Markov Property

If X_n is a Markov chain with transition probability p then by definition

$$P(X_{n+1} \in B | \mathcal{F}_n) = p(X_n, B)$$

In this section we will prove two extensions of the last equality in which $\{X_{n+1} \in B\}$ is replaced by a bounded function of the future, $h(X_n, X_{n+1}, \ldots)$, and n is replaced by a stopping time N. These results, especially the second, will be the keys to developing the theory of Markov chains.

As mentioned in Section 5.1, we can and will suppose that the X_n are the coordinate maps on sequence space

$$(\Omega_o, \mathcal{F}) = (S^{\{0,1,\ldots\}}, \mathcal{S}^{\{0,1,\ldots\}})$$

$\mathcal{F}_n = \sigma(X_0, X_1, \ldots, X_n)$, and for each initial distribution μ we have a measure P_μ defined by (1.2) that makes X_n a Markov chain with $P_\mu(X_0 \in A) = \mu(A)$. Define the shift operators $\theta_n : \Omega_o \to \Omega_o$ by $(\theta_n \omega)(m) = \omega(m + n)$.

(2.1) The Markov property. Let $Y : \Omega_o \to \mathbf{R}$ be bounded and measurable.

$$E_\mu(Y \circ \theta_n | \mathcal{F}_n) = E_{X_n} Y$$

Remark. Here the subscript μ on the left-hand side indicates that the conditional expectation is taken with respect to P_μ. The right-hand side is the function $\varphi(x) = E_x Y$ evaluated at $x = X_n$. To make the connection with the introduction of this section, let

$$Y(\omega) = h(\omega_0, \omega_1, \ldots)$$

We denote the function by Y, a letter usually used for random variables, because that's exactly what Y is, a measurable function defined on our probability space Ω_o.

Proof   We begin by proving the result in a special case and then use the $\pi - \lambda$ and monotone class theorems to get the general result. Let $\mathcal{A} = \{\omega : \omega_0 \in A_0, \ldots, \omega_m \in A_m\}$ and $g_0, \ldots g_n$ be bounded and measurable. Applying (1.6) with $f_k = 1_{A_k}$ for $k < m$, $f_m = 1_{A_m}g_0$, and $f_k = g_{k-m}$ for $m < k \leq m + n$ gives

$$E_\mu\left(\prod_{k=0}^{n} g_k(X_{m+k}); A\right) = \int_{A_0} \mu(dx_0) \int_{A_1} p(x_0, dx_1) \cdots \int_{A_m} p(x_{m-1}, dx_m)$$

$$\cdot g_0(x_m) \int p(x_m, dx_{m+1})g_1(x_{m+1})$$

$$\cdots \int p(x_{m+n-1}, dx_{m+n})g_n(x_{m+n})$$

$$= E_\mu\left(E_{X_m}\left(\prod_{k=0}^{n} g_k(X_k)\right); A\right)$$

The collection of sets for which the last formula holds is a λ-system, and the collection for which it has been proved is a π-system so using the $\pi - \lambda$ theorem ((4.2) in Chapter 1) shows that the last identity holds for all $A \in \mathcal{F}_m$.

Fix $A \in \mathcal{F}_m$ and let $\mathcal{H}$ be the collection of bounded measurable Y for which

$(*)$ $\qquad\qquad\qquad E_\mu(Y \circ \theta_m; A) = E_\mu(E_{X_m}Y; A)$

The last computation shows that $(*)$ holds when

$$Y(\omega) = \prod_{0 \leq k \leq n} g_k(\omega_k)$$

To finish the proof we will apply the monotone class theorem (1.4). Let $\mathcal{A}$ be the collection of sets of the form $\{\omega : \omega_0 \in A_0, \ldots, \omega_k \in A_k\}$. $\mathcal{A}$ is a π-system, so taking $g_k = 1_{A_k}$ shows (i) of (1.4) holds. $\mathcal{H}$ clearly has properties (ii) and (iii), so (1.4) implies that $\mathcal{H}$ contains the bounded functions measurable w.r.t $\sigma(\mathcal{A})$ and the proof is complete. $\square$

EXERCISE 2.1. Use the Markov property to show that if $A \in \sigma(X_0, \ldots, X_n)$ and $B \in \sigma(X_n, X_{n+1}, \ldots)$ then for any initial distribution μ

$$P_\mu(A \cap B|X_n) = P_\mu(A|X_n)P_\mu(B|X_n)$$

In words, the past and future are conditionally independent given the present. Hint: Write the left-hand side as $E_\mu(E_\mu(1_A 1_B|\mathcal{F}_n)|X_n)$.

The next two results illustrate the use of (2.1). We will see many other applications below.

(2.2) Chapman-Kolmogorov equation.

$$P_x(X_{m+n} = z) = \sum_y P_x(X_m = y)P_y(X_n = z)$$

Proof $P_x(X_{n+m} = z) = E_x(P_x(X_{n+m} = z|\mathcal{F}_m)) = E_x(P_{X_m}(X_n = z))$ by the Markov property (2.1) since $1_{(X_n=z)} \circ \theta_m = 1_{(X_{n+m}=z)}$. □

(2.3) Theorem. Let X_n be a Markov chain and suppose

$$P\left(\cup_{m=n+1}^{\infty}\{X_m \in B_m\}\,\middle|\, X_n\right) \geq \delta > 0 \quad \text{on } \{X_n \in A_n\}$$

Then $P(\{X_n \in A_n \text{ i.o.}\} - \{X_n \in B_n \text{ i.o.}\}) = 0$.

Remark. To quote Chung, "The intuitive meaning of the preceding theorem has been given by Doeblin as follows: if the chance of a pedestrian's getting run over is greater than $\delta > 0$ each time he crosses a certain street, then he will not be crossing it indefinitely (since he will be killed first)!"

Proof Let $\Lambda_n = \{X_{n+1} \in B_{n+1}\} \cup \{X_{n+2} \in B_{n+2}\} \cup \ldots$

$$\Lambda = \cap\Lambda_n = \{X_n \in B_n \text{ i.o.}\}$$

and $\Gamma = \{X_n \in A_n \text{ i.o.}\}$. Let $\mathcal{F}_n = \sigma(X_0, X_1, \ldots, X_n)$ and $\mathcal{F}_\infty = \sigma(\cup\mathcal{F}_n)$. Using the Markov property and the dominated convergence theorem for conditional expectations ((6.7) in Chapter 4)

$$E(1_{\Lambda_n}|X_n) = E(1_{\Lambda_n}|\mathcal{F}_n) \to E(1_\Lambda|\mathcal{F}_\infty) = 1_\Lambda$$

On Γ, the left-hand side is $\geq \delta$ i.o. This is only possible if $\Gamma \subset \Lambda$. □

EXERCISE 2.2. A state a is called absorbing if $P_a(X_1 = a) = 1$. Let $D = \{X_n = a \text{ for some } n \geq 1\}$ and let $h(x) = P_x(D)$. (i) Use (2.3) to conclude that $h(X_n) \to 0$ a.s. on D^c. Here a.s. means P_μ a.s. for any initial distribution μ. (ii) Obtain the result in Exercise 5.5 in Chapter 4 as a special case.

We are now ready for our second extension of the Markov property. Recall N is said to be a stopping time if $\{N = n\} \in \mathcal{F}_n$. As in Chapter 3 let

$$\mathcal{F}_N = \{A : A \cap \{N = n\} \in \mathcal{F}_n \text{ for all } n\}$$

be the information known at time N, and let

$$\theta_N \omega = \begin{cases} \theta_n \omega & \text{on } \{N = n\} \\ \Delta & \text{on } \{N = \infty\} \end{cases}$$

where Δ is an extra point that we add to Ω_o. In (2.4) and its applications we will explicitly restrict our attention to $\{N < \infty\}$, so the reader does not have to worry about the second part of the definition of θ_N.

(2.4) Strong Markov property. Suppose that for each n, $Y_n : \Omega \to \mathbf{R}$ is measurable and $|Y_n| \le M$ for all n. Then

$$E_\mu(Y_N \circ \theta_N | \mathcal{F}_N) = E_{X_N} Y_N \text{ on } \{N < \infty\}$$

where the right-hand side is $\varphi(x, n) = E_x Y_n$ evaluated at $x = X_N$, $n = N$.

Proof Let $A \in \mathcal{F}_N$.

$$E_\mu(Y_N \circ \theta_N ; A \cap \{N < \infty\}) = \sum_{n=0}^{\infty} E_\mu(Y_n \circ \theta_n ; A \cap \{N = n\})$$

Since $A \cap \{N = n\} \in \mathcal{F}_n$, using (2.2) now converts the right side into

$$\sum_{n=0}^{\infty} E_\mu(E_{X_n} Y_n ; A \cap \{N = n\}) = E_\mu(E_{X_N} Y_N ; A \cap \{N < \infty\})$$

Remark. The reader should notice that the proof is trivial. All we do is break things down according to the value of N, replace N by n, apply the Markov property (2.2), and reverse the process. This is the standard technique for proving results about stopping times.

The next example illustrates the use of (2.4), and explains why we want to allow the Y that we apply to the shifted path to depend on n.

(2.5) Reflection principle. Let $\xi_1, \xi_2, \ldots$ be independent and suppose that each ξ_m has a distribution that is symmetric about 0. Let $S_n = \xi_1 + \cdots + \xi_n$. If $a > 0$ then

$$P\left(\sup_{m \le n} S_m > a\right) \le 2P(S_n > a)$$

Remark. First, a trivial comment: the strictness of the inequality is not important. If the result holds for $>$ it holds for $\geq$ and vice versa.

A second more important one: we do the proof in two steps because that is how formulas like this are derived in practice. First, one computes intuitively and then figures out how to extract the desired formula from (2.4).

Proof in words First note that if Z has a distribuiton that is symmetric about 0 then

$$P(Z \geq 0) \geq P(Z > 0) + \frac{1}{2}P(Z = 0) = \frac{1}{2}$$

If we let $N = \inf\{m \leq n : S_m > a\}$ (with $\inf \emptyset = \infty$) then on $\{N < \infty\}$, $S_n - S_N$ is independent of S_N and has $P(S_n - S_N \geq 0) \geq 1/2$. So

$$P(S_n > a) \geq \frac{1}{2}P(N \leq n)$$

Formal Proof Let $Y_m(\omega) = 1$ if $m \leq n$ and $\omega_{n-m} > a$, $Y_m(\omega) = 0$ otherwise The definition of Y_m is chosen so that $(Y_N \circ \theta_N)(\omega) = 1$ if $\omega_n > a$ (and hence $N \leq n$), and $= 0$ otherwise. The strong Markov property implies

$$E_0(Y_N \circ \theta_N | \mathcal{F}_N) = E_{S_N} Y_N \quad \text{on } \{N < \infty\} = \{N \leq n\}$$

To evaluate the right-hand side, we note that if $y > a$ then

$$E_y Y_m = P_y(S_{n-m} > a) \geq P_y(S_{n-m} \geq y) \geq 1/2$$

So integrating over $\{N \leq n\}$ and using the definition of conditional expectation gives

$$\frac{1}{2}P(N \leq n) \leq E_0(E_0(Y_N \circ \theta_N | \mathcal{F}_N); N \leq n) = E_0(Y_N \circ \theta_N; N \leq n)$$

since $\{N \leq n\} \in \mathcal{F}_N$. Recalling that $Y_N \circ \theta_N = 1_{\{S_n > a\}}$, the last quantity

$$= E_0(1_{\{S_n > a\}}; N \leq n) = P_0(S_n > a)$$

since $\{S_n > a\} \subset \{N \leq n\}$. $\qquad\qquad\qquad\qquad\qquad\qquad\qquad$ $\square$

EXERCISES

The next five exercises concern the hitting times

$$\tau_A = \inf\{n \geq 0 : X_n \in A\} \qquad \tau_y = \tau_{\{y\}}$$
$$T_A = \inf\{n \geq 1 : X_n \in A\} \qquad T_y = T_{\{y\}}$$

To keep the two definitions straight note that the symbol τ is smaller than T. Some of the results below are valid for a general S but for simplicity.

We will suppose throughout that S is countable.

2.3. First entrance decomposition. Let $T_y = \inf\{n \geq 1 : X_n = y\}$. Show that

$$p^n(x,y) = \sum_{m=1}^{n} P_x(T_y = m)p^{n-m}(y,y)$$

2.4. Show that $\sum_{m=0}^{n} P_x(X_m = x) \geq \sum_{m=k}^{n+k} P_x(X_m = x)$.

2.5. Suppose that $S - C$ is finite and for each $x \in S - C$ $P_x(\tau_C < \infty) > 0$. Then there is an $N < \infty$ and $\epsilon > 0$ so that $P_y(\tau_C > kN) \leq (1 - \epsilon)^k$.

2.6. Let $h(x) = P_x(\tau_A < \tau_B)$. Suppose $A \cap B = \emptyset$, $S - (A \cup B)$ is finite, and $P_x(\tau_{A \cup B} < \infty) > 0$ for all $x \in S - (A \cup B)$. (i) Show that

$(*)$ $\qquad\qquad h(x) = \sum_y p(x,y)h(y) \quad \text{for } x \notin A \cup B$

(ii) Show that if h satisfies $(*)$ then $h(X(n \wedge \tau_{A \cup B}))$ is a martingale. (iii) Use this and Exercise 2.5 to conclude that $h(x) = P_x(\tau_A < \tau_B)$ is the only solution of $(*)$ that is 1 on A and 0 on B.

2.7. Let X_n be a Markov chain with $S = \{0, 1, \ldots, N\}$ and suppose that X_n is a martingale and $P_x(\tau_0 \wedge \tau_N < \infty) > 0$ for all x. (i) Show that 0 and N are absorbing states, i.e., $p(0,0) = p(N,N) = 1$. (ii) Show $P_x(\tau_N < \tau_0) = x/N$.

2.8. Genetics Chains. Suppose $S = \{0, 1, \ldots, N\}$ and consider

(i) $\qquad\qquad p(i,j) = \binom{N}{j}(i/N)^j (1 - i/N)^{N-j}$

(ii) $\qquad\qquad p(i,j) = \binom{2i}{j}\binom{2N - 2i}{N - j} \Big/ \binom{2N}{N}$

Show that these chains satisfy the hypotheses of Exercise 2.7.

2.9. In brother sister mating described in Exercise 1.4 AA, AA and aa, aa are absorbing states. Show that the number of A's in the pair is a martingale and use this to compute the probability of getting absorbed in AA, AA starting from each of the states.

2.10. Let $\tau_A = \inf\{n \geq 0 : X_n \in A\}$ and $g(x) = E_x\tau_A$. Suppose that $S - A$ is finite and for each $x \in S - A$, $P_x(\tau_A < \infty) > 0$. (i) Show that

$(*)$ $\qquad\qquad g(x) = 1 + \sum_y p(x,y)g(y) \quad \text{for } x \notin A$

(ii) Show that if g satisfies $(*)$, $g(X(n \wedge \tau_A)) + n \wedge \tau_A$ is a martingale. (iii) Use this to conclude that $g(x) = E_x \tau_A$ is the only solution of $(*)$ that is 0 on A.

2.11. Let $\xi_0, \xi_1, \ldots$ be i.i.d. $\in \{H, T\}$, taking each value with probability $1/2$, and let $X_n = (\xi_n, \xi_{n+1})$ be the Markov chain from Exercise 1.4. Let $N_1 = \inf\{n \geq 0 : (\xi_n, \xi_{n+1}) = (H, H)\}$. Use the results in the last exercise to compute EN_1. [No there is no missing subscript on E, but you will need to first to compute $g(x)$.]

2.12. Consider the Markov chain on $\{1, 2, \ldots, N\}$ with $p_{ij} = 1/(i-1)$ when $j < i$, $p_{11} = 1$ and $p_{ij} = 0$ otherwise. We claim that

$$E_k T_1 = 1 + 1/2 + \cdots + 1/(k-1)$$

Prove this by (i) using Exercise 2.10, OR (ii) letting $I_j = 1$ if X_n visits j, noticing that if $X_0 = k$, $T_1 = I_1 + \cdots + I_{k-1}$ where $I_1, I_2, \ldots, I_{k-1}$ are independent.

5.3. Recurrence and Transience

In this section and the next two, we will consider only Markov chains on a countable state space. Let $T_y^0 = 0$ and for $k \geq 1$ let

$$T_y^k = \inf\{n > T_y^{k-1} : X_n = y\}$$

T_y^k is the time of the kth return to y. The reader should note that $T_y^1 > 0$ so any visit at time 0 does not count. We adopt this convention so that if we let $T_y = T_y^1$ and $\rho_{xy} = P_x(T_y < \infty)$ then

(3.1) Theorem. $P_x(T_y^k < \infty) = \rho_{xy} \rho_{yy}^{k-1}$.

Intuitively, in order to make k visits to y we first have to go from x to y and then return $k - 1$ times to y.

Proof When $k = 1$ the result is trivial so we suppose $k \geq 2$. Let $Y(\omega) = 1$ if $\omega_n = y$ for some $n \geq 1$, $Y(\omega) = 0$ otherwise. If $N = T_y^{k-1}$ then $Y \circ \theta_N = 1$ if $T_y^k < \infty$. The strong Markov property (2.4) implies

$$E_x(Y \circ \theta_N | \mathcal{F}_N) = E_{X_N} Y \quad \text{on } \{N < \infty\}$$

On $\{N < \infty\}$, $X_N = y$, so the right-hand side is $P_y(T_y < \infty) = \rho_{yy}$, and it follows that

$$
\begin{aligned}
P_x(T_y^k < \infty) &= E_x(Y \circ \theta_N; N < \infty) \\
&= E_x(E_x(Y \circ \theta_N | \mathcal{F}_N); N < \infty) \\
&= E_x(\rho_{yy}; N < \infty) = \rho_{yy} P_x(T_y^{k-1} < \infty)
\end{aligned}
$$

The result now follows by induction.   □

A state y is said to be **recurrent** if $\rho_{yy} = 1$ and **transient** if $\rho_{yy} < 1$. If y is recurrent then (3.1) implies $P_y(T_y^k < \infty) = 1$ for all k, so $P_y(X_n = y \text{ i.o.}) = 1$.

EXERCISE 3.1. Suppose y is recurrent and for $k \geq 0$, let $R_k = T_y^k$ be the time of the kth return to y, and for $k \geq 1$ let $r_k = R_k - R_{k-1}$ be the kth interarrival time. Use the strong Markov property to conclude that under P_y, the vectors $v_k = (r_k, X_{R_{k-1}}, \ldots, X_{R_k-1})$, $k \geq 1$ are i.i.d.

If y is transient and we let $N(y) = \sum_{n=1}^{\infty} 1_{(X_n = y)}$ be the number of visits to y at positive times, then

(3.2)
$$E_x N(y) = \sum_{k=1}^{\infty} P_x(N(y) \geq k) = \sum_{k=1}^{\infty} P_x(T_y^k < \infty)$$
$$= \sum_{k=1}^{\infty} \rho_{xy}\rho_{yy}^{k-1} = \frac{\rho_{xy}}{1 - \rho_{yy}} < \infty$$

Combining the last computation with our result for recurrent states gives a result that generalizes (2.2) from Chapter 3.

(3.3) Theorem. y is recurrent if and only if $E_y N(y) = \infty$.

EXERCISE 3.2. Let $a \in S$, $f_n = P_a(T_a = n)$, and $u_n = P_a(X_n = a)$. (i) Show that $u_n = \sum_{1 \leq m \leq n} f_m u_{n-m}$. (ii) Let $u(s) = \sum_{n \geq 0} u_n s^n$, $f(s) = \sum_{n \geq 1} f_n s^n$, and show $u(s) = 1/(1 - f(s))$. Setting $s = 1$ gives (3.2) for $x = y = a$.

EXERCISE 3.3. Consider asymmetric simple random walk on $\mathbf{Z}$, i.e., we have $p(i, i+1) = p$, $p(i, i-1) = q = 1 - p$. In this case

$$p^{2m}(0,0) = \binom{2m}{m} p^m q^m \quad \text{and} \quad p^{2m+1}(0,0) = 0$$

(i) Use the Taylor series expansion for $h(x) = (1-x)^{-1/2}$ to show $u(s) = (1 - 4pqs^2)^{-1/2}$ and use the last exercise to conclude $f(s) = 1 - (1 - 4pqs^2)^{1/2}$. (ii) Set $s = 1$ to get the probability the random walk will return to 0 and check that this is the same as the answer given in Example 7.1 of Chapter 4.

The next result shows that recurrence is contagious.

(3.4) Theorem. If x is recurrent and $\rho_{xy} > 0$ then y is recurrent and $\rho_{yx} = 1$.

Proof We will first show $\rho_{yx} = 1$ by showing that if $\rho_{xy} > 0$ and $\rho_{yx} < 1$ then $\rho_{xx} < 1$. Let $K = \inf\{k : p^k(x, y) > 0\}$. There is a sequence $y_1, \ldots, y_{K-1}$ so that

$$p(x, y_1)p(y_1, y_2) \cdots p(y_{K-1}, y) > 0$$

Since K is minimal, $y_i \neq y$ for $1 \leq i \leq K - 1$. If $\rho_{yx} < 1$, we have

$$P_x(T_x = \infty) \geq p(x, y_1)p(y_1, y_2) \cdots p(y_{K-1}, y)(1 - \rho_{yx}) > 0$$

a contradiction. So $\rho_{yx} = 1$.

To prove that y is recurrent, observe that $\rho_{yx} > 0$ implies there is an L so that $p^L(y, x) > 0$. Now

$$p^{L+n+K}(y, y) \geq p^L(y, x)p^n(x, x)p^K(x, y)$$

Summing over n we see

$$\sum_{n=1}^{\infty} p^{L+n+K}(y, y) \geq p^L(y, x)p^K(x, y) \sum_{n=1}^{\infty} p^n(x, x) = \infty$$

so (3.3) implies y is recurrent. $\square$

EXERCISE 3.4. Use the strong Markov property to show that $\rho_{xz} \geq \rho_{xy}\rho_{yz}$.

The next fact will help us identify recurrent states in examples. First we need two definitions. C is **closed** if $x \in C$ and $\rho_{xy} > 0$ implies $y \in C$. The name comes from the fact that if C is closed and $x \in C$ then $P_x(X_n \in C) = 1$ for all n. D is **irreducible** if $x, y \in D$ implies $\rho_{xy} > 0$.

(3.5) Theorem. Let C be a finite closed set. Then C contains a recurrent state. If C is irreducible then all states in C are recurrent.

Proof In view of (3.4) it suffices to prove the first claim. Suppose it is false. Then for all $y \in C$, $\rho_{yy} < 1$ and $E_x N(y) = \rho_{xy}/(1 - \rho_{yy})$, but this is ridiculous since it implies

$$\infty > \sum_{y \in C} E_x N(y) = \sum_{y \in C} \sum_{n=1}^{\infty} p^n(x, y) = \sum_{n=1}^{\infty} \sum_{y \in C} p^n(x, y) = \sum_{n=1}^{\infty} 1$$

The first inequality follows from the fact that C is finite and the last equality from the fact that C is closed. $\square$

To illustrate the use of the last result consider

Example 3.1. Let X_n be a Markov chain with transition matrix

	1	2	3	4	5	6
1	0	1	0	0	0	0
2	.4	.6	0	0	0	0
3	.3	0	.4	.2	.1	0
4	0	0	0	.3	.7	0
5	0	0	0	.5	0	.5
6	0	0	0	.8	0	.2

Things become a little clearer if we represent the matrix as a graph with an edge from i to j if $i \neq j$ and $p(i, j) > 0$. (See Figure 5.3.1. To make the graph easier to draw we ignore the $p(i, i)$. They are irrelevant in computing ρ_{xy} for $x \neq y$.) It is easy to see that (when $x \neq y$) $\rho_{xy} > 0$ if and only if we can get from x to y on the graph.

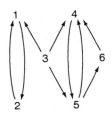

Figure 5.3.1

Looking at the graph we see that:

(i) $\rho_{34} > 0$ and $\rho_{43} = 0$ so 3 must be transient, or we would contradict (3.4).

(ii) $\{1, 2\}$ and $\{4, 5, 6\}$ are irreducible closed sets, so (3.5) implies these states are recurrent.

The last reasoning can be used to identify transient and recurrent states when S is finite since for $x \in S$ either: (i) there is a y with $\rho_{xy} > 0$ and $\rho_{yx} = 0$ and x must be transient, or (ii) $\rho_{xy} > 0$ implies $\rho_{yx} > 0$. In case (ii), Exercise 3.4 implies $C_x = \{y : \rho_{xy} > 0\}$ is an irreducible closed set. (If $y, z \in C_x$ then $\rho_{yz} \geq \rho_{yx}\rho_{xz} > 0$. If $\rho_{yw} > 0$ then $\rho_{xw} \geq \rho_{xy}\rho_{yw} > 0$ so $w \in C_x$.) So (3.5) implies x is recurrent.

EXERCISE 3.5. Show that in the Ehrenfest chain (Example 1.5) all states are recurrent.

Example 3.1 motivates the following:

(3.6) Decomposition theorem. Let $R = \{x : \rho_{xx} = 1\}$ be the recurrent states of a Markov chain. R can be written as $\cup_i R_i$ where each R_i is closed and irreducible.

Remark. This result shows that for the study of recurrent states we can without loss of generality consider a single irreducible closed set.

Proof If $x \in R$ let $C_x = \{y : \rho_{xy} > 0\}$. By (3.4) $C_x \subset R$ and if $y \in C_x$ then $\rho_{yx} > 0$. From this it follows easily that either $C_x \cap C_y = \emptyset$ or $C_x = C_y$. To prove the last claim suppose $C_x \cap C_y \neq \emptyset$. If $z \in C_x \cap C_y$ then $\rho_{xy} \geq \rho_{xz}\rho_{zy} > 0$, so if $w \in C_y$ we have $\rho_{xw} \geq \rho_{xy}\rho_{yw} > 0$ and it follows that $C_x \supset C_y$. Interchanging the roles of x and y gives $C_y \supset C_x$ and we have proved our claim. If we let R_i be a listing of the sets which appear as some C_x, we have the desired decomposition.   □

The rest of this section is devoted to examples. Specifically we concentrate on the question: How do we tell whether a state is recurrent or transient? Reasoning based on (3.4) works occasionally when S is infinite.

Example 3.2. Branching process. If the probability of no children is positive then $\rho_{k0} > 0$ and $\rho_{0k} = 0$ for $k \geq 1$, so (3.4) implies all states $k \geq 1$ are transient. The state 0 has $p(0,0) = 1$ and is recurrent. It is called an **absorbing state** to reflect the fact that once the chain enters 0, it remains there for all time.

If S is infinite and irreducible, all that (3.4) tells us is that either all the states are recurrent or all are transient, and we are left to figure out which case occurs.

Example 3.3. Renewal chain. Since $p(i, i-1) = 1$ for $i \geq 1$, it is clear that $\rho_{i0} = 1$ for all $i \geq 1$ and hence also for $i = 0$, i.e., 0 is recurrent. If we recall that $p(0,j) = f_{j+1}$ and suppose that $\{k : f_k > 0\}$ is unbounded then $\rho_{0i} > 0$ for all i and all states are recurrent. If $K = \sup\{k : f_k > 0\} < \infty$ then $\{0, 1, \ldots, K-1\}$ is an irreducible closed set of recurrent states and all states $k \geq K$ are transient.

Example 3.4. Birth and death chains on $\{0, 1, 2, \ldots\}$. Let

$$p(i, i+1) = p_i \quad p(i, i-1) = q_i \quad p(i, i) = r_i$$

where $q_0 = 0$. Let $N = \inf\{n : X_n = 0\}$. To analyze this example, we are going to define a function φ so that $\varphi(X_{N \wedge n})$ is a martingale. We start by setting

$\varphi(0) = 0$ and $\varphi(1) = 1$. For the martingale property to hold when $X_n = k \geq 1$ we must have

$$\varphi(k) = p_k \varphi(k+1) + r_k \varphi(k) + q_k \varphi(k-1)$$

Using $r_k = 1 - (p_k + q_k)$ we can rewrite the last equation as

$$q_k(\varphi(k) - \varphi(k-1)) = p_k(\varphi(k+1) - \varphi(k))$$

or
$$\varphi(k+1) - \varphi(k) = \frac{q_k}{p_k}(\varphi(k) - \varphi(k-1))$$

Here and in what follows, we suppose that $p_k, q_k > 0$ for $k \geq 1$. Otherwise the chain is not irreducible. Since $\varphi(1) - \varphi(0) = 1$, iterating the last result gives

$$\varphi(m+1) - \varphi(m) = \prod_{j=1}^{m} \frac{q_j}{p_j} \quad \text{for } m \geq 1$$

and
$$\varphi(n) = \sum_{m=0}^{n-1} \prod_{j=1}^{m} \frac{q_j}{p_j} \quad \text{for } n \geq 1$$

if we interpret the product as 1 when $m = 0$. Let $T_c = \inf\{n \geq 1 : X_n = c\}$. Now I claim that

(3.7) Theorem. If $a < x < b$ then

$$P_x(T_a < T_b) = \frac{\varphi(b) - \varphi(x)}{\varphi(b) - \varphi(a)} \qquad P_x(T_b < T_a) = \frac{\varphi(x) - \varphi(a)}{\varphi(b) - \varphi(a)}$$

Proof If we let $T = T_a \wedge T_b$ then $\varphi(X_{n \wedge T})$ is a bounded martingale and $T < \infty$ a.s. by Exercise 2.5, so $\varphi(x) = E_x \varphi(X_T)$ by (7.3) in Chapter 4. Since $X_T \in \{a, b\}$ a.s.,

$$\varphi(x) = \varphi(a) P_x(T_a < T_b) + \varphi(b)[1 - P_x(T_a < T_b)]$$

and solving gives the indicated formula.   □

Remark. The answer and the proof should remind the reader of Example 1.5 in Chapter 3 and Exercise 7.3 in Chapter 4. To help remember the formula observe that for any α and β if we let $\psi(x) = \alpha \varphi(x) + \beta$ then $\psi(X_{n \wedge T})$ is also a martingale and the answer we get using ψ must be the same. The last observation explains why the answer is a ratio of differences. To help remember which one, observe that the answer is 1 if $x = a$ and 0 if $x = b$.

Letting $a = 0$ and $b = M$ in (3.7) gives

$$P_x(T_0 > T_M) = \varphi(x)/\varphi(M)$$

Letting $M \to \infty$ and observing that $T_M \geq M - x$, P_x a.s. we have proved

(3.8) Theorem. 0 is recurrent if and only if $\varphi(M) \to \infty$ as $M \to \infty$, i.e.,

$$\varphi(\infty) \equiv \sum_{m=0}^{\infty} \prod_{j=1}^{m} \frac{q_j}{p_j} = \infty$$

If $\varphi(\infty) < \infty$ then $P_x(T_0 = \infty) = \varphi(x)/\varphi(\infty)$.

We will now see what (3.8) says about some concrete cases.

Example 3.5. Asymmetric simple random walk. Suppose $p_j = p$ and $q_j = 1 - p$ for $j \geq 1$. In this case

$$\varphi(n) = \sum_{m=0}^{n-1} \left(\frac{1-p}{p}\right)^m$$

From (3.8) it follows that 0 is recurrent if and only if $p \leq 1/2$, and if $p > 1/2$, then

$$P_x(T_0 < \infty) = \frac{\varphi(\infty) - \varphi(x)}{\varphi(\infty)} = \left(\frac{1-p}{p}\right)^x$$

EXERCISE 3.6. A gambler is playing roulette and betting \$1 on black each time. The probability she wins \$1 is 18/38 and the probability she loses \$1 is 20/38. (i) Calculate the probability that starting with \$20 she reaches \$40 before losing her money. (ii) Use the fact that $X_n + 2n/38$ is a martingale to calculate $E(T_{40} \wedge T_0)$.

Example 3.6. To probe the boundary between recurrence and transience suppose $p_j = 1/2 + \epsilon_j$ where $\epsilon_j \sim Cj^{-\alpha}$ as $j \to \infty$, and $q_j = 1 - p_j$. A little arithmetic shows

$$\frac{q_j}{p_j} = \frac{1/2 - \epsilon_j}{1/2 + \epsilon_j} = 1 - \frac{2\epsilon_j}{1/2 + \epsilon_j} \approx 1 - 4Cj^{-\alpha} \quad \text{for large } j$$

Case 1: $\alpha > 1$. It is easy to show that if $0 < \delta_j < 1$, then $\prod_j (1 - \delta_j) > 0$ if and only if $\sum_j \delta_j < \infty$, (see Exercise 3.5 in Chapter 4) so if $\alpha > 1$, $\prod_{j \leq k}(q_j/p_j) \downarrow$ a positive limit, and 0 is recurrent.

Case 2: $\alpha < 1$. Using the fact that $\log(1 - \delta) \sim -\delta$ as $\delta \to 0$ we see that

$$\log \prod_{j=1}^{k} \frac{q_j}{p_j} \sim - \sum_{j=1}^{k} 4Cj^{-\alpha} \sim -\frac{4C}{1-\alpha}k^{1-\alpha} \quad \text{as } k \to \infty$$

so for $k \geq K$

$$\prod_{j=1}^{k} \frac{q_j}{p_j} \leq \exp\left(\frac{-2Ck^{1-\alpha}}{1-\alpha}\right) \quad \text{and} \quad \sum_{k=0}^{\infty} \prod_{j=1}^{k} \frac{q_j}{p_j} < \infty$$

and hence 0 is transient.

Case 3: $\alpha = 1$. Repeating the argument for Case 2 shows

$$\log \prod_{j=1}^{k} \frac{q_j}{p_j} \sim -4C \log k$$

So if $C > 1/4$, 0 is transient and if $C < 1/4$, 0 is recurrent. The case $C = 1/4$ can go either way.

Example 3.7. $M/G/1$ **queue.** Let $\mu = \sum k a_k$ be the mean number of customers that arrive during one service time. We will now show that if $\mu > 1$ the chain is transient (i.e., all states are) but if $\mu \leq 1$ it is recurrent. For the case $\mu > 1$, we observe that if $\xi_1, \xi_2, \ldots$ are i.i.d. with $P(\xi_m = j) = a_{j+1}$ for $j \geq -1$ and $S_n = \xi_1 + \cdots + \xi_n$, then $X_0 + S_n$ and X_n behave the same until time $N = \inf\{n : X_0 + S_n = 0\}$. When $\mu > 1$, $E\xi_m = \mu - 1 > 0$, so $S_n \to \infty$ a.s., and $\inf S_n > -\infty$ a.s. It follows from the last observation that if x is large $P_x(N < \infty) < 1$, and the chain is transient.

To deal with the case $\mu \leq 1$ we observe that it follows from arguments in the last paragraph that $X_{n \wedge N}$ is a supermartingale. Let $T = \inf\{n : X_n \geq M\}$. Since $X_{n \wedge N}$ is a nonnegative supermartingale, using the optional stopping theorem ((7.6) in Chapter 4) at time $\tau = T \wedge N$, and observing $X_\tau \geq M$ on $\{T < N\}$, $X_\tau = 0$ on $\{N < T\}$ gives

$$x \geq MP_x(T < N)$$

Letting $M \to \infty$ shows $P_x(N < \infty) = 1$, so the chain is recurrent.

Remark. There is another way of seeing that the $M/G/1$ queue is transient when $\mu > 1$. If we consider the customers that arrive during a person's service time to be her children then we get a branching process. Results in Section

4.3 imply that when $\mu \leq 1$ the branching process dies out with probability one (i.e., the queue becomes empty), so the chain is recurrent. When $\mu > 1$, (3.9) in Chapter 4 implies $P_x(T_0 < \infty) = \rho^x$, where ρ is the smallest fixed point of the function

$$\varphi(\theta) = \sum_{k=0}^{\infty} a_k \theta^k$$

The next result encapsulates the techniques we used for birth and death chains and the $M/G/1$ queue.

(3.9) Theorem. Suppose S is irreducible, and φ is a nonnegative function with $E_x \varphi(X_1) \leq \varphi(x)$ for $x \notin F$, a finite set, and $\varphi(x) \to \infty$ as $x \to \infty$, i.e., $\{x : \varphi(x) \leq M\}$ is finite for any $M < \infty$, then the chain is recurrent.

Proof Let $\tau = \inf\{n > 0 : X_n \in F\}$. Our assumptions imply that $Y_n = \varphi(X_{n \wedge \tau})$ is a supermartingale. Let $T_M = \inf\{n > 0 : X_n \in F \text{ or } \varphi(X_n) > M\}$. Since $\{x : \varphi(x) \leq M\}$ is finite and the chain is irreducible, $T_M < \infty$ a.s. Using (7.6) in Chapter 4 now we see that

$$\varphi(x) \geq E_x \varphi(X_{T_M}) \geq M P_x(T_M < \tau)$$

since $\varphi(X_{T_M}) \geq M$ when $T_M < \tau$. Letting $M \to \infty$ we see that $P_x(\tau < \infty) = 1$ for all $x \notin F$. So $P_y(X_n \in F \text{ i.o.}) = 1$ for all $y \in S$, and since F is finite, $P_y(X_n = z \text{ i.o.}) = 1$ for some $z \in F$. $\qquad \square$

EXERCISE 3.7. Show that if we replace "$\varphi(x) \to \infty$" by "$\varphi(x) \to 0$" in the last exercise and assume that $\varphi(x) > 0$ for $x \in F$ then we can conclude that the chain is transient.

EXERCISE 3.8. Let X_n be a birth and death chain with $p_j - 1/2 \sim C/j$ as $j \to \infty$ and $q_j = 1 - p_j$. (i) Show that if we take $C < 1/4$ then we can pick $\alpha > 0$ so that $\varphi(x) = x^\alpha$ satifies the hypotheses of (3.9). (ii) Show that when $C > 1/4$ we can take $\alpha < 0$ and apply Exercise 3.7.

Remark. An advantage of the method of Exercise 3.8 over that of Example 3.6 is that it applies if we assume $P_x(|X_1 - x| \leq M) = 1$ and $E_x(X_1 - x) \sim 2C/x$.

EXERCISE 3.9. f is said to be **superharmonic** if $f(x) \geq \sum_y p(x,y) f(y)$, or equivalently $f(X_n)$ is a supermartingale. Suppose p is irreducible. Show that p is recurrent if and only if every nonnegative superharmonic function is constant.

EXERCISE 3.10. **M/M/∞ queue.** Consider a telephone system with an infinite number of lines. Let $X_n =$ the number of lines in use at time n, and suppose

$$X_{n+1} = \sum_{m=1}^{X_n} \xi_{n,m} + Y_{n+1}$$

where the $\xi_{n,m}$ are i.i.d. with $P(\xi_{n,m} = 1) = p$ and $P(\xi_{n,m} = 0) = 1 - p$, and Y_n is an independent i.i.d. sequence of Poisson mean λ r.v.'s. In words, for each conversation we flip a coin with probability p of heads to see if it continues for another minute. Meanwhile a Poisson mean λ number of conversations start between time n and $n + 1$. Use Exercise 3.9 with $\varphi(x) = x$ to show that the chain is recurrent for any $p < 1$.

5.4. Stationary Measures

A measure μ is said to be a **stationary measure** if

$$\sum_x \mu(x) p(x, y) = \mu(y)$$

The last equation says $P_\mu(X_1 = y) = \mu(y)$. Using the Markov property and induction it follows that $P_\mu(X_n = y) = \mu(y)$ for all $n \geq 1$. If μ is a probability measure, we call μ a **stationary distribution**, and it represents a possible equilibrium for the chain. That is, if X_0 has distribution μ then so does X_n for all $n \geq 1$. If we stretch our imagination a little, we can also apply this interpretation when μ is an infinite measure. (When the total mass is finite we can divide by $\mu(S)$ to get a stationary distribution.) Before getting into the theory, we consider some examples.

Example 4.1. Random walk. $S = \mathbf{Z}^d$. $p(x, y) = f(y - x)$ where $f(z) \geq 0$ and $\sum f(z) = 1$. In this case $\mu(x) \equiv 1$ is a stationary measure since

$$\sum_x p(x, y) = \sum_x f(y - x) = 1$$

A transition probability that has $\sum_x p(x, y) = 1$ is called **doubly stochastic**. This is obviously a necessary and sufficient condition for $\mu(x) \equiv 1$ to be a stationary measure.

Example 4.2. Asymmetric simple random walk. $S = \mathbf{Z}$.

$$p(x, x + 1) = p \qquad p(x, x - 1) = q = 1 - p$$

By the last example $\mu(x) \equiv 1$ is a stationary measure. When $p \neq q$, $\mu(x) = (p/q)^x$ is a second one. To check this, we observe that

$$\sum_x \mu(x)p(x,y) = \mu(y+1)p(y+1,y) + \mu(y-1)p(y-1,y)$$

$$= (p/q)^{y+1}q + (p/q)^{y-1}p = (p/q)^y[p+q] = (p/q)^y$$

Example 4.3. The Ehrenfest chain. $S = \{0,1,\ldots,r\}$.

$$p(k,k+1) = (r-k)/r \qquad p(k,k-1) = k/r$$

In this case $\mu(x) = 2^{-r}\binom{r}{x}$ is a stationary distribution. One can check this without pencil and paper by observing that μ corresponds to flipping r coins to determine which urn each ball is to be placed in, and the transitions of the chain correspond to picking a coin at random and turning it over. Alternatively, you can pick up your pencil and check that

$$\mu(k+1)p(k+1,k) + \mu(k-1)p(k-1,k) = \mu(k)$$

Example 4.4. Birth and death chains. $S = \{0,1,2,\ldots\}$

$$p(x,x+1) = p_x \quad p(x,x) = r_x \quad p(x,x-1) = q_x$$

with $q_0 = 0$ and $p(i,j) = 0$ otherwise. In this case there is the measure

$$\mu(x) = \prod_{k=1}^{x} \frac{p_{k-1}}{q_k}$$

which has

$$\mu(x)p(x,x+1) = p_x \prod_{k=1}^{x} \frac{p_{k-1}}{q_k} = \mu(x+1)p(x+1,x)$$

Since $p(x,y) = 0$ when $|x-y| > 1$ it follows that

(4.1) $$\mu(x)p(x,y) = \mu(y)p(y,x) \quad \text{for all } x,y$$

Summing over x gives

$$\sum_x \mu(x)p(x,y) = \mu(y)$$

so (4.1) is stronger than being a stationary measure. (4.1) asserts that the amount of mass that moves from x to y in one jump is exactly the same as the amount that moves from y to x. A measure μ that satisfies (4.1) is said to be a **reversible measure**. Since Examples 4.2 and 4.3 are birth and death chains they have reversible measures. In Example 4.1 (random walks) $\mu(x) \equiv 1$ is a reversible measure if and only if $p(x, y) = p(y, x)$.

The next exercise explains the name "reversible."

EXERCISE 4.1. Let μ be a stationary measure and suppose X_0 has "distribution" μ. Then $Y_m = X_{n-m}$, $0 \le m \le n$ is a Markov chain with initial measure μ and transition probability

$$q(x, y) = \mu(y)p(y, x)/\mu(x)$$

q is called the **dual transition probability**. If μ is a reversible measure then $q = p$.

Example 4.5. Random walks on graphs. A graph is described by giving a countable set of vertices S and an adjacency matrix a_{ij} that has $a_{ij} = 1$ if i and j are adjacent and 0 otherwise. To have an undirected graph with no loops we suppose $a_{ij} = a_{ji}$ and $a_{ii} = 0$. If we suppose that

$$\mu(i) = \sum_j a_{ij} < \infty \quad \text{and let} \quad p(i, j) = a_{ij}/\mu(i)$$

then p is a transition probability that corresponds to picking an edge at random and jumping to the other end. It is clear from the definition that

$$\mu(i)p(i, j) = a_{ij} = a_{ji} = \mu(j)p(j, i)$$

so μ is a reversible measure for p. A little thought reveals that if we assume only that

$$a_{ij} = a_{ji} \ge 0, \quad \mu(i) = \sum_j a_{ij} < \infty \quad \text{and} \quad p(i, j) = a_{ij}/\mu(i)$$

the same conclusion is valid. This is the most general example because if μ is a reversible measure for p we can let $a_{ij} = \mu(i)p(i, j)$.

Reviewing the last five examples might convince you that most chains have reversible measures. This is a false impression. The $M/G/1$ queue has no reversible measures because if $x > y + 1$, $p(x, y) = 0$ but $p(y, x) > 0$. The renewal chain has similar problems.

(4.2) Theorem. Suppose p is irreducible. A necessary and sufficient condition for the existence of a reversible measure is that (i) $p(x, y) > 0$ implies $p(y, x) > 0$ and (ii) for any loop $x_0, x_1, \ldots, x_n = x_0$ with $\prod_{1 \le i \le n} p(x_i, x_{i-1}) > 0$,

$$\prod_{i=1}^{n} \frac{p(x_{i-1}, x_i)}{p(x_i, x_{i-1})} = 1$$

Proof To prove the necessity of this **cycle condition**, due to Kolmogorov, we note that irreducibility implies that any stationary measure has $\mu(x) > 0$ for all x, so (4.1) implies (i) holds. To check (ii) note that (4.1) implies that for the sequences considered above

$$\prod_{i=1}^{n} \frac{p(x_{i-1}, x_i)}{p(x_i, x_{i-1})} = \prod_{i=1}^{n} \frac{\mu(x_i)}{\mu(x_{i-1})} = 1$$

To prove sufficiency, fix $a \in S$, set $\mu(a) = 1$ and if $x_0 = a$, $x_1, \ldots, x_n = x$ is a sequence with $\prod_{1 \le i \le n} p(x_i, x_{i-1}) > 0$, (irreducibility implies such a sequence will exist) we let

$$\mu(x) = \prod_{i=1}^{n} \frac{p(x_{i-1}, x_i)}{p(x_i, x_{i-1})}$$

The cycle condition guarantees that the last definition is independent of the path. To check (4.1) now, observe that if $p(y, x) > 0$ then adding $x_{n+1} = y$ to the end of a path to x we have

$$\mu(x) \frac{p(x, y)}{p(y, x)} = \mu(y)$$

□

Only special chains have reversible measures, but as the next result shows many Markov chains have stationary measures.

(4.3) Theorem. Let x be a recurrent state, and let $T = \inf\{n \ge 1 : X_n = x\}$. Then

$$\mu_x(y) = E_x \left(\sum_{n=0}^{T-1} 1_{\{X_n = y\}} \right) = \sum_{n=0}^{\infty} P_x(X_n = y, T > n)$$

defines a stationary measure.

Proof This is called the "cycle trick". The proof in words is simple. $\mu_x(y)$ is the expected number of visits to y in $\{0, \ldots, T-1\}$. $\mu_x p(y) \equiv \sum \mu_x(z) p(z, y)$ is the expected number of visits to y in $\{1, \ldots, T\}$, which is $= \mu_x(y)$ since

$X_T = X_0 = x$. To translate this intuition into a proof, let $\bar{p}_n(x, y) = P_x(X_n = y, T > n)$ and use Fubini's theorem to get

$$\sum_y \mu_x(y)p(y, z) = \sum_{n=0}^{\infty} \sum_y \bar{p}_n(x, y)p(y, z)$$

Case 1. $z \neq x$.

$$\sum_y \bar{p}_n(x, y)p(y, z) = \sum_y P_x(X_n = y, T > n, X_{n+1} = z)$$
$$= P_x(T > n + 1, X_{n+1} = z) = \bar{p}_{n+1}(x, z)$$

so

$$\sum_{n=0}^{\infty} \sum_y \bar{p}_n(x, y)p(y, z) = \sum_{n=0}^{\infty} \bar{p}_{n+1}(x, z) = \mu_x(z)$$

since $\bar{p}_0(x, z) = 0$.

Case 2. $z = x$.

$$\sum_y \bar{p}_n(x, y)p(y, x) = \sum_y P_x(X_n = y, T > n, X_{n+1} = x) = P_x(T = n + 1)$$

so

$$\sum_{n=0}^{\infty} \sum_y \bar{p}_n(x, y)p(y, x) = \sum_{n=0}^{\infty} P_x(T = n + 1) = 1 = \mu_x(x)$$

since $P_x(T = 0) = 0$. $\square$

Remark. If x is transient then we have $\mu_x p(z) \leq \mu_x(z)$ with equality for all $z \neq x$.

Technical Note. To show that we are not cheating, we should prove that $\mu_x(y) < \infty$ for all y. First observe that $\mu_x p = \mu_x$ implies $\mu_x p^n = \mu_x$ for all $n \geq 1$, and $\mu_x(x) = 1$, so if $p^n(y, x) > 0$ then $\mu_x(y) < \infty$. Since the last result is true for all n, we see that $\mu_x(y) < \infty$ whenever $\rho_{yx} > 0$, but this is good enough. By (3.3), when x is recurrent $\rho_{xy} > 0$ implies $\rho_{yx} > 0$ and it follows from the argument above that $\mu_x(y) < \infty$. If $\rho_{xy} = 0$ then $\mu_x(y) = 0$.

EXERCISE 4.2. (i) Use the construction in the proof of (4.3) to show that $\mu(j) = \sum_{k \geq j} f_{k+1}$ defines a stationary measure for the renewal chain (Example 1.4). (ii) Show that in this case the dual Markov chain defined in Exercise 4.1

represents the age of the item at use at time n, i.e., the amount of time since the last renewal $\leq n$.

(4.3) allows us to construct a stationary measure for each closed set of recurrent states. Conversely, we have

(4.4) Theorem. If p is irreducible and recurrent (i.e., all states are) then the stationary measure is unique up to constant multiples.

Proof   Let ν be a stationary measure and let $a \in S$.

$$\nu(z) = \sum_y \nu(y)p(y,z) = \nu(a)p(a,z) + \sum_{y \neq a} \nu(y)p(y,z)$$

Using the last identity to replace $\nu(y)$ on the right-hand side,

$$\nu(z) = \nu(a)p(a,z) + \sum_{y \neq a} \nu(a)p(a,y)p(y,z)$$
$$+ \sum_{x \neq a} \sum_{y \neq a} \nu(x)p(x,y)p(y,z)$$
$$= \nu(a)P_a(X_1 = z) + \nu(a)P_a(X_1 \neq a, X_2 = z)$$
$$+ P_\nu(X_0 \neq a, X_1 \neq a, X_2 = z)$$

Continuing in the obvious way we get

$$\nu(z) = \nu(a) \sum_{m=1}^n P_a(X_k \neq a, 1 \leq k < m, X_m = z)$$
$$+ P_\nu(X_j \neq a, 0 \leq j < n, X_n = z)$$

The last term is ≥ 0. Letting $n \to \infty$ gives $\nu(z) \geq \nu(a)\mu_a(z)$ where μ_a is the measure defined in (4.3) for $x = a$. It follows from (4.3) that μ is a stationary distribution with $\mu(a) = 1$. (Here we are summing from 1 to T rather than from 0 to $T - 1$.). To turn the $\geq$ in the last equation into $=$ we observe

$$\nu(a) = \sum_x \nu(x)p^n(x,a) \geq \nu(a) \sum_x \mu_a(x)p^n(x,a) = \nu(a)\mu_a(a) = \nu(a)$$

Since $\nu(x) \geq \nu(a)\mu_a(x)$ and the left- and right-hand sides are equal we must have $\nu(x) = \nu(a)\mu_a(x)$ whenever $p^n(x,a) > 0$. Since p is irreducible, it follows that $\nu(x) = \nu(a)\mu_a(x)$ for all $x \in S$ and the proof is complete. $\square$

(4.3) and (4.4) make a good team. (4.3) gives us a formula for a stationary distribution we call μ_x and (4.4) shows it is unique up to constant multiples. Together they allow us to derive a lot of formulas.

EXERCISE 4.3. Show that if p is irreducible and recurrent then

$$\mu_x(y)\mu_y(z) = \mu_x(z)$$

EXERCISE 4.4. Use (4.3) and (4.4) to show that for simple random walk, the expected number of visits to k between succesive visits to 0 is 1 for all k.

EXERCISE 4.5. Let $w_{xy} = P_x(T_y < T_x)$. Show that

$$\mu_x(y) = w_{xy}/w_{yx}$$

and use this to prove the result in the last exercise.

EXERCISE 4.6. **Another proof of (4.4).** Suppose p is irreducible and recurrent and let μ be the stationary measure constructed in (4.3). $\mu(x) > 0$ for all x and

$$q(x, y) = \mu(y)p(y, x)/\mu(x) \geq 0$$

defines a "dual" transition probability. (See Exercise 4.1.) (i) Show that q is irreducible and recurrent. (ii) Suppose $\nu(y) \geq \sum_x \nu(x)p(x, y)$ (i.e, ν is an **excessive measure**) and let $h(x) = \nu(x)/\mu(x)$. Verify that $h(y) \geq \sum q(y, x)h(x)$ and use Exercise 3.9 to conclude that h is constant, i.e., $\nu = c\mu$.

Remark. The last result is stronger than (4.4) since it shows that in the recurrent case any excessive measure is a constant multiple of one stationary measure. The remark after the proof of (4.3) shows that if p is irreducible and transient there is an excessive measure for each $x \in S$.

Having examined the existence and uniqueness of stationary measures, we turn our attention now to **stationary distributions**, i.e., probability measures π with $\pi p = \pi$. Stationary measures may exist for transient chains (e.g., random walks in $d \geq 3$) but

(4.5) **Theorem.** If there is a stationary distribution then all states y that have $\pi(y) > 0$ are recurrent.

Proof   Since $\pi p^n = \pi$, Fubini's theorem implies

$$\sum_x \pi(x) \sum_{n=1}^{\infty} p^n(x, y) = \sum_{n=1}^{\infty} \pi(y) = \infty$$

when $\pi(y) > 0$. Using (3.2) now gives

$$\infty = \sum_x \pi(x) \frac{\rho_{xy}}{1 - \rho_{yy}} \leq \frac{1}{1 - \rho_{yy}}$$

since $\rho_{xy} \leq 1$ and π is a probability measure. So $\rho_{yy} = 1$. $\qquad\square$

(4.6) Theorem. If p is irreducible and has stationary distribution π, then

$$\pi(x) = 1/E_x T_x$$

Remark. Recycling Chung's quote regarding (5.8) in Chapter 4, we note that the proof will make $\pi(x) = 1/E_x T_x$ obvious but it seems incredible that

$$\sum_x \frac{1}{E_x T_x} p(x, y) = \frac{1}{E_y T_y}$$

Proof Irreducibility implies $\pi(x) > 0$ so all states are recurrent by (4.5). From (4.3)

$$\mu_x(y) = \sum_{n=0}^{\infty} P_x(X_n = y, T_x > n)$$

defines a stationary measure with $\mu_x(x) = 1$ and Fubini's theorem implies

$$\sum_y \mu_x(y) = \sum_{n=0}^{\infty} P_x(T_x > n) = E_x T_x$$

By (4.4), the stationary measure is unique up to constant multiples so $\pi(x) = \mu_x(x)/E_x T_x$. Since $\mu_x(x) = 1$ by definition the desired result follows. $\qquad\square$

If a state x has $E_x T_x < \infty$, it is said to be **positive recurrent**. A recurrent state with $E_x T_x = \infty$ is said to be **null recurrent**. (5.1) will explain these names. The next result helps us identify positive recurrent states.

(4.7) Theorem. If p is irreducible then the following are equivalent:

(i) Some x is positive recurrent.

(ii) There is a stationary distribution.

(iii) All states are positive recurrent.

Proof   (i) implies (ii)   If x is positive recurrent then

$$\pi(y) = \sum_{n=0}^{\infty} P_x(X_n = y, T_x > n)/E_x T_x$$

defines a stationary distribution.

(ii) implies (iii)   (4.6) implies $\pi(y) = 1/E_y T_y$ and irreducibility tells us $\pi(y) > 0$ for all y, so $E_y T_y < \infty$.

(iii) implies (i)   Trivial.   □

EXERCISE 4.7. **Renewal chain.** Show that an irreducible renewal chain (Example 1.4) is positive recurrent (i.e., all the states are) if and only if $\mu = \sum k f_k < \infty$.

EXERCISE 4.8. Suppose p is irreducible and positive recurrent. Then $E_x T_y < \infty$ for all x, y.

EXERCISE 4.9. Suppose p is irreducible and has a stationary measure μ with $\sum_x \mu(x) = \infty$. Then p is not positive recurrent.

(4.7) shows that being positive recurrent is a **class property**. If it holds for one state in an irreducible set then it is true for all. Turning to our examples. Since $\mu(x) \equiv 1$ is a stationary measure Exercise 4.9 implies that random walks (Examples 4.1 and 4.2) are never positive recurrent. The Ehrenfest chain (Example 4.3) is positive recurrent. To see this note that the state space is finite so there is a stationary distribution and the conclusion follows from (4.7).

Birth and death chains (Example 4.4) have a stationary distribution if and only if

$$\sum_x \prod_{k=1}^{x} \frac{p_{k-1}}{q_k} < \infty$$

By (3.8), the chain is recurrent if and only if

$$\sum_{m=0}^{\infty} \prod_{j=1}^{m} \frac{q_j}{p_j} = \infty$$

When $p_j = p$ and $q_j = (1-p)$ for $j \geq 1$, there is a stationary distribution if and only if $p < 1/2$ and the chain is transient when $p > 1/2$. In Section 3 we probed the boundary between recurrence and transience by looking at examples with $p_j = 1/2 + \epsilon_j$ where $\epsilon_j \sim C\, j^{-\alpha}$ as $j \to \infty$ and $C, \alpha \in (0, \infty)$. Since $\epsilon_j \geq 0$ and hence $p_{j-1}/q_j \geq 1$ for large j, none of these chains have stationary distributions.

If we look at chains with $p_j = 1/2 - \epsilon_j$ then all we have done is interchange the roles of p and q and results from the last section imply that the chain is positive recurrent when $\alpha < 1$, or $\alpha = 1$ and $C > 1/4$.

Random walks on graphs (Example 4.5) are irreducible if and only if the graph is connected. Since $\mu(i) \geq 1$ in the connected case, we have positive recurrence if and only if the graph is finite.

EXERCISE 4.10. Compute the expected number of moves it takes a knight to return to its initial position if it starts in a corner of the chessboard, assuming there are no other pieces on the board, and each time it chooses a move at random from its legal moves. (Note: A chessboard is $\{0, 1, \ldots, 7\}^2$. A knight's move is L-shaped; two steps in one direction followed by one step in a perpindicular direction.)

Example 4.6. $M/G/1$ **queue.** Let $\mu = \sum k a_k$ be the mean number of customers that arrive during one service time. In Example 3.7 we showed that the chain is recurrent if and only if $\mu \leq 1$. We will now show that the chain is positive recurrent if and only if $\mu < 1$. First suppose that $\mu < 1$. When $X_n > 0$, the chain behaves like a random walk that has jumps with mean $\mu - 1$, so if $N = \inf\{n \geq 0 : X_n = 0\}$ then $X_{N \wedge n} - (\mu - 1)(N \wedge n)$ is a martingale. If $X_0 = x > 0$ then the martingale property implies

$$x = E_x X_{N \wedge n} + (1 - \mu) E_x(N \wedge n) \geq (1 - \mu) E_x(N \wedge n)$$

since $X_{N \wedge n} \geq 0$ and it follows that $E_x N \leq x/(1 - \mu)$.

To prove that there is equality observe that X_n decreases by at most one each time and for $x \geq 1$, $E_x T_{x-1} = E_1 T_0$, so $E_x N = cx$. To identify the constant observe that

$$E_1 N = 1 + \sum_{k=0}^{\infty} a_k E_k N$$

so $c = 1 + \mu c$ and $c = 1/(1 - \mu)$. If $X_0 = 0$ then $p(0, 0) = a_0 + a_1$ and $p(0, k - 1) = a_k$ for $k \geq 2$. By considering what happens on the first jump we see that (the first term may look wrong but recall $k - 1 = 0$ when $k = 1$)

$$E_0 T_0 = 1 + \sum_{k=1}^{\infty} a_k \frac{k-1}{1-\mu} = 1 + \frac{\mu - (1 - a_0)}{1 - \mu} = \frac{a_0}{1 - \mu} < \infty$$

This shows that the chain is positive recurrent if $\mu < 1$. To prove the converse observe that the arguments above show that if $E_0 T_0 < \infty$ then $E_k N < \infty$ for all k, $E_k N = ck$, and $c = 1/(1 - \mu)$, which is imposssible if $\mu \geq 1$.

The last result when combined with (4.4) and (4.6) allows us to conclude that the stationary distribution has $\pi(0) = (1 - \mu)/a_0$. This may not seem like much, but the equations in $\pi p = \pi$ are:

$$\pi(0) = \pi(0)(a_0 + a_1) + \pi(1)a_0$$
$$\pi(1) = \pi(0)a_2 + \pi(1)a_1 + \pi(2)a_0$$
$$\pi(2) = \pi(0)a_3 + \pi(1)a_2 + \pi(2)a_1 + \pi(3)a_0$$

or, in general, for $j \geq 1$

$$\pi(j) = \sum_{i=0}^{j+1} \pi(i)a_{j+1-i}$$

The equations have a "triangular" form, so knowing $\pi(0)$ we can solve for $\pi(1), \pi(2), \ldots$ The first expression,

$$\pi(1) = \pi(0)(1 - (a_0 + a_1))/a_0$$

is simple but the formulas get progressively messier, and there is no nice closed form solution.

Example 4.7. $M/M/\infty$ **queue.** In this chain, introduced in Exercise 3.9,

$$X_{n+1} = \sum_{m=1}^{X_n} \xi_{n,m} + Y_{n+1}$$

where $\xi_{n,m}$ are i.i.d. Bernoulli with mean p and Y_{n+1} is an independent Poisson mean λ. It follows from properties of the Poisson distribution that if X_n is Poisson with mean μ then X_{n+1} is Poisson with mean $\mu p + \lambda$. Setting $\mu = \mu p + \lambda$ we find that a Poisson distribution with mean $\mu = \lambda/(1 - p)$ is a stationary distribution.

There is a general result that handles Examples 4.6 and 4.7, and is useful in a number of other situations. This will be developed in the next two exercises.

EXERCISE 4.11. Let $X_n \geq 0$ be a Markov chain and suppose $E_x X_1 \leq x - \epsilon$ for $x > K$ where $\epsilon > 0$. Let $Y_n = X_n + n\epsilon$ and $\tau = \inf\{n : X_n \leq K\}$. $Y_{n \wedge \tau}$ is a positive supermartingale and the optional stopping theorem implies $E_x \tau \leq x/\epsilon$.

EXERCISE 4.12. Suppose that X_n has state space $\{0, 1, 2, \ldots\}$, the conditions of the last exercise hold when $K = 0$, and $E_0 X_1 < \infty$. Then 0 is positive recurrent. We leave it to the reader to formulate and prove a similar result when $K > 0$.

To close the section we will give a self-contained proof of

(4.8) Theorem. If p is irreducible and has a stationary distribution π then any other stationary measure is a multiple of π.

Remark. This result is a consequence of (4.5) and (4.4) but we find the method of proof amusing.

Proof Since p is irreducible, $\pi(x) > 0$ for all x. Let φ be a concave function that is bounded on $(0, \infty)$, e.g., $\varphi(x) = x/(x+1)$. Define the **entropy** of μ by

$$\mathcal{E}(\mu) = \sum_y \varphi\left(\frac{\mu(y)}{\pi(y)}\right) \pi(y)$$

The reason for the name will become clear during the proof.

$$\mathcal{E}(\mu p) = \sum_y \varphi\left(\sum_x \frac{\mu(x)p(x,y)}{\pi(y)}\right) \pi(y) = \sum_y \varphi\left(\sum_x \frac{\mu(x)}{\pi(x)} \cdot \frac{\pi(x)p(x,y)}{\pi(y)}\right) \pi(y)$$

$$\geq \sum_y \sum_x \varphi\left(\frac{\mu(x)}{\pi(x)}\right) \frac{\pi(x)p(x,y)}{\pi(y)} \pi(y)$$

since φ is concave and $\nu(x) = \pi(x)p(x,y)/\pi(y)$ is a probability distribution. Since the $\pi(y)$'s cancel and $\sum_y p(x,y) = 1$, the last expression $= \mathcal{E}(\mu)$ and we have shown $\mathcal{E}(\mu p) \geq \mathcal{E}(\mu)$, i.e., the entropy of an arbitrary initial measure μ is increased by an application of p.

If $p(x,y) > 0$ for all x and y, and $\mu p = \mu$, it follows that $\mu(x)/\pi(x)$ must be constant for otherwise there would be strict inequality in the application of Jensen's inequality. To get from the last special case to the general result, observe that if p is irreducible

$$\bar{p}(x,y) = \sum_{n=1}^{\infty} 2^{-n} p^n(x,y) > 0 \quad \text{for all } x, y$$

and $\mu p = \mu$ implies $\mu \bar{p} = \mu$. $\qquad\qquad\qquad\square$

5.5. Asymptotic Behavior

In this section we will investigate the asymptotic behavior of X_n and $p^n(x,y)$.

a. Convergence theorems

If y is transient $\sum_n p^n(x, y) < \infty$, so $p^n(x, y) \to 0$ as $n \to \infty$. To deal with the recurrent states, we let

$$N_n(y) = \sum_{m=1}^{n} 1_{\{X_m = y\}}$$

be the number of visits to y by time n.

(5.1) Theorem. Suppose y is recurrent. For any $x \in S$, as $n \to \infty$

$$\frac{N_n(y)}{n} \to \frac{1}{E_y T_y} 1_{\{T_y < \infty\}} \quad P_x\text{-a.s.}$$

Here $1/\infty = 0$.

Proof Suppose first that we start at y. Let $R(k) = \min\{n \geq 1 : N_n(y) = k\} =$ the time of the kth return to y. Let $t_k = R(k) - R(k-1)$ where $R(0) = 0$. Since we have assumed $X_0 = y$, $t_1, t_2, \ldots$ are i.i.d. and the strong law of large numbers implies

$$R(k)/k \to E_y T_y \quad \text{a.s.}$$

Since $R(N_n(y)) \leq n < R(N_n(y) + 1)$,

$$\frac{R(N_n(y))}{N_n(y)} \leq \frac{n}{N_n(y)} < \frac{R(N_n(y) + 1)}{N_n(y) + 1} \cdot \frac{N_n(y) + 1}{N_n(y)}$$

Letting $n \to \infty$, and recalling $N_n(y) \to \infty$ a.s. since y is recurrent, we have

$$\frac{n}{N_n(y)} \to E_y T_y \quad \text{a.s.}$$

To generalize now to $x \neq y$, observe that if $T_y = \infty$ then $N_n(y) = 0$ for all n and hence

$$N_n(y)/n \to 0 \quad \text{on } \{T_y = \infty\}$$

The strong Markov property implies that conditional on $\{T_y < \infty\}$, $t_2, t_3, \ldots$ are i.i.d. and have $P_x(t_k = n) = P_y(T_y = n)$ so

$$R(k)/k = t_1/k + (t_2 + \cdots + t_k)/k \to 0 + E_y T_y \quad \text{a.s.}$$

Repeating the proof for the case $x = y$ shows

$$N_n(y)/n \to 1/E_y T_y \quad \text{a.s. on } \{T_y < \infty\}$$

and combining this with the result for $\{T_y = \infty\}$ completes the proof. $\qquad\square$

Remark. (5.1) should help explain the terms positive and null recurrent. If we start from x then in the first case the asymptotic fraction of time spent at x is positive and in the second case it is 0.

Since $0 \leq N_n(y)/n \leq 1$, it follows from the bounded convergence theorem that $E_x N_n(y)/n \to E_x(1_{\{T_y < \infty\}}/E_y T_y)$ so

$$(5.2) \qquad \frac{1}{n} \sum_{m=1}^{n} p^m(x, y) \to \rho_{xy}/E_y T_y$$

The last result was proved for recurrent y but also holds for transient y, since in that case $E_y T_y = \infty$ and the limit is 0, since $\sum_m p^m(x, y) < \infty$.

(5.2) shows that the sequence $p^n(x, y)$ always converges in the Cesaro sense. The next example shows that $p^n(x, y)$ need not converge.

Example 5.1.

$$p = \begin{pmatrix} 0 & 1 \\ 1 & 0 \end{pmatrix} \quad p^2 = \begin{pmatrix} 1 & 0 \\ 0 & 1 \end{pmatrix} \quad p^3 = p, \quad p^4 = p^2, \ldots$$

A similar problem also occurs in the Ehrenfest chain. In that case if X_0 is even, then X_1 is odd, X_2 is even, $\ldots$ so $p^n(x, x) = 0$ unless n is even. It is easy to construct examples with $p^n(x, x) = 0$ unless n is a multiple of 3 or 17 or $\ldots$

(5.5) below will show that this "periodicity" is the only thing that can prevent the convergence of the $p^n(x, y)$. First, we need a definition and two preliminary results. Let x be a recurrent state, let $I_x = \{n \geq 1 : p^n(x, x) > 0\}$, and let d_x be the greatest common divisor of I_x. d_x is called the **period** of x. The first result says that the period is a class property.

(5.3) Lemma. If $\rho_{xy} > 0$ then $d_y = d_x$.

Proof Let K and L be such that $p^K(x, y) > 0$ and $p^L(y, x) > 0$. (x is recurrent so $\rho_{yx} > 0$.)

$$p^{K+L}(y, y) \geq p^L(y, x) p^K(x, y) > 0$$

so d_y divides $K + L$, abbreviated $d_y|(K + L)$. Let n be such that $p^n(x, x) > 0$.

$$p^{K+n+L}(y, y) \geq p^L(y, x) p^n(x, x) p^K(x, y) > 0$$

so $d_y|(K+n+L)$, and hence $d_y|n$. Since $n \in I_x$ is arbitrary, $d_y|d_x$. Interchanging the roles of y and x gives $d_x|d_y$, and hence $d_x = d_y$. $\qquad\square$

The next result implies that $I_x \supset \{m \cdot d_x : m \geq m_0\}$. (Apply (5.4) to $p^{d(x)}$.)

(5.4) Lemma. If $d_x = 1$ then $p^m(x, x) > 0$ for $m \geq m_0$.

Proof by example Suppose $4, 7 \in I_x$. $p^{m+n}(x, x) \geq p^m(x, x)p^n(x, x)$ so I_x is closed under addition, i.e., if $m, n \in I_x$ then $m + n \in I_x$. A little calculation shows that in the example

$$I_x \supset \{4, \quad 7, 8, \quad 11, 12, \quad 14, 15, 16, \quad 18, 19, 20, 21, \ldots\}$$

so the result is true with $m_0 = 18$. (Once I_x contains four consecutive integers, it will contain all the rest.)

Proof Our first goal is to prove that I_x contains two consecutive integers. Let n_0, $n_0 + k \in I_x$. If $k = 1$ we are done. If not, then since the greatest common divisor of I_x is 1, there is an $n_1 \in I_x$ so that k is not a divisor of n_1. Write $n_1 = mk + r$ with $0 < r < k$. Since I_x is closed under addition, $(m + 1)(n_0 + k) > (m + 1)n_0 + n_1$ are both in I_x. Their difference is

$$k(m + 1) - n_1 = k - r < k$$

Repeating the last argument (at most k times) we eventually arrive at a pair of consecutive integers $N, N + 1 \in I_x$. It is now easy to show that the result holds for $m_0 = N^2$. Let $m \geq N^2$ and write $m - N^2 = kN + r$ with $0 \leq r < N$. Then

$$m = r + N^2 + kN = r(1 + N) + (N - r + k)N \in I_x \qquad \square$$

(5.5) Convergence theorem. Suppose p is irreducible, aperiodic (i.e., all states have $d_x = 1$), and has stationary distribution π. Then as $n \to \infty$, $p^n(x, y) \to \pi(y)$.

Proof Let $S^2 = S \times S$. Define a transition probability $\bar{p}$ on $S \times S$ by

$$\bar{p}((x_1, y_1), (x_2, y_2)) = p(x_1, x_2)p(y_1, y_2)$$

i.e., each coordinate moves independently. Our first step is to check that $\bar{p}$ is irreducible. This may seem like a silly thing to do first, but this is the only step that requires aperiodicity. Since p is irreducible, there are K, L, so that $p^K(x_1, x_2) > 0$ and $p^L(y_1, y_2) > 0$. From (5.4) it follows that if M is large $p^{L+M}(x_2, x_2) > 0$ and $p^{K+M}(y_2, y_2) > 0$, so

$$\bar{p}^{K+L+M}((x_1, y_1), (x_2, y_2)) > 0$$

Our second step is to observe that since the two coordinates are independent $\bar{\pi}(a,b) = \pi(a)\pi(b)$ defines a stationary distribution for $\bar{p}$, and (4.5) implies that for $\bar{p}$ all states are recurrent. Let (X_n, Y_n) denote the chain on $S \times S$, and let T be the first time that this chain hits the diagonal $\{(y,y) : y \in S\}$. Let $T_{(x,x)}$ be the hitting time of (x,x). Since $\bar{p}$ is irreducible and recurrent, $T_{(x,x)} < \infty$ a.s. and hence $T < \infty$ a.s. The final step is to observe that on $\{T \le n\}$, the two coordinates X_n and Y_n have the same distribution. By considering the time and place of the first intersection and then using the Markov property.

$$P(X_n = y, T \le n) = \sum_{m=1}^{n} \sum_x P(T = m, X_m = x, X_n = y)$$
$$= \sum_{m=1}^{n} \sum_x P(T = m, X_m = x)P(X_n = y|X_m = x)$$
$$= \sum_{m=1}^{n} \sum_x P(T = m, Y_m = x)P(Y_n = y|Y_m = x)$$
$$= P(Y_n = y, T \le n)$$

To finish up we observe that

$$P(X_n = y) = P(Y_n = y, T \le n) + P(X_n = y, T > n)$$
$$\le P(Y_n = y) + P(X_n = y, T > n)$$

and similarly $P(Y_n = y) \le P(X_n = y) + P(Y_n = y, T > n)$. So

$$|P(X_n = y) - P(Y_n = y)| \le P(X_n = y, T > n) + P(Y_n = y, T > n)$$

and summing over y gives

$$\sum_y |P(X_n = y) - P(Y_n = y)| \le 2P(T > n)$$

If we let $X_0 = x$ and let Y_0 have the stationary distribution π, then Y_n has distribution π and it follows that

$$\sum_y |p^n(x,y) - \pi(y)| \le 2P(T > n) \to 0$$

proving the desired result. If we recall the definition of the total variation distance given in Section 2.6, the last conclusion can be written as

$$\|p^n(x,\cdot) - \pi(\cdot)\| \le P(T > n) \to 0 \qquad \square$$

At first glance it may seem strange to prove the convergence theorem by running independent copies of the chain. An approach that is slightly more complicated, but explains better what is happening is to define

$$q((x_1, y_1), (x_2, y_2)) = \begin{cases} p(x_1, x_2)p(y_1, y_2) & \text{if } x_1 \neq y_1 \\ p(x_1, x_2) & \text{if } x_1 = y_1,\ x_2 = y_2 \\ 0 & \text{otherwise} \end{cases}$$

In words, the two coordinates move independently until they hit and then move together. It is easy to see from the definition that each coordinate is a copy of the original process. If T' is the hitting time of the diagonal for the new chain (X_n', Y_n'), then $X_n' = Y_n'$ on $T' \leq n$, so it is clear that

$$\sum_y |P(X_n' = y) - P(Y_n' = y)| \leq 2\, P(X_n' \neq Y_n') = 2P(T' > n)$$

On the other hand T and T' have the same distribution so $P(T' > n) \to 0$ and the conclusion follows as before. The technique used in the last proof is called **coupling**. Generally this term refers to building two sequences X_n and Y_n on the same space to conclude that X_n converges in distribution by showing $P(X_n \neq Y_n) \to 0$ or more generally that for some metric ρ, $\rho(X_n, Y_n) \to 0$ in probability.

Having completed the proof of (5.5) we pause to show that coupling can be fun and it doesn't always happen.

Example 5.2. A coupling card trick. The following demonstration used by E.B. Dynkin in his probability class is a variation of a card trick that appeared in Scientific American. The instructor asks a student to write 100 random digits from 0 to 9 on the blackboard. Another student chooses one of the first 10 numbers and does not tell the instructor. If that digit is 7 say she counts 7 places along the list, notes the digit at that location, and continues the process. If the digit is 0 she counts 10. A possible sequence is underlined on the list below:

$$3\ 4\ \underline{7}\ 8\ 2\ 3\ 7\ 5\ 6\ \underline{1}\ \underline{6}\ 4\ 6\ 5\ 7\ 8\ \underline{3}\ 1\ 5\ \underline{3}\ 0\ 7\ \underline{9}\ 2\ 3 \ldots$$

The trick is that, without knowing the student's first digit, the instructor can point to her final stopping position. To this end, he picks the first digit, and forms his own sequence in the same manner as the student and announces his stopping position. He makes an error if the coupling time is larger than 100. Numerical computations done by one of Dynkin's graduate students show that the probability of error is approximately .026

Example 5.3. There is a transition probability that is irreducible, aperiodic, and recurrent but has the property that 2 independent particles need not meet.

Let S_n be a modified two dimensional simple random walk that stays where it is with probability 1/5 and jumps to each of its four neighbors with probability 1/5 each. Let $T = \inf\{n \geq 1 : S_n = (0,0)\}$ and $f_k = P(T = k)$. Let p be the transition probability with

$$p(0,j) = f_{j+1}, \quad p(i,i-1) = 1, \quad p(i,j) = 0 \text{ otherwise}$$

p is the renewal chain corresponding to the distribution f, so p is recurrent. $f_k > 0$ for all k so p is irreducible and aperiodic. Let S_n^1 and S_n^2 be independent copies of the random walk starting from $S_0^1 \neq S_0^2$, and let $X_n^i = \inf\{m - n : m \geq n, S_m^i = (0,0)\}$. From our discussion of the renewal chain in Example 1.4, it follows that X_n^1 and X_n^2 are independent Markov chains with transition probability p. It is easy to see that if $X_n^1 = X_n^2 = j$ then $X_{n+j}^1 = X_{n+j}^2 = 0$, and at this time the four dimensional random walk (S_n^1, S_n^2) is at $(0,0,0,0)$. Since

$$P((S_n^1, S_n^2) = (0,0,0,0) \quad \text{for some } n \geq 1) < 1$$

we have proved our claim.

(5.5) applies almost immediately to the examples considered in Section 1. The $M/G/1$ queue has $a_k > 0$ for all $k \geq 0$ so if $\mu < 1$, $P_x(X_n = y) \to \pi(y)$ for any $x, y \geq 0$. The same result holds for the renewal chain provided $\{k : f_k > 0\}$ is unbounded (for irreducibility) and its greatest common divisor is 1. In this case $P_x(X_n = 0) \to \pi(0) = 1/\nu$ where $\nu = \sum k f_k$ is the mean time between renewals.

EXERCISE 5.1. Historically the first chain for which (5.5) was proved was the **Bernoulli-Laplace model of diffusion**. Suppose 2 urns, which we will call left and right, have m balls each. b (which we will assume is $\leq m$) balls are black and $2m - b$ are white. At each time we pick one ball from each urn and interchange them. Let the state at time n be the number of black balls in the left urn. Compute the transition probability for this chain, find its stationary distribution, and use (5.5) to conclude that the chain approaches equilibrium as $n \to \infty$.

And now for something completely different:

Example 5.4. Shuffling cards. The state of a deck of n cards can be represented by a permutation, $\pi(i)$ giving the location of the ith card. Consider the following method of mixing the deck up. The top card is removed and inserted under one of the $n - 1$ cards that remain. I claim that by following the bottom card of the deck we can see that it takes about $n \log n$ moves to mix up the deck. This card stays at the bottom until the first time (T_1) a card is inserted below it. It is easy to see that when the kth card is inserted below the original bottom card (at time T_k), all $k!$ arrangements of the cards below are equally likely, so at time $\tau_n = T_{n-1} + 1$ all $n!$ arrangements are equally likely. If we let $T_0 = 0$ and $t_k = T_k - T_{k-1}$ for $1 \le k \le n-1$, then these r.v.'s are independent and t_k has a geometric distribution with success probability $k/(n-1)$. These waiting times are the same as the ones in the coupon collector's problem (Example 5.3 in Chapter 1), so $\tau_n/(n \log n) \to 1$ in probability as $n \to \infty$. For more on card shuffling, see Aldous and Diaconis (1986).

Example 5.5. Random walk on the hypercube. Consider $\{0,1\}^d$ as a graph with edges connecting each pair of points that differ in only one coordinate. Let X_n be a random walk on $\{0,1\}^d$ that stays put with probability $1/2$ and jumps to one of its d neighbors with probability $1/2d$ each. Let Y_n be another copy of the chain in which Y_0 (and hence Y_n, $n \ge 1$) is uniformly distributed on $\{0,1\}^d$. We construct a coupling of X_n and Y_n by letting $U_1, U_2, \ldots$ be uniform on $\{1, 2, \ldots, 2d\}$. At time n the jth coordinates of X and Y are each set equal to 1 if $U_n = 2j - 1$ and are each set equal to 0 if $U_n = 2j$. The other coordinates are unchanged. Let $T_d = \inf\{m : \{U_1, \ldots, U_m\} = \{1, 2, \ldots, 2d\}\}$. When $n \ge T_d$, $X_n = Y_n$. Results for the coupon collectors problem (Example (5.10) in Chapter 1) show that $T_d/(d \log d) \to 1$ in probability as $d \to \infty$.

*b. Periodic case

(5.6) Lemma. Suppose p is irreducible, recurrent, and all states have period d. Fix $x \in S$ and for each $y \in S$ let $K_y = \{n \ge 1 : p^n(x,y) > 0\}$. (i) There is an $r_y \in \{0, 1, \ldots, d-1\}$ so that if $n \in K_y$ then $n = r_y \bmod d$, i.e., the difference $n - r_y$ is a multiple of d. (ii) Let $S_r = \{y : r_y = r\}$ for $0 \le r < d$. If $y \in S_i$, $z \in S_j$, and $p^n(y,z) > 0$ then $n = (j - i) \bmod d$. (iii) $S_0, S_1, \ldots, S_{d-1}$ are irreducible classes for p^d, and all states have period 1.

Proof   (i) Let $m(y)$ be such that $p^{m(y)}(y,x) > 0$. If $n \in K_y$ then $p^{n+m(y)}(x,x)$ is positive so $d|(n+m)$. Let $r_y = (d - m(y)) \bmod d$. (ii) Let m, n be such that $p^n(y,z)$, $p^m(x,y) > 0$. Since $p^{n+m}(x,z) > 0$ it follows from (i) that $n + m = j \bmod d$. Since $m = i \bmod d$, the result follows. The irreducibility in (iii) follows immediately from (ii). The aperiodicity follows from the definition of the period as the g.c.d. $\{x : p^n(x,x) > 0\}$. □

A partition of the state space $S_0, S_1, \ldots, S_d$ satisfying (ii) in (5.6) is called a **cyclic decomposition** of the state space. Except for the choice of the set to put first, it is unique. (Pick an $x \in S$. It lies in some S_j but once the value of j is known, irreducibility and (ii) allow us to calculate all the sets.)

EXERCISE 5.2. Find the decomposition for the Markov chain with transition probability

	1	2	3	4	5	6	7
1	0	0	0	.5	.5	0	0
2	.3	0	0	0	0	0	.7
3	0	0	0	0	0	0	1
4	0	0	1	0	0	0	0
5	0	0	1	0	0	0	0
6	0	1	0	0	0	0	0
7	0	0	0	.4	0	.6	0

(5.7) Convergence theorem, periodic case. Suppose p is irreducible, has a stationary distribution π, and all states have period d. Let $x \in S$, and let $S_0, S_1, \ldots, S_d$ be the cyclic decomposition of the state space with $x \in S_0$. If $y \in S_r$ then

$$\lim_{m \to \infty} p^{md+r}(x, y) = \pi(y)d$$

Proof If $y \in S_0$ then using (iii) in (5.6) and applying (5.5) to p^d shows

$$\lim_{m \to \infty} p^{md}(x, y) \text{ exists}$$

To identify the limit we note that (5.2) implies

$$\frac{1}{n} \sum_{m=1}^{n} p^m(x, y) \to \pi(y)$$

and (ii) of (5.6) implies $p^m(x, y) = 0$ unless $d|m$, so the limit in the first display must be $\pi(y)d$. If $y \in S_r$ with $1 \le r < d$ then

$$p^{md+r}(x, y) = \sum_{z \in S_r} p^r(x, z)p^{md}(z, y)$$

Since $y, z \in S_r$ it follows from the first case in the proof that $p^{md}(z, y) \to \pi(y)d$ as $m \to \infty$. $p^{md}(z, y) \le 1$, and $\sum_z p^r(x, z) = 1$, so (5.7) follows from the dominated convergence theorem. $\square$

*c. Tail σ-field

Let $\mathcal{F}'_n = \sigma(X_{n+1}, X_{n+2}, \ldots)$ and $\mathcal{T} = \cap_n \mathcal{F}'_n$ be the tail σ-field. The next result is due to Orey. The proof we give is from Blackwell and Freedman (1964).

(5.8) Theorem. Suppose p is irreducible, recurrent, and all states have period d, $\mathcal{T} = \sigma(\{X_0 \in S_r\} : 0 \le r < d)$

Remark. To be precise, if μ is any initial distribution and $A \in \mathcal{T}$ then there is an r so that $A = \{X_0 \in S_r\}$ P_μ-a.s.

Proof We build up to the general result in three steps.

Case 1. Suppose $P(X_0 = x) = 1$. Let $T_0 = 0$ and for $n \ge 1$ let $T_n = \inf\{m > T_{n-1} : X_m = x\}$ be the time of the nth return to x. Let

$$V_n = (X(T_{n-1}), \ldots, X(T_n - 1))$$

The vectors V_n are i.i.d. by Exercise 3.1, and the tail σ-field is contained in the exchangeable field of the V_n, so the Hewitt-Savage 0-1 law ((1.1) in Chapter 3, proved there for r.v's taking values in a general measurable space) implies that $\mathcal{T}$ is trivial in this case.

Case 2. Suppose that the initial distribution is concentrated on one cyclic class, say S_0. If $A \in \mathcal{T}$ then $P_x(A) \in \{0,1\}$ for each x by case 1. If $P_x(A) = 0$ for all $x \in S_0$ then $P_\mu(A) = 0$. Suppose $P_y(A) > 0$, and hence $= 1$, for some $y \in S_0$. Let $z \in S_0$. Since p^d is irreducible and aperiodic on S_0, there is an n so that $p^n(z,y) > 0$ and $p^n(y,y) > 0$. If we write $1_A = 1_B \circ \theta_n$ then the Markov property implies

$$1 = P_y(A) = E_y(E_y(1_B \circ \theta_n | \mathcal{F}_n)) = E_y(E_{X_n} 1_B)$$

so $P_y(B) = 1$. Another application of the Markov property gives

$$P_z(A) = E_z(E_{X_n} 1_B) \ge p^n(z, y) > 0$$

so $P_z(A) = 1$, and since $z \in S_0$ is arbitrary $P_\mu(A) = 1$.

General Case. From case 2 we see that $P(A|X_0 = y) \equiv 1$ or $\equiv 0$ on each cyclic class. This implies that either $\{X_0 \in S_r\} \subset A$ or $\{X_0 \in S_r\} \cap A = \emptyset$ P_μ a.s. Conversely, it is clear that $\{X_0 \in S_r\} = \{X_{nd} \in S_r \text{ i.o.}\} \in \mathcal{T}$ and the proof is complete. $\square$

The next result will help us identify the tail σ-field in transient examples.

(5.9) Theorem. Suppose X_0 has initial distribution μ. The equations

$$h(X_n, n) = E_\mu(Z|\mathcal{F}_n) \quad \text{and} \quad Z = \lim_{n \to \infty} h(X_n, n)$$

set up a 1-1 correspondence between bounded $Z \in \mathcal{T}$, and bounded **space-time harmonic functions**, i.e., bounded $h : S \times \{0,1,\ldots\} \to \mathbf{R}$, so that $h(X_n, n)$ is a martingale.

Proof   Let $Z \in \mathcal{T}$, write $Z = Y_n \circ \theta_n$, and let $h(x, n) = E_x Y_n$.

$$E_\mu(Z|\mathcal{F}_n) = E_\mu(Y_n \circ \theta_n|\mathcal{F}_n) = h(X_n, n)$$

by the Markov property, so $h(X_n, n)$ is a martingale. Conversely if $h(X_n, n)$ is a bounded martingale, using (2.10) and (5.6) from Chapter 4 shows $h(X_n, n) \to Z \in \mathcal{T}$ as $n \to \infty$, and $h(X_n, n) = E_\mu(Z|\mathcal{F}_n)$. $\qquad\square$

EXERCISE 5.3. A random variable Z with $Z = Z \circ \theta$, and hence $= Z \circ \theta_n$ for all n, is called **invariant**. Show there is a 1-1 correspondence between bounded invariant random variables and bounded harmonic functions. We will have more to say about invariant r.v.'s in Section 6.1.

Example 5.6. Simple random walk in d dimensions. We begin by constructing a coupling for this process. Let $i_1, i_2, \ldots$ be i.i.d. uniform on $\{1, \ldots, d\}$. Let $\xi_1, \xi_2, \ldots$ and $\eta_1, \eta_2, \ldots$ be i.i.d. uniform on $\{-1, 1\}$. Let e_j be the jth unit vector. Construct a coupled pair of d-dimensional simple random walks by

$$X_n = X_{n-1} + e(i_n)\xi_n$$

$$Y_n = \begin{cases} Y_{n-1} + e(i_n)\xi_n & \text{if } X_{n-1}^{i_n} = Y_{n-1}^{i_n} \\ Y_{n-1} + e(i_n)\eta_n & \text{if } X_{n-1}^{i_n} \neq Y_{n-1}^{i_n} \end{cases}$$

In words, the coordinate that changes is always the same in the two walks and once they agree in one coordinate, future movements in that direction are the same. It is easy to see that if $X_0^i - Y_0^i$ is even for $1 \leq i \leq d$ then the two random walks will hit with probability one.

Let $L_0 = \{z \in \mathbf{Z}^d : z^1 + \cdots + z^d \text{ is even }\}$ and $L_1 = \mathbf{Z}^d - L_0$. Although we have only defined the notion for the recurrent case it should be clear that L_0, L_1 is the cyclic decomposition of the state space for simple random walk. If $S_n \in L_i$ then $S_{n+1} \in L_{1-i}$ and p^2 is irreducible on each L_i. To couple two random walks starting from $x, y \in L_i$, let them run independently until the first time all the coordinate differences are even, and then use the last coupling. In the remaining case $x \in L_0$, $y \in L_1$ coupling is impossible.

The next result should explain our interest in coupling two d dimensional simple random walks.

(5.10) **Theorem.** For d dimensional simple random walk,

$$\mathcal{T} = \sigma(\{X_0 \in L_i\}, i = 1, 2)$$

Proof Let $x, y \in L_i$, and let X_n, Y_n be a realization of the coupling defined above for $X_0 = x$ and $Y_0 = y$. Let $h(x, n)$ be a bounded space-time harmonic function. The martingale property implies $h(x, 0) = E_x h(X_n, n)$. If $|h| \leq C$, it follows from the coupling that

$$|h(x, 0) - h(y, 0)| = |Eh(X_n, n) - Eh(Y_n, n)| \leq 2CP(X_n \neq Y_n) \to 0$$

so $h(x, 0)$ is constant on L_0 and L_1. Applying the last result to $h'(x, m) = h(x, n + m)$ we see that $h(x, n) = a_n^i$ on L_i. The martingale property implies $a_n^i = a_{n+1}^{1-i}$ and the desired result follows from (5.9).

Example 5.7. Ornstein's coupling. Let $p(x, y) = f(y - x)$ be the transition probability for an irreducible aperiodic random walk on $\mathbf{Z}$. To prove that the tail σ-field is trivial, pick M large enough so that the random walk generated by the probability distribution $f_M(x)$ with $f_M(x) = c_M f(x)$ for $|x| \leq M$ and $f_M(x) = 0$ for $|x| > M$ is irreducible and aperiodic. Let $Z_1, Z_2, \ldots$ be i.i.d. with distribution f and let $W_1, W_2, \ldots$ be i.i.d. with distribution f_M. Let $X_n = X_{n-1} + Z_n$ for $n \geq 1$. If $X_{n-1} = Y_{n-1}$ we set $X_n = Y_n$. Otherwise we let

$$Y_n = \begin{cases} Y_{n-1} + Z_n & \text{if } |Z_n| > m \\ Y_{n-1} + W_n & \text{if } |Z_n| \leq m \end{cases}$$

In words, the big jumps are taken in parallel and the small jumps are independent. The recurrence of 1 dimensional random walks with mean 0 implies $P(X_n \neq Y_n) \to 0$. Repeating the proof of (5.10) we see that $\mathcal{T}$ is trivial.

The tail σ-field in (5.10) is essentially the same as in (5.8). To get a more interesting $\mathcal{T}$, we look at

Example 5.8. Random walk on a tree. To facilitate definitions, we will consider the system as a random walk on a group with 3 generators a, b, c which have $a^2 = b^2 = c^2 = e$, the identity element. To form the random walk, let $\xi_1, \xi_2, \ldots$ be i.i.d. with $P(\xi_n = x) = 1/3$ for $x = a, b, c$, and let $X_n = X_{n-1}\xi_n$. (This is equivalent to a random walk on the tree in which each vertex has degree 3 but the algebraic formulation is convenient for computations.) Let L_n be the length of the word X_n when it has been reduced as much as possible, with $L_n = 0$ if $X_n = e$. The reduction can be done as we go along. If the last letter of X_{n-1} is the same as ξ_n, we erase it, otherwise we add the new letter. It is easy to see that L_n is a Markov chain with a transition probability that has $p(0, 1) = 1$ and

$$p(j, j - 1) = 1/3 \qquad p(j, j + 1) = 2/3 \quad \text{for } j \geq 1$$

As $n \to \infty$, $L_n \to \infty$. From this it follows easily that the word X_n has a limit in the sense that the ith letter X_n^i stays the same for large n. Let X_∞ be the limiting word, i.e., $X_\infty^i = \lim X_n^i$. $\mathcal{T} \supset \sigma(X_\infty^i, i \geq 1)$ but it is easy to see that this is not all. If $S_0 =$ the words of even length, and $S_1 = S_0^c$ then $X_n \in S_i$ implies $X_{n+1} \in S_{1-i}$, so $\{X_0 \in S_0\} \in \mathcal{T}$. Can the reader prove that we have now found all of $\mathcal{T}$? As Fermat once said "I have a proof but it won't fit in the margin."

Remark. This time the solution does not involve elliptic curves but uses "h-paths." See Furstenburg (1970) or decode the following: "Condition on the exit point (the infinite word). Then the resulting RW is an h-process, which moves closer to the boundary with probability 2/3 and farther with probability 1/3 (1/6 each to the two possiblities). Two such random walks couple, provided they have same parity." The quote is from Robin Pemantle, who says he consulted Itai Benajamini and Yuval Peres.

EXERCISES

5.4. M/G/1 queue. Let $\xi_1, \xi_2, \ldots$ be i.i.d. with $P(\xi_m = k) = a_{k+1}$ for $k \geq -1$. Let $S_n = x + \xi_1 + \cdots + \xi_n$ where $x \geq 0$, and let

$$X_n = S_n + \left(\min_{m \leq n} S_m \right)^-$$

(i) Show that X_n has the same distribution as the $M/G/1$ queue (Example 1.5) starting from $X_0 = x$. (ii) Use this to conclude that if $\mu = \sum k a_k < 1$ then as $n \to \infty$

$$\frac{1}{n} |\{m \leq n : X_{m-1} = 0, \xi_n = -1\}| \to (1 - \mu) \quad \text{a.s.}$$

This gives a roundabout way of getting the result $\pi(0) = (1 - \mu)/a_0$ proved in Example 4.6.

5.5. Strong law for additive functionals. Suppose p is irreducible and has stationary distribution π. Let f be a function that has $\sum |f(y)| \pi(y) < \infty$. Let T_x^k be the time of the kth return to x. (i) Show that

$$V_k^f = f(X(T_x^k)) + \cdots + f(X(T_x^{k+1} - 1)), \quad k \geq 1 \text{ are i.i.d.}$$

with $E|V_k^f| < \infty$. (ii) Let $K_n = \inf\{k : T_x^k \geq n\}$ and show that

$$\frac{1}{n} \sum_{m=1}^{K_n} V_m^f \to \frac{E V_1^f}{E_x T_x^1} = \sum f(y) \pi(y) \quad P_\mu - \text{a.s.}$$

(iii) Show that $\max_{1 \leq m \leq n} V_m^{|f|}/n \to 0$ and conclude

$$\frac{1}{n} \sum_{m=1}^{n} f(X_m) \to \sum_{y} f(y)\pi(y) \quad P_\mu - \text{a.s.}$$

for any initial distribution μ.

5.6. Central limit theorem for additive functionals. Suppose in addition to the conditions in the Exercise 5.5 that $\sum f(y)\pi(y) = 0$, and $E_x(V_k^{|f|})^2 < \infty$. (i) Use the random index central limit theorem (Exercise 4.6 in Chapter 2) to conclude that for any initial distribution μ

$$\frac{1}{\sqrt{n}} \sum_{m=1}^{K_n} V_m^f \Rightarrow c\chi \quad \text{under } P_\mu$$

(ii) Show that $\max_{1 \leq m \leq n} V_m^{|f|}/\sqrt{n} \to 0$ in probability, and conclude

$$\frac{1}{\sqrt{n}} \sum_{m=1}^{n} f(X_m) \Rightarrow c\chi \quad \text{under } P_\mu$$

5.7. Ratio Limit Theorems. (5.1) does not say much in the null recurrent case. To get a more informative limit theorem, suppose that y is recurrent and m is the (unique up to constant multiples) stationary measure on $C_y = \{z : \rho_{yz} > 0\}$. Let $N_n(z) = |\{m \leq n : X_n = z\}|$. Break up the path at successive returns to y and show that $N_n(z)/N_n(y) \to m(z)/m(y)$ P_x-a.s. for all $x, z \in C_y$. Note that $n \to N_n(z)$ is increasing so this is much easier than the previous problem.

5.8. We got (5.2) from (5.1) by taking expected value. This does not work for the ratio in the previous exercise so we need another approach. Suppose $z \neq y$. (i) Let $\bar{p}_n(x, z) = P_x(X_n = z, T_y > n)$ and decompose $p^m(x, z)$ according to the value of $J = \sup\{j \in [1, m] : X_j = y\}$ to get

$$\sum_{m=1}^{n} p^m(x, z) = \sum_{m=1}^{n} \bar{p}_m(x, z) + \sum_{j=1}^{n-1} p^j(x, y) \sum_{k=1}^{n-j} \bar{p}_k(y, z)$$

(ii) Show that

$$\sum_{m=1}^{n} p^m(x, z) \Big/ \sum_{m=1}^{n} p^m(x, y) \to \frac{m(z)}{m(y)}$$

5.9. Show that if S is finite and p is irreducible and aperiodic then there is an m so that $p^m(x, y) > 0$ for all x, y.

5.10. Show that if S is finite, p is irreducible and aperiodic, and T is the coupling time defined in the proof of (5.5) then $P(T > n) \le Cr^n$ for some $r < 1$ and $C < \infty$. So the convergence to equilibrium occurs exponentially rapidly in this case. Hint: first consider the case in which $p(x, y) > 0$ for all x and y, and reduce the general case to this one by looking at a power of p.

5.11. For any transition matrix p, define

$$\alpha_n = \sup_{i,j} \frac{1}{2} \sum_k |p^n(i, k) - p^n(j, k)|$$

The $1/2$ is there because for any i and j we can define r.v.'s X and Y so that $P(X = k) = p^n(i, k)$, $P(Y = k) = p^n(j, k)$, and

$$P(X \ne Y) = (1/2) \sum_k |p^n(i, k) - p^n(j, k)|$$

Show that $\alpha_{m+n} \le \alpha_n \alpha_m$. Here you may find the coupling interpretation may help you from getting lost in the algebra.

Remark. Using (9.1) in Chapter 1 we can conclude that

$$\frac{1}{n} \log \alpha_n \to \inf_{m \ge 1} \frac{1}{m} \log \alpha_m$$

so if $\alpha_m < 1$ for some m it approaches 0 exponentially fast.

*5.6. General State Space

In this section we will generalize the results from the last three sections to a collection of Markov chains with uncountable state space called Harris chains. The developments here are motivated by three ideas. First, the proofs in the last two sections work if there is one point in the state space that the chain hits with probability one. (Think, for example, about the construction of the stationary measure in (4.3).) Second, a recurrent Harris chain can be modified to contain such a point. Third, the collection of Harris chains is a comfortable level of generality — broad enough to contain a large number of interesting examples, yet restrictive enough to allow for a rich theory.

We say that a Markov chain X_n is a **Harris chain** if we can find sets $A, B \in \mathcal{S}$, a function q with $q(x, y) \ge \epsilon > 0$ for $x \in A$, $y \in B$, and a probability measure ρ concentrated on B so that:

(i) If $\tau_A = \inf\{n \ge 0 : X_n \in A\}$, then $P_z(\tau_A < \infty) > 0$ for all $z \in S$.

(ii) If $x \in A$ and $C \subset B$ then $p(x, C) \geq \int_C q(x, y)\, \rho(dy)$.

To explain the definition we turn to some examples:

Example 6.1. Countable state space. If S is countable and there is a point a with $\rho_{xa} > 0$ for all x (a condition slightly weaker than irreducibility) then we can take $A = \{a\}$, $B = \{b\}$ where b is any state with $p(a, b) > 0$, $\mu = \delta_b$ the point mass at b, and $q(a, b) = p(a, b)$.

Conversely, if S is countable and (A', B') is a pair for which (i) and (ii) hold then we can without loss of generality reduce B' to a single point b. Having done this, if we set $A = \{b\}$, pick c so that $p(b, c) > 0$ and set $B = \{c\}$ then (i) and (ii) hold with A and B both singletons.

Example 6.2. Chains with continuous densities. Suppose $X_n \in \mathbf{R}^d$ is a Markov chain with a transition probability that has $p(x, dy) = p(x, y)\, dy$ where $(x, y) \to p(x, y)$ is continuous. Pick (x_0, y_0) so that $p(x_0, y_0) > 0$. Let A and B be open sets around x_0 and y_0 that are small enough so that $p(x, y) \geq \epsilon > 0$ on $A \times B$. If we let $\rho(C) = |B \cap C|/|B|$, where $|B|$ is the Lebesgue measure of B then (ii) holds. If (i) holds, then X_n is a Harris chain.

For concrete examples, consider

(a) **Diffusion processes** is a large class of examples that lie outside the scope of this book but is too important to ignore. Specifically, if the generator of X has Hölder continuous coefficients satisfying suitable growth conditions. (See Dynkin (1960) Appendix) then $P(X_1 \in dy) = p(x, y)\, dy$ and p satisfies the conditions above.

(b) **ARMAP's.** Let $\xi_1, \xi_2, \ldots$ be i.i.d. and $V_n = \theta V_{n-1} + \xi_n$. V_n is called an **autoregressive moving average process** or **armap** for short. We call V_n a **smooth armap** if the distribution of ξ_n has a continuous density g. In this case $p(x, dy) = g(y - \theta x)\, dy$ with $(x, y) \to g(y - \theta x)$ continuous.

In the analyzing the behavior of ARMAP's there are a number of cases to consider depending on the nature of the support of ξ_n. We call V_n a **simple armap** if the density function for ξ_n is positive for at all points in $\mathbf{R}$. In this case we can take $A = B = [-1/2, 1/2]$ with $\rho =$ the restriction of Lebesgue measure.

(c) The **discrete Ornstein-Uhlenbeck process** is a special case of (a) and (b). Let $\xi_1, \xi_2, \ldots$ be i.i.d. standard normals and let $V_n = \theta V_{n-1} + \xi_n$. The Ornstein-Uhlenbeck process is a diffusion process $\{V_t, t \in [0, \infty)\}$ that models the velocity of a particle suspended in a liquid. (See, e.g., Breiman (1968)

Section 16.1.) Looking at V_t at integer times (and dividing by a constant to make the variance 1) gives a Markov chain with the indicated distributions.

Example 6.3. GI/G/1 queue, or storage model. Let $\xi_1, \xi_2, \ldots$ be i.i.d. and define W_n inductively by $W_n = (W_{n-1} + \xi_n)^+$. If $P(\xi_n < 0) > 0$ then we can take $A = B = \{0\}$ and (i) and (ii) hold. To explain the first name in the title, consider a queueing system in which customers arrive at times of a renewal process, i.e., at times $0 = T_0 < T_1 < T_2 \ldots$ with $\zeta_n = T_n - T_{n-1}$, $n \geq 1$ i.i.d. Let η_n, $n \geq 0$, be the amount of service time the nth customer requires and let $\xi_n = \eta_{n-1} - \zeta_n$. I claim that W_n is the amount of time the nth customer has to wait to enter service. To see this notice that the $(n-1)$th customer adds η_{n-1} to the server's workload and if the server is busy at all times in $[T_{n-1}, T_n)$ he reduces his workload by ζ_n. If $W_{n-1} + \eta_{n-1} < \zeta_n$ then the server has enough time to finish his work and the next arriving customer will find an empty queue.

The second name in the title refers to the fact that W_n can be used to model the contents of a storage facility. For an intuitive description consider water reservoirs. We assume that rainstorms occur at times of a renewal process $\{T_n : n \geq 1\}$, that the nth rainstorm contributes an amount of water η_n, and that water is consumed at constant rate c. If we let $\zeta_n = T_n - T_{n-1}$ as before, and $\xi_n = \eta_{n-1} - c\zeta_n$ then W_n gives the amount of water in the reservoir just before the nth rainstorm.

History Lesson. Doeblin was the first to prove results for Markov chains on general state space. He supposed that there was an n so that $p^n(x, C) \geq \epsilon \rho(C)$ for all $x \in S$ and $C \subset S$. See Doob (1956) Section V.5 for an account of his results. Harris (1956) generalized Doeblin's result by observing that it was enough to have a set A so that (i) holds and the chain viewed on A ($Y_k = X(T_A^k)$ where $T_A^k = \inf\{n > T_A^{k-1} : X_n \in A\}$ and $T_A^0 = 0$) satisfies Doeblin's condition. Our formulation as well as most of the proofs in this section follows Athreya and Ney (1978). For a nice description of the "traditional approach," see Revuz (1984).

Given a Harris chain on $(S, \mathcal{S})$, we will construct a Markov chain $\bar{X}_n$ with transition probability $\bar{p}$ on $(\bar{S}, \bar{\mathcal{S}})$ where $\bar{S} = S \cup \{\alpha\}$ and $\bar{\mathcal{S}} = \{B, B \cup \{\alpha\} : B \in \mathcal{S}\}$. The aim, as advertised earlier, is to manufacture a point α that the process hits with probability 1 in the recurrent case.

If $x \in S - A$ $\bar{p}(x, C) = p(x, C)$ for $C \in \mathcal{S}$

If $x \in A$ $\bar{p}(x, \{\alpha\}) = \epsilon$

             $\bar{p}(x, C) = p(x, C) - \epsilon \, \rho(C)$ for $C \in \mathcal{S}$

If $x = \alpha$ $\bar{p}(\alpha, D) = \int \rho(dx) \bar{p}(x, D)$ for $D \in \bar{\mathcal{S}}$

Intuitively, $\bar{X}_n = \alpha$ corresponds to X_n being distributed on B according to

ρ. Here and in what follows we will reserve A and B for the special sets that occur in the definition and use C and D for generic elements of $\mathcal{S}$. We will often simplify notation by writing $\bar{p}(x, \alpha)$ instead of $\bar{p}(x, \{\alpha\})$, $\mu(\alpha)$ instead of $\mu(\{\alpha\})$, etc.

Our next step is to prove three technical lemmas, (6.1)–(6.3), that will help us develop the theory below. Define a transition probability v by

$$v(x, \{x\}) = 1 \quad \text{if} \quad x \in S \qquad v(\alpha, C) = \rho(C)$$

In words, V leaves mass in S alone but returns the mass at α to S and distributes it according to ρ.

(6.1) Lemma. $v\bar{p} = \bar{p}$ and $\bar{p}v = p$.

Proof Before giving the proof we would like to remind the reader that measures multiply the transition probability on the left, i.e., in the first case we want to show $\mu v \bar{p} = \mu \bar{p}$. If we first make a transition according to v and then one according to $\bar{p}$, this amounts to one transition according to $\bar{p}$, since only mass at α is affected by v and

$$\bar{p}(\alpha, D) = \int \rho(dx) \bar{p}(x, D)$$

The second equality also follows easily from the definition. In words if $\bar{p}$ acts first and then v, then v returns the mass at α to where it came from.   □

From (6.1) it follows easily that we have:

(6.2) Lemma. Let Y_n be an inhomogeneous Markov chain with $p_{2k} = v$ and $p_{2k+1} = \bar{p}$. Then $\bar{X}_n = Y_{2n}$ is a Markov chain with transition probability $\bar{p}$ and $X_n = Y_{2n+1}$ is a Markov chain with transition probability p.

(6.2) shows that there is an intimate relationship between the asymptotic behavior of X_n and of $\bar{X}_n$. To quantify this we need a definition. If f is a bounded measurable function on S, let $\bar{f} = vf$, i.e., $\bar{f}(x) = f(x)$ for $x \in S$ and $\bar{f}(\alpha) = \int f \, d\rho$.

(6.3) Lemma. If μ is a probability measure on $(S, \mathcal{S})$ then

$$E_\mu f(X_n) = E_\mu \bar{f}(\bar{X}_n)$$

Proof Observe that if X_n and $\bar{X}_n$ are constructed as in (6.2), and $P(\bar{X}_0 \in S) = 1$ then $X_0 = \bar{X}_0$ and X_n is obtained from $\bar{X}_n$ by making a transition according to v.   □

(6.1)–(6.3) will allow us to obtain results for X_n from those for $\bar{X}_n$. We turn now to the task of generalizing the results of Sections 5.3–5.5 to $\bar{X}_n$. To facilitate comparison with the results for countable state space, we will break this section into four subsections, the first three of which correspond to Sections 5.3–5.5. In the fourth subsection we take an in depth look at Example 6.3. Before developing the theory we will give one last example that explains why some of the statements are messy.

Example 6.4. Perverted O.U. process. Take the discrete O.U. processs of Example 6.2 and modify the transition probability at the integers $x \geq 2$ so that

$$p(x, \{x+1\}) = 1 - x^{-2}$$
$$p(x, A) = x^{-2}|A| \quad \text{for } A \subset (0,1)$$

p is the transition probability of a Harris chain but

$$P_2(X_n = n + 2 \text{ for all } n) > 0$$

I can sympathize with the reader who thinks that such chains will not arise "in application" but it seems easier (and better) to adapt the theory to include them than to modify the assumptions to exclude them.

a. Recurrence and transience

We begin with the dichotomy between recurrence and transience. Let $R = \inf\{n \geq 1 : \bar{X}_n = \alpha\}$. If $P_\alpha(R < \infty) = 1$ then we call the chain **recurrent**, otherwise we call it **transient**. Let $R_1 = R$ and for $k \geq 2$, let $R_k = \inf\{n > R_{k-1} : \bar{X}_n = \alpha\}$ be the time of the kth return to α. The strong Markov property implies $P_\alpha(R_k < \infty) = P_\alpha(R < \infty)^k$ so $P_\alpha(\bar{X}_n = \alpha \text{ i.o.}) = 1$ in the recurrent case and $= 0$ in the transient case. It is easy to generalize (3.3) to the current setting.

EXERCISE 6.1. $\bar{X}_n$ is recurrent if and only if $\sum_{n=1}^{\infty} \bar{p}^n(\alpha, \alpha) = \infty$.

The next result generalizes (3.4).

(6.4) **Theorem.** Let $\lambda(C) = \sum_{n=1}^{\infty} 2^{-n} \bar{p}^n(\alpha, C)$. In the recurrent case if $\lambda(C) > 0$ then $P_\alpha(\bar{X}_n \in C \text{ i.o.}) = 1$. For λ-a.e. x, $P_x(R < \infty) = 1$.

Proof The first conclusion follows from (2.3). For the second let $D = \{x : P_x(R < \infty) < 1\}$ and observe that if $p^n(\alpha, D) > 0$ for some n then

$$P_\alpha(\bar{X}_m = \alpha \text{ i.o.}) \leq \int \bar{p}^n(\alpha, dx) P_x(R < \infty) < 1 \qquad \square$$

Remark. Example 6.4 shows that we cannot expect to have $P_x(R < \infty) = 1$ for all x. To see that even when the state space is countable, we need not hit every point starting from α do

EXERCISE 6.2. If X_n is a recurrent Harris chain on a countable state space then S can only have one irreducible set of recurrent states but may have a nonempty set of transient states. For a concrete example consider a branching process in which the probability of no children $p_0 > 0$ and set $A = B = \{0\}$.

EXERCISE 6.3. Suppose X_n is a recurrent Harris chain. Show that if (A', B') is another pair satisfying the conditions of the definition then (6.4) implies $P_\alpha(\bar{X}_n \in A'$ i.o.$) = 1$, so the recurrence or transience does not depend on the choice of (A, B).

As in Section 6.3, we need special methods to determine whether an example is recurrent or transient.

EXERCISE 6.4. In the $GI/G/1$ queue, the waiting time W_n and the random walk $S_n = X_0 + \xi_1 + \cdots + \xi_n$ agree until $N = \inf\{n : S_n < 0\}$, and at this time $W_N = 0$. Use this observation to show that Example 6.3 is recurrent when $E\xi_n \le 0$ and transient when $E\xi_n > 0$.

EXERCISE 6.5. Let V_n be a simple smooth armap with $E|\xi_i| < \infty$. Show that if $\theta < 1$ then $E_x|V_1| \le |x|$ for $|x| \ge M$. Use this and ideas from the proof of (3.9) to show that the chain is recurrent in this case.

EXERCISE 6.6. Let V_n be an armap (not necessarily smooth or simple) and suppose $\theta > 1$. Let $\gamma \in (1, \theta)$ and observe that if $x > 0$ then $P_x(V_1 < \gamma x) \le C/((\theta - \gamma)x)$, so if x is large $P_x(V_n \ge \gamma^n x$ for all $n) > 0$.

Remark. In the case $\theta = 1$ the chain V_n discussed in the last two exercises is a random walk with mean 0 and hence recurrent.

EXERCISE 6.7. In the discrete O.U. process X_{n+1} is normal with mean θX_n and variance 1. What happens to the recurrence and transience if instead Y_{n+1} is normal with mean 0 and variance $\beta^2 |Y_n|$?

b. Stationary measures

(6.5) Theorem. In the recurrent case, there is a stationary measure.

Proof Let $R = \inf\{n \geq 1 : \bar{X}_n = \alpha\}$, and let

$$\bar{\mu}(C) = E_\alpha \left(\sum_{n=0}^{R-1} 1_{\{\bar{X}_n \in C\}} \right) = \sum_{n=0}^{\infty} P_\alpha(\bar{X}_n \in C, R > n)$$

Repeating the proof of (4.3) shows that $\bar{\mu}\bar{p} = \bar{\mu}$. If we let $\mu = \bar{\mu}v$ then it follows from (6.1) that $\bar{\mu}v\,p = \bar{\mu}\bar{p}v = \bar{\mu}v$, so $\mu\,p = \mu$.

EXERCISE 6.8. Let $G_{k,\delta} = \{x : \bar{p}^k(x, \alpha) \geq \delta\}$. Show that $\bar{\mu}(G_{k,\delta}) \leq 2k/\delta$ and use this to conclude that $\bar{\mu}$ and hence μ is σ-finite.

EXERCISE 6.9. Let λ be the measure defined in (6.4). Show that $\bar{\mu} \ll \lambda$ and $\lambda \ll \bar{\mu}$.

EXERCISE 6.10. Let V_n be an armap (not necessarily smooth or simple) with $\theta < 1$ and $E \log^+ |\xi_n| < \infty$. Show that $\sum_{m \geq 0} \theta^m \xi_m$ converges a.s. and defines a stationary distribution for V_n.

To investigate uniqueness of the stationary measure we begin with:

(6.6) Lemma. If ν is a σ-finite stationary measure for p, then $\nu(A) < \infty$ and $\bar{\nu} = \nu\bar{p}$ is a stationary measure for $\bar{p}$ with $\bar{\nu}(\alpha) < \infty$.

Proof We will first show that $\nu(A) < \infty$. If $\nu(A) = \infty$ then part (ii) of the definition implies $\nu(C) = \infty$ for all sets C with $\rho(C) > 0$. If $B = \cup_i B_i$ with $\nu(B_i) < \infty$ then $\rho(B_i) = 0$ by the last observation and $\rho(B) = 0$ by countable subadditivity, a contradiction. So $\nu(A) < \infty$ and $\bar{\nu}(\alpha) = \nu\bar{p}(\alpha) = \epsilon\nu(A) < \infty$. Using the fact that $\nu\,p = \nu$, we find

$$\nu\bar{p}(C) = \nu(C) - \epsilon\nu(A)\rho(B \cap C)$$

the last subtraction being well defined since $\nu(A) < \infty$, and it follows that $\bar{\nu}v = \nu$. To check $\bar{\nu}\bar{p} = \bar{\nu}$, we observe that (6.1) and the last result imply $\bar{\nu}\bar{p} = \bar{\nu}v\bar{p} = \nu\bar{p} = \bar{\nu}$. □

(6.7) Theorem. Suppose p is recurrent. If ν is a σ-finite stationary measure then $\nu = \bar{\nu}(\alpha)\mu$ where μ is the measure constructed in the proof of (6.5).

Proof By (6.6) it suffices to prove that if $\bar{\nu}$ is a stationary measure for $\bar{p}$ with $\bar{\nu}(\alpha) < \infty$ then $\bar{\nu} = \bar{\nu}(\alpha)\bar{\mu}$. Repeating the proof of (4.4) with $a = \alpha$, it is easy to show that $\bar{\nu}(C) \geq \bar{\nu}(\alpha)\bar{\mu}(C)$. Continuing to compute as in that proof:

$$\bar{\nu}(\alpha) = \int \bar{\nu}(dx)\bar{p}^n(x, \alpha) \geq \bar{\nu}(\alpha) \int \bar{\mu}(dx)\bar{p}^n(x, \alpha) = \bar{\nu}(\alpha)\bar{\mu}(\alpha) = \bar{\nu}(\alpha)$$

Let $S_n = \{x : p^n(x, \alpha) > 0\}$. By assumption $\cup_n S_n = S$. If $\bar{\nu}(D) > \bar{\nu}(\alpha)\bar{\mu}(D)$ for some D, then $\bar{\nu}(D \cap S_n) > \bar{\nu}(\alpha)\bar{\mu}(D \cap S_n)$ and it follows that $\bar{\nu}(\alpha) > \bar{\nu}(\alpha)$ a contradiction.

c. Convergence theorem

We say that a recurrent Harris chain X_n is **aperiodic** if g.c.d. $\{n \geq 1 : p^n(\alpha, \alpha) > 0\} = 1$. This occurs, for example, if we can take $A = B$ in the definition for then $p(\alpha, \alpha) > 0$.

(6.8) Theorem. Let X_n be an aperiodic recurrent Harris chain with stationary distribution π. If $P_x(R < \infty) = 1$ then as $n \to \infty$,

$$\|p^n(x, \cdot) - \pi(\cdot)\| \to 0$$

Note. Here $\| \, \|$ denotes the total variation distance between the measures. (6.4) guarantees that π a.e. x satisfies the hypothesis.

Proof In view of (6.3) it suffices to prove the result for $\bar{p}$. We begin by observing that the existence of a stationary probability measure and the uniqueness result in (6.7) implies that the measure constructed in (6.5) has $E_\alpha R = \bar{\mu}(S) < \infty$. As in the proof of (5.5) we let X_n and Y_n be independent copies of the chain with initial distributions δ_x and π respectively, and let $\tau = \inf\{n \geq 0 : X_n = Y_n = \alpha\}$. For $m \geq 0$ let S_m (resp. T_m) be the times at which X_n (resp. Y_n) visit α for the $(m+1)$th time. $S_m - T_m$ is a random walk with mean 0 steps, so $M = \inf\{m \geq 1 : S_m = T_m\} < \infty$ a.s. and it follows that this is true for τ as well. The computations in the proof of (5.5) show $|P(X_n \in C) - P(Y_n \in C)| \leq P(\tau > n)$. Since this is true for all C, $\|p^n(x, \cdot) - \pi(\cdot)\| \leq P(\tau > n)$ and the proof is complete. $\qquad\square$

EXERCISE 6.11. Use Exercise 6.1 and imitate the proof of (4.5) to show that a Harris chain with a stationary distribution must be recurrent.

EXERCISE 6.12. Show that an armap with $\theta < 1$ and $E \log^+ |\xi_n| < \infty$ converges in distribution as $n \to \infty$. Hint: Recall the construction of π in Exercise 6.10.

d. GI/G/1 queue

For the rest of the section we will concentrate on the $GI/G/1$ queue. Let $\xi_1, \xi_2, \ldots$ be i.i.d., let $W_n = (W_{n-1} + \xi_n)^+$ and let $S_n = \xi_1 + \cdots + \xi_n$. Recall $\xi_n = \eta_{n-1} - \zeta_n$ where the η's are service times, ζ's are the interarrival times, and suppose $E\xi_n < 0$ so that Exercise 6.11 implies there is a stationary distribution.

EXERCISE 6.13. Let $m_n = \min(S_0, S_1, \ldots, S_n)$ where S_n is the random walk defined above. (i) Show that $S_n - m_n =_d W_n$. (ii) Let $\xi'_m = \xi_{n+1-m}$ for $1 \le m \le n$. Show that $S_n - m_n = \max(S'_0, S'_1, \ldots, S'_n)$. (iii) Conclude that as $n \to \infty$ we have $W_n \Rightarrow M \equiv \max(S'_0, S'_1, S'_2, \ldots)$.

Explicit formulas for the distribution of M are in general difficult to obtain. However, this can be done if either the arrival or service distribution is exponential. One reason for this is

EXERCISE 6.14. Suppose $X, Y \ge 0$ are independent and $P(X > x) = e^{-\lambda x}$. Show that $P(X - Y > x) = ae^{-\lambda x}$ where $a = P(X - Y > 0)$.

Example 6.5. Exponential service time. Suppose $P(\eta_n > x) = e^{-\beta x}$ and $E\zeta_n > E\eta_n$. Let $T = \inf\{n : S_n > 0\}$ and $L = S_T$, setting $L = -\infty$ if $T = \infty$. The lack of memory property of the exponential distribution implies that $P(L > x) = re^{-\beta x}$ where $r = P(T < \infty)$. To compute the distribution of the maximum, M, let $T_1 = T$ and let $T_k = \inf\{n > T_{k-1} : S_n > S_{T_{k-1}}\}$ for $k \ge 2$. (1.3) in Chapter 3 implies that if $T_k < \infty$ then $S(T_{k+1}) - S(T_k) =_d L$ and is independent of $S(T_k)$. Using this and breaking things down according to the value of $K = \inf\{k : L_{k+1} = -\infty\}$ we see that for $x > 0$ the density function

$$P(M = x) = \sum_{k=1}^{\infty} r^k (1-r) e^{-\beta x} \beta^k x^{k-1}/(k-1)! = \beta r(1-r) e^{-\beta x(1-r)}$$

To complete the calculation we need to calculate r. To do this, let

$$\varphi(\theta) = E \exp(\theta \xi_n) = E \exp(\theta \eta_{n-1}) E \exp(-\theta \zeta_n)$$

which is finite for $0 < \theta < \beta$ since $\zeta_n \ge 0$ and η_{n-1} has an exponential distribution. It is easy to see that

$$\varphi'(0) = E\xi_n < 0 \qquad \lim_{\theta \uparrow \beta} \varphi(\theta) = \infty$$

so there is a $\theta \in (0, \beta)$ with $\varphi(\theta) = 1$. Exercise 7.6 in Chapter 4 implies $\exp(\theta S_n)$ is a martingale. (4.1) in Chapter 4 implies $1 = E \exp(\theta S_{T \wedge n})$. Letting $n \to \infty$ and noting that $(S_n | T = n)$ has an exponential distribution and $S_n \to -\infty$ on $\{T = \infty\}$ we have

$$1 = r \int_0^\infty e^{\theta x} \beta e^{-\beta x} \, dx = \frac{r\beta}{\beta - \theta}$$

Example 6.6. Poisson arrivals. Suppose $P(\zeta_n > x) = e^{-\alpha x}$ and $E\zeta_n > E\eta_n$. Let $\bar{S}_n = -S_n$. Reversing time as in (ii) of Exercise 6.14, we see (for $n \ge 1$)

$$P\left(\max_{0 \le k < n} \bar{S}_k < \bar{S}_n \in A\right) = P\left(\min_{1 \le k \le n} \bar{S}_k > 0, \bar{S}_n \in A\right)$$

Let $\psi_n(A)$ be the common value of the last two expression and let $\psi(A) = \sum_{n \geq 0} \psi_n(A)$. $\psi_n(A)$ is the probability the random walk reaches a new maximum (or ladder height, see Example 1.4 in Chapter 3) in A at time n, so $\psi(A)$ is the number of ladder points in A with $\psi(\{0\}) = 1$. Letting the random walk take one more step

$$P\left(\min_{1 \leq k \leq n} \bar{S}_k > 0, \bar{S}_{n+1} \leq x\right) = \int F(x - z)\, d\psi_n(z)$$

The last identity is valid for $n = 0$ if we interpret the left-hand side as $F(x)$. Let $\tau = \inf\{n \geq 1 : \bar{S}_n \leq 0\}$ and $x \leq 0$. Integrating by parts on the right-hand side and then summing over $n \geq 0$ gives

(6.9)
$$P(\bar{S}_\tau \leq x) = \sum_{n=0}^{\infty} P\left(\min_{1 \leq k \leq n} \bar{S}_k > 0, \bar{S}_{n+1} \leq x\right)$$
$$= \int_{y \leq x} \psi[0, x - y]\, dF(y)$$

The limit $y \leq x$ comes from the fact that $\psi((-\infty, 0)) = 0$.

Let $\bar{\xi}_n = \bar{S}_n - \bar{S}_{n-1} = -\xi_n$. Exercise 6.14 implies $P(\bar{\xi}_n > x) = ae^{-\alpha x}$. Let $\bar{T} = \inf\{n : \bar{S}_n > 0\}$. $E\bar{\xi}_n > 0$ so $P(\bar{T} < \infty) = 1$. Let $J = \bar{S}_{\bar{T}}$. As in the previous example, $P(J > x) = e^{-\alpha x}$. Let $V_n = J_1 + \cdots + J_n$. V_n is a rate α Poisson process, so $\psi[0, x - y] = 1 + \alpha(x - y)$ for $x - y \geq 0$. Using (6.9) now and integrating by parts gives

(6.10)
$$P(\bar{S}_\tau \leq x) = \int_{y \leq x} (1 + \alpha(x - y))\, dF(y)$$
$$= F(x) + \alpha \int_{-\infty}^{x} F(y)\, dy \qquad \text{for } x \leq 0$$

Since $P(\bar{S}_n = 0) = 0$ for $n \geq 1$, $-\bar{S}_\tau$ has the same distribution as S_T where $T = \inf\{n : S_n > 0\}$. Combining this with part (ii) of Exercise 6.13 gives a "formula" for $P(M > x)$. Straightforward but somewhat tedious calculations show that if $B(s) = E\exp(-s\eta_n)$ then

$$E\exp(-sM) = \frac{(1 - \alpha \cdot E\eta)s}{s - \alpha + \alpha B(s)}$$

a result known as the **Pollaczek-Khintchine formula**. The computations we omitted can be found in Billingsley (1979) on p. 277 or several times in Feller Vol. II (1971).

6 Ergodic Theorems

X_n, $n \geq 0$, is said to be a stationary sequence if for each $k \geq 1$ it has the same distribution as the shifted sequence X_{n+k}, $n \geq 0$. The basic fact about these sequences, called the ergodic theorem, is that if $E|f(X_0)| < \infty$ then

$$(2.1) \qquad \lim_{n \to \infty} \frac{1}{n} \sum_{m=0}^{n-1} f(X_m) \quad \text{exists a.s.}$$

If X_n is ergodic (intuitively, it is not a mixture of two other stationary sequences) then the limit is $Ef(X_0)$. Sections 6.1 and 6.2 develop the theory needed to prove the ergodic theorem. The remaining five sections of this chapter develop various complements and with the exception of Section 6.7, which gives applications of a result proved in Section 6.6, can be read in any order. In Section 6.3 we apply the ergodic theorem to study the recurrence of stationary sequences. In Section 6.4 we study "mixing", an asymptotic independence property stronger than ergodicity. In Section 6.5, we discuss entropy and give a proof of the Shannon-McMillan-Breiman theorem for X_n's taking values in a finite set. In Section 6.6, we prove the subadditive ergodic theorem. As five examples in Section 6.6 and four applications in Section 6.7 should indicate, this result is a useful generalization of the ergodic theorem.

6.1. Definitions and Examples

$X_0, X_1, \ldots$ is said to be a **stationary sequence** if for every k, the sequence $X_k, X_{k+1}, \ldots$ has the same distribution, i.e., for each n, $(X_0, \ldots, X_n)$ and $(X_k, \ldots, X_{k+n})$ have the same distribution. We begin by giving four examples that will be our constant companions.

Example 1.1. $X_0, X_1, \ldots$ are i.i.d.

Example 1.2. Let X_n be a Markov chain with transition probability $p(x, A)$ and stationary distribution π, i.e., $\pi(A) = \int \pi(dx)\, p(x, A)$. If X_0 has distribution π then $X_0, X_1, \ldots$ is a stationary sequence. A special case to keep in mind for counterexamples is the chain with state space $S = \{0,1\}$ and transition probability $p(x, \{1 - x\}) = 1$. In this case the stationary distribution has $\pi(0) = \pi(1) = 1/2$ and $(X_0, X_1, \ldots) = (0, 1, 0, 1, \ldots)$ or $(1, 0, 1, 0, \ldots)$ with probability $1/2$ each.

Example 1.3. Rotation of the circle. Let $\Omega = [0, 1)$, $\mathcal{F} =$ Borel subsets, $P =$ Lebesgue measure. Let $\theta \in (0, 1)$ and for $n \geq 0$ let $X_n(\omega) = (\omega + n\theta)$ mod 1, where x mod $1 = x - [x]$, $[x]$ being the greatest integer $\leq x$. To see the reason for the name map $[0, 1)$ into $\mathbf{C}$ by $x \to \exp(2\pi i x)$. This example is a special case of the last one. Let $p(x, \{y\}) = 1$ if $y = (x + \theta)$ mod 1.

To make new examples from old we can use

(1.1) Theorem. If $X_0, X_1, \ldots$ is a stationary sequence and $g : \mathbf{R}^{\{0,1,\ldots\}} \to \mathbf{R}$ is measurable then $Y_k = g(X_k, X_{k+1}, \ldots)$ is a stationary sequence.

Proof If $x \in \mathbf{R}^{\{0,1,\ldots\}}$, let $g_k(x) = g(x_k, x_{k+1}, \ldots)$ and if $B \in \mathcal{R}^{\{0,1,\ldots\}}$ let

$$A = \{x : (g_0(x), g_1(x), \ldots) \in B\}$$

To check stationarity now we observe:

$$
\begin{aligned}
P(\omega : (Y_0, Y_1, \ldots) \in B) &= P(\omega : (X_0, X_1, \ldots) \in A) \\
&= P(\omega : (X_k, X_{k+1}, \ldots) \in A) \\
&= P(\omega : (Y_k, Y_{k+1}, \ldots) \in B)
\end{aligned}
$$

$\square$

Example 1.4. Bernoulli shift. $\Omega = [0, 1)$, $\mathcal{F} =$ Borel subsets, $P =$ Lebesgue measure. $Y_0(\omega) = \omega$ and for $n \geq 1$ let $Y_n(\omega) = (2\,Y_{n-1}(\omega))$ mod 1. This example is a special case of (1.1). Let $X_0, X_1, \ldots$ be i.i.d. with $P(X_i = 0) = P(X_i = 1) = 1/2$, and let

$$g(x) = \sum_{i=0}^{\infty} x_i 2^{-(i+1)}$$

The name comes from the fact that multiplying by 2 shifts the X's to the left. This example is also a special case of Example 1.2. Let $p(x, \{y\}) = 1$ if $y = (2x)$ mod 1.

Examples 1.3 and 1.4 are special cases of the following situation.

Example 1.5. Let $(\Omega, \mathcal{F}, P)$ be a probability space. A measurable map $\varphi : \Omega \to \Omega$ is said to be **measure preserving** if $P(\varphi^{-1}A) = P(A)$ for all $A \in$

$\mathcal{F}$. Let φ^n be the nth iterate of φ defined inductively by $\varphi^n = \varphi(\varphi^{n-1})$ for $n \geq 1$ where $\varphi^0(\omega) = \omega$. We claim that if $X \in \mathcal{F}$ then $X_n(\omega) = X(\varphi^n \omega)$ defines a stationary sequence. To check this, let $B \in \mathcal{R}^{n+1}$ and $A = \{\omega : (X_0(\omega), \ldots, X_n(\omega)) \in B\}$. Then

$$P((X_k, \ldots, X_{k+n}) \in B) = P(\varphi^k \omega \in A) = P(\omega \in A) = P((X_0, \ldots, X_n) \in B)$$

The last example is more than an important example. In fact, it is the only example! If $Y_0, Y_1, \ldots$ is a stationary sequence taking values in a nice space, Kolmogorov's extension theorem ((7.1) in the Appendix) allows us to construct a measure P on sequence space $(S^{\{0,1,\ldots\}}, \mathcal{S}^{\{0,1,\ldots\}})$, so that the sequence $X_n(\omega) = \omega_n$ has the same distribution as that of $\{Y_n, n \geq 0\}$. If we let φ be the shift operator, i.e., $\varphi(\omega_0, \omega_1, \ldots) = (\omega_1, \omega_2, \ldots)$, and let $X(\omega) = \omega_0$ then φ is measure preserving and $X_n(\omega) = X(\varphi^n \omega)$.

In some situations (e.g., in the proof of (3.3) below) it is useful to observe:

(1.2) Theorem. Any stationary sequence $\{X_n \ , \ n \geq 0\}$ can be embedded in a two sided stationary sequence $\{Y_n : n \in \mathbf{Z}\}$.

Proof We observe that

$$P(Y_{-m} \in A_0, \ldots, Y_n \in A_{m+n}) = P(X_0 \in A_0, \ldots, X_{m+n} \in A_{m+n})$$

is a consistent set of finite dimensional distributions, so a trivial generalization of the Kolmogorov extension theorem implies there is a measure P on $(S^{\mathbf{Z}}, \mathcal{S}^{\mathbf{Z}})$ so that the variables $Y_n(\omega) = \omega_n$ have the desired distributions. $\square$

In view of the observations above, it suffices to give our definitions and prove our results in the setting of Example 1.5. Thus our basic set-up consists of

$$\begin{aligned} (\Omega, \mathcal{F}, P) \quad &\text{a probability space} \\ \varphi \quad &\text{a map which preserves } P \\ X_n(\omega) = X(\varphi^n \omega) \quad &\text{where } X \text{ is a random variable} \end{aligned}$$

We will now give some important definitions. Here and in what follows we assume φ is measure preserving. A set $A \in \mathcal{F}$ is said to be **invariant** if $\varphi^{-1} A = A$. (Here, as usual, two sets are considered to be equal if their symmetric difference has probability 0.) Some authors call A **almost invariant** if $P(A \triangle \varphi^{-1}(A)) = 0$. We call such sets invariant and call B **invariant in the strict sense** if $B = \varphi^{-1}(B)$.

EXERCISE 1.1. Show that the class of invariant events $\mathcal{I}$ is a σ-field, and $X \in \mathcal{I}$ if and only if X is **invariant**, i.e., $X \circ \varphi = X$ a.s.

EXERCISE 1.2. (i) Let A be any set, let $B = \cup_{n=0}^{\infty} \varphi^{-n}(A)$. Show $\varphi^{-1}(B) \subset B$. (ii) Let B be any set with $\varphi^{-1}(B) \subset B$ and let $C = \cap_{n=0}^{\infty} \varphi^{-n}(B)$. Show that $\varphi^{-1}(C) = C$. (iii) Show that A is almost invariant if and only if there is a C invariant in the strict sense with $P(A \Delta C) = 0$.

A measure preserving transformation on $(\Omega, \mathcal{F}, P)$ is said to be **ergodic** if $\mathcal{I}$ is trivial, i.e., for every $A \in \mathcal{I}$, $P(A) \in \{0, 1\}$. If φ is not ergodic then the space can be split into two sets A and A^c each having positive measure so that $\varphi(A) = A$ and $\varphi(A^c) = A^c$. In words, φ is not "irreducible."
To investigate further the meaning of ergodicity we turn to our examples.

For an **i.i.d. sequence** (Example 1.1), we begin by observing that if $\Omega = \mathbf{R}^{\{0,1,\dots\}}$ and φ is the shift operator then an invariant set A has $\{\omega : \omega \in A\} = \{\omega : \varphi\omega \in A\} \in \sigma(X_1, X_2, \dots)$. Iterating gives

$$A \in \cap_{n=1}^{\infty} \sigma(X_n, X_{n+1}, \dots) = \mathcal{T}, \quad \text{the tail } \sigma\text{-field}$$

so $\mathcal{I} \subset \mathcal{T}$. For an i.i.d. sequence, Kolmogorov's 0-1 law implies $\mathcal{T}$ is trivial, so $\mathcal{I}$ is trivial and the sequence is ergodic (i.e., when the corresponding measure is put on sequence space $\Omega = \mathbf{R}^{\{0,1,2,\dots\}}$ the shift is).

Turning to **Markov chains** (Example 1.2), suppose the state space S is countable and the stationary distribution has $\pi(x) > 0$ for all $x \in S$. By (4.5) and (3.6) in Chapter 5, all states are recurrent, and we can write $S = \cup R_i$, where the R_i are disjoint irreducible closed sets. If $X_0 \in R_i$ then with probability one, $X_n \in R_i$ for all $n \geq 1$ so $\{\omega : X_0(\omega) \in R_i\} \in \mathcal{I}$. The last observation shows that if the Markov chain is not irreducible then the sequence is not ergodic. To prove the converse, observe that if $A \in \mathcal{I}$, $1_A \circ \theta_n = 1_A$ where $\theta_n(\omega_0, \omega_1, \dots) = (\omega_n, \omega_{n+1}, \dots)$. So if we let $\mathcal{F}_n = \sigma(X_0, \dots, X_n)$, the shift invariance of 1_A and the Markov property imply

$$E_\pi(1_A | \mathcal{F}_n) = E_\pi(1_A \circ \theta_n | \mathcal{F}_n) = h(X_n)$$

where $h(x) = E_x 1_A$. Lévy's 0-1 law implies that the left-hand side converges to 1_A as $n \to \infty$. If X_n is irreducible and recurrent then for any $y \in S$, the right-hand side $= h(y)$ i.o., so either $h(x) \equiv 0$ or $h(x) \equiv 1$, and $P_\pi(A) \in \{0, 1\}$. This example also shows that $\mathcal{I}$ and $\mathcal{T}$ may be different. When the transition probability p is irreducible $\mathcal{I}$ is trivial, but if all the states have period $d > 1$, $\mathcal{T}$ is not. In (5.8) of Chapter 5 we showed that if $S_0, \dots, S_{d-1}$ is the cyclic decomposition of S then $\mathcal{T} = \sigma(\{X_0 \in S_r\} : 0 \leq r < d)$.

EXERCISE 1.3. Give an example of an ergodic measure preserving transformation T on $(\Omega, \mathcal{F}, P)$, so that T^2 is not ergodic.

Rotation of the circle (Example 1.3) is not ergodic if $\theta = m/n$ where $m < n$ are positive integers. If B is a Borel subset of $[0, 1/n)$ and

$$A = \cup_{k=0}^{n-1}(B + k/n)$$

then A is invariant. Conversely, if θ is irrational then φ is ergodic. To prove this we need a fact from Fourier analysis. If f is a measurable function on $[0, 1)$ with $\int f^2(x)\, dx < \infty$ then f can be written as

$$f(x) = \sum_k c_k e^{2\pi i k x}$$

where the equality is in the sense that as $K \to \infty$

$$\sum_{k=-K}^{K} c_k e^{2\pi i k x} \to f(x) \text{ in } L^2[0, 1)$$

and this is possible for only one choice of the coefficients c_k :

$$c_k = \int f(x) e^{-2\pi i k x}\, dx$$

Now

$$f(\varphi(x)) = \sum_k c_k e^{2\pi i k(x+\theta)} = \sum_k (c_k e^{2\pi i k\theta}) e^{2\pi i k x}$$

The uniqueness of the coefficients c_k implies that $f(\varphi(x)) = f(x)$ if and only if

$$c_k(e^{2\pi i k\theta} - 1) = 0$$

If θ is irrational this implies $c_k = 0$ for $k \neq 0$, so f is constant. Applying the last result to $f = 1_A$ with $A \in \mathcal{I}$ shows that $A = \emptyset$ or $[0, 1)$ a.s.

EXERCISE 1.4. **A direct proof of ergodicity.** (i) Show that if θ is irrational, $x_n = n\theta \mod 1$ is dense in $[0,1)$. Hint: all the x_n are distinct, so for any $N < \infty$, $|x_n - x_m| \leq 1/N$ for some $m < n \leq N$. (ii) Use Exercise 3.1 in the Appendix to show that if A is a Borel set with $|A| > 0$, then for any $\delta > 0$ there is an interval $J = [a, b)$ so that $|A \cap J| > (1 - \delta)|J|$. (iii) Combine this with (i) to conclude $P(A) = 1$.

Finally the **Bernoulli shift** (Example 1.4) is ergodic. To prove this, we recall that the stationary sequence $Y_n(\omega) = \varphi^n(\omega)$ can be represented as

$$Y_n = \sum_{m=0}^{\infty} 2^{-(m+1)} X_{n+m}$$

where $X_0, X_1, \ldots$ are i.i.d. with $P(X_k = 1) = P(X_k = 0) = 1/2$, and use the following fact:

(1.3) Theorem. Let $g : \mathbf{R}^{\{0,1,\ldots\}} \to \mathbf{R}$ be measurable. If $X_0, X_1, \ldots$ is an ergodic stationary sequence and then $Y_k = g(X_k, X_{k+1}, \ldots)$ is ergodic.

Proof Suppose $X_0, X_1, \ldots$ is defined on sequence space with $X_n(\omega) = \omega_n$. If B has $\{\omega : (Y_0, Y_1, \ldots) \in B\} = \{\omega : (Y_1, Y_2, \ldots) \in B\}$ then $A = \{\omega : (Y_0, Y_1, \ldots) \in B\}$ is shift invariant. □

Remark. The proofs of (1.1) and (1.3) generalize easily to functions $g : \mathbf{R}^{\mathbf{Z}} \to \mathbf{R}$ of a two-sided stationary sequence X_n, $n \in \mathbf{Z}$. An example of the use of this generalization is the following: Let ξ_n, $n \in \mathbf{Z}$ be i.i.d. with $E\xi_n = 0$ and $E\xi_n^2 < \infty$. Let c_n, $n \in \mathbf{Z}$, be constants with $\sum c_n^2 < \infty$. (8.3) in Chapter 1 implies that $X_n = \sum_m c_{n-m}\xi_m$, $n \geq 0$, converges a.s., generalizations of (1.1) and (1.3) imply that it is stationary and ergodic.

EXERCISES

1.5. Use Fourier analysis as in Example 1.3 to prove that Example 1.4 is ergodic.

1.6. **Continued fractions.** Let $\varphi(x) = 1/x - [1/x]$ for $x \in (0,1)$ and $A(x) = [1/x]$ where $[1/x] =$ the largest integer $\leq 1/x$. $a_n = A(\varphi^n x)$, $n = 0, 1, 2, \ldots$ gives the continued fraction representation of x, i.e.,

$$x = 1/(a_0 + 1/(a_1 + 1/(a_2 + 1/\ldots)))$$

Show that φ preserves

$$\mu(A) = \frac{1}{\log 2} \int_A \frac{dx}{1+x} \quad \text{for } A \subset (0,1)$$

Remark. In his (1959) monograph Kac claimed that it was "entirely trivial" to check that φ is ergodic but retracted his claim in a later footnote. We leave it to the reader to construct a proof or look up the answer in Ryll-Nardzewski (1951). Chapter 9 of Lévy (1937) is devoted to this topic and is still interesting reading today.

1.7. **Independent blocks.** Let $X_1, X_2, \ldots$ be a stationary sequence. Let $n < \infty$ and let $Y_1, Y_2, \ldots$ be a sequence so that $(Y_{nk+1}, \ldots, Y_{n(k+1)})$, $k \geq 0$ are i.i.d. and $(Y_1, \ldots, Y_n) = (X_1, \ldots, X_n)$. Finally let ν be uniformly distributed on $\{1, 2, \ldots, n\}$ and let $Z_m = Y_{\nu+m}$ for $m \geq 1$. Show that Z is stationary and ergodic.

6.2. Birkhoff's Ergodic Theorem

Throughout this section φ is a measure preserving transformation on $(\Omega, \mathcal{F}, P)$. We begin by proving a result which is usually referred to as:

(2.1) The ergodic theorem. For any $X \in L^1$,

$$\frac{1}{n} \sum_{m=0}^{n-1} X(\varphi^m \omega) \to E(X|\mathcal{I}) \quad \text{a.s. and in } L^1$$

This result due to Birkhoff (1931) is sometimes called the pointwise or individual ergodic theorem because of the a.s. convergence in the conclusion. The proof we give is based on an odd integration inequality due to Yosida and Kakutani (1939).

(2.2) Maximal ergodic lemma. Let $X_j(\omega) = X(\varphi^j \omega)$, $S_k(\omega) = X_0(\omega) + \ldots + X_{k-1}(\omega)$, and $M_k(\omega) = \max(0, S_1(\omega), \ldots, S_k(\omega))$. Then $E(X; M_k > 0) \geq 0$.

Proof We follow Garsia (1965). The proof is not intuitive but none of the steps are difficult. If $j \leq k$ then $M_k(\varphi \omega) \geq S_j(\varphi \omega)$, so adding $X(\omega)$ gives

$$X(\omega) + M_k(\varphi \omega) \geq X(\omega) + S_j(\varphi \omega) = S_{j+1}(\omega)$$

and rearranging we have

$$X(\omega) \geq S_{j+1}(\omega) - M_k(\varphi \omega) \text{ for } j = 1, \ldots, k$$

Trivially $X(\omega) \geq S_1(\omega) - M_k(\varphi \omega)$, since $S_1(\omega) = X(\omega)$ and $M_k(\varphi \omega) \geq 0$. Therefore

$$E(X(\omega); M_k > 0) \geq \int_{\{M_k > 0\}} \max(S_1(\omega), \ldots, S_k(\omega)) - M_k(\varphi \omega) \, dP$$

$$= \int_{\{M_k > 0\}} M_k(\omega) - M_k(\varphi \omega) \, dP$$

Now $M_k(\omega) = 0$ and $M_k(\varphi \omega) \geq 0$ on $\{M_k > 0\}^c$, so the last expression is

$$\geq \int M_k(\omega) - M_k(\varphi \omega) \, dP = 0$$

since φ is measure preserving. $\square$

Proof of (2.1) $E(X|\mathcal{I})$ is invariant under φ (see Exercise 1.1) so letting $X' = X - E(X|\mathcal{I})$ we can assume without loss of generality that $E(X|\mathcal{I}) = 0$. Let $\bar{X} = \limsup S_n/n$, let $\epsilon > 0$, and let $D = \{\omega : \bar{X}(\omega) > \epsilon\}$. Our goal is to prove that $P(D) = 0$. $\bar{X}(\varphi\omega) = \bar{X}(\omega)$, so $D \in \mathcal{I}$. Let

$$X^*(\omega) = (X(\omega) - \epsilon)1_D(\omega) \qquad S_n^*(\omega) = X^*(\omega) + \ldots + X^*(\varphi^{n-1}\omega)$$
$$M_n^*(\omega) = \max(0, S_1^*(\omega), \ldots, S_n^*(\omega)) \qquad F_n = \{M_n^* > 0\}$$
$$F = \cup_n F_n = \left\{\sup_{k \geq 1} S_k^*/k > 0\right\}$$

Since $X^*(\omega) = (X(\omega) - \epsilon)1_D(\omega)$ and $D = \{\limsup S_k/k > \epsilon\}$, it follows that

$$F = \left\{\sup_{k \geq 1} S_k/k > \epsilon\right\} \cap D = D$$

(2.2) implies that $E(X^*; F_n) \geq 0$. Since $E|X^*| \leq E|X| + \epsilon < \infty$, the dominated convergence theorem implies $E(X^*; F_n) \to E(X^*; F)$ and it follows that $E(X^*; F) \geq 0$. The last conclusion looks innocent, but $F = D \in \mathcal{I}$ so it implies

$$0 \leq E(X^*; D) = E(X - \epsilon; D) = E(E(X|\mathcal{I}); D) - \epsilon P(D) = -\epsilon P(D)$$

since $E(X|\mathcal{I}) = 0$. The last inequality implies that

$$0 = P(D) = P(\limsup S_n/n > \epsilon)$$

and since $\epsilon > 0$ is arbitrary, it follows that $\limsup S_n/n \leq 0$. Applying the last result to $-X$ shows that $S_n/n \to 0$ a.s.

To prove that convergence occurs in L^1 let

$$X_M'(\omega) = X(\omega)1_{(|X(\omega)| \leq M)} \quad \text{and} \quad X_M''(\omega) = X(\omega) - X_M'(\omega)$$

The part of the ergodic theorem we have proved implies

$$\frac{1}{n} \sum_{m=0}^{n-1} X_M'(\varphi^m \omega) \to E(X_M'|\mathcal{I}) \quad \text{a.s.}$$

Since X_M' is bounded, the bounded convergence theorem implies

$$E\left|\frac{1}{n} \sum_{m=0}^{n-1} X_M'(\varphi^m \omega) - E(X_M'|\mathcal{I})\right| \to 0$$

To handle X_M'' we observe

$$E\left|\frac{1}{n}\sum_{m=0}^{n-1}X_M''(\varphi^m\omega)\right| \le \frac{1}{n}\sum_{m=0}^{n-1}E|X_M''(\varphi^m\omega)| = E|X_M''|$$

and $E|E(X_M''|\mathcal{I})| \le EE(|X_M''||\mathcal{I}) = E|X_M''|$. So

$$E\left|\frac{1}{n}\sum_{m=0}^{n-1}X_M''(\varphi^m\omega) - E(X_M''|\mathcal{I})\right| \le 2E|X_M''|$$

and it follows that

$$\limsup_{n\to\infty} E\left|\frac{1}{n}\sum_{m=0}^{n-1}X(\varphi^m\omega) - E(X|\mathcal{I})\right| \le 2E|X_M''|$$

As $M \to \infty$, $E|X_M''| \to 0$ by the dominated convergence theorem, so we have completed the proof of (2.1).   □

EXERCISE 2.1. Show that if $X \in L^p$ with $p > 1$ then the convergence in (2.1) occurs in L^p.

EXERCISE 2.2. (i) Show that if $g_n(\omega) \to g(\omega)$ a.s. and $E(\sup_k |g_k(\omega)|) < \infty$, then

$$\lim_{n\to\infty}\frac{1}{n}\sum_{m=0}^{n-1}g_m(\varphi^m\omega) = E(g|\mathcal{I})\quad \text{a.s.}$$

(ii) Show that if we suppose only that $g_n \to g$ in L^1 we get L^1 convergence.

Before turning to examples, we would like to prove a useful result that is a simple consequence of (2.2):

(2.3) **Wiener's maximal inequality.** Let $X_j(\omega) = X(\varphi^j\omega)$, $S_k(\omega) = X_0(\omega) + \cdots + X_{k-1}(\omega)$, $A_k(\omega) = S_k(\omega)/k$, and $D_k = \max(A_1,\ldots,A_k)$. If $\alpha > 0$ then

$$P(D_k > \alpha) \le \alpha^{-1}E|X|$$

Proof   Let $B = \{D_k > \alpha\}$. Applying (2.2) to $X' = (X - \alpha)1_B$, with $X_j'(\omega) = X'(\varphi^j\omega)$, $S_k' = X_0'(\omega) + \cdots + X_{k-1}'$, and $M_k' = \max(0, S_1',\ldots,S_k')$ gives $E(X'; M_k' > 0) \ge 0$. Since $\{M_k' > 0\} = \{D_k > \alpha\} \equiv B$, it follows that

$$E|X| \ge \int_B X\,dP \ge \alpha\int_B 1_B\,dP = \alpha P(B)$$ □

EXERCISE 2.3. Use (2.3) and the truncation argument at the end of the proof of (2.1) to conclude that if (2.1) holds for bounded r.v.'s then it holds whenever $E|X| < \infty$.

Our next step is to see what (2.1) says about our examples.

Example 2.1. i.i.d. sequences. Since $\mathcal{I}$ is trivial, the ergodic theorem implies that

$$\frac{1}{n} \sum_{m=0}^{n-1} X_m \to EX_0 \quad \text{a.s. and in } L^1$$

The a.s. convergence is the strong law of large numbers.

Remark. We can prove the L^1 convergence in the law of large numbers without invoking the ergodic theorem. To do this note that

$$\frac{1}{n} \sum_{m=1}^{n} X_m^+ \to EX^+ \quad \text{a.s.} \qquad E\left(\frac{1}{n} \sum_{m=1}^{n} X_m^+\right) = EX^+$$

and use (5.2) in Chapter 4 to conclude that $\frac{1}{n} \sum_{m=1}^{n} X_m^+ \to EX^+$ in L^1. A similar result for the negative part and the triangle inequality now give the desired result.

Example 2.2. Markov chains. Let X_n be an irreducible Markov chain on a countable state space which has a stationary distribution π. Let f be a function with

$$\sum_x |f(x)|\pi(x) < \infty$$

In Section 1 we showed that $\mathcal{I}$ is trivial, so applying the ergodic theorem to $f(X_0(\omega))$ gives

$$\frac{1}{n} \sum_{m=0}^{n-1} f(X_m) \to \sum_x f(x)\pi(x) \quad \text{a.s. and in } L^1$$

For another proof, see Exercise 5.5 in Chapter 5.

Example 2.3. Rotation of the circle. $\Omega = [0,1)$ $\varphi(\omega) = (\omega + \theta) \bmod 1$. Suppose that $\theta \in (0,1)$ is irrational so Example 1.3 implies that $\mathcal{I}$ is trivial. If we set $X(\omega) = 1_A(\omega)$, with A a Borel subset of $[0,1)$, then the ergodic theorem implies

$$\frac{1}{n} \sum_{m=0}^{n-1} 1_{(\varphi^m \omega \in A)} \to |A| \quad \text{a.s.}$$

where $|A|$ denotes the Lebesgue measure of A. The last result for $\omega = 0$ is usually called **Weyl's equidistribution theorem**, although Bohl and Sierpinski should also get credit. For the history and a nonprobabilistic proof, see Hardy and Wright (1959) p. 390–393.

To recover the number theoretic result we will now show that

(2.4) Theorem. If $A = [a, b)$ then the exceptional set is $\emptyset$.

Proof Let $A_k = [a + 1/k, b - 1/k)$. If $b - a > 2/k$, the ergodic theorem implies

$$\frac{1}{n} \sum_{m=0}^{n-1} 1_{A_k}(\varphi^m \omega) \to b - a - \frac{2}{k}$$

for $\omega \in \Omega_k$ with $P(\Omega_k) = 1$. Let $G = \cap \Omega_k$ where the intersection is over integers k with $b - a > 2/k$. $P(G) = 1$ so G is dense in $[0,1)$. If $x \in [0, 1)$ and $\omega_k \in G$ with $|\omega_k - x| < 1/k$ then $\varphi^m \omega_k \in A_k$ implies $\varphi^m x \in A$ so

$$\liminf_{n \to \infty} \frac{1}{n} \sum_{m=0}^{n-1} 1_A(\varphi^m x) \geq b - a - \frac{2}{k}$$

for all large enough k. Noting that k is arbitrary and applying similar reasoning to A^c shows

$$\frac{1}{n} \sum_{m=0}^{n-1} 1_A(\varphi^m x) \to b - a \qquad \square$$

Example 2.4. Benford's law. As Gelfand first observed, the equidistribution theorem says something interesting about 2^m. Let $\theta = \log_{10} 2$, $1 \leq k \leq 9$, and $A_k = [\log_{10} k, \log_{10} k + 1)$ where $\log_{10} y$ is the logarithm of y to the base 10. Taking $x = 0$ in the last result we have

$$\frac{1}{n} \sum_{m=0}^{n-1} 1_A(\varphi^m 0) \to \log_{10} \left(\frac{k+1}{k} \right)$$

A little thought reveals that the first digit of 2^m is k if and only if $m\theta$ mod $1 \in A_k$. Taking $k = 1$ for example we have shown that the asymptotic fraction of time 1 is the first digit of 2^m is $\log_{10} 2 = .3010$.

The limit distribution on $\{1, \ldots, 9\}$ is called Benford's (1938) law, although it was discovered by Newcomb (1881). As Raimi (1976) explains, in many tables the observed frequency with which k appears as a first digit is approximately $\log_{10}((k + 1)/k)$. He mentions powers of two as an example. Two other data sets that fit this very well are (i) the street addresses of the first 342 person in American Men Of Science, 1938, and (ii) the kilowatt hours of 1243 electric bills for October 1969 from Honiara in the British Solomon Islands. We leave it to the reader to figure out why Benford's law should appear in the last two situations.

Example 2.5.   Bernoulli shift. $\Omega = [0,1)$, $\varphi(\omega) = (2\omega) \bmod 1$. Let $i_1, \ldots, i_k \in \{0, 1\}$, let $r = i_1 2^{-1} + \cdots + i_k 2^{-k}$, and let $X(\omega) = 1$ if $r \leq \omega < r + 2^{-k}$. In words, $X(\omega) = 1$ if the first k digits of the binary expansion of ω are $i_1, \ldots, i_k$. The ergodic theorem implies that

$$\frac{1}{n} \sum_{m=0}^{n-1} X(\varphi^m \omega) \to 2^{-k} \quad \text{a.s.}$$

i.e., in almost every $\omega \in [0, 1)$ the pattern $i_1, \ldots, i_k$ occurs with its expected frequency. Since there are only a countable number of patterns (of finite length) it follows that almost every $\omega \in [0, 1)$ is **normal**, i.e., all patterns occur with their expected frequency. This is the binary version of Borel's (1909) normal number theorem.

6.3.   Recurrence

In this section we will study the recurrence properties of stationary sequences. Our first result is an application of the ergodic theorem. Let $X_1, X_2, \ldots$ be a stationary sequence taking values in $\mathbf{R}^d$, let $S_k = X_1 + \cdots + X_k$, let $A = \{S_k \neq 0$ for all $k \geq 1\}$, and let $R_n = |\{S_1, \ldots, S_n\}|$ be the number of points visited at time n. Kesten, Spitzer, and Whitman (see Spitzer (1964) p.40) proved the next result when the X_i are i.i.d. In that case $\mathcal{I}$ is trivial so the limit is $P(A)$.

(3.1) Theorem. As $n \to \infty$, $R_n/n \to E(1_A | \mathcal{I})$ a.s.

Proof Suppose $X_1, X_2, \ldots$ are constructed on $(\mathbf{R}^d)^{\{0, 1, \cdots\}}$ with $X_n(\omega) = \omega_n$, and let φ be the shift operator. It is clear that

$$R_n \geq \sum_{m=1}^{n} 1_A(\varphi^m \omega)$$

since the right-hand side $= |\{m : 1 \leq m \leq n, S_\ell \neq S_m \text{ for all } \ell > m\}|$. Using the ergodic theorem now gives

$$\liminf_{n \to \infty} R_n/n \geq E(1_A | \mathcal{I}) \quad \text{a.s.}$$

To prove the opposite inequality, let $A_k = \{S_1 \neq 0, S_2 \neq 0, \ldots, S_k \neq 0\}$. It is clear that

$$R_n \leq k + \sum_{m=1}^{n-k} 1_{A_k}(\varphi^m \omega)$$

since the sum on the right-hand side $= |\{m : 1 \leq m \leq n - k, S_\ell \neq S_m$ for $m < \ell \leq m + k\}|$. Using the ergodic theorem now gives

$$\limsup_{n \to \infty} R_n/n \leq E(1_{A_k}|\mathcal{I})$$

As $k \uparrow \infty$, $A_k \downarrow A$, so the monotone convergence theorem for conditional expectations, (1.1c) in Chapter 4, implies

$$E(1_{A_k}|\mathcal{I}) \downarrow E(1_A|\mathcal{I}) \quad \text{as } k \uparrow \infty$$

and the proof is complete. $\square$

EXERCISE 3.1. Let $g_n = P(S_1 \neq 0, \ldots, S_n \neq 0)$ for $n \geq 1$ and $g_0 = 1$. Show that $ER_n = \sum_{m=1}^n g_{m-1}$.

From (3.1) we get a result about the recurrence of random walks with stationary increments that is (for integer valued random walks) a generalization of the Chung-Fuchs theorem ((2.7) in Chapter 3).

(3.2) Theorem. Let $X_1, X_2, \ldots$ be a stationary sequence taking values in $\mathbf{Z}$ with $E|X_i| < \infty$. Let $S_n = X_1 + \cdots + X_n$, and let $A = \{S_1 \neq 0, S_2 \neq 0, \ldots\}$. (i) If $E(X_1|\mathcal{I}) = 0$ then $P(A) = 0$. (ii) If $P(A) = 0$ then $P(S_n = 0 \text{ i.o.}) = 1$.

Remark. In words, mean zero implies recurrence. The condition $E(X_1|\mathcal{I}) = 0$ is needed to rule out trivial examples that have mean 0 but are a combination of a sequence with positive and negative means, e.g., $P(X_n = 1$ for all $n) = P(X_n = -1$ for all $n) = 1/2$.

Proof of (i)   If $E(X_1|\mathcal{I}) = 0$ then the ergodic theorem implies $S_n/n \to 0$ a.s. Now

$$\limsup_{n \to \infty} \left(\max_{1 \leq k \leq n} |S_k|/n \right) = \limsup_{n \to \infty} \left(\max_{K \leq k \leq n} |S_k|/n \right) \leq \left(\max_{k \geq K} |S_k|/k \right)$$

for any K and the right-hand side $\downarrow 0$ as $K \uparrow \infty$. The last conclusion leads easily to

$$\lim_{n \to \infty} \left(\max_{1 \leq k \leq n} |S_k| \right) \Big/ n = 0 = \lim_{n \to \infty} \left(\min_{1 \leq k \leq n} |S_k| \right) \Big/ n$$

Since

$$R_n \leq 1 + \max_{1 \leq k \leq n} S_k - \min_{1 \leq k \leq n} S_k$$

it follows that $R_n/n \to 0$ and (3.1) implies $P(A) = 0$.

Proof of (ii)   Let $F_j = \{S_i \neq 0 \text{ for } i < j, S_j = 0\}$ and $G_{j,k} = \{S_{j+i} - S_j \neq 0 \text{ for } i < k, S_{j+k} - S_j = 0\}$. $P(A) = 0$ implies that $\sum P(F_k) = 1$. Stationarity implies $P(G_{j,k}) = P(F_k)$, and for fixed j the $G_{j,k}$ are disjoint, so $\cup_k G_{j,k} = \Omega$ a.s. It follows that

$$\sum_k P(F_j \cap G_{j,k}) = P(F_j) \quad \text{and} \quad \sum_{j,k} P(F_j \cap G_{j,k}) = 1$$

On $F_j \cap G_{j,k}$, $S_j = 0$ and $S_{j+k} = 0$, so we have shown $P(S_n = 0$ at least two times $) = 1$. Repeating the last argument shows $P(S_n = 0$ at least k times$) = 1$ for all k, and the proof is complete. $\qquad\square$

EXERCISE 3.2.   Imitate the proof of (i) in (3.2) to show that if we assume $P(X_i > 1) = 0$ and $EX_i > 0$ in addition to the hypotheses of (3.2) then $P(A) = EX_i$.

Remark. You have proved the last result twice for asymmetric simple random walk (Exercise 1.13 in Chapter 3, Exercise 7.3 in Chapter 4). For more general random walks, the conclusion is new. It is interesting to note that we can use martingale theory to prove a result for random walks that do not skip over integers on the way down.

EXERCISE 3.3.   Suppose $X_1, X_2, \ldots$ are i.i.d. with $P(X_i < -1) = 0$ and $EX_i > 0$. If $P(X_i = -1) > 0$ there is a unique $\theta < 0$ so that $E \exp(\theta X_i) = 1$. Let $S_n = X_1 + \cdots + X_n$ and $N = \inf\{n : S_n < 0\}$. Show that $\exp(\theta S_n)$ is a martingale and use the optional stopping theorem to conclude that $P(N < \infty) = e^\theta$.

Extending the reasoning in the proof of part (ii) of (3.2) gives a result of Kac (1947b). Let $X_0, X_1, \ldots$ be a stationary sequence taking values in $(S, \mathcal{S})$. Let $A \in \mathcal{S}$, let $T_0 = 0$, and for $n \geq 1$, let $T_n = \inf\{m > T_{n-1} : X_m \in A\}$ be the time of the nth return to A.

(3.3) **Theorem.** If $P(X_n \in A \text{ at least once}) = 1$, then under $P(\cdot | X_0 \in A)$, $t_n = T_n - T_{n-1}$ is a stationary sequence with $E(T_1 | X_0 \in A) = 1/P(X_0 \in A)$.

Remark. If X_n is an irreducible Markov chain on a countable state space S starting from its stationary distribution π, and $A = \{x\}$, then (3.3) says

$E_x T_x = 1/\pi(x)$, which is (4.6) in Chapter 5. (3.3) extends that result to an arbitrary $A \subset S$ and drops the assumption that X_n is a Markov chain.

Proof We first show that under $P(\cdot|X_0 \in A)$, $t_1, t_2, \ldots$ is stationary. To cut down on ...'s, we will only show that

$$P(t_1 = m, t_2 = n | X_0 \in A) = P(t_2 = m, t_3 = n | X_0 \in A)$$

It will be clear that the same proof works for any finite dimensional distribution. Our first step is to extend $\{X_n, n \geq 0\}$ to a two sided stationary sequence $\{X_n, n \in \mathbf{Z}\}$ using (1.2). Let $C_k = \{X_{-1} \notin A, \ldots, X_{-k+1} \notin A, X_{-k} \in A\}$.

$$\left(\cup_{k=1}^K C_k\right)^c = \{X_k \notin A \text{ for } -K \leq k \leq -1\}$$

The last event has the same probability as $\{X_k \notin A \text{ for } 1 \leq k \leq K\}$, so letting $K \to \infty$ we get $P\left(\cup_{k=1}^{\infty} C_k\right) = 1$. To prove the desired stationarity, we let $I_{j,k} = \{i \in [j, k] : X_i \in A\}$ and observe that

$$\begin{aligned}
P(t_2 = m, t_3 = n, X_0 \in A) &= \sum_{\ell=1}^{\infty} P(X_0 \in A, t_1 = \ell, t_2 = m, t_3 = n) \\
&= \sum_{\ell=1}^{\infty} P(I_{0,\ell+m+n} = \{0, \ell, \ell+m, \ell+m+n\}) \\
&= \sum_{\ell=1}^{\infty} P(I_{-\ell, m+n} = \{-\ell, 0, m, m+n\}) \\
&= \sum_{\ell=1}^{\infty} P(C_\ell, X_0 \in A, t_1 = m, t_2 = n)
\end{aligned}$$

To complete the proof, we compute

$$E(t_1 | X_0 \in A) = \sum_{k=1}^{\infty} P(t_1 \geq k | X_0 \in A) = P(X_0 \in A)^{-1} \sum_{k=1}^{\infty} P(t_1 \geq k, X_0 \in A)$$

$$= P(X_0 \in A)^{-1} \sum_{k=1}^{\infty} P(C_k) = 1/P(X_0 \in A)$$

since the C_k are disjoint and their union has probability 1. $\square$

In the next two exercises we continue to use the notation of (3.3).

EXERCISE 3.4. Show that if $P(X_n \in A \text{ at least once}) = 1$ and $A \cap B = \emptyset$ then

$$E\left(\sum_{1 \leq m \leq T_1} 1_{(X_m \in B)} \middle| X_0 \in A\right) = \frac{P(X_0 \in B)}{P(X_0 \in A)}$$

When $A = \{x\}$ and X_n is a Markov chain, this is the "cycle trick" for defining a stationary measure. (See (4.3) in Chapter 5.)

EXERCISE 3.5. Consider the special case in which $X_n \in \{0,1\}$, and let $\bar{P} = P(\cdot|X_0 = 1)$. Here $A = \{1\}$ and so $T_1 = \inf\{m > 0 : X_m = 1\}$. Show $P(T_1 = n) = \bar{P}(T_1 \geq n)/\bar{E}T_1$. When $t_1, t_2, \ldots$ are i.i.d. this reduces to the formula for the first waiting time in a stationary renewal process.

In checking the hypotheses of Kac's theorem, a result Poincaré proved in 1899 is useful. First we need a definition. Let $T_A = \inf\{n \geq 1 : \varphi^n(\omega) \in A\}$.

(3.4) Theorem. Suppose $\varphi : \Omega \to \Omega$ preserves P, that is, $P \circ \varphi^{-1} = P$. (i) $T_A < \infty$ a.s. on A, that is, $P(\omega \in A, T_A = \infty) = 0$. (ii) $\{\varphi^n(\omega) \in A \text{ i.o.}\} \supset A$. (iii) If φ is ergodic and $P(A) > 0$ then $P(\varphi^n(\omega) \in A \text{ i.o.}) = 1$.

Remark. Note that in (i) and (ii) we assume only that φ is measure preserving. Extrapolating from Markov chain theory, the conclusions can be "explained" by noting that: (i) the existence of a stationary distribution implies the sequence is recurrent, and (ii) since we start in A we do not have to assume irreducibility. Conclusion (iii) is, of course, a consequence of the ergodic theorem, but as the self-contained proof below indicates, it is a much simpler fact.

Proof Let $B = \{\omega \in A, T_A = \infty\}$. A little thought shows that if $\omega \in \varphi^{-m}B$ then $\varphi^m(\omega) \in A$ but $\varphi^n(\omega) \notin A$ for $n > m$, so the $\varphi^{-m}B$ are pairwise disjoint. The fact that φ is measure preserving implies $P(\varphi^{-m}B) = P(B)$ so we must have $P(B) = 0$ (or P would have infinite mass). To prove (ii), note that for any k, φ^k is measure preserving, so (i) implies

$$0 = P(\omega \in A, \varphi^{nk}(\omega) \notin A \text{ for all } n \geq 1)$$
$$\geq P(\omega \in A, \varphi^m(\omega) \notin A \text{ for all } m \geq k)$$

Since the last probability is 0 for all k, (ii) follows. Finally for (iii), note that $B \equiv \{\omega : \varphi^n(\omega) \in A \text{ i.o.}\}$ is invariant and $\supset A$ by (b), so $P(B) > 0$ and it follows from ergodicity that $P(B) = 1$. $\qquad\square$

*6.4. Mixing

A measure preserving transformation φ on $(\Omega, \mathcal{F}, P)$ is called **mixing** if for all measurable sets A and B

(4.1) $$\lim_{n \to \infty} P(A \cap \varphi^{-n}B) = P(A)P(B)$$

A sequence X_n, $n \geq 0$, is said to be **mixing** if the corresponding shift on sequence space is. To see that mixing implies ergodicity, observe that if A is invariant then (4.1) implies $P(A) = P(A)^2$, i.e., $P(A) \in \{0,1\}$. In the other direction, ergodicity implies

$$\frac{1}{n} \sum_{m=0}^{n-1} 1_B(\varphi^m \omega) \to P(B) \quad \text{a.s.}$$

so integrating over A and using the bounded convergence theorem gives

(4.2) $$\frac{1}{n} \sum_{m=0}^{n-1} P(A \cap \varphi^{-m} B) \to P(A)P(B)$$

i.e., (4.1) holds in a Cesaro sense. To see that (4.2) is equivalent to ergodicity, we note that the ergodic theorem implies

$$\frac{1}{n} \sum_{m=0}^{n-1} 1_B(\varphi^m \omega) \to E(1_B | \mathcal{I}) \quad \text{a.s.}$$

Integrating over A using the bounded convergence theorem and using (4.2) we have

$$\int_A E(1_B | \mathcal{I}) \, dP = P(A)P(B)$$

Since this holds for all A, $E(1_B | \mathcal{I}) = P(B)$. Since that holds for all B, $\mathcal{I}$ must be trivial.

As usual when we meet a new concept, we turn to our examples to see what it means. To handle Examples 1.1 and 1.2, we use a general result. For this result we suppose that φ is the shift operator on sequence space, and $X_n(\omega) = \omega_n$. Let $\mathcal{F}'_n = \sigma(X_n, X_{n+1}, \dots)$ and $\mathcal{T} = \cap_n \mathcal{F}'_n$.

(4.3) Theorem. If $\mathcal{T}$ is trivial then φ is mixing and

(*) $$\lim_{n \to \infty} \sup_B |P(A \cap \varphi^{-n} B) - P(A)P(B)| = 0$$

Conversely, if (*) holds $\mathcal{T}$ is trivial.

Proof Let $C = \varphi^{-n} B \in \mathcal{F}'_n$.

$$|P(A \cap C) - P(A)P(C)| = \left| \int_C (1_A - P(A)) \, dP \right|$$

$$= \left| \int_C P(A | \mathcal{F}'_n) - P(A) \, dP \right|$$

$$\leq \int |P(A | \mathcal{F}'_n) - P(A)| \, dP \to 0$$

since $P(A|\mathcal{F}'_n) \to P(A)$ in L^1 by (6.3) in Chapter 4.

Conversely if $A \in \mathcal{T}$ has $P(A) \in (0,1)$ and we write $A = \varphi^{-n}B_n$ (which is possible since $A \in \mathcal{F}'_n$) then $P(B_n) = P(A)$ so

$$|P(A \cap \varphi^{-n}B_n) - P(A)P(B_n)| = P(A) - P(A)^2$$

and $(*)$ is false. $\qquad\qquad\qquad\qquad\qquad\qquad\qquad\qquad\qquad\qquad\qquad\qquad$ $\square$

Combining (4.3), the definition of mixing, and (4.2) we have

$\mathcal{T}$ is trivial iff	$\sup_B |P(A \cap \varphi^{-n}B) - P(A)P(B)| \to 0$	for all A
φ is mixing iff	$P(A \cap \varphi^{-n}B) - P(A)P(B) \to 0$	for all A, B
φ is ergodic iff	$\frac{1}{n}\sum_{m=0}^{n-1} P(A \cap \varphi^{-m}B) \to P(A)P(B)$	for all A, B

Example 4.1. i.i.d. sequences are mixing since $\mathcal{T}$ is trivial.

Example 4.2. Markov chains. Suppose the state space S is countable, p is irreducible, and has stationary distribution π. If p is aperiodic, $\mathcal{T}$ is trivial by (5.8) in Chapter 5 and the sequence is mixing. Conversely if p has period d and cyclic decomposition $S_0, S_1, \ldots, S_{d-1}$

$$\lim_{n \to \infty} P(X_0 \in S_0, X_{nd+1} \in S_0) = 0 \neq \pi(S_0)^2$$

so the sequence is not mixing.

Example 4.3. Rotation of the circle. Let $\Omega = [0,1)$ and $\varphi(\omega) = (\omega + \theta)$ mod 1, where θ is irrational. To prove that this transformation is not mixing, we begin by observing that $n\theta$ mod 1 is dense. (Use Exercise 1.3 or recall (2.4) shows that for intervals the exceptional set in the ergodic theorem is $\emptyset$). Since $n\theta$ mod 1 is dense, there is a sequence $n_k \to \infty$ with $n_k\theta$ mod $1 \to 1/2$. Let $A = B = [0, 1/3)$. If k is large then $A \cap \varphi^{-n_k}B = \emptyset$, so $P(A \cap \varphi^{-n_k}B)$ does not converge to 1/9 and the transformation is not mixing.

To treat our last example, we will use

(4.4) Lemma. Let $\mathcal{A}$ be a π-system. If

$$(\star) \qquad\qquad \lim_{n \to \infty} P(A \cap \varphi^{-n}B) = P(A)P(B)$$

holds for $A, B \in \mathcal{A}$ then $(\star)$ holds for $A, B \in \sigma(\mathcal{A})$.

Proof As the reader can probably guess, we are going to use the $\pi - \lambda$ theorem ((4.2) in Chapter 1). Fix $A \in \mathcal{A}$ and let $\mathcal{L}$ be the collection of B for which $(\star)$

holds. By assumption $\mathcal{L} \supset \mathcal{A}$. To begin to check the assumptions of the $\pi - \lambda$ theorem, we note that $\Omega \in \mathcal{L}$. Let $B_1 \supset B_2$ be in $\mathcal{L}$, $\varphi^{-n}(B_1 - B_2) = \varphi^{-n}B_1 - \varphi^{-n}B_2$ so

$$P(A \cap \varphi^{-n}(B_1 - B_2)) = P(A \cap \varphi^{-n}B_1) - P(A \cap \varphi^{-n}B_2)$$

and

$$\lim_{n \to \infty} P(A \cap \varphi^{-n}(B_1 - B_2)) = P(A)P(B_1) - P(A)P(B_2) = P(A)P(B_1 - B_2)$$

To finish the proof that $\mathcal{L}$ is a λ-system, let $B_k \in \mathcal{L}$ with $B_k \uparrow B$.

$$|P(A \cap \varphi^{-n}B) - P(A \cap \varphi^{-n}B_k)| \le |P(\varphi^{-n}B) - P(\varphi^{-n}B_k)| = P(B) - P(B_k)$$

since φ^n preserves P.

At this point we have checked that $\mathcal{L}$ is a λ-system so the $\pi - \lambda$ theorem implies that $\mathcal{L} \supset \sigma(\mathcal{L})$. In other words, if $A \in \mathcal{A}$ then $(\star)$ holds for $B \in \sigma(\mathcal{A})$. Fixing $B \in \sigma(\mathcal{A})$, the reader can now repeat the last argument to show that $(\star)$ holds for $A, B \in \sigma(\mathcal{A})$. Indeed, the proof is simpler since you do not have to worry about what φ^{-n} does. $\qquad \square$

Example 4.4. Bernoulli shift. $\Omega = [0, 1)$, $\varphi(\omega) = 2\omega \mod 1$. Let $\mathcal{F}_n = \sigma([k2^{-n}, (k+1)2^{-n}); 0 \le k < 2^n)$ and $\mathcal{A} = \cup \mathcal{F}_n$. If $A \in \mathcal{F}_n$, $B \in \mathcal{A}$, and then A and $\varphi^{-m}B$ are independent for $m \ge n$, i.e., $P(A \cap \varphi^{-m}B) = P(A)P(\varphi^{-m}B) = P(A)P(B)$ since φ is measure preserving. So (4.3) implies φ is mixing.

Up to this point all our examples either have a trivial tail field (Examples 4.1, 4.2, and 4.4) or are not mixing (Example 4.3). To complete the picture we close the section with some examples that are mixing but have nontrivial tail fields.

Example 4.5. Let X_n, $n \in \mathbf{Z}$, be i.i.d. with $P(X_n = 1) = P(X_n = 0) = 1/2$ and let

$$Z_n(\omega) = \sum_{m=0}^{\infty} 2^{-(m+1)} X_{n-m}(\omega)$$

Z_n is mixing but $\mathcal{T} = \sigma(X_n : n \in \mathbf{Z})$.

Proof Since all the X_k, $k \le n$, can be recovered by expanding Z_n in its binary decimal representation, the claim about $\mathcal{T}$ is clear. To prove that Z_n is mixing, we use (4.4) with $\mathcal{A} = $ sets of the form $\{\omega : Z_i(\omega) \in G_i, 0 \le i \le k\}$ where the G_i are open intervals. Now

$$Z_n = \sum_{m=0}^{n-k-1} 2^{-(m+1)} X_{n-m} + 2^{-(n-k)} Z_k$$

so if $\mathcal{F}_k = \sigma(X_j : j \leq k)$ and G_0 is an open interval

$$P(Z_n \in G_0 | \mathcal{F}_k) = \mu_{n-k-1}(G_0 - 2^{-(n-k)}Z_k)$$

where μ_M is the distribution of $\sum_{0 \leq m \leq M} 2^{-(m+1)}X_{-m}$ and $G_0 - c = \{x - c : x \in G_0\}$.

Thinking about the binary digits of a number chosen at random from $[0,1)$ it is easy to see that as $M \to \infty$, $\mu_M \Rightarrow \mu_\infty$, the uniform distribution on $(0,1)$. Since the uniform distribution has no atoms, so for a.e. ω

$$\mu_{n-k-1}(G_0 - 2^{-(n-k)}Z_k) \to \mu_\infty(G_0)$$

Integrating over $A \in \mathcal{F}_k$ and using the bounded convergence theorem and definition of conditional expectation

$$P(Z_n \in G_0, A) \to \mu_\infty(G_0)P(A)$$

Letting $B = \{\omega : Z_0(\omega) \in G_0\}$ we have shown

$$P(A \cap \varphi^{-n}B) \to P(A)P(B) \quad \text{for } A \in \mathcal{F}_k$$

The last argument generalizes easily to $B = \{\omega : Z_i(\omega) \in G_i, 0 \leq i \leq k\}$. This checks (4.4) and completes the proof. $\qquad \square$

Our last example, borrowed from Lasota and MacKay (1985), is named for its famous cousin: geodesic flow on a compact Riemannian manifold M with negative curvature, e.g., the two hole torus with a suitable metric. For more on these examples see Anosov (1963), (1967).

Example 4.6.   Anosov map. $\Omega = [0,1)^2$, $\varphi(x, y) = (x + y, x + 2y)$. By drawing a picture it is easy to see that φ maps Ω $1 - 1$ onto itself. Since the Jacobian

$$J = \det \begin{pmatrix} 1 & 1 \\ 1 & 2 \end{pmatrix} = 1$$

φ preserves Lebesgue measure. Since φ is invertible the entire sequence can be recovered from one term and T is far from trivial. We will now show that φ is mixing. To begin we observe that an induction argument shows

$$\varphi^n(x, y) = (a_{2n-2}x + a_{2n-1}y, a_{2n-1}x + a_{2n}y)$$

where the a_n are the **Fibonacci numbers** given by $a_0 = a_1 = 1$ and $a_{n+1} = a_n + a_{n-1}$ for $n \geq 1$. (To help check this note that $a_{2n-2} + 2a_{2n-1} + a_{2n} = a_{2n} + a_{2n+1}$ and then subtract a_{2n} from each side.) Let

$$f(x, y) = \exp(2\pi i(px + qy)) \qquad g(x, y) = \exp(2\pi i(rx + sy))$$

Since $\int_{[0,1]} \exp(2\pi i kx)dx = 0$ unless $k = 0$, it follows that

$$\int_0^1 \int_0^1 f(x,y)g(\varphi^n(x,y))\, dx\, dy = 0$$

unless

(a) $\qquad ra_{2n-2} + sa_{2n-1} + p = 0 \qquad ra_{2n-1} + sa_{2n} + q = 0$

Now the difference equation $b_{n+1} - b_n - b_{n-1} = 0$ has a two parameter family of solutions given by

$$b_n = C_1 \left(\frac{1+\sqrt{5}}{2}\right)^n + C_2 \left(\frac{1-\sqrt{5}}{2}\right)^n$$

so the Fibonacci numbers are given by

(b) $\qquad a_n = \dfrac{1}{\sqrt{5}} \left(\dfrac{1+\sqrt{5}}{2}\right)^{n+1} - \dfrac{1}{\sqrt{5}} \left(\dfrac{1-\sqrt{5}}{2}\right)^{n+1} \qquad$ for $n \geq 0$

To see this note that $b_n = \lambda^n$ is a solution of $b_{n+1} - b_n - b_{n-1} = 0$ if and only if $\lambda^2 - \lambda - 1 = 0$ and any solution is determined by the values of b_0 and b_1.

From (b) we see that

$$\lim_{n\to\infty} \frac{a_{2n}}{a_{2n-1}} = \lim_{n\to\infty} \frac{a_{2n-1}}{a_{2n-2}} = \frac{1+\sqrt{5}}{2}$$

Since $(1+\sqrt{5})/2$ is irrational, (a) cannot hold for infinitely many n unless $p = q = r = s = 0$. The last result implies that if

(c)
$$f(x,y) = \sum_{j=1}^{k} a_j \exp(2\pi i(p_j x + q_j y))$$

$$g(x,y) = \sum_{j=1}^{\ell} b_j \exp(2\pi i(r_j x + s_j y))$$

then as $n \to \infty$

$$\int_0^1 \int_0^1 f(x,y)g(\varphi^n(x,y))\, dx\, dy \to 0$$

To finish the proof, we observe

(4.5) Lemma. Since the f and g for which (c) holds are dense in

$$L_0^2 = \{f : \int_0^1 \int_0^1 f(x, y)\, dx\, dy = 0, \quad \int_0^1 \int_0^1 f^2(x, y)\, dx\, dy < \infty\}$$

it follows that φ is mixing.

Proof Let

$$< h, k > = \int_0^1 \int_0^1 h(x, y)k(x, y)\, dx\, dy$$

and $\|h\|_2 = < h, h >^{1/2}$. Adding and subtracting $< h, g \circ \varphi^n >$ then using the Cauchy-Schwarz inequality gives

$$
\begin{aligned}
| < f, g \circ \varphi^n > - < h, k \circ \varphi^n > | &\le | < f, g \circ \varphi^n > - < h, g \circ \varphi^n > | \\
&\quad + | < h, g \circ \varphi^n > - < h, k \circ \varphi^n > | \\
&\le \|f - h\|_2 \|g \circ \varphi^n\|_2 + \|h\|_2 \|(g - k) \circ \varphi^n\|_2 \\
&= \|f - h\|_2 \|g\|_2 + \|h\|_2 \|(g - k)\|_2
\end{aligned}
$$

Suppose now that $h = 1_A - P(A)$, $k = 1_B - P(B) \in L_0^2$, pick f and g so that (c) holds and $\|h - f\|_2$, $\|k - g\|_2$ are $< \epsilon$. The last inequality implies

$$\limsup_{n \to \infty} |P(A \cap \varphi^{-n} B) - P(A)P(B)| = \limsup_{n \to \infty} | < h, k \circ \varphi^n > |$$

$$\le \epsilon(\|k\|_2 + \epsilon) + \epsilon\|h\|_2$$

and since ϵ is arbitrary the proof is complete. $\qquad\square$

*6.5. Entropy

Throughout this section we will suppose that X_n, $n \in \mathbf{Z}$, is an ergodic stationary sequence taking values in a finite set S. Let

$$p(x_0, \ldots, x_{n-1}) = P(X_0 = x_0, \ldots, X_{n-1} = x_{n-1})$$

and

$$p(x_n | x_{n-1}, \ldots, x_0) = P(X_n = x_n | X_{n-1} = x_{n-1}, \ldots, X_0 = x_0)$$

whenever the conditioning event has positive probability. Define random variables

$$p(X_0, \ldots, X_n) \quad \text{and} \quad p(X_n | X_{n-1}, \ldots, X_0)$$

by setting $x_j = X_j(\omega)$ in the corresponding definitions. Since

$$P(p(X_0, \ldots, X_n) = 0) = 0$$

the conditional probability makes sense a.s.

(5.1) **The Shannon-McMillan-Breiman theorem** asserts that

$$-\frac{1}{n} \log \ p(X_0, \ldots, X_{n-1}) \to H \quad \text{a.s.}$$

where $H = \lim_{n \to \infty} E\{-\log p(X_n | X_{n-1}, \ldots, X_0)\}$ is the **entropy rate** of X_n.

Remark. The three names indicate the evolution of the theorem: Shannon (1948), McMillan (1953), Breiman (1957). Our proof follows Algoet and Cover (1988).

Proof   The first step is to prove that the limit defining H exists. To do this, we observe that if $\mathcal{F}_n = \sigma(X_{-1}, \ldots, X_{-n})$ and

$$Y_n = p(x_0 | X_{-1}, \ldots, X_{-n}) = P(X_0 = x_0 | \mathcal{F}_n)$$

then Y_n is a bounded martingale, so as $n \to \infty$

$$Y_n \to Y_\infty \equiv p(x_0 | X_{-1}, X_{-2}, \ldots) \quad \text{a.s. and in } L^1$$

Since S is finite and $x \log x$ is bounded on $[0, 1]$, it follows that

(a) $\qquad H_k \equiv E(-\log p(X_0 | X_{-1}, \ldots, X_{-k}))$

$$= E\left(-\sum_x p(x | X_{-1}, \ldots, X_{-k}) \log p(x | X_{-1}, \ldots, X_{-k})\right)$$

$$\to E\left(-\sum_x p(x | X_{-1}, X_{-2}, \ldots) \log p(x | X_{-1}, X_{-2}, \ldots)\right) \equiv H$$

as $k \to \infty$.

Our second step is to find something related to the quantity of interest that converges to H. Elementary conditional probabilities give

$$p(X_0, \ldots, X_{n-1} | X_{-1}, \ldots, X_{-k}) = \prod_{m=0}^{n-1} p(X_m | X_{m-1}, \ldots, X_{-k})$$

and (5.7) in Chapter 4 implies

$$p(X_0, \ldots, X_{n-1}|X_{-1}, X_{-2}, \ldots) \equiv \lim_{k \to \infty} p(X_0, \ldots, X_{n-1}|X_{-1}, \ldots, X_{-k})$$

So we have

$$p(X_0, \ldots, X_{n-1}|X_{-1}, X_{-2}, \ldots) = \prod_{m=0}^{n-1} p(X_m|X_{m-1}, X_{m-2}, \ldots)$$

Taking logs gives

$$-\frac{1}{n} \log p(X_0, \ldots, X_{n-1}|X_{-1}, X_{-2}, \ldots)$$

(b)
$$= -\frac{1}{n} \sum_{m=0}^{n-1} \log p(X_m|X_{m-1}, X_{m-2}, \ldots)$$

$$\to E(-\log p(X_0|X_{-1}, \ldots)) = H$$

by the ergodic theorem applied to $F(\omega) = -\log\ p(\omega_0|\omega_{-1}, \ldots)$.

For the other side of our sandwich, we define the k step Markovian approximation

$$p^k(X_0, \ldots, X_{n-1}) = p(X_0, \ldots, X_{k-1}) \prod_{m=k}^{n-1} p(X_m|X_{m-1}, \ldots, X_{m-k})$$

for $k < n$, and observe that another application of the ergodic theorem gives

(c) $-\dfrac{1}{n} \log p^k(X_0, \ldots, X_{n-1}) \to E(-\log p(X_0|X_{-1}, \ldots, X_{-k})) = H_k$

To put the sandwich together we let

$$A_n^k = p^k(X_0, \ldots, X_{n-1})$$
$$B_n = p(X_0, \ldots, X_{n-1})$$
$$C_n = p(X_0, \ldots, X_{n-1}|X_{-1}, X_{-2}, \ldots)$$
$$W_n^1 = A_n^k/B_n \qquad W_n^2 = B_n/C_n$$

and note that

(d)
$$-\frac{1}{n} \log A_n^k = -\frac{1}{n} \log W_n^1 - \frac{1}{n} \log B_n$$

(e)
$$-\frac{1}{n} \log B_n = -\frac{1}{n} \log W_n^2 - \frac{1}{n} \log C_n$$

To get from (b) and (c) to the desired result we use:

(f) **Lemma.** If $W_n \geq 0$ and $EW_n \leq 1$ then $\limsup_{n\to\infty} n^{-1}\log W_n \leq 0$ a.s.

Proof If $\epsilon > 0$ then

$$P(n^{-1}\log W_n \geq \epsilon) = P(W_n \geq e^{\epsilon n}) \leq e^{-\epsilon n}$$

by Chebyshev's inequality. $\sum \exp^{-\epsilon n} < \infty$, so the Borel-Cantelli lemma implies that $\limsup n^{-1}\log W_n \leq \epsilon$. ϵ is arbitrary, so the proof is complete. $\square$

To check that $EW_n^i \leq 1$ we observe that

$$EW_n^1 = \sum_x \frac{p^k(x_0,\ldots,x_{n-1})}{p(x_0,\ldots,x_{n-1})} p(x_0,\ldots,x_{n-1}) = 1$$

For the second we observe that

$$E\,\frac{p(X_0,\ldots,X_{n-1})}{p(X_0,\ldots,X_{n-1}|X_{-1},\ldots,X_{-k})} = E\,\frac{p(X_{-k},\ldots,X_{-1})p(X_0,\ldots,X_{n-1})}{p(X_{-k},\ldots,X_{n-1})}$$

and $\sum_x p(x_{-k},\ldots,x_{-1})p(x_0,\ldots,x_{n-1}) = 1$.

Letting $k \to \infty$ and using Fatou's lemma gives $EW_n^2 \leq 1$. Applying (f) now and using (c), (d), (e), and (b) gives

$$H_k \geq \limsup_{n\to\infty} n^{-1}\log p(X_0,\ldots,X_{n-1})$$
$$\geq \liminf_{n\to\infty} n^{-1}\log p(X_0,\ldots,X_{n-1}) \geq H$$

(a) implies $H_k \to H$ as $k \to \infty$, and the proof of (5.1) is complete. $\square$

Example 5.1. $X_0, X_1, \ldots$ are i.i.d. In this case

$$p(X_0,\ldots,X_{n-1}) = \prod_{m=0}^{n-1} p(X_m)$$

so the strong law of large numbers implies

$$-\frac{1}{n}\log p(X_0,\ldots,X_{n-1}) \to -E\log p(X_0) = -\sum_x p(x)\log p(x)$$

For a concrete example, suppose $P(X_i = 0) = 2/3$ and $P(X_i = 1) = 1/3$. In this case

$$H = \frac{2}{3}\log\frac{3}{2} + \frac{1}{3}\log 3 \approx .6365$$

Example 5.2. $X_0, X_1, \ldots$ is a Markov chain. In this case $H = H_1$ so

$$H = E\{-\log p(X_0|X_{-1})\} = \sum_{x,y} \pi(x) p(x,y)\{-\log p(x,y)\}$$

where p is the transition probability and π is the stationary distribution. For a concrete example suppose

$$p = \begin{pmatrix} .5 & .5 \\ 1 & 0 \end{pmatrix} \qquad \pi = (2/3, 1/3)$$

In this case, $H = \frac{2}{3}\log 2 \approx .4621$.

Comparing the values of H in the two examples we see that although they have the same marginal distributions, the i.i.d. sequence is "more random" than the Markov chain. To explain what the values of H tell us about the stationary sequence, we state a result that is merely a reformulation of (5.1).

(5.2) Asymptotic equipartition property. Let $a = |S|$. Given $\epsilon > 0$ there is an N_ϵ so that for $n \geq N_\epsilon$, the a^n possible outcomes of $(X_0, \ldots, X_{n-1})$ can be divided into two classes: (i) a class B_n with probability $< \epsilon$, (ii) a class G_n in which each outcome $(x_0, x_1, \ldots, x_{n-1})$ has

$$|-H - n^{-1}\log p(x_0, \ldots, x_{n-1})| < \epsilon$$

Since the outcomes in G_n have probability $\exp(-(H \pm \epsilon)n)$, it follows that the number of outcomes in G_n is $\exp((H \pm \epsilon)n)$. A trivial but informative special case occurs when X_n is an i.i.d. sequence with $P(X_n = x) = 1/a$ for $x \in S$. In this case $H = \log a$, so we need almost all the outcomes to capture $1 - \epsilon$ of the probability mass. Since all outcomes have equal probability the last conclusion should not be too surprising.

Breiman's proof. If we let $g_n(\omega) = -\log p(\omega_0|\omega_{-1}, \ldots, \omega_{-n})$ then we can use Exercise 2.2 to prove (5.1). To check $E(\sup g_n) < \infty$, let

$$A_k = \left\{ g_k > \lambda, \sup_{j < k} g_j \leq \lambda \right\}$$

and observe

$$P(A_j) = \sum_{\omega \in A_j} p(\omega_{-j}, \ldots, \omega_0) \leq e^{-\lambda} \sum_{\omega \in A_j} p(\omega_{-j}, \ldots, \omega_{-1}) \leq a e^{-\lambda}$$

where $a = |S|$ since for each ω_0 the sum is ≤ 1.

To apply the result in Exercise 2.2, we note

$$\frac{1}{n}\sum_{m=0}^{n-1} g_m(\varphi^m\omega) = -\frac{1}{n}\sum_{m=0}^{n-1}\log p(\omega_m|\omega_{m-1},\ldots,\omega_0)$$

$$= -\frac{1}{n}\log p(\omega_0,\ldots,\omega_{m-1})$$

and we have assumed X_n is ergodic so

$$E(g|\mathcal{I}) = E(-\log p(\omega_0|\omega_{-1},\omega_{-2},\ldots))$$

*6.6. A Subadditive Ergodic Theorem

In this section we will prove Liggett's (1985) version of Kingman's (1968)

(6.1) Subadditive ergodic theorem. Suppose $X_{m,n}$, $0 \leq m < n$ satisfy:

(i) $X_{0,m} + X_{m,n} \geq X_{0,n}$

(ii) $\{X_{nk,(n+1)k}, n \geq 1\}$ is a stationary sequence for each k.

(iii) The distribution of $\{X_{m,m+k}, k \geq 1\}$ does not depend on m.

(iv) $EX_{0,1}^+ < \infty$ and for each n, $EX_{0,n} \geq \gamma_0 n$ where $\gamma_0 > -\infty$.

Then

(a) $\lim_{n\to\infty} EX_{0,n}/n = \inf_m EX_{0,m}/m \equiv \gamma$

(b) $X = \lim_{n\to\infty} X_{0,n}/n$ exists a.s. and in L^1, so $EX = \gamma$.

(c) If all the stationary sequences in (ii) are ergodic then $X = \gamma$ a.s.

Remark. Kingman assumed (iv), but instead of (i)-(iii) he assumed that $X_{\ell,m} + X_{m,n} \geq X_{\ell,n}$ for all $\ell < m < n$ and that the distribution of $\{X_{m+k,n+k}, 0 \leq m < n\}$ does not depend on k. In two of the four applications in Section 6.7 these stronger conditions do not hold.

Before giving the proof, which is somewhat lengthy, we will consider several examples for motivation. Since the validity of (ii) and (iii) in each case is clear, we will only check (i) and (iv). The first example shows that (6.1) contains the ergodic theorem (5.1) as a special case.

Example 6.1.   Stationary sequences. Suppose $\xi_1, \xi_2, \ldots$ is a stationary sequence with $E|\xi_k| < \infty$, and let $X_{m,n} = \xi_{m+1} + \cdots + \xi_n$. Then $X_{0,n} = X_{0,m} + X_{m,n}$ and (iv) holds.

Example 6.2.   Range of random walk. Suppose $\xi_1, \xi_2, \ldots$ is a stationary sequence and let $S_n = \xi_1 + \cdots + \xi_n$. Let $X_{m,n} = |\{S_{m+1}, \ldots, S_n\}|$. It is clear that $X_{0,m} + X_{m,n} \geq X_{0,n}$. $0 \leq X_{0,n} \leq n$, so (iv) holds. Applying (6.1) now gives $X_{0,n}/n \to X$ a.s. and in L^1 but it does not tell us what the limit is.

EXERCISE 6.1. Suppose $\xi_1, \xi_2, \ldots$ is ergodic in Example 6.2. Use (c) and (a) of (6.1) to conclude that $|\{S_1, \ldots, S_n\}|/n \to P(\text{ no return to } 0)$.

Example 6.3.   Records from "improving populations." Ballerini and Resnick (1985,1987). Let $\xi_1, \xi_2, \ldots$ be a stationary sequence and let $\zeta_n = \xi_n + cn$ where $c > 0$. Let

$$X_{m,n} = |\{\ell : m < \ell \leq n \text{ and } \zeta_\ell > \zeta_k \text{ for all } m < k < \ell\}|$$

It is clear that $X_{0,m} + X_{m,n} \geq X_{0,n}$. $0 \leq X_{0,n} \leq n$ so (iv) holds. Applying (6.1) now gives that $X_{0,n}/n \to X$ a.s. and in L^1. To identify the limit extend ξ_n, $n \geq 0$, to $\{\xi_n : n \in \mathbf{Z}\}$ using (1.2), and let $\zeta_n = \xi_n + cn$, $n \in \mathbf{Z}$. Let $Y_m = 1$ if $\zeta_m > \zeta_k$ for all $k < m$; $Y_m = 0$ otherwise. An easy extension of (1.1) implies Y_m is a stationary sequence, so the ergodic theorem implies

$$(Y_1 + \cdots + Y_n)/n \to Y \quad \text{a.s. and in } L^1$$

To see that $X = Y$ observe that $X \geq Y$ but

$$EX_{0,n}/n = \frac{1}{n} \sum_{m=1}^{n} P(\zeta_0 > \zeta_k \text{ for } 0 > k > -m) \to EY$$

as $n \to \infty$, so $EX = EY$ and hence $X = Y$ a.s.

The analysis in the last paragraph could be applied to identify the limit in Example 6.2. If we let $Y_m = 1$ when $S_k \neq S_m$ for all $k > m$ then we get a stationary sequence that has the same limit as $X_{0,n}/n$. (This was the key to the proof in Section 6.3.) Kingman's original proof shows that we can always identify the limit by this method. That is, we can write

$$X_{m,n} = \sum_{k=m+1}^{n} Y_k + Z_{m,n}$$

where Y_m, $m \geq 1$, is a stationary sequence, and $Z_{m,n} \geq 0$ is a subadditive process with $EZ_{0,n}/n \to 0$. We will prove (6.1) without proving Kingman's "decomposition theorem."

Example 6.4. Longest common subsequences. Let $X_1, X_2, X_3, \ldots$ and $Y_1, Y_2, Y_3, \ldots$ be ergodic stationary sequences. Let $L_{m,n} = \max\{K : X_{i_k} = Y_{j_k}$ for $1 \leq k \leq K$ where $m < i_1 < i_2 \ldots < i_K \leq n$ and $m < j_1 < j_2 \ldots < j_K \leq n\}$. It is clear that

$$L_{0,m} + L_{m,n} \geq L_{0,n}$$

so $X_{m,n} = -L_{m,n}$ is subadditive. $0 \leq L_{0,n} \leq n$ so (iv) holds. Applying (6.1) now we conclude that

$$L_{0,n}/n \to \gamma = \sup_{m \geq 1} E(L_m/m)$$

EXERCISE 6.2. Suppose that in the last exercise $X_1, X_2, \ldots$ and $Y_1, Y_2, \ldots$ are i.i.d. and take the values 0 and 1 with probability $1/2$ each. (a) Compute EL_1 and $EL_2/2$ to get lower bounds on γ. (b) Show $\gamma < 1$ by computing the expected number of i and j sequences of length $K = an$ with the desired property.

Remark. Chvatal and Sankoff (1975) have shown $.727273 \leq \gamma \leq .866595$

Our final example shows that the convergence in (a) of (6.1) may occur arbitrarily slowly.

Example 6.5. Suppose $X_{m,m+k} = f(k) \geq 0$ where $f(k)/k$ is decreasing.

$$X_{0,n} = f(n) = m\frac{f(n)}{n} + (n-m)\frac{f(n)}{n}$$
$$\leq m\frac{f(m)}{m} + (n-m)\frac{f(n-m)}{n-m} = X_{0,m} + X_{m,n}$$

The examples above should provide enough motivation for now. In Section 6.7, we will give four more applications of (6.1).

Proof of (6.1) There are four steps. The first, second, and fourth date back to Kingman (1968). The half dozen proofs of subadditive ergodic theorems that exist all do the crucial third step in a different way. Here we use the approach of S. Leventhal (1988), who in turn based his proof on Katznelson and Weiss (1982).

Step 1. The first thing to check is that $E|X_{0,n}| \leq Cn$. To do this we note that (i) implies $X_{0,m}^+ + X_{m,n}^+ \geq X_{0,n}^+$. Repeatedly using the last inequality and invoking (iii) gives $EX_{0,n}^+ \leq nEX_{0,1}^+ < \infty$. Since $|x| = 2x^+ - x$ it follows from (iv) that

$$E|X_{0,n}| \leq 2EX_{0,n}^+ - EX_{0,n} \leq Cn < \infty$$

Let $a_n = EX_{0,n}$. (i) and (iii) imply that

(6.2) $$a_m + a_{n-m} \geq a_n$$

From this it follows easily that

(6.3) $$a_n/n \to \inf_{m \geq 1} a_m/m \equiv \gamma$$

To prove this, we observe that the liminf is clearly $\geq \gamma$, so all we have to show is that the limsup $\leq a_m/m$ for any m. The last fact is easy, for if we write $n = km + \ell$ with $0 \leq \ell < m$ then repeated use of (6.2) gives $a_n \leq ka_m + a_\ell$. Dividing by $n = km + \ell$ gives

$$\frac{a_n}{n} \leq \frac{km}{km+\ell} \cdot \frac{a_m}{m} + \frac{a_\ell}{n}$$

Letting $n \to \infty$ and recalling $0 \leq \ell < m$ gives (6.3) and proves (a) in (1).

Remark. Chvatal and Sankoff (1975) attribute (6.3) to Fekete (1923).

Step 2. Making repeated use of (i) we get

$$X_{0,n} \leq X_{0,km} + X_{km,n}$$
$$X_{0,n} \leq X_{0,(k-1)m} + X_{(k-1)m,km} + X_{km,n}$$

and so on until the first term on the right is $X_{0,m}$. Dividing by $n = km + \ell$ then gives

(6.4) $$\frac{X_{0,n}}{n} \leq \frac{k}{km+\ell} \cdot \frac{X_{0,m} + \cdots + X_{(k-1)m,km}}{k} + \frac{X_{km,n}}{n}$$

Using (ii) and the ergodic theorem now gives that

$$\frac{X_{0,m} + \cdots + X_{(k-1)m,km}}{k} \to A_m \quad \text{a.s. and in } L^1$$

where $A_m = E(X_{0,m}|\mathcal{I}_m)$ and the subscript indicates that $\mathcal{I}_m$ is the shift invariant σ-field for the sequence $X_{(k-1)m,km}$, $k \geq 1$. The exact formula for the limit is not important but we will need to know later that $EA_m = EX_{0,m}$.

If we fix ℓ and let $\epsilon > 0$ then (iii) implies

$$\sum_{k=1}^{\infty} P(X_{km,km+\ell} > (km + \ell)\epsilon) = \sum_{k=1}^{\infty} P(X_{0,\ell} > k\epsilon) < \infty$$

since $EX_{0,\ell}^+ < \infty$ by the result at the beginning of Step 1. The last two observations imply

(6.5)
$$\overline{X} \equiv \limsup_{n \to \infty} X_{0,n}/n \leq A_m/m$$

Taking expected values now gives $E\overline{X} \leq E(X_{0,m}/m)$ and taking the infimum over m we have $E\overline{X} \leq \gamma$. Note that if all the stationary sequences in (ii) are ergodic we have $\overline{X} \leq \gamma$.

Remark. If (i)-(iii) hold, $EX_{0,1}^+ < \infty$, and $\inf EX_{0,m}/m = -\infty$, then it follows from the last argument that as $X_{0,n}/n \to -\infty$ a.s. as $n \to \infty$.

Step 3. The next step is to let

$$\underline{X} = \liminf_{n \to \infty} X_{0,n}/n$$

and show that $E\underline{X} \geq \gamma$. Since $\infty > EX_{0,1} \geq \gamma \geq \gamma_0 > -\infty$, and we have shown in Step 2 that $E\overline{X} \leq \gamma$, it will follow that $\underline{X} = \overline{X}$, i.e., the limit of $X_{0,n}/n$ exists a.s. Let

$$\underline{X}_m = \liminf_{n \to \infty} X_{m,m+n}/n$$

(i) implies

$$X_{0,m+n} \leq X_{0,m} + X_{m,m+n}$$

Dividing both sides by n and letting $n \to \infty$ gives $\underline{X} \leq \underline{X}_m$ a.s. However, (iii) implies that $\underline{X}_m$ and $\underline{X}$ have the same distribution so $\underline{X} = \underline{X}_m$ a.s.

Let $\epsilon > 0$ and let $Z = \epsilon + (\underline{X} \vee -M)$. Since $\underline{X} \leq \overline{X}$ and $E\overline{X} \leq \gamma < \infty$ by Step 2, $E|Z| < \infty$. Let

$$Y_{m,n} = X_{m,n} - (n - m)Z$$

Y satisfies (i)-(iv) (since $Z_{m,n} = -(n - m)Z$ does) and has

$$\underline{Y} \equiv \liminf_{n \to \infty} Y_{0,n}/n \leq -\epsilon$$

Let $T_m = \min\{n \geq 1 : Y_{m,m+n} \leq 0\}$. (iii) implies $T_m =_d T_0$ and

$$E(Y_{m,m+1}; T_m > N) = E(Y_{0,1}; T_0 > N)$$

(6.6) implies that $P(T_0 < \infty) = 1$, so we can pick N large enough so that

$$E(Y_{0,1}; T_0 > N) \le \epsilon$$

Let

$$S_m = \begin{cases} T_m & \text{on } \{T_m \le N\} \\ 1 & \text{on } \{T_m > N\} \end{cases}$$

(This is not a stopping time but there is nothing special about stopping times for a stationary sequence!) Let

$$\xi_m = \begin{cases} 0 & \text{on } \{T_m \le N\} \\ Y_{m,m+1} & \text{on } \{T_m > N\} \end{cases}$$

Since $Y(m, m+T_m) \le 0$ always and we have $S_m = 1$, $Y_{m,m+1} > 0$ on $\{T_m > N\}$ we have $Y(m, m + S_m) \le \xi_m$ and $\xi_m \ge 0$. Let $R_0 = 0$ and for $k \ge 1$ let $R_k = R_{k-1} + S(R_{k-1})$. Let $K = \max\{k : R_k \le n\}$. From (i) it follows that

$$Y(0, n) \le Y(R_0, R_1) + \cdots + Y(R_{K-1}, R_K) + Y(R_K, n)$$

Since $\xi_m \ge 0$ and $n - R_K \le N$ the last quantity is

$$\le \sum_{m=0}^{n-1} \xi_m + \sum_{j=1}^{N} |Y_{n-j,n-j+1}|$$

(Here we have used (i) on $Y(R_K, n)$.) Dividing both sides by n, taking expected values and letting $n \to \infty$ gives

$$\limsup_{n \to \infty} EY_{0,n}/n \le E\xi_0 \le E(Y_{0,1}; T_0 > N) \le \epsilon$$

It follows from (a) and the definition of $Y_{0,n}$ that

$$\gamma = \lim_{n \to \infty} EX_{0,n}/n \le 2\epsilon + E(\underline{X} \vee -M)$$

Since $\epsilon > 0$ and M are arbitrary, it follows that $E\underline{X} \ge \gamma$ and Step 3 is complete.

Step 4. It only remains to prove convergence in L^1. Let $\Gamma_m = A_m/m$ be the limit in (6.5), recall $E\Gamma_m = E(X_{0,m}/m)$, and let $\Gamma = \inf \Gamma_m$. Observing that $|z| = 2z^+ - z$ (consider two cases $z \ge 0$ and $z < 0$) we can write

$$E|X_{0,n}/n - \Gamma| = 2E(X_{0,n}/n - \Gamma)^+ - E(X_{0,n}/n - \Gamma) \le 2E(X_{0,n}/n - \Gamma)^+$$

since

$$E(X_{0,n}/n) \geq \gamma = \inf E\Gamma_m \geq E\Gamma$$

Using the trivial inequality $(x+y)^+ \leq x^+ + y^+$ and noticing $\Gamma_m \geq \Gamma$ now gives

$$E(X_{0,n}/n - \Gamma)^+ \leq E(X_{0,n}/n - \Gamma_m)^+ + E(\Gamma_m - \Gamma)$$

Now $E\Gamma_m \to \gamma$ as $m \to \infty$ and $E\Gamma \geq E\bar{X} \geq E\underline{X} \geq \gamma$ by steps 2 and 3, so $E\Gamma = \gamma$ and it follows that $E(\Gamma_m - \Gamma)$ is small if m is large. To bound the other term observe (i) implies

$$E(X_{0,n}/n - \Gamma_m)^+ \leq E\left(\frac{X(0,m) + \cdots + X((k-1)m, km)}{km + \ell} - \Gamma_m\right)^+$$
$$+ E\left(\frac{X(km, n)}{n}\right)^+$$

The second term $= E(X_{0,\ell}^+/n) \to 0$ as $n \to \infty$. For the first we observe $y^+ \leq |y|$ and the ergodic theorem implies

$$E\left|\frac{X(0,m) + \cdots + X((k-1)m, km)}{k} - \Gamma_m\right| \to 0$$

so the proof of (6.1) is complete. $\qquad\qquad\qquad\qquad\qquad\qquad\qquad\qquad\square$

*6.7. Applications

In this section we will give four applications of our subadditive ergodic theorem (6.1). These examples are independent of each other and can be read in any order. In the last two we encounter situations to which Liggett's version applies but Kingman's version does not.

Example 7.1.   Products of random matrices. Suppose $A_1, A_2, \ldots$ is a stationary sequence of $k \times k$ matrices with positive entries and let $\alpha_{m,n}(i,j) = (A_{m+1} \cdots A_n)(i,j)$, i.e., the entry in row i of column j of the product. It is clear that

$$\alpha_{0,m}(1,1)\alpha_{m,n}(1,1) \leq \alpha_{0,n}(1,1)$$

so if we let $X_{m,n} = -\log \alpha_{m,n}(1,1)$ then $X_{0,m} + X_{m,n} \geq X_{0,n}$. To check (iv) we observe that

$$\prod_{m=1}^{n} A_m(1,1) \leq \alpha_{0,n}(1,1) \leq k^{n-1} \prod_{m=1}^{n} \left(\sup_{i,j} A_m(i,j)\right)$$

or taking logs

$$-\sum_{m=1}^{n} \log A_m(1,1) \geq X_{0,n} \geq -(n \log k) - \sum_{m=1}^{n} \log \left(\sup_{i,j} A_m(i,j) \right)$$

So if $E \log A_m(1,1) > -\infty$ then $EX_{0,1}^+ < \infty$, and if

$$E \log \left(\sup_{i,j} A_m(i,j) \right) < \infty$$

then $EX_{0,n}^- \leq \gamma_0 n$. If we observe that

$$P \left(\log \left(\sup_{i,j} A_m(i,j) \right) \geq x \right) \leq \sum_{i,j} P \left(\log A_m(i,j) \geq x \right)$$

we see that it is enough to assume that

(*) $E| \log A_m(i,j)| < \infty$ for all i,j

When (*) holds, applying (6.1) gives $X_{0,n}/n \to X$ a.s. Using the strict positivity of the entries it is easy to improve that result to

(7.1) $\dfrac{1}{n} \log \alpha_{0,n}(i,j) \to -X$ a.s. for all i,j

a result first proved by Furstenberg and Kesten (1960).

The key to the proof above was the fact that $\alpha_{0,n}(1,1)$ was supermultiplicative. An alternative approach is to let

$$\|A\| = \max_i \sum_j |A(i,j)| = \max\{\|xA\|_1 : \|x\|_1 = 1\}$$

where $(xA)_j = \sum_i x_i A(i,j)$ and $\|x\|_1 = |x_1| + \cdots + |x_k|$. From the second definition it is clear that $\|AB\| \leq \|A\| \cdot \|B\|$, so if we let

$$\beta_{m,n} = \|A_{m+1} \cdots A_n\|$$

and $Y_{m,n} = \log \beta_{m,n}$, then $Y_{m,n}$ is subadditive. It is easy to use (7.1) to show that

$$\dfrac{1}{n} \log \|A_{m+1} \cdots A_n\| \to -X$ \text{a.s.}$$

where X is the limit of $X_{0,n}/n$. To see the advantage in having two proofs of the same result, we observe that if $A_1, A_2, \ldots$ is an i.i.d. sequence then X is constant and we can get upper and lower bounds by observing

$$\sup_{m \geq 1} (E \log \alpha_{0,m})/m = -X = \inf_{m \geq 1} (E \log \beta_{0,m})/m$$

Remark. Oseleděc (1968) proved a result which gives the asymptotic behavior of all of the eigenvalues of A. As Raghunathan (1979) and Ruelle (1979) have observed, this result can also be obtained from (6.1). See Krengel (1985) or the papers cited for details. Furstenberg and Kesten (1960) and later Ishitani (1977) have proved central limit theorems:

$$(\log \alpha_{0,n}(1,1) - \mu n)/n^{1/2} \Rightarrow \sigma \chi$$

where χ has the standard normal distribution. For more about products of random matrices, see Cohen, Kesten, and Newman (1985).

Example 7.2. Increasing sequences in random permutations. Let π be a permutation of $\{1, 2, \ldots, n\}$ and let $\ell(\pi)$ be the length of the longest increasing sequence in π. That is, the largest k for which there are integers $i_1 < i_2 \ldots < i_k$ so that $\pi(i_1) < \pi(i_2) < \ldots < \pi(i_k)$. Hammersley (1970) attacked this problem by putting a rate one Poisson process in the plane, and for $s < t \in [0, \infty)$, letting $Y_{s,t}$ denote the length of the longest increasing path lying in the square $R_{s,t}$ with vertices (s, s), (s, t), (t, t), and (t, s). That is, the largest k for which there are points (x_i, y_i) in the Poisson process with $s < x_1 < \ldots < x_k < t$ and $s < y_1 < \ldots < y_k < t$. It is clear that $Y_{0,m} + Y_{m,n} \leq Y_{0,n}$. Applying (6.1) to $-Y_{0,n}$ shows

$$Y_{0,n}/n \to \gamma \equiv \sup_{m \geq 1} EY_{0,m}/m \quad \text{a.s.}$$

For each k, $Y_{nk,(n+1)k}$, $n \geq 0$ is i.i.d. so the limit is constant. We will show that $\gamma < \infty$ in Exercise 7.3.

To get from the result about the Poisson process back to the random permutation problem, let $\tau(n)$ be the smallest value of t for which there are n points in $R_{0,t}$. Let the n points in $R_{0,\tau(n)}$ be written as (x_i, y_i) where $0 < x_1 < x_2 \ldots < x_n \leq \tau(n)$ and let π_n be the unique permutation of $\{1, 2, \ldots, n\}$ so that $y_{\pi_n(1)} < y_{\pi_n(2)} \ldots < y_{\pi_n(n)}$. It is clear that $Y_{0,\tau(n)} = \ell(\pi_n)$. An easy argument shows

(7.2) Lemma. $\tau(n)/\sqrt{n} \to 1$ a.s.

Proof Let S_n be the number of points in $R_{0,\sqrt{n}}$. $S_n - S_{n-1}$ are independent Poisson r.v.'s with mean 1, so the strong law of large numbers implies $S_n/n \to 1$

a.s. If $\epsilon > 0$ then for large n, $S_{n(1-\epsilon)} < n < S_{n(1+\epsilon)}$ and hence $\sqrt{(1-\epsilon)n} \leq \tau(n) \leq \sqrt{(1+\epsilon)n}$. □

It follows from (7.2) and the monotonicity of $m \to Y_{0,m}$ that

$$n^{-1/2}\ell(\pi_n) \to \gamma \quad \text{a.s.}$$

Hammersley (1970) has a proof that $\pi/2 \leq \gamma \leq e$, and Kingman (1973) shows that $1.59 < \gamma < 2.49$. (See Exercises 7.2 and 7.3.) Subsequent work on the random permutation problem, see Logan and Shepp (1977) and Vershik and Kerov (1977), has shown that $\gamma = 2$.

EXERCISE 7.1. Given a rate one Poisson process in $[0,\infty) \times [0,\infty)$, let (X_1, Y_1) be the point that minimizes $x+y$. Let (X_2, Y_2) be the point in $[X_1, \infty) \times [Y_1, \infty)$ that minimizes $x + y$, and so on. Use this construction to show that $\gamma \geq (8/\pi)^{1/2} > 1.59$.

EXERCISE 7.2. Let π_n be a random permuation of $\{1,\dots,n\}$ and let J_k^n be the number of subsets of $\{1,\dots n\}$ of size k so that the associated $\pi_n(j)$ form an increasing subsequence. Compute $E J_k^n$ and take $k \sim \alpha n^{1/2}$ to conclude $\gamma \leq e$.

Remark. Kingman improved this by observing that $\ell(\pi_n) \geq \ell$ then $J_k^n \geq \binom{\ell}{k}$. Using this with the bound on $E J_k^n$ and taking $\ell \sim \beta n^{1/2}$ and $k \sim \alpha n^{1/2}$, he showed $\gamma < 2.49$.

Example 7.3. Age-dependent branching processes. This is a variation of the branching process introduced in Section 4.3 in which each individual lives for an amount of time with distribution F before producing k offspring with probability p_k. The description of the process is completed by supposing that the process starts with one individual in generation 0 who is born at time 0, and when this particle dies, its offspring start independent copies of the original process.

Suppose $p_0 = 0$, let $X_{0,m}$ be the birth time of the first member of generation m, and let $X_{m,n}$ be the time lag necessary for that individual to have an offspring in generation n. (In case of ties pick an individual at random from those in generation m born at time $X_{0,m}$.) It is clear that $X_{0,n} \leq X_{0,m} + X_{m,n}$. Since $X_{0,n} \geq 0$, (iv) holds if we assume F has finite mean. Applying (6.1) now, it follows that

$$X_{0,n}/n \to \gamma \quad \text{a.s.}$$

The limit is constant because the sequences $\{X_{nk,(n+1)k}, n \geq 0\}$ are i.i.d.

Remark. The inequality $X_{\ell,m} + X_{m,n} \geq X_{\ell,n}$ is false when $\ell > 0$, because if we call i_m the individual that determines the value of $X_{m,n}$ for $n > m$, then i_m may not be a descendant of i_ℓ.

As usual one has to use other methods to identify the constant. Let $t_1, t_2, \ldots$ be i.i.d. with distribution F, let $T_n = t_1 + \cdots + t_n$, and $\mu = \sum k p_k$. Let $Z_n(an)$ be the number of individuals in generation n born by time an. Each individual in generation n has probability $P(T_n \leq an)$ to be born by time an and the times are independent of the offspring numbers so

$$EZ_n(an) = EE(Z_n(an)|Z_n) = E(Z_n P(T_n \leq an)) = \mu^n P(T_n \leq an)$$

By results in Section 1.9, $n^{-1} \log P(T_n \leq an) \to -c(a)$ as $n \to \infty$. If $\log \mu - c(a) < 0$ then Chebyshev's inequality and the Borel-Cantelli lemma imply $P(Z_n(an) \text{ i.o.}) = 0$. Conversely, if $EZ_n(an) > 1$ for some n then we can define a supercritical branching process Y_m that consists of the offspring in generation mn that are descendants of individuals in Y_{m-1} in generation $(m-1)n$ which are born less than an units of time after their parents. This shows that with positive probability $X_{0,mn} \leq mna$ for all m. Combining the last two observations with the fact that $c(a)$ is strictly increasing gives

$$\gamma = \inf\{a : \log \mu - c(a) > 0\}$$

The last result is from Biggins (1977). See his (1978) and (1979) papers for extensions and refinements. Kingman (1975) has an approach to the problem via martingales:

EXERCISE 7.3. Let $\varphi(\theta) = E \exp(-\theta t_i)$ and

$$Y_n = (\mu\varphi(\theta))^{-n} \sum_{i=1}^{Z_n} \exp(-\theta T_n(i))$$

where the sum is over individuals in generation n and $T_n(i)$ is the ith person's birth time. Show that Y_n is a nonnegative martingale and use this to conclude that if $\exp(-\theta a)/\mu\varphi(\theta) > 1$ then $P(X_{0,n} \leq an) \to 0$. A little thought reveals that this bound is the same as the answer in the last exercise.

Example 7.4. First passage percolation. Consider $\mathbf{Z}^d$ as a graph with edges connecting each $x, y \in \mathbf{Z}^d$ with $|x - y| = 1$. Assign an independent nonnegative random variable $\tau(e)$ to each edge that represents the time required to traverse the edge going in either direction. If e is the edge connecting x and

y, let $\tau(x, y) = \tau(y, x) = \tau(e)$. If $x_0 = x, x_1, \ldots, x_n = y$ is a path from x to y, i.e., a sequence with $|x_m - x_{m-1}| = 1$ for $1 \leq m \leq n$, we define the **travel time** for the path to be $\tau(x_0, x_1) + \cdots + \tau(x_{n-1}, x_n)$. Define the **passage time** from x to y, $t(x, y) = $ the infimum of the travel times over all paths from x to y. Let $z \in \mathbf{Z}^d$ and let $X_{m,n} = t(mu, nu)$, where $u = (1, 0, \ldots, 0)$.

Clearly $X_{0,m} + X_{m,n} \geq X_{0,n}$. $X_{0,n} \geq 0$ so if $E\tau(x, y) < \infty$ then (iv) holds, and (6.1) implies that $X_{0,n}/n \to X$ a.s. To see that the limit is constant, enumerate the edges in some order $e_1, e_2, \ldots$ and observe that X is measurable with respect to the tail σ-field of the i.i.d. sequence $\tau(e_1), \tau(e_2), \ldots$

Remark. It is not hard to see that the assumption of finite first moment can be weakened. If τ has distribution F with

(*)
$$\int_0^\infty (1 - F(x))^{2d} \, dx < \infty$$

i.e., the minimum of $2d$ independent copies has finite mean, then by finding $2d$ disjoint paths from 0 to $u = (1, 0, \ldots, 0)$, one concludes that $E\tau(0, u) < \infty$ and (6.1) can be applied. The condition (*) is also necessary for $X_{0,n}/n$ to converge to a finite limit. If (*) fails and Y_n is the minimum of $t(e)$ over all the edges from nu then

$$\limsup_{n \to \infty} X_{0,n}/n \geq \limsup_{n \to \infty} Y_n/n = \infty \quad \text{a.s.}$$

Above we considered the **point-to-point passage time**. A second object of interest is the **point-to-line passage time**:

$$a_n = \inf\{t(0, x) : x_1 = n\}$$

Unfortunately it does not seem to be possible to embed this sequence in a subadditive family. To see the difficulty let $\bar{t}(0, x)$ be infimum of travel times over paths from 0 to x that lie in $\{y : y_1 \geq 0\}$, let

$$\bar{a}_m = \inf\{\bar{t}(0, x) : x_1 = m\}$$

and let x^m be a point at which the infimum is achieved. (We leave to the reader the highly nontrivial task of proving that such a point exists; see Smythe and Wierman (1978) for a proof.) If we let $\bar{a}_{m,n}$ be the infimum of travel times over all paths which start at x^m, stay in $\{y : y_1 \geq m\}$, and end on $\{y : y_1 = n\}$, then $\bar{a}_{m,n}$ is independent of $\bar{a}_m$ and

$$\bar{a}_m + \bar{a}_{m,n} \geq \bar{a}_n$$

The last inequality is true without the half-space restriction but the independence is not and without the half-space restriction we cannot get the stationarity properties needed to apply (4.1).

Remark. The family $\bar{a}_{m,n}$ is another example where $\bar{a}_{\ell,m} + \bar{a}_{m,n} \geq \bar{a}_{\ell,n}$ need not hold for $\ell > 0$.

A second approach to limit theorems for a_m is to prove a result about the set of points which can be reached by time t: $\xi_t = \{x : t(0, x) \leq t\}$. Cox and Durrett (1981) have shown

(7.3) Theorem. For any passage time distribution F with $F(0) = 0$ there is a convex set A so that for any $\epsilon > 0$ we have with probability one

$$\xi_t \subset (1 + \epsilon)tA \text{ for all } t \text{ sufficiently large}$$

and $|\xi_t^\epsilon \cap (1 - \epsilon)tA \cap \mathbf{Z}^d|/t^d \to 0$ as $t \to \infty$.

Ignoring the boring details of how to state things precisely, the last result says $\xi_t/t \to A$ a.s. It implies that $a_n/n \to \gamma$ a.s. where $\gamma = 1/\sup\{x_1 : x \in A\}$. (Use the convexity and reflection symmetry of A.) When the distribution has finite mean (or satisfies the weaker condition in the remark above), γ is the limit of $t(0, nu)/n$. Without any assumptions $t(0, nu)/n \to \gamma$ in probability. For more details see the paper cited above. Kesten (1986) and (1987) are good sources for more about first passage percolation.

EXERCISE 7.4. **Oriented first passage percolation.** Consider a graph with vertices $\{(m, n) \in \mathbf{Z}^2 : m+n \text{ is even and } n \leq 0\}$, and oriented edges connecting (m, n) to $(m - 1, n - 1)$ and (m, n) to $(m - 1, n - 1)$. Assign i.i.d. exponential mean one r.v.'s to each edge. Thinking of the number on edge e as giving the time it takes water to travel down the edge, define $t(m, n) = $ the time at which the fluid first reaches (m, n), and $a_n = \inf\{t(m, -n)\}$. Show that as $n \to \infty$, a_n/n converges to a limit γ a.s.

EXERCISE 7.5. Continuing with the set-up in the last exercise. (i) Show $\gamma \leq 1/2$ by considering a_1. (ii) Get a positive lower bound on γ by looking at the expected number of paths down to $\{(m, -n) : -n \leq m \leq n\}$ with passage time $\leq an$ and using results from Section 1.9.

Remark. If we replace the graph in Exercise 7.4 by a binary tree then we get a problem equivalent to the first birth problem (Example 7.3) for $p_2 = 2$, $P(t_i > x) = e^{-x}$. In that case the lower bound obtained by the methods of part (ii) Exercise 7.5 was sharp but in this case it is not.

7 Brownian Motion

Brownian motion is a process of tremedous practical and theoretical significance. It originated (a) as a model of the phenomenon observed by Robert Brown in 1828 that "pollen grains suspended in water perform a continual swarming motion," and (b) in Bachelier's (1900) work as a model of the stock market. These are just two of many systems that Brownian motion has been used to model. On the theoretical side, Brownian motion is a Gaussian Markov process with stationary independent increments. It lies in the intersection of three important classes of processes and is a fundamental example in each theory.

The first part of this chapter develops properties of Brownian motion. In Section 7.1 we define Brownian motion and investigate continuity properties of its paths. In Section 7.2, we prove the Markov property and a related 0-1 law. In Section 7.3, we define stopping times and prove the strong Markov property. In Section 7.4 we take a close look at the zero set of Brownian motion. In Section 7.5, we introduce some martingales associated with Brownian motion and use them to compute the distribution of T and B_T for some stopping times T.

The second part of this chapter applies Brownian motion to some of the problems considered in Chapters 1 and 2. In Section 7.6, we embed random walks into Brownian motion to prove Donsker's theorem, a far-reaching generalization of the central limit theorem. In Section 7.7, we extend Donsker's theorem to martingales satisfying "Lindeberg-Feller conditions" and to weakly dependent stationary sequences. In Section 7.8 we show that the discrepancy between the empirical distribution and the true distribution when suitably magnified converges to Brownian bridge. In Section 7.9, we prove laws of the iterated logarithm for Brownian motion and random walks with finite variance. The last three sections depend on Section 7.6 but are independent of each other and can be read in any order.

7.1. Definition and Construction

A one-dimensional **Brownian motion** is a real valued process B_t, $t \geq 0$ that has the following properties:

(a) If $t_0 < t_1 < \ldots < t_n$ then $B(t_0), B(t_1) - B(t_0), \ldots, B(t_n) - B(t_{n-1})$ are independent.

(b) If $s, t \geq 0$ then

$$P(B(s+t) - B(s) \in A) = \int_A (2\pi t)^{-1/2} \exp(-x^2/2t)\,dx$$

(c) With probability one, $t \to B_t$ is continuous.

(a) says that B_t has independent increments. (b) says that the increment $B(s+t) - B(s)$ has a normal distribution with mean 0 and variance t. (c) is self-explanatory.

Thinking of Brown's pollen grain (c) is certainly reasonable. (a) and (b) can be justified by noting that the movement of the pollen grain is due to the net effect of the bombardment of millions of water molecules, so by the central limit theorem the displacement in any one interval should have a normal distribution, and the displacements in two disjoint intervals should be independent.

Two immediate consequences of the definition that will be useful many times are:

(1.1) **Translation invariance.** $\{B_t - B_0, t \geq 0\}$ is independent of B_0 and has the same distribution as a Brownian motion with $B_0 = 0$.

Proof   Let $\mathcal{A}_1 = \sigma(B_0)$ and $\mathcal{A}_2$ be the events of the form $\{B(t_1) - B(t_0) \in A_1, \ldots, B(t_n) - B(t_{n-1}) \in A_n\}$. The $\mathcal{A}_i$ are π-systems that are independent so the desired result follows from (4.2) in Chapter 1. $\square$

(1.2) **The Brownian scaling relation.** If $B_0 = 0$ then for any $t > 0$,

$$\{B_{st}, s \geq 0\} \stackrel{d}{=} \{t^{1/2} B_s, s \geq 0\}$$

To be precise, the two families of r.v.'s have the same finite dimensional distributions, i.e., if $s_1 < \ldots < s_n$ then

$$(B_{s_1 t}, \ldots, B_{s_n t}) \stackrel{d}{=} (t^{1/2} B_{s_1}, \ldots t^{1/2} B_{s_n})$$

Proof   To check this when $n = 1$, we note that $t^{1/2}$ times a normal with mean 0 and variance s is a normal with mean 0 and variance st. The result for $n > 1$ follows from independent increments. $\square$

A second equivalent definition of Brownian motion starting from $B_0 = 0$, that we will occasionally find useful is that B_t, $t \geq 0$, is a real valued process satisfying

(a') $B(t)$ is a **Gaussian process** (i.e., all its finite dimensional distributions are multivariate normal).

(b') $EB_s = 0$ and $EB_s B_t = s \wedge t$.

(c') With probability one, $t \to B_t$ is continuous.

It is easy to see that (a) and (b) imply (a'). To get (b') from (a) and (b) suppose $s < t$ and write

$$EB_s B_t = E(B_s^2) + E(B_s(B_t - B_s)) = s$$

The converse is even easier. (a') and (b') specify the finite dimensional distributions of B_t, which by the last calculation must agree with the ones defined in (a) and (b).

The first question that must be addressed in any treatment of Brownian motion is, "Is there a process with these properties?" The answer is "Yes," of course, or this chapter would not exist. For pedagogical reasons we will pursue an approach that leads to a dead end and then retreat a little to rectify the difficulty. Fix an $x \in \mathbf{R}$ and for each $0 < t_1 < \ldots < t_n$ define a measure on $\mathbf{R}^n$ by

$$\mu_{x,t_1,\ldots,t_n}(A_1 \times \ldots \times A_n) = \int_{A_1} dx_1 \cdots \int_{A_n} dx_n \prod_{m=1}^n p_{t_m - t_{m-1}}(x_{m-1}, x_m)$$

where $A_i \in \mathcal{R}$, $x_0 = x$, $t_0 = 0$, and

$$p_t(a, b) = (2\pi t)^{-1/2} \exp(-(b-a)^2/2t)$$

From the formula above it is easy to see that for fixed x the family μ is a consistent set of finite dimensional distributions (f.d.d.'s), that is, if $\{s_1, \ldots, s_{n-1}\} \subset \{t_1, \ldots, t_n\}$ and $t_j \notin \{s_1, \ldots, s_{n-1}\}$ then

$$\mu_{x,s_1,\ldots,s_{n-1}}(A_1 \times \cdots \times A_{n-1}) = \mu_{x,t_1,\ldots,t_n}(A_1 \times \cdots \times A_{j-1} \times \mathbf{R} \times A_j \times \cdots \times A_{n-1})$$

This is clear when $j = n$. To check the equality when $1 \leq j < n$, it is enough to show that

$$\int p_{t_j - t_{j-1}}(x, y) p_{t_{j+1} - t_j}(y, z) \, dy = p_{t_{j+1} - t_{j-1}}(x, z)$$

By translation invariance, we can without loss of generality assume $x = 0$, but all this says is that the sum of independent normals with mean 0 and variances $t_j - t_{j-1}$ and $t_{j+1} - t_j$ has a normal distribution with mean 0 and variance $t_{j+1} - t_{j-1}$.

With the consistency of f.d.d.'s verified we get our first construction of Brownian motion:

(1.3) Theorem. Let $\Omega_o = \{$functions $\omega : [0,\infty) \to \mathbf{R}\}$ and $\mathcal{F}_o$ be the σ-field generated by the finite dimensional sets $\{\omega : \omega(t_i) \in A_i$ for $1 \le i \le n\}$ where $A_i \in \mathcal{R}$. For each $x \in \mathbf{R}$, there is a unique probability measure ν_x on $(\Omega_o, \mathcal{F}_o)$ so that $\nu_x\{\omega : \omega(0) = x\} = 1$ and when $0 < t_1 < \ldots < t_n$

$$(*) \qquad \nu_x\{\omega : \omega(t_i) \in A_i\} = \mu_{x,t_1,\ldots,t_n}(A_1 \times \cdots \times A_n)$$

This follows from a generalization of Kolmogorov's extension theorem ((7.1) in the Appendix). We will not bother with the details since at this point we are at the dead end referred to above. If $C = \{\omega : t \to \omega(t)$ is continuous$\}$ then $C \notin \mathcal{F}_o$, that is, C is not a measurable set. The easiest way of proving $C \notin \mathcal{F}_o$ is to do

EXERCISE 1.1. $A \in \mathcal{F}_o$ if and only if there is a sequence of times $t_1, t_2, \ldots$ in $[0, \infty)$ and a $B \in \mathcal{R}^{\{1,2,\ldots\}}$ so that $A = \{\omega : (\omega(t_1), \omega(t_2), \ldots) \in B\}$. In words, all events in $\mathcal{F}_o$ depend on only countably many coordinates.

The above problem is easy to solve. Let $\mathbf{Q}_2 = \{m2^{-n} : m, n \ge 0\}$ be the **dyadic rationals**. If $\Omega_q = \{\omega : \mathbf{Q}_2 \to \mathbf{R}\}$ and $\mathcal{F}_q$ is the σ-field generated by the finite dimensional sets, then enumerating the rationals $q_1, q_2, \ldots$ and applying Kolmogorov's extension theorem shows that we can construct a probability ν_x on $(\Omega_q, \mathcal{F}_q)$ so that $\nu_x\{\omega : \omega(0) = x\} = 1$ and $(*)$ in (1.3) holds when the $t_i \in \mathbf{Q}_2$. To extend B_t to a process defined on $[0, \infty)$ we will show:

(1.4) Theorem. Let $T < \infty$ and $x \in \mathbf{R}$. ν_x assigns probability one to paths $\omega : \mathbf{Q}_2 \to \mathbf{R}$ that are uniformly continuous on $\mathbf{Q}_2 \cap [0, T]$.

Remark. It will take quite a bit of work to prove (1.4). Before taking on that task, we will attend to the last measure theoretic detail: we tidy things up by moving our probability measures to $(C, \mathcal{C})$ where $C = \{$continuous $\omega : [0, \infty) \to \mathbf{R}\}$ and $\mathcal{C}$ is the σ-field generated by the the coordinate maps $t \to \omega(t)$. To do this, we observe that the map ψ that takes a uniformly continuous point in Ω_q to its unique continuous extension in C is measurable, and we set

$$P_x = \nu_x \circ \psi^{-1}$$

Our construction guarantees that $B_t(\omega) = \omega_t$ has the right finite dimensional distributions for $t \in \mathbf{Q}_2$. Continuity of paths and a simple limiting argument shows that this is true when $t \in [0, \infty)$. Finally, the reader should note that, as in the case of Markov chains, we have one set of random variables $B_t(\omega) = \omega(t)$, and a family of probability measures P_x, $x \in \mathbf{R}$, so that under P_x, B_t is a Brownian motion with $P_x(B_0 = x) = 1$.

Proof of (1.4)   By (1.1) and (1.2), we can without loss of generality suppose $B_0 = 0$ and prove the result for $T = 1$. In this case, part (b) of the definition and the scaling relation (1.2) imply

$$E_0(|B_t - B_s|)^4 = E_0|B_{t-s}|^4 = C(t-s)^2$$

where $C = E|B_1|^4 < \infty$. From the last observation we get the desired uniform continuity by using the following result due to Kolmogorov.

(1.5) Theorem.   Suppose $E|X_s - X_t|^\beta \le K|t - s|^{1+\alpha}$ where α, $\beta > 0$. If $\gamma < \alpha/\beta$ then with probability one there is a contant $C(\omega)$ so that

$$|X(q) - X(r)| \le C|q - r|^\gamma \quad \text{for all } q, r \in \mathbf{Q}_2 \cap [0, 1]$$

Proof   Let $\gamma < \alpha/\beta$, $\eta > 0$, $I_n = \{(i, j) : 0 \le i \le j \le 2^n, 0 < j - i \le 2^{n\eta}\}$ and $G_n = \{|X(j2^{-n}) - X(i2^{-n})| \le ((j-i)2^{-n})^\gamma$ for all $(i, j) \in I_n\}$. Since $a^\beta P(|Y| > a) \le E|Y|^\beta$ we have

$$P(G_n^c) \le \sum_{(i,j) \in I_n} ((j-i)2^{-n})^{-\beta\gamma} E|X(j2^{-n}) - X(i2^{-n})|^\beta$$

Using our assumption and then noticing the number of $(i, j) \in I_n$ is $\le 2^n 2^{n\eta}$, we see that the right-hand side is (for the second step note $\alpha - \beta\gamma > 0$)

$$\le K \sum_{(i,j) \in I_n} ((j-i)2^{-n})^{-\beta\gamma+1+\alpha} \le K2^n 2^{n\eta}(2^{n\eta}2^{-n})^{-\beta\gamma+1+\alpha} = K2^{-n\lambda}$$

where $\lambda = (1 - \eta)(1 + \alpha - \beta\gamma) - (1 + \eta)$. Since $\gamma < \alpha/\beta$, we can pick η small enough so that $\lambda > 0$. To complete the proof now we will show

(1.6) Lemma.   Let $A = 3 \cdot 2^{(1-\eta)\gamma}/(1 - 2^{-\gamma})$. On $H_N = \cap_{n=N}^\infty G_n$ we have $|X(q) - X(r)| \le A|t - s|^\gamma$ for $q, r \in \mathbf{Q}_2 \cap [0, 1]$ with $|q - r| < 2^{-(1-\eta)N}$.

(1.5) implies (1.6)   The trivial inequality

$$P(H_N^c) \le \sum_{n=N}^\infty P(G_n^c) \le K \sum_{n=N}^\infty 2^{-n\lambda} = K2^{-N\lambda}/(1 - 2^{-\lambda})$$

implies that $|X(q) - X(r)| \leq A|q - r|^{\gamma}$ for $q, r \in \mathbf{Q}_2$ with $|q - r| < \delta(\omega)$. To extend this to $q, r \in \mathbf{Q}_2 \cap [0, 1]$ let $s_0 = q < s_1 < \ldots < s_n = r$ with $|s_i - s_{i-1}| < \delta(\omega)$ and use the triangle inequality to conclude $|X(q) - X(r)| \leq C(\omega)|q - r|^{\gamma}$ where $C(\omega) = 1 + \delta(\omega)^{-1}$.

Proof of (1.6) Let $q, r \in \mathbf{Q}_2 \cap [0, 1]$ with $0 < r - q < 2^{-(1-\eta)N}$. Pick $m \geq N$ so that

$$2^{-(m+1)(1-\eta)} \leq r - q < 2^{-m(1-\eta)}$$

and write

$$r = j2^{-m} + 2^{-r(1)} + \cdots + 2^{-r(\ell)}$$
$$q = i2^{-m} - 2^{-q(1)} - \cdots - 2^{-q(k)}$$

where $m < r(1) < \cdots < r(\ell)$ and $m < q(1) < \cdots < q(k)$. Now $0 < r - q < 2^{-m(1-\eta)}$, so $(j - i) < 2^{m\eta}$ and it follows that on H_N

(a)
$$|X(i2^{-m}) - X(j2^{-m})| \leq ((2^{m\eta})2^{-m})^{\gamma}$$

On H_N it follows from the triangle inequality that

(b)
$$|X(q) - X(i2^{-m})| \leq \sum_{h=1}^{k}(2^{-q(h)})^{\gamma} \leq \sum_{h=m}^{\infty}(2^{-\gamma})^h \leq C_{\gamma}2^{-\gamma m}$$

where $C_{\gamma} = 1/(1 - 2^{-\gamma}) > 1$. Repeating the last computation shows

(c)
$$|X(r) - X(j2^{-m})| \leq C_{\gamma}2^{-\gamma m}$$

Combining (a)–(c) gives

$$|X(q) - X(r)| \leq 3C_{\gamma}2^{-\gamma m(1-\eta)} \leq 3C_{\gamma}2^{(1-\eta)\gamma}|r - q|^{\gamma}$$

since $2^{-m(1-\eta)} \leq 2^{1-\eta}|r - q|$. This completes the proof of (1.6) and hence of (1.5) and (1.4). $\qquad \square$

The scaling relation (1.2) implies

$$E|B_t - B_s|^{2m} = C_m|t - s|^m \quad \text{where} \quad C_m = E|B_1|^{2m}$$

so using (1.5) with $\beta = 2m$, $\alpha = m - 1$ and letting $m \to \infty$ gives a result of Wiener (1923).

(1.7) Theorem. Brownian paths are Hölder continuous with exponent γ for any $\gamma < 1/2$.

It is easy to show:

(1.8) **Theorem.** With probability one, Brownian paths are not Lipschitz continuous (and hence not differentiable) at any point.

Remark. The nondifferentiability of Brownian paths was discovered by Paley, Wiener, and Zygmund (1933). Paley died in 1933 at the age of 26 in a skiing accident while the paper was in press. The proof we are about to give is due to Dvoretsky, Erdös, and Kakutani (1961).

Proof   Fix a constant $C < \infty$ and let $A_n = \{\omega :$ there is an $s \in [0,1]$ so that $|B_t - B_s| \le C|t - s|$ when $|t - s| \le 3/n\}$. For $1 \le k \le n - 2$ let

$$Y_{k,n} = \max\left\{\left|B\left(\frac{k+j}{n}\right) - B\left(\frac{k+j-1}{n}\right)\right| : j = 0, 1, 2\right\}$$

$$B_n = \{\text{ at least one } Y_{k,n} \le 5C/n\}$$

The triangle inequality implies $A_n \subset B_n$. The worst case is $s = 1$. We pick $k = n - 2$ and observe

$$\left|B\left(\frac{n-3}{n}\right) - B\left(\frac{n-2}{n}\right)\right| \le \left|B\left(\frac{n-3}{n}\right) - B(1)\right| + \left|B(1) - B\left(\frac{n-2}{n}\right)\right|$$

$$\le C(3/n + 2/n)$$

Using $A_n \subset B_n$ and the scaling relation (1.2) gives

$$P(A_n) \le P(B_n) \le nP(|B(1/n)| \le 5C/n)^3 = nP(|B(1)| \le 5C/n^{1/2})^3$$

$$\le n\{(10C/n^{1/2}) \cdot (2\pi)^{-1/2}\}^3$$

since $\exp(-x^2/2) \le 1$. Letting $n \to \infty$ shows $P(A_n) \to 0$. Noticing $n \to A_n$ is increasing shows $P(A_n) = 0$ for all n and completes the proof. $\square$

EXERCISE 1.2.   Looking at the proof of (1.8) carefully shows that if $\gamma > 5/6$ then B_t is not Hölder continuous with exponent γ at any point in $[0,1]$. Show by considering k increments instead of 3 that the last conclusion is true for all $\gamma > 1/2 + 1/k$.

The next result is more evidence that the sample paths of Brownian motion behave locally like $\sqrt{t}$.

EXERCISE 1.3.   Fix t and let $\Delta_{m,n} = B(tm2^{n-1}) - B(t(m-1)2^{n-1})$. Compute

$$E\left(\sum_{m \le 2^n} \Delta_{m,n}^2 - t\right)^2$$

and use Borel-Cantelli to conclude that $\sum_{m\leq 2^n} \Delta_{m,n}^2 \to t$ a.s. as $n \to \infty$.

Remark. The last result is true if we consider a sequence of partitions $\Pi_1 \subset \Pi_2 \subset \ldots$ with mesh $\to 0$. See Freedman (1971a) p. 42-46. However, the true quadratic variation, defined as the sup over all partitions, is ∞.

7.2. Markov Property, Blumenthal's 0-1 Law

Intuitively the Markov property says "if $s \geq 0$ then $B(t+s) - B(s)$, $t \geq 0$ is a Brownian motion that is independent of what happened before time s." The first step in making this into a precise statement is to explain what we mean by "what happened before time s." The first thing that comes to mind is

$$\mathcal{F}_s^o = \sigma(B_r : r \leq s)$$

For reasons that will become clear as we go along it is convenient to replace $\mathcal{F}_s^o$ by.

$$\mathcal{F}_s^+ = \cap_{t>s}\mathcal{F}_t^o$$

The fields $\mathcal{F}_s^+$ are nicer because they are **right continuous**:

$$\cap_{t>s}\mathcal{F}_t^+ = \cap_{t>s}(\cap_{u>t}\mathcal{F}_u^o) = \cap_{u>s}\mathcal{F}_u^o = \mathcal{F}_s^+$$

In words the $\mathcal{F}_s^+$ allow us an "infinitesimal peek at the future," i.e., $A \in \mathcal{F}_s^+$ if it is in $\mathcal{F}_{s+\epsilon}^o$ for any $\epsilon > 0$. If $f(u) > 0$ for all $u > 0$ then the random variable

$$\limsup_{t\downarrow s} \frac{B_t - B_s}{f(t-s)}$$

is measurable with respect to $\mathcal{F}_s^+$ but not $\mathcal{F}_s^o$. We will see below that there are no interesting examples, i.e., $\mathcal{F}_s^+$ and $\mathcal{F}_s^o$ are the same (up to null sets).

To state the Markov property we need some notation. Recall that we have a family of measures P_x, $x \in \mathbf{R}$, on $(C, \mathcal{C})$ so that under P_x, $B_t(\omega) = \omega(t)$ is a Brownian motion starting at x. For $s \geq 0$ we define the **shift transformation** $\theta_s : C \to C$ by

$$(\theta_s\omega)(t) = \omega(s+t) \quad \text{for } t \geq 0$$

In words, we cut off the part of the path before time s and then shift the path so that time s becomes time 0.

(2.1) **Markov property.** If $s \geq 0$ and Y is bounded and $\mathcal{C}$ measurable then for all $x \in \mathbf{R}$

$$E_x(Y \circ \theta_s | \mathcal{F}_s^+) = E_{B_s} Y$$

where the right-hand side is the function $\varphi(x) = E_x Y$ evaluated at $x = B_s$.

Proof   By the definition of conditional expectation, what we need to show is that

(∗) $$E_x(Y \circ \theta_s; A) = E_x(E_{B_s}Y; A) \quad \text{for all } A \in \mathcal{F}_s^+$$

We will begin by proving the result for a carefully chosen special case and then use the monotone class theorem (MCT) to get the general case. Suppose $Y(\omega) = \prod_{1 \le m \le n} f_m(\omega(t_m))$ where $0 < t_1 < \ldots < t_n$ and the f_m are bounded and measurable. Let $0 < h < t_1$, let $0 < s_1 \ldots < s_k \le s + h$, and let $A = \{\omega : \omega(s_j) \in A_j, 1 \le j \le k\}$ where $A_j \in \mathcal{R}$ for $1 \le j \le k$. From the definition of Brownian motion it follows that

$$E(Y \circ \theta_s; A) = \int_{A_1} dx_1 \, p_{s_1}(x, x_1) \int_{A_2} dx_2 \, p_{s_2-s_1}(x_1, x_2) \cdots$$
$$\int_{A_k} dx_k \, p_{s_k-s_{k-1}}(x_{k-1}, x_k) \int dy \, p_{s+h-s_k}(x_k, y)\varphi(y, h)$$

where

$$\varphi(y, h) = \int dy_1 \, p_{t_1-h}(y, y_1)f_1(y_1) \ldots \int dy_n \, p_{t_n-t_{n-1}}(y_{n-1}, y_n)f_n(y_n)$$

For more details see the proof of (1.6) in Chapter 5, which applies without change here. Using that identity on the right hand side we have

(∗∗) $$E(Y \circ \theta_s; A) = E(\varphi(B_{s+h}, h); A)$$

The last equality holds for all finite dimensional sets A so the $\pi - \lambda$ theorem ((4.2) in Chapter 1) implies that it is valid for all $A \in \mathcal{F}_{s+h}^o \supset \mathcal{F}_s^+$.

It is easy to see by induction on n that

$$\psi(y_1) = f_1(y_1) \int dy_2 \, p_{t_2-t_1}(y_1, y_2)f_2(y_2)$$
$$\ldots \int dy_n \, p_{t_n-t_{n-1}}(y_{n-1}, y_n)f_n(y_n)$$

is bounded and measurable. Letting $h \downarrow 0$ and using the dominated convergence theorem shows that if $x_h \to x$ then

$$\varphi(x_h, h) = \int dy_1 \, p_{t_1-h}(x_h, y_1)\psi(y_1) \to \varphi(x, 0)$$

as $h \downarrow 0$. Using (**) and the bounded convergence theorem now gives

$$E_x(Y \circ \theta_s; A) = E_x(\varphi(B_s, 0); A)$$

for all $A \in \mathcal{F}_s^+$. This shows that (*) holds for $Y = \prod_{1 \le m \le n} f_m(\omega(t_m))$ and the f_m are bounded and measurable.

The desired conclusion now follows from the monotone class theorem, (1.4) in Chapter 5. Let $\mathcal{H} =$ the collection of bounded functions for which (*) holds. $\mathcal{H}$ clearly has properties (ii) and (iii). Let $\mathcal{A}$ be the collection of sets of the form $\{\omega : \omega(t_j) \in A_j\}$ where $A_j \in \mathcal{R}$. The speical case treated above shows (i) holds and the desired conclusion follows. $\square$

The next two exercises give typical applications of the Markov property. In Section 7.4 we will use these equalities to compute the distributions of L and R.

EXERCISE 2.1. Let $T_0 = \inf\{s > 0 : B_s = 0\}$ and let $R = \inf\{t > 1 : B_t = 0\}$. R is for right or return. Use the Markov property at time 1 to get

$$(2.2) \qquad P_x(R > 1 + t) = \int p_1(x, y) P_y(T_0 > t)\, dy$$

EXERCISE 2.2. Let $T_0 = \inf\{s > 0 : B_s = 0\}$ and let $L = \sup\{t \le 1 : B_t = 0\}$. L is for left or last. Use the Markov property at time $0 < t < 1$ to conclude

$$(2.3) \qquad P_0(L \le t) = \int p_t(0, y) P_y(T_0 > 1 - t)\, dy$$

The reader will see many applications of the Markov property below, so we turn our attention now to a "triviality" that has surprising consequences. Since

$$E_x(Y \circ \theta_s | \mathcal{F}_s^+) = E_{B(s)} Y \in \mathcal{F}_s^o$$

it follows from (1.1) in Chapter 5 that

$$E_x(Y \circ \theta_s | \mathcal{F}_s^+) = E_x(Y \circ \theta_s | \mathcal{F}_s^o)$$

From the last equation it is a short step to

(2.4) Theorem. If $Z \in \mathcal{C}$ is bounded then for all $s \ge 0$ and $x \in \mathbf{R}$,

$$E_x(Z | \mathcal{F}_s^+) = E_x(Z | \mathcal{F}_s^o)$$

Proof   As in the proof of (2.1) it suffices to prove the result when

$$Z = \prod_{m=1}^{n} f_m(B(t_m))$$

and the f_m are bounded and measurable. In this case Z can be written as $X(Y \circ \theta_s)$ where $X \in \mathcal{F}_s^o$ and Y is $\mathcal{C}$ measurable so

$$E_x(Z|\mathcal{F}_s^+) = X E_x(Y \circ \theta_s|\mathcal{F}_s^+) = X E_{B_s} Y \in \mathcal{F}_s^o$$

and the proof is complete.   □

If we let $Z \in \mathcal{F}_s^+$ then (2.4) implies $Z = E_x(Z|\mathcal{F}_s^o) \in \mathcal{F}_s^o$, so the two σ-fields are the same up to null sets. At first glance, this conclusion is not exciting. The fun starts when we take $s = 0$ in (2.4) to get

(2.5) Blumenthal's 0-1 law. If $A \in \mathcal{F}_0^+$ then for all $x \in \mathbf{R}$, $P_x(A) \in \{0, 1\}$.

Proof   Using $A \in \mathcal{F}_0^+$, (2.5), and $\mathcal{F}_0^o = \sigma(B_0)$ is trivial under P_x gives

$$1_A = E_x(1_A|\mathcal{F}_0^+) = E_x(1_A|\mathcal{F}_0^o) = P_x(A)   P_x \text{ a.s.}$$

This shows that the indicator function 1_A is a.s. equal to the number $P_x(A)$ and the result follows.   □

In words, the last result says that the **germ field**, $\mathcal{F}_0^+$, is trivial. This result is very useful in studying the local behavior of Brownian paths.

(2.6) Theorem. If $\tau = \inf\{t \geq 0 : B_t > 0\}$ then $P_0(\tau = 0) = 1$.

Proof $P_0(\tau \leq t) \geq P_0(B_t > 0) = 1/2$ since the normal distribution is symmetric about 0. Letting $t \downarrow 0$ we conclude

$$P_0(\tau = 0) = \lim_{t \downarrow 0} P_0(\tau \leq t) \geq 1/2$$

so it follows from (2.5) that $P_0(\tau = 0) = 1$.   □

Once Brownian motion must hit $(0, \infty)$ immediately starting from 0, it must also hit $(-\infty, 0)$ immediately. Since $t \to B_t$ is continuous, this forces:

(2.7) Theorem. If $T_0 = \inf\{t > 0 : B_t = 0\}$ then $P_0(T_0 = 0) = 1$.

A corollary of (2.7) is

EXERCISE 2.3. If $a < b$, there is a local maximum of B_t in (a, b). So the set of local maxima of B_t is a dense set.

Another typical application of (2.5) is

EXERCISE 2.4. (i) Suppose $f(t) > 0$ for all $t > 0$. Use (2.6) to conclude that $\limsup_{t \downarrow 0} B(t)/f(t) = c$, P_0 a.s. where $c \in [0, \infty]$ is a constant. (ii) Show that when $f(t) = \sqrt{t}$ then $c = \infty$, so with probability one Brownian paths are not Hölder continuous of order $1/2$ at 0.

Remark. Let $\mathcal{H}_\gamma(\omega)$ be the set of times at which the path $\omega \in C$ is Hölder continuous of order γ. (1.7) shows that $P(\mathcal{H}_\gamma = [0, \infty)) = 1$ for $\gamma < 1/2$. Exercise 1.2 shows that $P(\mathcal{H}_\gamma = \emptyset) = 1$ for $\gamma > 1/2$. The last exercise shows $P(t \in \mathcal{H}_{1/2}) = 0$ for each t, but B. Davis (1983) has shown $P(\mathcal{H}_{1/2} \neq \emptyset) = 1$.

(2.5) concerns the behavior of B_t as $t \to 0$. By using a trick we can use this result to get information about the behavior as $t \to \infty$.

(2.8) Theorem. If B_t is a Brownian motion starting at 0 then so is the process defined by $X_0 = 0$ and $X_t = tB(1/t)$ for $t > 0$.

Proof   Here we will check the second definition of Brownian motion. To do this we note: (i) If $0 < t_1 < \ldots < t_n$ then $(X(t_1), \ldots, X(t_n))$ has a multivariate normal distribution with mean 0. (ii) $EX_s = 0$ and if $s < t$ then

$$E(X_s X_t) = st E(B(1/s)B(1/t)) = s$$

For (iii) we note that X is clearly continuous at $t \neq 0$. To handle $t = 0$ we note that X has the same finite dimensional distributions as Brownian motion and then use (1.4).   □

Direct proof of continuity.   We begin by observing that the strong law of large numbers implies $B_n/n \to 0$ as $n \to \infty$ through the integers. To handle values in between integers we note that Kolmogorov's inequality, (8.2) in Chapter 1 implies

$$P\left(\sup_{0 < k \leq 2^m} |B(n + k2^{-m}) - B_n| > n^{2/3} \right) \leq n^{-4/3} E(B_{n+1} - B_n)^2$$

Letting $m \to \infty$ we have

$$P\left(\sup_{u \in [n, n+1]} |B_u - B_n| > n^{2/3} \right) \leq n^{-4/3}$$

Since $\sum_n n^{-4/3} < \infty$, the Borel-Cantelli lemma implies $B_u/u \to 0$ as $u \to \infty$. Taking $u = 1/t$, we have $X_t \to 0$ as $t \to 0$. $\square$

(2.8) allows us to relate the behavior of B_t as $t \to \infty$ and as $t \to 0$. Combining this idea with Blumenthal's 0-1 law leads to a very useful result. Let

$$\mathcal{F}'_t = \sigma(B_s : s \geq t) = \text{ the future at time } t$$
$$\mathcal{T} = \cap_{t \geq 0} \mathcal{F}'_t = \text{ the tail } \sigma\text{-field.}$$

(2.9) **Theorem.** If $A \in \mathcal{T}$ then either $P_x(A) \equiv 0$ or $P_x(A) \equiv 1$.

Remark. Notice that this is stronger than the conclusion of Blumenthal's 0-1 law (2.5). The examples $A = \{\omega : \omega(0) \in D\}$ show that for A in the germ σ-field $\mathcal{F}_0^+$ the value of $P_x(A)$, $1_D(x)$ in this case, may depend on x.

ABC Notes. In the remark we used D is the definition of A to avoid conflict with B for Brownian motion and C for constant. D may remind the reader of the unit disk but then we would have had to skip E for expected value, F for function, and G for open set.

Proof Since the tail σ-field of B is the same as the germ σ-field for X, it follows that $P_0(A) \in \{0, 1\}$. To improve this to the conclusion given observe that $A \in \mathcal{F}'_1$, so A can be written as $1_D \circ \theta_1$. Applying the Markov property gives

$$P_x(A) = E_x(1_D \circ \theta_1) = E_x(E_x(1_D \circ \theta_1 | \mathcal{F}_1)) = E_x(E_{B_1} 1_D)$$
$$= \int (2\pi)^{-1/2} \exp(-(y-x)^2/2) P_y(D) \, dy$$

Taking $x = 0$ we see that if $P_0(A) = 0$ then $P_y(D) = 0$ for a.e. y with respect to Lebesgue measure, and using the formula again shows $P_x(A) = 0$ for all x. To handle the case $P_0(A) = 1$ observe that $A^c \in \mathcal{T}$ and $P_0(A^c) = 0$, so the last result implies $P_x(A^c) = 0$ for all x. $\square$

The next result is a typical application of (2.9).

(2.10) **Theorem.** Let B_t be a one dimensional Brownian motion starting at 0 then with probability 1,

$$\limsup_{t \to \infty} B_t/\sqrt{t} = \infty \qquad \liminf_{t \to \infty} B_t/\sqrt{t} = -\infty$$

Proof Let $K < \infty$. By Exercise 6.6 in Chapter 1, and scaling

$$P_0(B_n/\sqrt{n} \geq K \text{ i.o.}) \geq \limsup_{n \to \infty} P_0(B_n \geq K\sqrt{n}) = P_0(B_1 \geq K) > 0$$

so the 0–1 law (2.9) implies the probability is 1. Since K is arbitrary this proves the first result. The second one follows from symmetry. $\qquad\square$

From (2.10), translation invariance, and the continuity of Brownian paths it follows that we have

(2.11) Theorem. Let B_t be a one dimensional Brownian motion and let $A = \cap_n \{B_t = 0$ for some $t \geq n\}$. Then $P_x(A) = 1$ for all x.

In words, one dimensional Brownian motion is recurrent. For any starting point x, it will return to 0 "infinitely often," i.e., there is a sequence of times $t_n \uparrow \infty$ so that $B_{t_n} = 0$. We have to be careful with the interpretation of the phrase in quotes since starting from 0, B_t will hit 0 infinitely many times by time $\epsilon > 0$.

Last rites. With our discussion of Blumenthal's 0–1 law complete, the distinction between $\mathcal{F}_s^+$ and $\mathcal{F}_s^o$ is no longer important, so we will make one final improvement in our σ-fields, and remove the superscripts. Let

$$\mathcal{N}_x = \{A : A \subset D \text{ with } P_x(D) = 0\}$$
$$\mathcal{F}_s^x = \sigma(\mathcal{F}_s^+ \cup \mathcal{N}_x)$$
$$\mathcal{F}_s = \cap_x \mathcal{F}_s^x$$

$\mathcal{N}_x$ are the **null sets** and $\mathcal{F}_s^x$ are the completed σ-fields for P_x. Since we do not want the filtration to depend on the initial state we take the intersection of all the σ-fields.

7.3. Stopping Times, Strong Markov Property

Generalizing a definition in Section 3.1, we call a random variable S taking values in $[0, \infty]$ a **stopping time** if for all $t \geq 0$, $\{S < t\} \in \mathcal{F}_t$. In the last definition we have obviously made a choice between $\{S < t\}$ and $\{S \leq t\}$. This makes a big difference in discrete time but none in continuous time (for a right continuous filtration $\mathcal{F}_t$) :

If $\{S \leq t\} \in \mathcal{F}_t$ then $\{S < t\} = \cup_n \{S \leq t - 1/n\} \in \mathcal{F}_t$.

If $\{S < t\} \in \mathcal{F}_t$ then $\{S \leq t\} = \cap_n \{S < t + 1/n\} \in \mathcal{F}_t$.

The first conclusion requires only that $t \to \mathcal{F}_t$ is increasing. The second relies on the fact that $t \to \mathcal{F}_t$ is right continuous. (3.2) and (3.3) below show that when checking something is a stopping time it is nice to know that the two definitions are equivalent.

(3.1) Theorem. If G is an open set and $T = \inf\{t \geq 0 : B_t \in G\}$ then T is a stopping time.

Proof   Since G is open and $t \to B_t$ is continuous $\{T < t\} = \cup_{q<t}\{B_q \in G\}$ where the union is over all rational q, so $\{T < t\} \in \mathcal{F}_t$. Here, we need to use the rationals to get a countable union, and hence a measurable set.   □

(3.2) Theorem. If T_n is a sequence of stopping times and $T_n \downarrow T$ then T is a stopping time.

Proof $\{T < t\} = \cup_n\{T_n < t\}$.   □

(3.3) Theorem. If T_n is a sequence of stopping times and $T_n \uparrow T$ then T is a stopping time.

Proof $\{T \leq t\} = \cap_n\{T_n \leq t\}$.   □

(3.4) Theorem. If K is a closed set and $T = \inf\{t \geq 0 : B_t \in K\}$ then T is a stopping time.

Proof   Let $G_n = \cup\{(x - 1/n, x + 1/n) : x \in K\}$ and let $T_n = \inf\{t \geq 0 : B_t \in G_n\}$. Since G_n is open, it follows from (3.1) that T_n is a stopping time. I claim that as $n \uparrow \infty$, $T_n \uparrow T$. To prove this notice that $T \geq T_n$ for all n, so $\lim T_n \leq T$. To prove $T \leq \lim T_n$ we can suppose that $T_n \uparrow t < \infty$. Since $B(T_n) \in \bar{G}_n$ for all n and $B(T_n) \to B(t)$, it follows that $B(t) \in K$ and $T \leq t$. □

EXERCISE 3.1.   Let S be a stopping time and let $S_n = ([2^n S] + 1)/2^n$ where $[x] =$ the largest integer $\leq x$. That is,

$$S_n = (m+1)2^{-n} \text{ if } m2^{-n} \leq S < (m+1)2^{-n}$$

In words, we stop at the first time of the form $k2^{-n}$ after S (i.e., $> S$). From the verbal description it should be clear that S_n is a stopping time. Prove that it is.

EXERCISE 3.2.   If S and T are stopping times, then $S \wedge T = \min\{S, T\}$, $S \vee T = \max\{S, T\}$, and $S + T$ are also stopping times. In particular, if $t \geq 0$, then $S \wedge t$, $S \vee t$, and $S + t$ are stopping times.

EXERCISE 3.3.   Let T_n be a sequence of stopping times. Show that

$$\sup_n T_n, \quad \inf_n T_n, \quad \limsup_n T_n, \quad \liminf_n T_n$$

are stopping times.

(3.1) and (3.4) will take care of all the hitting times we will consider. Our next goal is to state and prove the strong Markov property. To do this, we need to generalize two definitions from Section 3.1. Given a nonnegative random variable $S(\omega)$ we define the random shift θ_S which "cuts off the part of ω before $S(\omega)$ and then shifts the path so that time $S(\omega)$ becomes time 0." In symbols, we set

$$(\theta_S \omega)(t) = \begin{cases} \omega(S(\omega) + t) & \text{on } \{S < \infty\} \\ \Delta & \text{on } \{S = \infty\} \end{cases}$$

where Δ is an extra point we add to C. As in Section 5.2, we will usually explicitly restrict our attention to $\{S < \infty\}$ so the reader does not have to worry about the second half of the definition.

The second quantity $\mathcal{F}_S$, "the information known at time S," is a little more subtle. We defined

$$\mathcal{F}_s = \cap_{\epsilon > 0} \, \sigma(B_{t \wedge (s+\epsilon)}, t \geq 0)$$

so by analogy we should set

$$\mathcal{F}_S = \cap_{\epsilon > 0} \, \sigma(B_{t \wedge (S+\epsilon)}, t \geq 0)$$

The definition we will now give is less transparent but easier to work with.

$$\mathcal{F}_S = \{A : A \cap \{S \leq t\} \in \mathcal{F}_t \text{ for all } t \geq 0\}$$

In words, this makes the reasonable demand that the part of A that lies in $\{S \leq t\}$ should be measurable with respect to the information available at time t. Again we have made a choice between $\leq t$ and $< t$ but as in the case of stopping times, this makes no difference and it is useful to know that the two definitions are equivalent.

EXERCISE 3.4. Show that when $\mathcal{F}_t$ is right continuous, the last definition is unchanged if we replace $\{S \leq t\}$ by $\{S < t\}$.

For practice with the definition of $\mathcal{F}_S$ do

EXERCISE 3.5. Let S be a stopping time, let $A \in \mathcal{F}_S$, and let

$$R = \begin{cases} S & \text{on } A \\ \infty & \text{on } A^c \end{cases}$$

Show that R is a stopping time.

EXERCISE 3.6. Let S and T be stopping times.
(i) $\{S < t\}$, $\{S > t\}$, $\{S = t\}$ are in $\mathcal{F}_S$.
(ii) $\{S < T\}$, $\{S > T\}$, and $\{S = T\}$ are in $\mathcal{F}_S$ (and in $\mathcal{F}_T$).

Most of the properties of $\mathcal{F}_N$ derived in Section 3.1 carry over to continuous time. Two that will be useful below are:

(3.5) Theorem. If $S \le T$ are stopping times then $\mathcal{F}_S \subset \mathcal{F}_T$.

Proof   If $A \in \mathcal{F}_S$ then $A \cap \{T \le t\} = (A \cap \{S \le t\}) \cap \{T \le t\} \in \mathcal{F}_t$. $\square$

(3.6) Theorem. If $T_n \downarrow T$ are stopping times then $\mathcal{F}_T = \cap \mathcal{F}(T_n)$.

Proof   (3.5) implies $\mathcal{F}(T_n) \supset \mathcal{F}_T$ for all n. To prove the other inclusion, let $A \in \cap \mathcal{F}(T_n)$. Since $A \cap \{T_n < t\} \in \mathcal{F}_t$ and $T_n \downarrow T$, it follows that $A \cap \{T < t\} \in \mathcal{F}_t$. $\square$

The last result allows us to prove something that is obvious from the verbal definition.

EXERCISE 3.7. $B_S \in \mathcal{F}_S$, i.e., the value of B_S is measurable with respect to the information known at time S! To prove this let $S_n = ([2^n S] + 1)/2^n$ be the stopping times defined in Exercise 3.1. Show $B(S_n) \in \mathcal{F}_{S_n}$ then let $n \to \infty$ and use (3.6).

We are now ready to state the strong Markov property, which says that the Markov property holds at stopping times.

(3.7) Strong Markov property. Let $(s, \omega) \to Y(s, \omega)$ be bounded and $\mathcal{R} \times \mathcal{C}$ measurable. If S is a stopping time then for all $x \in \mathbf{R}$

$$E_x(Y_S \circ \theta_S | \mathcal{F}_S) = E_{B(S)} Y_S \text{ on } \{S < \infty\}$$

where the right-hand side is the function $\varphi(x, t) = E_x Y_t$ evaluated at $x = B(S)$, $t = S$.

Proof   We first prove the result under the assumption that there is a sequence of times $t_n \uparrow \infty$, so that $P_x(S < \infty) = \sum P_x(S = t_n)$. In this case the proof is basically the same as the proof of (2.4) in Chapter 5. We break things down according to the value of S, apply the Markov property and put the pieces back together. If we let $Z_n = Y_{t_n}(\omega)$ and $A \in \mathcal{F}_S$ then

$$E_x(Y_S \circ \theta_S; A \cap \{S < \infty\}) = \sum_{n=1}^{\infty} E_x(Z_n \circ \theta_{t_n}; A \cap \{S = t_n\})$$

Now if $A \in \mathcal{F}_S$, $A \cap \{S = t_n\} = (A \cap \{S \leq t_n\}) - (A \cap \{S \leq t_{n-1}\}) \in \mathcal{F}_{t_n}$, so it follows from the Markov property that the above sum is

$$= \sum_{n=1}^{\infty} E_x(E_{B(t_n)} Z_n; A \cap \{S = t_n\}) = E_x(E_{B(S)} Y_S; A \cap \{S < \infty\})$$

To prove the result in general we let $S_n = ([2^n S] + 1)/2^n$ be the stopping time defined in Exercise 3.1. To be able to let $n \to \infty$ we restrict our attention to Y's of the form

$$(*) \qquad\qquad Y_s(\omega) = f_0(s) \prod_{m=1}^{n} f_m(\omega(t_m))$$

where $0 < t_1 < \ldots < t_n$ and $f_0, \ldots, f_n$ are bounded and continuous. If f is bounded and continuous then the dominated convergence theorem implies that

$$x \to \int dy\, p_t(x, y) f(y)$$

is continuous. From this and induction it follows that

$$\varphi(x, s) = E_x Y_s = f_0(s) \int dy_1\, p_{t_1}(x, y_1) f(y_1)$$

$$\ldots \int dy_n\, p_{t_n - t_{n-1}}(y_{n-1}, y_n) f(y_n)$$

is bounded and continuous.

Having assembled the necessary ingredients we can now complete the proof. Let $A \in \mathcal{F}_S$. Since $S \leq S_n$, (3.5) implies $A \in \mathcal{F}(S_n)$. Applying the special case of (3.7) proved above to S_n and observing that $\{S_n < \infty\} = \{S < \infty\}$ gives

$$E_x(Y_{S_n} \circ \theta_{S_n}; A \cap \{S < \infty\}) = E_x(\varphi(B(S_n), S_n); A \cap \{S < \infty\})$$

Now as $n \to \infty$, $S_n \downarrow S$, $B(S_n) \to B(S)$, $\varphi(B(S_n), S_n) \to \varphi(B(S), S)$ and

$$Y_{S_n} \circ \theta_{S_n} \to Y_S \circ \theta_S$$

so the bounded convergence theorem implies that (3.7) holds when Y has the form given in $(*)$.

To complete the proof now we will apply the monotone class theorem. As in the proof of (2.1), we let $\mathcal{H}$ be the collection of Y for which

$$E_x(Y_S \circ \theta_S; A) = E_x(E_{B(S)}Y_S; A) \quad \text{for all } A \in \mathcal{F}_S$$

and it is easy to see that (ii) and (iii) hold. This time however, we take $\mathcal{A}$ to be the sets of the form $A = \{\omega : \omega(s_j) \in G_j, 1 \le j \le k\}$ where the G_j are open sets. To verify (i) we note that if $K_j = G_j^c$ and $f_j^n(x) = 1 \wedge n\rho(x, K_j)$ where $\rho(x, K) = \inf\{|x = y| : y \in K\}$ then f_j^n are continuous functions with $f_j^n \uparrow 1_{G_j}$ as $n \uparrow \infty$. The facts that

$$Y_n(\omega) = \prod_{j=1}^{k} f_j^n(\omega(s_j)) \in \mathcal{H}$$

and (iii) holds for $\mathcal{H}$ imply that $1_A \in \mathcal{H}$. This verifies (i) in the monotone class theorem and completes the proof of (3.7). $\square$

The next section is devoted to applications of the strong Markov property.

7.4. Maxima and Zeros

In this section we will use the strong Markov property to derive properties of the zero set $\{t : B_t = 0\}$, the hitting times $T_a = \inf\{t : B_t = a\}$ and $\max_{0 \le s \le t} B_s$. This is the tip of an iceberg that is bigger than the one the Titanic ran into.

Example 4.1. Zeros of Brownian motion. Let $R_t = \inf\{u > t : B_u = 0\}$ and let $T_0 = \inf\{u > 0 : B_u = 0\}$. Now (2.10) implies $P_x(R_t < \infty) = 1$, so $B(R_t) = 0$ and the strong Markov property and (2.9) imply

$$P_x(T_0 \circ \theta_{R_t} > 0 | \mathcal{F}_{R_t}) = P_0(T_0 > 0) = 0$$

Taking expected value of the last equation we see that

$$P_x(T_0 \circ \theta_{R_t} > 0 \text{ for some rational } t) = 0$$

From this it follows that if a point $u \in \mathcal{Z}(\omega) \equiv \{t : B_t(\omega) = 0\}$ is isolated on the left (i.e., there is a rational $t < u$ so that $(t, u) \cap \mathcal{Z}(\omega) = \emptyset$) then it is, with probability one, a decreasing limit of points in $\mathcal{Z}(\omega)$. This shows that the closed set $\mathcal{Z}(\omega)$ has no isolated points and hence must be uncountable. For the last step see Hewitt and Stromberg (1969), page 72.

If we let $|\mathcal{Z}(\omega)|$ denote the Lebesgue measure of $\mathcal{Z}(\omega)$ then Fubini's theorem implies

$$E_x(|\mathcal{Z}(\omega)| \cap [0,T]) = \int_0^T P_x(B_t = 0)\, dt = 0$$

So $\mathcal{Z}(\omega)$ is a set of measure zero.

Example 4.2. Hitting times have independent increments.

(4.1) Theorem. Under P_0, $\{T_a, a \geq 0\}$ has stationary independent increments.

Proof The first step is to notice that if $0 < a < b$ then

$$T_b \circ \theta_{T_a} = T_b - T_a,$$

so if f is bounded and measurable, the strong Markov property (3.7) and translation invariance imply

$$E_0\left(f(T_b - T_a)\,|\,\mathcal{F}_{T_a}\right) = E_0\left(f(T_b) \circ \theta_{T_a}\,|\,\mathcal{F}_{T_a}\right)$$
$$= E_a f(T_b) = E_0 f(T_{b-a})$$

To show that the increments are independent, let $a_0 < a_1 \ldots < a_n$, and let f_i, $1 \leq i \leq n$ be bounded and measurable. Conditioning on $\mathcal{F}_{T_{a_{n-1}}}$ and using the preceding calculation we have

$$E_0\left(\prod_{i=1}^n f_i(T_{a_i} - T_{a_{i-1}})\right)$$
$$= E_0\left(\prod_{i=1}^{n-1} f_i(T_{a_i} - T_{a_{i-1}}) E_0(f_n(T_{a_n} - T_{a_{n-1}})|\mathcal{F}_{T_{a_{n-1}}})\right)$$
$$= E_0\left(\prod_{i=1}^{n-1} f_i(T_{a_i} - T_{a_{i-1}})\right) E_0 f(T_{a_n - a_{n-1}})$$

By induction it follows that

$$E_0\left(\prod_{i=1}^n f_i(T_{a_i} - T_{a_{i-1}})\right) = \prod_{i=1}^n E_0 f_i(T_{a_i - a_{i-1}})$$

which implies the desired conclusion. $\square$

The scaling relation (1.2) implies

$$(4.2) \qquad\qquad T_a \stackrel{d}{=} a^2 T_1$$

Combining (4.1) and (4.2), and using (7.15) from Chapter 2 we see that T_a has a stable law with index $\alpha = 1/2$. Since $T_a \geq 0$, the skewness parameter $\kappa = 1$. For a derivation that does not rely on the fact that (7.13) gives all the stable laws, combine Example 6.6 below with Exercise 7.8 in Chapter 2.

Without knowing the theory mentioned in the previous paragraph, it is easy to determine the Laplace transform $\varphi_a(\lambda) = E_0 \exp(-\lambda T_a)$. To do this, we start by observing that (4.1) implies

$$\varphi_x(\lambda)\varphi_y(\lambda) = \varphi_{x+y}(\lambda)$$

It follows easily from this that

(4.3)
$$\varphi_a(\lambda) = \exp(-ac(\lambda))$$

Proof Let $c(\lambda) = -\log \varphi_1(\lambda)$ so (4.3) holds when $a = 1$. Using the previous identity with $x = y = 2^{-m}$ and induction gives the result for $a = 2^{-m}$, $m \geq 1$. Then letting $x = k2^{-m}$ and $y = 2^{-m}$ we get the result for $a = (k+1)2^{-m}$ with $k \geq 1$. Finally, to extend to $a \in [0, \infty)$ note that $a \to \varphi_a(\lambda)$ is decreasing. $\square$

To identify $c(\lambda)$ we observe that (4.2) implies

$$E \exp(-T_a) = E \exp(-a^2 T_1)$$

so $ac(1) = c(a^2)$, i.e., $c(\lambda) = c\sqrt{\lambda}$. In the next section (see Exercise 5.2), we will show

(4.4)
$$E_0(\exp(-\lambda T_a)) = \exp(-a\sqrt{2\lambda})$$

so $c = \sqrt{2}$.

Our next goal is to compute the distribution of the hitting times T_a. This application of the strong Markov property shows why we want to allow the function Y that we apply to the shifted path to depend on the stopping time S.

Example 4.3. Reflection principle. Let $a > 0$ and let $T_a = \inf\{t : B_t = a\}$. Then

(4.5)
$$P_0(T_a < t) = 2P_0(B_t \geq a)$$

Intuitive proof We observe that if B_s hits a at some time $s < t$ then the strong Markov property implies that $B_t - B(T_a)$ is independent of what happened before time T_a. The symmetry of the normal distribution and $P_a(B_u = a) = 0$ for $u > 0$ then imply

(4.6)
$$P_0(T_a < t, B_t > a) = \frac{1}{2}P_0(T_a < t)$$

Rearranging the last equation and using $\{B_t > a\} \subset \{T_a < t\}$ gives

$$P_0(T_a < t) = 2P_0(T_a < t, B_t > a) = 2P_0(B_t > a)$$

Proof To make the intuitive proof rigorous we only have to prove (4.6). To extract this from the strong Markov property (3.7), we let

$$Y_s(\omega) = \begin{cases} 1 & \text{if } s < t, \, \omega(t-s) > a \\ 0 & \text{otherwise} \end{cases}$$

We do this so that if we let $S = \inf\{s < t : B_s = a\}$ with $\inf \emptyset = \infty$ then

$$Y_S(\theta_S \omega) = \begin{cases} 1 & \text{if } S < t, \, B_t > a \\ 0 & \text{otherwise} \end{cases}$$

and the strong Markov property implies

$$E_0(Y_S \circ \theta_S | \mathcal{F}_S) = \varphi(B_S, S) \quad \text{on } \{S < \infty\} = \{T_a < t\}$$

where $\varphi(x, s) = E_x Y_s$. $B_S = a$ on $\{S < \infty\}$ and $\varphi(a, s) = 1/2$ if $s < t$, so taking expected values gives

$$P_0(T_a < t, B_t \geq a) = E_0(Y_S \circ \theta_S ; S < \infty)$$
$$= E_0(E_0(Y_S \circ \theta_S | \mathcal{F}_S); S < \infty) = E_0(1/2 ; T_a < \infty)$$

which proves (4.6). $\square$

EXERCISE 4.1. (i) Generalize the proof of (4.6) to conclude that if $u < v \leq a$ then

(4.7) $$P_0(T_a < t, u < B_t < v) = P_0(2a - v < B_t < 2a - u)$$

(ii) Use (i) to derive the joint density

$$P_0(M_t = a, B_t = x) = \frac{2(2a - x)}{\sqrt{2\pi t^3}} e^{-(2a-x)^2/2t}$$

Using (4.5) we can compute the probability density of T_a. We begin by noting that

$$P(T_a \leq t) = 2 \, P_0(B_t \geq a) = 2 \int_a^\infty (2\pi t)^{-1/2} \exp(-x^2/2t) dx$$

then change variables $x = (t^{1/2}a)/s^{1/2}$ to get

(4.8)
$$P_0(T_a \le t) = 2 \int_t^0 (2\pi t)^{-1/2} \exp(-a^2/2s) \left(-t^{1/2}a/2s^{3/2}\right) ds$$
$$= \int_0^t (2\pi s^3)^{-1/2} a \exp(-a^2/2s) ds$$

Using the last formula we can compute

Example 4.4. The distribution of $L = \sup\{t \le 1 : B_t = 0\}$. By (2.3)

$$P_0(L \le s) = \int_{-\infty}^{\infty} p_s(0, x) P_x(T_0 > 1 - s) dx$$
$$= 2 \int_0^{\infty} (2\pi s)^{-1/2} \exp(-x^2/2s) \int_{1-s}^{\infty} (2\pi r^3)^{-1/2} x \exp(-x^2/2r) dr dx$$
$$= \frac{1}{\pi} \int_{1-s}^{\infty} (sr^3)^{-1/2} \int_0^{\infty} x \exp(-x^2(r+s)/2rs) dx dr$$
$$= \frac{1}{\pi} \int_{1-s}^{\infty} (sr^3)^{-1/2} rs/(r+s) dr$$

Our next step is to let $t = s/(r+s)$ to convert the integral over $r \in [1-s, \infty)$ into one over $t \in [0, s]$. $dt = -s/(r+s)^2 dr$, so to make the calculations easier we first rewrite the integral as

$$= \frac{1}{\pi} \int_{1-s}^{\infty} \left(\frac{(r+s)^2}{rs}\right)^{1/2} \frac{s}{(r+s)^2} dr$$

and then change variables to get

(4.9)
$$P_0(L \le s) = \frac{1}{\pi} \int_0^s (t(1-t))^{-1/2} dt = \frac{2}{\pi} \arcsin(\sqrt{s})$$

The arcsin may remind the reader of the limit theorem for $L_{2n} = \sup\{m \le 2n : S_m = 0\}$ given in (3.6) of Chapter 3. We will see in Section 7.6 that our new result is a consequence of the old one.

EXERCISE 4.2. Show that $R = \inf\{t > 1 : B_t = 0\}$ has probability density

$$P_0(R = 1 + t) = 1/(\pi t^{1/2}(1 + t))$$

Example 4.5. Lévy's modulus of continuity

$$\text{osc}(\delta) = \sup\{|B_s - B_t| : s, t \in [0, 1], |t - s| < \delta\}$$

(4.10) **Theorem.** With probability 1,

$$\limsup_{\delta \to 0} \operatorname{osc}(\delta)/(\delta \log(1/\delta))^{1/2} \le 6$$

Remark. The constant 6 is not the best possible because the end of the proof is sloppy. Lévy (1937) showed

$$\limsup_{\delta \to 0} \operatorname{osc}(\delta)/(\delta \log(1/\delta))^{1/2} = \sqrt{2}$$

See McKean (1969) p. 14-16 or Itô and McKean (1965) p. 36-38 where a sharper result due to Chung, Erdös and Sirao (1959) is proved. In contrast if we look at the behavior at a single point, (9.5) below shows

$$\limsup_{t \to 0} |B_t|/\sqrt{2t \log \log(1/t)} = 1 \quad \text{a.s.}$$

Proof   Let $I_{m,n} = [m2^{-n}, (m+1)2^{-n}]$, and $\Delta_{m,n} = \sup\{|B_t - B(m2^{-n})| : t \in I_{m,n}\}$. From (4.5) and the scaling relation it follows that

$$P(\Delta_{m,n} \ge a2^{-n/2}) \le 4P(B(2^{-n}) \ge a2^{-n/2})$$
$$= 4P(B(1) \ge a) \le 4\exp(-a^2/2)$$

by (1.3) in Chapter 1 if $a \ge 1$. If $\epsilon > 0$, $b = 2(1 + \epsilon)(\log 2)$, and $a_n = (bn)^{1/2}$ then the last result implies

$$P(\Delta_{m,n} \ge a_n 2^{-n/2} \text{ for some } m \le 2^n) \le 2^n \cdot 4\exp(-bn/2) = 4 \cdot 2^{-n\epsilon}$$

so the Borel-Cantelli lemma implies that if $n \ge N(\omega)$, $\Delta_{m,n} \le (bn)^{1/2}2^{-n/2}$. Now if $s \in I_{m,n}$, $s < t$ and $|s - t| < 2^{-n}$ then $t \in I_{m,n}$ or $I_{m+1,n}$. I claim that in either case the triangle inequality implies

$$|B_t - B_s| \le 3(bn)^{1/2}2^{-n/2}$$

To see this note that the worst case is $t \in I_{m+1,n}$, but even in this case

$$|B_t - B_s| \le |B_t - B((m+1)2^{-n})|$$
$$+ |B((m+1)2^{-n}) - B(m2^{-n})| + |B(m2^{-n}) - B_s|$$

It follows from the last estimate that for $2^{-(n+1)} \le \delta < 2^{-n}$

$$\operatorname{osc}(\delta) \le 3(bn)^{1/2}2^{-n/2} \le 3(b\log_2(1/\delta))^{1/2}(2\delta)^{1/2} = 6((1+\epsilon)\delta \log(1/\delta))^{1/2}$$

Recall $b = 2(1 + \epsilon) \log 2$ and observe $\exp((\log 2)(\log_2 1/\delta)) = 1/\delta$.   □

7.5. Martingales

At the end of Section 4.7 we used martingales to study the hitting times of random walks. The same methods can be used on Brownian motion once we prove

(5.1) Theorem. Let X_t be a right continuous martingale adapted to a right continuous filtration. If T is a bounded stopping time then $EX_T = EX_0$.

Proof Let n be an integer so that $P(T \leq n - 1) = 1$. As in the proof of the strong Markov property let $T_m = ([2^m T] + 1)/2^m$. $Y_k^m = X(k2^{-m})$ is a martingale with respect to $\mathcal{F}_k^m = \mathcal{F}(k2^{-m})$ and $S_m = 2^m T_m$ is a stopping time for $(Y_k^m, \mathcal{F}_k^m)$ so by Exercise 4.3 in Chapter 4

$$X(T_m) = Y_{S_m}^m = E(Y_{n2^m}^m | \mathcal{F}_{S_m}^m) = E(X_n | \mathcal{F}(T_m))$$

As $m \uparrow \infty$, $X(T_m) \to X(T)$ by right continuity and $\mathcal{F}(T_m) \downarrow \mathcal{F}(T)$ by (3.6), so it follows from (6.3) in Chapter 4 that

$$X(T) = E(X_n | \mathcal{F}(T))$$

Taking expected values now gives $EX(T) = EX_n = EX_0$, since X_n is a martingale.   □

(5.2) Theorem. B_t is a martingale w.r.t. the σ-fields $\mathcal{F}_t$ defined in Section 3.

Note: We will use these σ-fields in (5.4), (5.6), and (5.8) but will not mention them explicitly in the statements.

Proof The Markov property implies that

$$E_x(B_t | \mathcal{F}_s) = E_{B_s}(B_{t-s}) = B_s$$

since symmetry implies $E_y B_u = y$ for all $u \geq 0$.   □

From (5.2) it follows immediately that we have

(5.3) Theorem. If $a < x < b$ then $P_x(T_a < T_b) = (b - x)/(b - a)$.

Proof Let $T = T_a \wedge T_b$. (2.10) implies that $T < \infty$ a.s. Using (5.1) and (5.2) it follows that $x = E_x B(T \wedge t)$. Letting $t \to \infty$ and using the bounded convergence theorem it follows that

$$x = aP_x(T_a < T_b) + b(1 - P_x(T_a < T_b))$$

Solving for $P_x(T_a < T_b)$ now gives the desired result. □

Example 5.1. Optimal doubling in Backgammon (Keeler and Spencer (1975)). In our idealization, backgammon is a Brownian motion starting at 1/2 run until it hits 1 or 0, and B_t is the probability you will win given the events up to time t. Initially the "doubling cube" sits in the middle of the board and either player can "double" = tell the other player to play on for twice the stakes or give up and pay the current wager. If a player accepts the double (i.e., decides to play on), she gets possession of the doubling cube and is the only one who can offer the next double.

A doubling strategy is given by two numbers $b < 1/2 < a$, i.e., offer a double when $B_t \geq a$ and give up if the other player doubles and $B_t < b$. It is not hard to see that for the optimal strategy $b^* = 1 - a^*$ and that when $B_t = b^*$ accepting and giving up must have the same payoff. If you accept when your probability of winning is b^* then you lose 2 dollars when your probability hits 0 but you win 2 dollars when your probability of winning hits a^*, since at that moment you can double and the other player gets the same payoff if they give up or play on. If giving up or playing on at b^* is to have the same payoff we must have

$$-1 = \frac{b^*}{a^*} \cdot 2 + \frac{a^* - b^*}{a^*} \cdot (-2)$$

Writing $b^* = c$ and $a^* = 1 - c$ and solving we have

$$-(1 - c) = 2c - 2(1 - 2c) \qquad 1 = 5c$$

so $b^* = 1/5$ and $a^* = 4/5$. □

(5.4) Theorem. $B_t^2 - t$ is a martingale.

Proof $E_x(B_t^2|\mathcal{F}_s) = E_x(B_s^2 + 2B_s(B_t - B_s) + (B_t - B_s)^2|\mathcal{F}_s)$
$$= B_s^2 + 2B_s E_x(B_t - B_s|\mathcal{F}_s) + E_x((B_t - B_s)^2|\mathcal{F}_s)$$
$$= B_s^2 + 0 + (t - s)$$

since $B_t - B_s$ is independent of $\mathcal{F}_s$ and has mean 0 and variance $t - s$. □

(5.5) Theorem. Let $T = \inf\{t : B_t \notin (a, b)\}$ where $a < 0 < b$.

$$E_0 T = -ab$$

Proof (5.1) and (5.4) imply $E_0(B^2(T \wedge t)) = E_0(T \wedge t)$. Letting $t \to \infty$ and using the monotone convergence theorem gives $E_0(T \wedge t) \uparrow E_0 T$. Using the bounded convergence theorem and (5.3) we have

$$E_0 B^2(T \wedge t) \to E_0 B_T^2 = a^2 \frac{b}{b-a} + b^2 \frac{-a}{b-a} = ab \frac{a-b}{b-a} = -ab \qquad \square$$

(5.6) Theorem. $\exp(\theta B_t - (\theta^2 t/2))$ is a martingale.

Proof
$$E_x(\exp(\theta B_t)|\mathcal{F}_s) = \exp(\theta B_s) E(\exp(\theta(B_t - B_s))|\mathcal{F}_s)$$
$$= \exp(\theta B_s) \exp(\theta^2(t-s)/2)$$

since $B_t - B_s$ is independent of $\mathcal{F}_s$ and has a normal distribution with mean 0 and variance $t - s$. $\qquad \square$

(5.7) Theorem. Let $T = \inf\{t : B_t \notin (-a, a)\}$.

$$E_0 \exp(-\lambda T) = 1/\cosh(a\sqrt{2\lambda})$$

Proof (5.1) and (5.6) imply that $1 = E_0 \exp(\theta B(T \wedge t) - \theta^2(T \wedge t))$. Letting $t \to \infty$ and using the bounded convergence theorem gives

$$1 = E_0 \exp(\theta B_T - \theta^2 T/2)$$

Symmetry implies that $P(B_T = a) = P(B_T = -a) = 1/2$ and that $B(T)$ is independent of T so
$$1 = \cosh(\theta a) E_0 \exp(-\theta^2 T/2)$$

Setting $\theta = \sqrt{2\lambda}$ now gives the desired result. $\qquad \square$

If you don't like the "symmetry implies that" then

EXERCISE 5.1. Derive (5.7) by showing that $\exp(-\theta^2 t/2) \cosh(\theta B_t)$ is a martingale.

(5.8) Theorem. $B_t^3 - 3t B_t$, $B_t^4 - 6t B_t^2 + 3t^2$, ... are martingales.

Proof The result in (5.6) can be written as

$$E(\exp(\theta B_t - \theta^2 t/2); A) = E(\exp(\theta B_s - \theta^2 s/2); A) \quad \text{for } A \in \mathcal{F}_s$$

Let $f(x, \theta) = \exp(\theta x - \theta^2 t/2)$, let $f_k(x, \theta)$ be the k the derivative w.r.t. θ, and let $h_k(x) = f_k(x, 0)$. Differentiating the last identity k times (referring to Section

A.9 for the justification) and setting $\theta = 0$ we see that $h_k(B_t)$ is a martingale. We have seen $h_1(B_t)$ and $h_2(B_t)$ but the other ones are new:

k	$h_k(\theta)$	Martingale
1	$(x - \theta t)f(\theta)$	B_t
2	$\{(x - \theta t)^2 - t\}f(\theta)$	$B_t^2 - t$
3	$\{(x - \theta t)^3 - 3t(x - \theta t)\}f(\theta)$	$B_t^3 - 3tB_t$
4	$\{(x - \theta t)^4 - 6t(x - \theta t)^2 + 3t^2\}f(\theta)$	$B_t^4 - 6tB_t^2 + 3t^2$

(5.9) Theorem. Let $T = \inf\{t : B_t \notin (-a, a)\}$.

$$ET^2 = 5a^4/3$$

Proof (5.1) and (5.8) imply $E(B(T \wedge t)^4 - 6(T \wedge t)B(T \wedge t)^2) = -3E(T \wedge t)^2$. From (5.5) we know that $ET = a^2 < \infty$. Letting $t \to \infty$, using the dominated convergence theorem on the left-hand side, and the monotone convergence theorem on the right gives

$$a^4 - 6a^2 ET = -3E(T^2)$$

Plugging in $ET = a^2$ gives the desired result. $\square$

EXERCISES

5.2. The point of this exercise is to get information about the amount of time it takes Brownian motion with drift $-b$, $X_t \equiv B_t - bt$ to hit level a. Let $\tau = \inf\{t : B_t = a + bt\}$ where $a > 0$. (i) Use the martingale $\exp(\theta B_t - \theta^2 t/2)$ with $\theta = b + (b^2 + 2\lambda)^{1/2}$ to show

$$E_0 \exp(-\lambda\tau) = \exp(-a\{b + (b^2 + 2\lambda)^{1/2}\})$$

Setting $b = 0$ gives formula (4.4) as promised. Letting $\lambda \to 0$

$$P_0(\tau < \infty) = \exp(-2ab)$$

5.3. Let $\sigma = \inf\{t : B_t \notin (a, b)\}$ and let $\lambda > 0$. Use the strong Markov property to show

$$E_x \exp(-\lambda T_a) = E_x(e^{-\lambda\sigma}; T_a < T_b) + E_x(e^{-\lambda\sigma}; T_b < T_a)E_b \exp(-\lambda T_a)$$

(ii) Interchange the roles of a and b to get a second equation, use (4.4), and solve to get

$$E_x(e^{-\lambda T}; T_a < T_b) = \sinh(\sqrt{2\lambda}(b - x))/\sinh(\sqrt{2\lambda}(b - a))$$
$$E_x(e^{-\lambda T}; T_b < T_a) = \sinh(\sqrt{2\lambda}(x - a))/\sinh(\sqrt{2\lambda}(b - a))$$

5.4. If $U = \inf\{t : B_t \notin (a,b)\}$, where $a < 0 < b$ and $a \neq -b$, then U and B_U^2 are not independent so we cannot calculate EU^2 as we did in the proof of (5.9). Use the Cauchy-Schwarz inequality to estimate $E(UB_U^2)$ and conclude $EU^2 \leq C\, E(B_U^4)$ where C is independent of a and b.

5.5. Let $u(t,x)$ be a function that satisfies

$$(*) \qquad \frac{\partial u}{\partial t} + \frac{1}{2}\frac{\partial^2 u}{\partial x^2} = 0 \quad\text{and}\quad \left|\frac{\partial^2 u}{\partial x^2}(t,x)\right| \leq C_T \exp(x^2/(t+\epsilon)) \quad\text{for } t \leq T$$

Show that $u(t, B_t)$ is a martingale by checking

$$\frac{\partial}{\partial t} p_t(x,y) = \frac{1}{2}\frac{\partial^2}{\partial y^2} p_t(x,y)$$

interchanging $\partial/\partial t$ and $\int$, and then integrating by parts twice to show

$$\frac{\partial}{\partial t} E_x u(t, B_t) = \int \frac{\partial}{\partial t}(p_t(x,y)u(t,y))\, dy = 0$$

Examples of functions that satisfy $(*)$ are $\exp(\theta x - \theta^2 t/2)$, x, $x^2 - t$, ...

5.6. Find a martingale of the form $B_t^6 - atB_t^4 + bt^2B_t^2 - ct^3$ and use it to compute the third moment of $T = \inf\{t : B_t \notin (-a,a)\}$. Note: you can differentiate $\exp(\theta x - \theta^2 t/2)$ six times, but it is easier to get a, b, and c from the last exercise.

5.7. Show that $(1+t)^{-1/2}\exp(B_t^2/(1+t))$ is a martingale and use this to conclude that $\limsup_{t\to\infty} B_t/(2t\log t)^{1/2} \leq 1$ a.s.

7.6. Donsker's Theorem

Let $X_1, X_2, \ldots$ be i.i.d. with $EX = 0$ and $EX^2 = 1$, and let $S_n = X_1 + \cdots + X_n$. In this section we will show that as $n \to \infty$, $S(nt)/n^{1/2}$, $0 \leq t \leq 1$ converges in distribution to B_t a Brownian motion starting from $B_0 = 0$. We will say precisely what the last sentence means below. The key to its proof is

(6.1) Skorokhod's representation theorem. If $EX = 0$ and $EX^2 < \infty$ then there is a stopping time T for Brownian motion so that $B_T =_d X$ and $ET = EX^2$.

Remark. The Brownian motion in the statement and all the Brownian motions in this section have $B_0 = 0$.

Proof Suppose first that X is supported on $\{a, b\}$ where $a < 0 < b$. Since $EX = 0$, we must have

$$P(X = a) = \frac{b}{b - a} \qquad P(X = b) = \frac{-a}{b - a}$$

If we let $T = T_{a,b} = \inf\{t : B_t \notin (a, b)\}$ then (5.3) implies $B_T =_d X$ and (5.5) tells us that

$$ET = -ab = EB_T^2$$

To treat the general case we will write $F(x) = P(X \le x)$ as a mixture of two point distributions with mean 0. Let

$$c = \int_{-\infty}^0 (-u)\, dF(u) = \int_0^\infty v\, dF(v)$$

If φ is bounded and $\varphi(0) = 0$ then

$$c \int \varphi(x)\, dF(x) = \left(\int_0^\infty \varphi(v)\, dF(v) \right) \int_{-\infty}^0 (-u) dF(u)$$
$$+ \left(\int_{-\infty}^0 \varphi(u)\, dF(u) \right) \int_0^\infty v\, dF(v)$$
$$= \int_0^\infty dF(v) \int_{-\infty}^0 dF(u)\, (v\varphi(u) - u\varphi(v))$$

So

$$\int \varphi(x)\, dF(x) = c^{-1} \int_0^\infty dF(v) \int_{-\infty}^0 dF(u)(v - u) \left\{ \frac{v}{v - u}\varphi(u) + \frac{-u}{v - u}\varphi(v) \right\}$$

The last equation gives the desired mixture. If we let $(U, V) \in \mathbf{R}^2$ have

(6.2)
$$P\{(U, V) = (0, 0)\} = F(\{0\})$$
$$P((U, V) \in A) = c^{-1} \iint_{(u,v)\in A} dF(u)\, dF(v)\, (v - u)$$

for $A \subset (-\infty, 0) \times (0, \infty)$ and define probability measures by $\mu_{0,0}(\{0\}) = 1$ and

$$\mu_{u,v}(\{u\}) = \frac{v}{v - u} \qquad \mu_{u,v}(\{v\}) = \frac{-u}{v - u} \quad \text{for } u < 0 < v$$

then

$$\int \varphi(x)\, dF(x) = E\left(\int \varphi(x)\, \mu_{U,V}(dx) \right)$$

We proved the last formula when $\varphi(0) = 0$, but it is easy to see that it is true in general. Letting $\varphi \equiv 1$ in the last equation shows that the measure defined in (6.2) has total mass 1.

From the calculations above it follows that if we have (U, V) with distribution given in (6.2) and an independent Brownian motion defined on the same space then $B(T_{U,V}) =_d X$. Sticklers for detail will notice that $T_{U,V}$ is not a stopping time for B_t since (U, V) is independent of the Brownian motion. This is not a serious problem since if we condition on $U = u$ and $V = v$ then $T_{u,v}$ is a stopping time and this is good enough for all the calculations below. For instance, to compute $E(T_{U,V})$ we observe

$$E(T_{U,V}) = E\{E(T_{U,V}|(U,V))\} = E(-UV)$$

by (5.5). (6.2) implies

$$E(-UV) = \int_{-\infty}^{0} dF(u)(-u) \int_{0}^{\infty} dF(v)v(v-u)c^{-1}$$

$$= \int_{-\infty}^{0} dF(u)(-u)\left\{-u + \int_{0}^{\infty} dF(v)c^{-1}v^2\right\}$$

since

$$c = \int_{0}^{\infty} v\, dF(v) = \int_{-\infty}^{0} (-u)\, dF(u)$$

Using the second expression for c now gives

$$E(T_{U,V}) = E(-UV) = \int_{-\infty}^{0} u^2 dF(u) + \int_{0}^{\infty} v^2 dF(v) = EX^2 \qquad \square$$

EXERCISE 6.1. Use Exercise 5.4 to conclude that $E(T_{U,V}^2) \leq C E X^4$.

Remark. One can embed distributions in Brownian motion without adding random variables to the probability space: see Dubins (1968), (1969) or Sheu (1986).

From (6.1) it is only a small step to

(6.3) Theorem. Let $X_1, X_2, \ldots$ be i.i.d. with a distribution F which has mean 0 and variance 1, and let $S_n = X_1 + \ldots + X_n$. There is a sequence of stopping times $T_0 = 0, T_1, T_2, \ldots$ such that $S_n =_d B(T_n)$ and $T_n - T_{n-1}$ are independent and identically distributed.

Proof   Let $(U_1, V_1), (U_2, V_2), \ldots$ be i.i.d. and have distribution given in (6.2) and let B_t be an independent Brownian motion. Let $T_0 = 0$ and for $n \geq 1$ let

$$T_n = \inf\{t \geq T_{n-1} : B_t - B(T_{n-1}) \notin (U_n, V_n)\} \qquad \square$$

As a corollary of (6.3) we get

(6.4) Central limit theorem. Under the hypotheses of (6.3), $S_n/\sqrt{n} \Rightarrow \chi$, where χ has the standard normal distribution.

Proof   If we let $W_n(t) = B(nt)/\sqrt{n} =_d B_t$ by Brownian scaling, then

$$S_n/\sqrt{n} \stackrel{d}{=} B(T_n)/\sqrt{n} = W_n(T_n/n)$$

The weak law of large numbers implies that $T_n/n \to 1$ in probability. It should be clear from this that $S_n/\sqrt{n} \Rightarrow B_1$. To fill in the details, let $\epsilon > 0$, pick δ so that

$$P(|B_t - B_1| > \epsilon \text{ for some } t \in (1 - \delta, 1 + \delta)) < \epsilon/2$$

then pick N large enough so that for $n \geq N$, $P(|T_n/n - 1| > \delta) < \epsilon/2$. The last two estimates imply that for $n \geq N$

$$P(|W_n(T_n/n) - W_n(1)| > \epsilon) < \epsilon$$

Since ϵ is arbitrary it follows that $W_n(T_n/n) - W_n(1) \to 0$ in probability. Applying the converging together lemma, Exercise 2.10 in Chapter 2, with $X_n = W_n(1)$ and $Z_n = W_n(T_n/n)$ the desired result follows. $\qquad \square$

Our next goal is to prove a strengthening of (6.4) that allows us to obtain limit theorems for functionals of $\{S_m : 0 \leq m \leq n\}$, e.g., $\max_{0 \leq m \leq n} S_m$ or $|\{m \leq n : S_m > 0\}|$. Let $C[0,1] = \{\text{continuous } \omega : [0,1] \to \mathbf{R}\}$. When equipped with the norm $\|\omega\| = \sup\{|\omega(s)| : s \in [0,1]\}$, $C[0,1]$ becomes a complete separable metric space. To fit $C[0,1]$ into the framework of Section 2.9 we want our measures defined on $\mathcal{B} =$ the σ-field generated by the open sets. Fortunately,

(6.5) Lemma. $\mathcal{B}$ is the same as $\mathcal{C}$ the σ-field generated by the finite dimensional sets $\{\omega : \omega(t_i) \in A_i\}$.

Proof   Observe that if ξ is a given continuous function

$$\{\omega : \|\omega - \xi\| \leq \epsilon - 1/n\} = \cap_q \{\omega : |\omega(q) - \xi(q)| \leq \epsilon - 1/n\}$$

where the intersection is over all rationals in $[0,1]$. Letting $n \to \infty$ shows $\{\omega : \|\omega - \xi\| < \epsilon\} \in \mathcal{C}$ and $\mathcal{B} \subset \mathcal{C}$. To prove the reverse inclusion observe that if the A_i are open the finite dimensional set $\{\omega : \omega(t_i) \in A_i\}$ is open so the $\pi - \lambda$ theorem implies $\mathcal{B} \supset \mathcal{C}$. $\qquad \square$

A sequence of probability measures μ_n on $C[0,1]$ is said to **converge weakly** to a limit μ if for all bounded continuous functions $\varphi : C[0,1] \to \mathbf{R}$, $\int \varphi \, d\mu_n \to \int \varphi \, d\mu$. Let $\mathbf{N}$ be the nonnegative integers and let

$$S(u) = \begin{cases} S_k & \text{if } u = k \in \mathbf{N} \\ \text{linear on } [k, k+1] & \text{for } k \in \mathbf{N} \end{cases}$$

We will prove:

(6.6) **Donsker's theorem.** Under the hypotheses of (6.3), $S(n\cdot)/\sqrt{n} \Rightarrow B(\cdot)$, i.e., the associated measures on $C[0,1]$ converge weakly.

To motivate ourselves for the proof we will begin by extracting several corollaries. The key to each one is a consequence of (9.1) in Chapter 2.

(6.7) **Theorem.** If $\psi : C[0,1] \to \mathbf{R}$ has the property that it is continuous P_0-a.s. then

$(*)$ $$\psi(S(n\cdot)/\sqrt{n}) \Rightarrow \psi(B(\cdot))$$

Example 6.1. Let $\psi(\omega) = \omega(1)$. In this case $\psi : C[0,1] \to \mathbf{R}$ is continuous and $(*)$ is the central limit theorem.

Example 6.2. Let $\psi(\omega) = \max\{\omega(t) : 0 \le t \le 1\}$. Again $\psi : C[0,1] \to \mathbf{R}$ is continuous. This time $(*)$ says

$$\max_{0 \le m \le n} S_m/\sqrt{n} \Rightarrow M_1 \equiv \max_{0 \le t \le 1} B_t$$

To complete the picture, we observe that by (4.5) the distribution of the right-hand side is

$$P_0(M_1 \ge a) = P_0(T_a \le 1) = 2P_0(B_1 \ge a)$$

EXERCISE 6.2. Suppose S_n is one dimensional simple random walk and let

$$R_n = 1 + \max_{m \le n} S_m - \min_{m \le n} S_m$$

be the number of points visited by time n. Show that $R_n/\sqrt{n} \Rightarrow$ a limit.

Example 6.3. Let $\psi(\omega) = \sup\{t \le 1 : \omega(t) = 0\}$. This time ψ is not continuous, for if ω_ϵ has $\omega_\epsilon(0) = 0$, $\omega_\epsilon(1/3) = 1$, $\omega_\epsilon(2/3) = \epsilon$, $\omega(1) = 2$, and linear on each interval $[j, (j+1)/3]$ then $\psi(\omega_0) = 2/3$ but $\psi(\omega_\epsilon) = 0$ for $\epsilon > 0$. It is easy to see that if $\psi(\omega) < 1$ and $\omega(t)$ has positive and negative values in each interval $(\psi(\omega) - \delta, \psi(\omega))$ then ψ is continuous at ω. By arguments in Example 3.1, the last set has P_0 measure 1. (If the zero at $\psi(\omega)$ was isolated on the left, it would not be isolated on the right.) It follows that

$$\sup\{m \le n : S_{m-1} \cdot S_m \le 0\}/n \Rightarrow L = \sup\{t \le 1 : B_t = 0\}$$

The distribution of L is given in (4.2). The last result shows that the arcsine law, (3.4) in Chapter 3, proved for simple random walks holds when the mean is 0 and variance is finite.

Example 6.4. Let $\psi(\omega) = |\{t \in [0,1] : \omega(t) > a\}|$. The point $\omega \equiv a$ shows that ψ is not continuous but it is easy to see that ψ is continuous at paths ω with $|\{t \in [0,1] : \omega(t) = a\}| = 0$. Fubini's theorem implies that

$$E_0|\{t \in [0,1] : B_t = a\}| = \int_0^1 P_0(B_t = a)\, dt = 0$$

so ψ is continuous P_0-a.s. With a little work $(*)$ implies

$$|\{m \le n : S_m > a\sqrt{n}\}|/n \Rightarrow |\{t \in [0,1] : B_t > a\}|$$

Proof Application of $(*)$ gives that for any a,

$$|\{t \in [0,1] : S(nt) > a\sqrt{n}\}| \Rightarrow |\{t \in [0,1] : B_t > a\}|$$

To convert this into a result about $|\{m \le n : S_m > a\sqrt{n}\}|$ we note that on $\{\max_{m \le n} |X_m| \le \epsilon\sqrt{n}\}$, which by Chebyshev's inequality has a probability $\to 1$, we have

$$|\{t \in [0,1] : S(nt) > (a+\epsilon)\sqrt{n}\}| \le \frac{1}{n}|\{m \le n : S_m > a\sqrt{n}\}|$$
$$\le |\{t \in [0,1] : S(nt) > (a-\epsilon)\sqrt{n}\}|$$

Combining this with the first conclusion of the proof and using the fact that $b \to |\{t \in [0,1] : B_t > b\}|$ is continuous at $b = a$ with probability one, one arrives easily at the desired conclusion. $\square$

To compute the distribution of $|\{t \in [0,1] : B_t > 0\}|$ observe that we proved in Section 3.4 that if $S_n =_d -S_n$ and $P(S_m = 0 \text{ for all } m \ge 1) = 0$, e.g., the

X_i have a symmetric continuous distribution, then the left-hand side converges to the arcsine law, so the right-hand side has that distribution and is the limit for any random walk with mean 0 and finite variance. The last argument uses an idea called the "invariance principle" that originated with Erdös and Kac (1946,1947): the asymptotic behavior of functionals of S_n should be the same as long as the central limit theorem applies. Our final application is from the original paper of Donsker (1951). Erdös and Kac (1946) give the limit distribution for the case $k = 2$.

Example 6.5. Let $\psi(\omega) = \int_{[0,1]} \omega(t)^k dt$ where $k > 0$ is an integer. ψ is continuous so applying (6.7) gives

$$\int_0^1 (S(nt)/\sqrt{n})^k \, dt \Rightarrow \int_0^1 B_t^k \, dt$$

To convert this into a result about the original sequence, we begin by observing that if $x < y$ with $|x - y| \le \epsilon$ and $|x|, |y| \le M$ then

$$|x^k - y^k| \le \int_x^y \frac{|z|^{k+1}}{k+1} \, dz \le \frac{\epsilon M^{k+1}}{k+1}$$

From this it follows that on

$$G_n(M) = \left\{ \max_{m \le n} |X_m| \le M^{-(k+2)} \sqrt{n}, \ \max_{m \le n} |S_m \le M\sqrt{n} \right\}$$

we have

$$\left| \int_0^1 (S(nt)/\sqrt{n})^k \, dt - n^{-1-(k/2)} \sum_{m=1}^n S_m^k \right| \le \frac{1}{(k+1)M}$$

For fixed M it follows from Chebyshev's inequality, Example 6.2, and (2.4) in Chapter 2 that

$$\liminf_{n \to \infty} P(G_n(M)) \ge P\left(\max_{0 \le t \le 1} |B_t| < M \right)$$

The right-hand side is close to 0 if M is large so

$$\int_0^1 (S(nt)/\sqrt{n})^k \, dt - n^{-1-(k/2)} \sum_{m=1}^n S_m^k \to 0$$

in probability and it follows from the converging together lemma (Exercise 2.10 in Chapter 2) that

$$n^{-1-(k/2)} \sum_{m=1}^n S_m^k \Rightarrow \int_0^1 B_t^k \, dt$$

It is remarkable that the last result holds under the assumption that $EX_i = 0$ and $EX_i^2 = 1$, i.e., we do not need to assume that $E|X_i^k| < \infty$.

EXERCISE 6.3. When $k = 1$ the last result says that if $X_1, X_2, \ldots$ are i.i.d. with $EX_i = 0$ and $EX_i^2 = 1$ then

$$n^{-3/2} \sum_{m=1}^{n} (n + 1 - m) X_m \Rightarrow \int_0^1 B_t\, dt$$

(i) Show that the right-hand side has a normal distribution with mean 0 and variance 1/3. (ii) Deduce this result from the Lindeberg-Feller theorem.

Proof of (6.6) To simplify the proof and prepare for generalizations in the next section, let $X_{n,m}$, $1 \leq m \leq n$, be a triangular array of random variables, $S_{n,m} = X_{n,1} + \cdots + X_{n,m}$ and suppose $S_{n,m} = B(\tau_m^n)$. Let

$$S_{n,(u)} = \begin{cases} S_{n,m} & \text{if } u = m \in \{0, 1, \ldots, n\} \\ \text{linear for } u \in [m-1, m] & \text{when } m \in \{1, \ldots, n\} \end{cases}$$

(6.8) Lemma. If $\tau_{[ns]}^n \to s$ in probability for each $s \in [0, 1]$ then

$$\|S_{n,(n\cdot)} - B(\cdot)\| \to 0 \quad \text{in probability}$$

To make the connection with the original problem, let $X_{n,m} = X_m/\sqrt{n}$ and define $\tau_1^n, \ldots, \tau_n^n$ so that $(S_{n,1}, \ldots, S_{n,n}) =_d (B(\tau_1^n), \ldots, B(\tau_n^n))$. If $T_1, T_2, \ldots$ are the stopping times defined in (6.3), Brownian scaling implies $\tau_m^n =_d T_m/n$, so the hypothesis of (6.8) is satisfied.

Proof The fact that B has continuous paths (and hence uniformly continuous on $[0,1]$) implies that if $\epsilon > 0$ then there is a $\delta > 0$ so that $1/\delta$ is an integer and

(a) $P(|B_t - B_s| < \epsilon \text{ for all } 0 \leq s \leq 1, |t - s| < 2\delta) > 1 - \epsilon$

The hypothesis of (6.8) implies that if $n \geq N_\delta$ then

$$P(|\tau_{[nk\delta]}^n - k\delta| < \delta \quad \text{for } k = 1, 2, \ldots, 1/\delta) \geq 1 - \epsilon$$

Since $m \to \tau_m^n$ is increasing, it follows that if $s \in ((k-1)\delta, k\delta)$

$$\tau_{[ns]}^n - s \geq \tau_{[n(k-1)\delta]}^n - k\delta$$
$$\tau_{[ns]}^n - s \leq \tau_{[nk\delta]}^n - (k+1)\delta$$

so if $n \geq N_\delta$,

(b)
$$P\left(\sup_{0 \leq s \leq 1} |\tau^n_{[ns]} - s| < 2\delta\right) \geq 1 - \epsilon$$

When the events in (a) and (b) occur

(c)
$$|S_{n,m} - B_{m/n}| < \epsilon \text{ for all } m \leq n$$

To deal with $t = (m + \theta)/n$ with $0 < \theta < 1$ we observe that

$$|S_{n,(nt)} - B_t| \leq (1 - \theta)|S_{n,m} - B_{m/n}| + \theta|S_{n,m+1} - B_{(m+1)/n}|$$
$$+ (1 - \theta)|B_{m/n} - B_t| + \theta|B_{(m+1)/n} - B_t|$$

Using (c) on the first two terms and (a) on the last two, we see that if $n \geq N_\delta$ and $1/n < 2\delta$ then $\|S_{n,(n\cdot)} - B(\cdot)\| < 2\epsilon$ with probability $\geq 1 - 2\epsilon$. Since ϵ is arbitrary the proof of (6.8) is complete. $\qquad\square$

To get (6.6) now we have to show:

(6.9) **Lemma.** If φ is bounded and continuous $E\varphi(S_{n,(n\cdot)}) \to E\varphi(B(\cdot))$.

Proof For fixed $\epsilon > 0$ let $G_\delta = \{\omega : \text{if } \|\omega - \omega'\| < \delta \text{ then } |\varphi(\omega) - \varphi(\omega')| < \epsilon\}$. Since φ is continuous, $G_\delta \uparrow C[0,1]$ as $\delta \downarrow 0$. Let $\Delta = \|S_{n,(n\cdot)} - B(\cdot)\|$. The desired result now follows from (6.8) and the trivial inequality

$$|E\varphi(S_{n,(n\cdot)}) - \varphi(B(\cdot))| \leq \epsilon + (2 \sup |\varphi(\omega)|)\{P(G^c_\delta) + P(\Delta \geq \delta)\} \qquad \square$$

To accomodate our final example we need a trivial generalization of (6.6). Let $C[0,\infty) = \{\text{continuous } \omega : [0,\infty) \to \mathbf{R}\}$ and let $\mathcal{C}[0,\infty)$ be the σ-field generated by the finite dimensional sets. Given a probability measure μ on $C[0,\infty)$, there is a corresponding measure $\pi_M\mu$ on $C[0,M] = \{\text{continuous } \omega : [0,M] \to \mathbf{R}\}$ (with $\mathcal{C}[0,M]$ the σ-field generated by the finite dimensional sets) obtained by "cutting off the paths at time M." Let $(\psi_M\omega)(t) = \omega(t)$ for $t \in [0,M]$ and let $\pi_M\mu = \mu \circ \psi_M^{-1}$. We say that a sequence of probability measures μ_n on $C[0,\infty)$ converges weakly to μ if for all M, $\pi_M\mu_n$ converges weakly to $\pi_M\mu$ on $C[0,M]$, the last concept being defined by a trivial extension of the definitions for $M = 1$. With these definitions it is easy to conclude

(6.10) **Theorem.** $S(n\cdot)/\sqrt{n} \Rightarrow B(\cdot)$, i.e., the associated measures on $C[0,\infty)$ converge weakly.

Proof By definition all we have to show is that weak convergence occurs on $C[0, M]$ for all $M < \infty$. The proof of (6.6) works in the same way when 1 is replaced by M. $\qquad\square$

Example 6.6. Let $N_n = \inf\{m : S_m \geq \sqrt{n}\}$ and $T_1 = \inf\{t : B_t \geq 1\}$. Since $\psi(\omega) = T_1(\omega) \wedge 1$ is continuous P_0 a.s. on $C[0, 1]$ and the distribution of T_1 is continuous, it follows from (6.7) that for $0 < t < 1$

$$P(N_n \leq nt) \to P(T_1 \leq t)$$

Repeating the last argument with 1 replaced by M and using (6.10) shows that the last conclusion holds for all t.

*7.7. CLT's for Dependent Variables

In this section we will prove central limit theorems for some dependent sequences. First, by embedding of martingales in Brownian motion we will prove a Lindeberg-Feller theorem for martingales, (7.3). Then using an idea of Gordin (1969) we will use the result for martingales to get a CLT for stationary sequences, (7.6). The condition in (7.6) may look difficult to check, but we show that it is implied by the usual mixing conditions. We begin with the central limit theorems for

a. Martingales

(7.1) Theorem. If S_n is a martingale with $S_0 = 0$, and B_t is a Brownian motion, there is a sequence of stopping times $0 = T_0 \leq T_1 \leq T_2 \ldots$ for the Brownian motion so that

$$(S_0, S_1, \ldots, S_k) \overset{d}{=} (B(T_0), B(T_1), \ldots, B(T_k)) \quad \text{for all } k \geq 0$$

Remark. This is due to Strassen (1967), see Theorem 4.3.

Proof We include $S_0 = 0 = B(T_0)$ only for the sake of starting the induction argument. Suppose we have $(S_0, \ldots, S_{k-1}) =_d (B(T_0), \ldots, B(T_{k-1}))$ for some $k \geq 1$. The strong Markov property implies that $\{B(T_{k-1}+t) - B(T_{k-1}) : t \geq 0\}$ is a Brownian motion that is independent of $\mathcal{F}(T_{k-1})$. Let $\mu_k(s_0, \ldots, s_{k-1}; \cdot)$ be a regular conditional distribution of $S_k - S_{k-1}$ given $S_j = s_j$, $0 \leq j \leq k$, that is, for each Borel set A

$$P(S_k - S_{k-1} \in A | S_0, \ldots, S_{k-1}) = \mu_k(S_0, \ldots, S_{k-1}; A)$$

By Exercises 1.16 and 1.14 in Chapter 4 this exists and we have

$$0 = E(S_k - S_{k-1}|S_0, \ldots, S_{k-1}) = \int x\mu_k(S_0, \ldots, S_{k-1}; dx)$$

so the mean of the conditional distribution is 0 almost surely. Using (6.1) now we see that for almost every $\hat{S} \equiv (S_0, \ldots, S_{k-1})$ there is a stopping time $\tau_{\hat{S}}$ (for $\{B(T_{k-1} + t) - B(T_{k-1}) : t \geq 0\}$) so that

$$B(T_{k-1} + \tau_{\hat{S}}) - B(T_{k-1}) \overset{d}{=} \mu_k(S_1, \ldots, S_{k-1}; \cdot)$$

If we let $T_k = T_{k-1} + \tau_{\hat{S}}$ then $(S_0, \ldots, S_k) \overset{d}{=} (B(T_0), \ldots, B(T_k))$ and the result follows by induction.

Remark. While the details of the proof are fresh in the reader's mind, we would like to observe that if $E(S_k - S_{k-1})^2 < \infty$ then

$$E(\tau_{\hat{S}}|S_0, \ldots, S_{k-1}) = \int x^2 \mu_k(S_0, \ldots, S_{k-1}; dx)$$

since $B_t^2 - t$ is a martingale and $\tau_{\hat{S}}$ is the exit time from a randomly chosen interval $(S_{k-1} + U_k, S_{k-1} + V_k)$.

Our first step toward the promised Lindeberg-Feller theorem is to prove a result of Freedman (1971a). We say that $X_{n,m}$, $\mathcal{F}_{n,m}$, $1 \leq m \leq n$, is a **martingale difference array** if $X_{n,m} \in \mathcal{F}_{n,m}$ and $E(X_{n,m}|\mathcal{F}_{n,m-1}) = 0$ for $1 \leq m \leq n$, where $\mathcal{F}_{n,0} = \{\emptyset, \Omega\}$. Let

$$V_{n,k} = \sum_{1 \leq m \leq k} E(X_{n,m}^2|\mathcal{F}_{n,m-1})$$

(7.2) **Theorem.** Suppose $\{X_{n,m}, \mathcal{F}_{n,m}\}$ is a martingale difference array and let $S_{n,m} = X_{n,1} + \ldots + X_{n,m}$.

If (i) $|X_{n,m}| \leq \epsilon_n$ for all m with $\epsilon_n \to 0$, and

(ii) for each t, $V_{n,[nt]} \to t$ in probability,

then $S_{n,(n\cdot)} \Rightarrow B(\cdot)$.

Let $\mathcal{N} = \{0, 1, 2, \ldots\}$. Here and throughout the section $S_{n,(n\cdot)}$ denotes the linear interpolation of $S_{n,m}$ defined by

$$S_{n,(u)} = \begin{cases} S_k & \text{if } u = k \in N \\ \text{linear for } u \in [k, k+1] & \text{when } k \in N \end{cases}$$

and $B(\cdot)$ is a Brownian motion with $B_0 = 0$.

Proof (ii) implies $V_{n,n} \to 1$ in probability. By stopping each sequence at the first time $V_{n,k} > 2$ and setting the later $X_{n,m} = 0$, we can suppose without loss of generality that $V_{n,n} \leq 2 + \epsilon_n^2$ for all n. By (7.1) we can find stopping times $T_{n,1}, \ldots, T_{n,n}$ so that $(S_{n,1}, \ldots, S_{n,n}) =_d (B(T_{n,1}), \ldots, B(T_{n,n}))$. By (6.8) it suffices to show that $T_{n,[nt]} \to t$ in probability for each $t \in [0, 1]$. To do this we let $t_{n,m} = T_{n,m} - T_{n,m-1}$ (with $T_{n,0} = 0$) observe that by the remark after the proof of (7.1), $E(t_{n,m}|\mathcal{F}_{n,m-1}) = E(X_{n,m}^2|\mathcal{F}_{n,m-1})$. The last observation and hypothesis (ii) imply

$$\sum_{m=1}^{[nt]} E(t_{n,m}|\mathcal{F}_{n,m-1}) \to t \quad \text{in probability}$$

To get from this to $T_{n,[nt]} \to t$ in probability, we observe

$$E\left(\sum_{m=1}^{[nt]} t_{n,m} - E(t_{n,m}|\mathcal{F}_{n,m-1})\right)^2 = E\sum_{m=1}^{[nt]} \{t_{n,m} - E(t_{n,m}|\mathcal{F}_{n,m-1})\}^2$$

by the orthogonality of martingale increments ((4.6) in Chapter 4). Now

$$E(\{t_{n,m} - E(t_{n,m}|\mathcal{F}_{n,m-1})\}^2|\mathcal{F}_{n,m-1}) \leq E(t_{n,m}^2|\mathcal{F}_{n,m-1})$$
$$\leq CE(X_{n,m}^4|\mathcal{F}_{n,m-1}) \leq C\epsilon_n^2 E(X_{n,m}^2|\mathcal{F}_{n,m-1})$$

by Exercise 5.4 and assumption (i). Summing over n, taking expected values, and recalling we have assumed $V_{n,n} \leq 2 + \epsilon_n^2$, it follows that

$$E\left(\sum_{m=1}^{[nt]} t_{n,m} - E(t_{n,m}|\mathcal{F}_{n,m-1})\right)^2 \to C\epsilon_n^2 EV_{n,n} \to 0$$

Unscrambling the definitions we have shown $E(T_{n,[nt]} - V_{n,[nt]})^2 \to 0$, so Chebyshev's inequality implies $P(|T_{n,[nt]} - V_{n,[nt]}| > \epsilon) \to 0$, and using (ii) now completes the proof. $\square$

Remark. We can get rid of assumption (ii) in (7.2) by putting the martingale on its "natural time scale." Let $X_{n,m}$, $\mathcal{F}_{n,m}$, $1 \leq m < \infty$, be a martingale difference array with $|X_{n,m}| \leq \epsilon_n$ and $\epsilon_n \to 0$. and suppose that $V_{n,m} \to \infty$ as $m \to \infty$. Define $B_n(t)$ by requiring $B_n(V_{n,m}) = S_{n,m}$ and $B_n(t)$ is linear on each interval $[V_{n,m-1}, V_{n,m}]$. A generalization of the proof of (7.2) shows that $B_n(\cdot) \Rightarrow B(\cdot)$. See Freedman (1971a), p. 89–93 for details.

With (7.2) established, a truncation argument gives us:

(7.3) Lindeberg-Feller theorem for martingales. Suppose $X_{n,m}$, $\mathcal{F}_{n,m}$, $1 \leq m \leq n$ is a martingale difference array.

If (i) $V_{n,[nt]} \to t$ in probability for all $t \in [0,1]$ and

(ii) $\sum_{m \leq n} E(X_{n,m}^2 1_{(|X_{n,m}|>\epsilon)}|\mathcal{F}_{n,m-1}) \to 0$ in probability,

then $S_{n,(n \cdot)} \Rightarrow B(\cdot)$.

Remark. Literally dozens of papers have been written proving results like this. See Hall and Heyde (1980) for some history. Here we follow Durrett and Resnick (1978).

Proof   The first step is to truncate so that we can apply (7.2). Let

$$\hat{V}_n(\epsilon) = \sum_{m=1}^{n} E(X_{n,m}^2 1_{(|X_{n,m}|>\epsilon_n)}|\mathcal{F}_{n,m-1})$$

(a) If $\epsilon_n \to 0$ slowly enough then $\epsilon_n^{-2} \hat{V}_n(\epsilon_n) \to 0$ in probability.

Remark. The ϵ_n^{-2} in front is so that we can conclude

$$\sum_{m \leq n} P(|X_{n,m}| > \epsilon_n|\mathcal{F}_{n,m-1}) \to 0 \text{ in probability}$$

Proof of (a)   Let N_m be chosen so that $P(m^2 \hat{V}_n(1/m) > 1/m) \leq 1/m$ for $n \geq N_m$. Let $\epsilon_n = 1/m$ for $n \in [N_m, N_{m+1})$ and $\epsilon_n = 1$ if $n < N_1$. If $\delta > 0$ and $1/m < \delta$ then for $n \in [N_m, N_{m+1})$

$$P(\epsilon_n^{-2} \hat{V}_n(\epsilon_n) > \delta) \leq P(m^2 \hat{V}_n(1/m) > 1/m) \leq 1/m \qquad \square$$

Let $\hat{X}_{n,m} = X_{n,m} 1_{(|X_{n,m}|>\epsilon_n)}$, $\bar{X}_{n,m} = X_{n,m} 1_{(|X_{n,m}|\leq\epsilon_n)}$ and

$$\tilde{X}_{n,m} = \bar{X}_{n,m} - E(\bar{X}_{n,m}|\mathcal{F}_{n,m-1})$$

Our next step is to show:

(b) If we define $\tilde{S}_{n,(n \cdot)}$ in the obvious way then (7.2) implies $\tilde{S}_{n,(n \cdot)} \Rightarrow B(\cdot)$.

Proof of (b) Since $|\tilde{X}_{n,m}| \leq 2\epsilon_n$ we only have to check (ii) in (7.2). To do this, we observe that the conditional variance formula ((4.7) in Chapter 4) implies

$$E(\tilde{X}_{n,m}^2|\mathcal{F}_{n,m-1}) = E(\bar{X}_{n,m}^2|\mathcal{F}_{n,m-1}) - E(\bar{X}_{n,m}|\mathcal{F}_{n,m-1})^2$$

For the first term we observe

$$E(\bar{X}_{n,m}^2|\mathcal{F}_{n,m-1}) = E(X_{n,m}^2|\mathcal{F}_{n,m-1}) - E(\hat{X}_{n,m}^2|\mathcal{F}_{n,m-1})$$

For the second we observe that $E(X_{n,m}|\mathcal{F}_{n,m-1}) = 0$ implies

$$E(\bar{X}_{n,m}|\mathcal{F}_{n,m-1})^2 = E(\hat{X}_{n,m}|\mathcal{F}_{n,m-1})^2 \leq E(\hat{X}_{n,m}^2|\mathcal{F}_{n,m-1})$$

by Jensen's inequality, so it follows from (a) and (i) that

$$\sum_{m=1}^{[nt]} E(\tilde{X}_{n,m}^2|\mathcal{F}_{n,m-1}) \to t \quad \text{for all } t \in [0,1] \qquad \square$$

Having proved (b), it remains to estimate the difference between $S_{n,(n\cdot)}$ and $\tilde{S}_{n,(n\cdot)}$. On $\{|X_{n,m}| \leq \epsilon_n \text{ for all } 1 \leq m \leq n\}$ we have

(c)
$$\|S_{n,(n\cdot)} - \tilde{S}_{n,(n\cdot)}\| \leq \sum_{m=1}^{n} |E(\bar{X}_{n,m}|\mathcal{F}_{n,m-1})|$$

To handle the right-hand side, we observe

(d)
$$\sum_{m=1}^{n} |E(\bar{X}_{n,m}|\mathcal{F}_{n,m-1})| = \sum_{m=1}^{n} |E(\hat{X}_{n,m}|\mathcal{F}_{n,m-1})|$$
$$\leq \sum_{m=1}^{n} E(|\hat{X}_{n,m}||\mathcal{F}_{n,m-1})$$
$$\leq \epsilon_n^{-1} \sum_{m=1}^{n} E(\hat{X}_{n,m}^2|\mathcal{F}_{n,m-1}) \to 0$$

in probability by (a). To complete the proof now, it suffices to show

(e)
$$P(|X_{n,m}| > \epsilon_n \text{ for some } m \leq n) \to 0$$

for with (d) and (c) this implies $\|S_{n,(n\cdot)} - \tilde{S}_{n,(n\cdot)}\| \to 0$ in probability. The proof of (7.2) constructs a Brownian motion with $\|\tilde{S}_{n,(n\cdot)} - B(\cdot)\| \to 0$, so the desired result follows from the triangle inequality and (6.9).

To prove (e) we will use Lemma 3.5 of Dvoretsky (1972).

(f) If A_n is adapted to $\mathcal{G}_n$ then for any nonnegative $\delta \in \mathcal{G}_0$,

$$P\left(\cup_{m=1}^n A_m | \mathcal{G}_0\right) \leq \delta + P\left(\left.\sum_{m=1}^n P(A_m|\mathcal{G}_{m-1}) > \delta \right| \mathcal{G}_0\right)$$

Proof of (f)   We proceed by induction. When $n = 1$ the conclusion says

$$P(A_1|\mathcal{G}_0) \leq \delta + P(P(A_1|\mathcal{G}_0) > \delta|\mathcal{G}_0)$$

This is obviously true on $\Omega_- \equiv \{P(A_1|\mathcal{G}_0) \leq \delta\}$ and also on $\Omega_+ \equiv \{P(A_1|\mathcal{G}_0) > \delta\} \in \mathcal{G}_0$ since on Ω_+

$$P(P(A_1|\mathcal{G}_0) > \delta|\mathcal{G}_0) = 1 \geq P(A_1|\mathcal{G}_0)$$

To prove the result for n sets, observe that by the last argument the inequality is trivial on Ω_+. Let $B_m = A_m \cap \Omega_-$. Since $\Omega_- \in \mathcal{G}_0 \subset \mathcal{G}_{m-1}$, $P(B_m|\mathcal{G}_{m-1}) = P(A_m|\mathcal{G}_{m-1})$ on Ω_-. (See Exercise 1.1. in Chapter 4.) Applying the result for $n - 1$ sets with $\gamma = \delta - P(B_1|\mathcal{G}_0) \geq 0$

$$P\left(\cup_{m=2}^n B_m | \mathcal{G}_1\right) \leq \gamma + P\left(\left.\sum_{m=2}^n P(B_m|\mathcal{G}_{m-1}) > \gamma \right| \mathcal{G}_1\right)$$

Taking conditional expectation w.r.t. $\mathcal{G}_0$ and noting $\gamma \in \mathcal{G}_0$

$$P\left(\cup_{m=2}^n B_m | \mathcal{G}_0\right) \leq \gamma + P\left(\left.\sum_{m=1}^n P(B_m|\mathcal{G}_{m-1}) > \delta \right| \mathcal{G}_0\right)$$

$\cup_{2 \leq m \leq n} B_m = (\cup_{2 \leq m \leq n} A_m) \cap \Omega_-$ and another use of Exercise 1.1 from Chapter 4 shows

$$\sum_{1 \leq m \leq n} P(B_m|\mathcal{G}_{m-1}) = \sum_{1 \leq m \leq n} P(A_m|\mathcal{G}_{m-1})$$

on Ω_-. So on Ω_-

$$P\left(\cup_{m=2}^n A_m | \mathcal{G}_0\right) \leq \delta - P(A_1|\mathcal{G}_0) + P\left(\sum_{m=1}^n P(A_m|\mathcal{G}_{m-1}) > \delta|\mathcal{G}_0\right)$$

The result now follows from

$$P\left(\cup_{m=1}^n A_m | \mathcal{G}_0\right) \leq P(A_1|\mathcal{G}_0) + P\left(\cup_{m=2}^n A_m | \mathcal{G}_0\right)$$

To see this, let $C = \cup_{2 \leq m \leq n} A_m$, observe that $1_{A_1 \cup C} \leq 1_{A_1} + 1_C$, and use the monotonicity of conditional expectations. $\square$

Proof of (e) Let $A_m = \{|X_{n,m}| > \epsilon_n\}$, $\mathcal{G}_m = \mathcal{F}_{n,m}$, and let δ be a positive number. (f) implies

$$P(|X_{n,m}| > \epsilon_n \text{ for some } m \leq n) \leq \delta + P\left(\sum_{m=1}^{n} P(|X_{n,m}| > \epsilon_n | \mathcal{F}_{n,m-1}) > \delta\right)$$

To estimate the right-hand side we observe "Chebyshev's inequality" (Exercise 1.3 in Chapter 4) implies

$$\sum_{m=1}^{n} P(|X_{n,m}| > \epsilon_n | \mathcal{F}_{n,m-1}) \leq \epsilon_n^{-2} \sum_{m=1}^{n} E(\hat{X}_{n,m}^2 | \mathcal{F}_{n,m-1}) \to 0$$

so $\limsup_{n \to \infty} P(|X_{n,m}| > \epsilon_n \text{ for some } m \leq n) \leq \delta$. Since δ is arbitrary the proof of (e) and hence of (7.3) is complete. $\square$

For applications it is useful to have a result for a single sequence.

(7.4) Martingale central limit theorem. Suppose X_n, $\mathcal{F}_n$, $n \geq 1$, is a martingale difference sequence and let $V_k = \sum_{1 \leq n \leq k} E(X_n^2 | \mathcal{F}_{n-1})$. If (i) $V_k/k \to \sigma^2 > 0$ in probability and

(ii) $n^{-1} \sum_{m \leq n} E(X_m^2 1_{(|X_m| > \epsilon \sqrt{n})}) \to 0$ then $S_{(n \cdot)}/\sqrt{n} \Rightarrow \sigma B(\cdot)$.

Proof Let $X_{n,m} = X_m/\sigma\sqrt{n}$, $\mathcal{F}_{n,m} = \mathcal{F}_m$. Changing notation and letting $k = nt$, our first assumption becomes (i) of (7.3). To check (ii) of (7.3) observe that

$$E\sum_{m=1}^{n} E(X_{n,m}^2 1_{(|X_{n,m}| > \epsilon)} | \mathcal{F}_{n,m-1}) = \sigma^{-2} n^{-1} \sum_{m=1}^{n} E(X_m^2 1_{(|X_m| > \epsilon\sqrt{n})}) \to 0$$

EXERCISE 7.1. **Chain dependent random variables** (Kielson and Wishart (1964), O'Brien (1974)). Let ζ_n be a Markov chain with finite state space S, irreducible transition probability $p(i, j)$, and stationary distribution π_i. The random variables $\{X_n, n \geq 1\}$ are said to be chain dependent if for $\mathcal{G}_n = \sigma(\zeta_m, X_m, m \leq n)$ and $n \geq 0$ we have

$$P(\zeta_{n+1} = j, X_{n+1} \leq x | \mathcal{G}_n) = p(\zeta_n, j) H_{\zeta_n}(x)$$

where $H_1, \ldots, H_n$ are given distribution functions. Intuitively ζ_n is the "environment", which may be in one of a finite number of states, and the distribution of the random variable we observe depends upon the value of ζ_n. Show that if $\int x\,dH_i = 0$ and $\int x^2\,dH_i = \sigma_i^2 < \infty$ then X_n is a martingale difference sequence and

$$n^{-1/2}S_{(n\cdot)} \Rightarrow \sigma B(\cdot) \quad \text{where} \quad \sigma^2 = \sum_{m=1}^{M} \pi_m \sigma_m^2$$

The unappetizing assumption $\int x\,dH_i = 0$ for all i will be dropped in Exercise 7.3.

To get from our result about martingales to our result about

b. Stationary sequences

we begin by considering the intersection of the two cases.

(7.5) Theorem. Suppose X_n, $n \in \mathbf{Z}$, is an ergodic stationary sequence of square integrable martingale differences, i.e., $\sigma^2 = EX_n^2 < \infty$ and $E(X_n|\mathcal{F}_{n-1}) = 0$ where $\mathcal{F}_n = \sigma(X_m, m \le n)$. Let $S_n = X_1 + \cdots + X_n$. Then

$$S_{(n\cdot)}/n^{1/2} \Rightarrow \sigma B(\cdot)$$

Remark. This result was discovered independently by Billingsley (1961) and Ibragimov (1963).

Proof $u_n \equiv E(X_n^2|\mathcal{F}_{n-1})$ can be written as $\varphi(X_{n-1}, X_{n-2}, \ldots)$ so (1.3) in Chapter 6 implies u_n is stationary and ergodic, and the ergodic theorem implies

$$n^{-1}\sum_{m=1}^{n} u_m \to Eu_0 = EX_0^2 \quad \text{a.s.}$$

The last conclusion shows that (i) of (7.4) holds. To verify (ii), we observe

$$n^{-1}\sum_{m=1}^{n} E(X_m^2 1_{(|X_m|>\epsilon\sqrt{n})}) = E(X_0^2 1_{(|X_0|>\epsilon\sqrt{n})}) \to 0$$

by the dominated convergence theorem. $\square$

We come now to the promised central limit theorem for stationary sequences. Our proof is based on Scott (1973), but the key is an idea of Gordin

(1969). In some applications of this result, see e.g. Exercise 7.2, it is convenient to use σ-fields richer than $\mathcal{F}_m = \sigma(X_n, n \leq m)$. So we adopt abstract conditions that satisfy the demands of martingale theory and ergodic theory:

(i) $n \to \mathcal{F}_n$ is increasing, and $X_n \in \mathcal{F}_n$

(ii) $\theta^{-1}\mathcal{F}_m = \mathcal{F}_{m+1}$

(7.6) **Theorem.** Suppose X_n, $n \in \mathbf{Z}$ is an ergodic stationary sequence with $EX_n = 0$. Let $\mathcal{F}_n$ satisfy (i), (ii), and

$$\sum_{n \geq 1} \|E(X_0|\mathcal{F}_{-n})\|_2 < \infty$$

where $\|Y\|_2 = (EY^2)^{1/2}$. Let $S_n = X_1 + \cdots + X_n$. Then $S_{(n\cdot)}/\sqrt{n} \Rightarrow \sigma B(\cdot)$ where

$$\sigma^2 = EX_0^2 + 2\sum_{n=1}^{\infty} EX_0X_n$$

and the series in the definition converges absolutely.

Proof Suppose X_n, $n \in \mathbf{Z}$ is defined on sequence space $(\mathbf{R}^{\mathbf{Z}}, \mathcal{R}^{\mathbf{Z}}, P)$ with $X_n(\omega) = \omega_n$ and let $(\theta^n\omega)(m) = \omega(m + n)$. Let

$$H_n = \{Y \in \mathcal{F}_n \text{ with } EY^2 < \infty\}$$
$$K_n = \{Y \in H_n \text{ with } E(YZ) = 0 \text{ for all } Z \in H_{n-1}\}$$

Geometrically, $H_0 \supset H_{-1} \supset H_{-2} \ldots$ is a sequence of subspaces of L^2 and K_n is the orthogonal complement of H_{n-1} in H_n. If Y is a random variable, let $(\theta^nY)(\omega) = Y(\theta^n\omega)$. Generalizing from the example $Y = f(X_{-j}, \ldots, X_k)$, which has $\theta^nY = f(X_{n-j}, \ldots, X_{n+K})$, it is easy to see that if $Y \in H_k$ then $\theta^nY \in H_{k+n}$, and hence if $Y \in K_j$ then $\theta^nY \in K_{n+j}$.

If X_0 happened to be in K_0 then we would be happy since then $X_n = \theta^nX_0 \in K_n$ for all n, and taking $Z = 1_A \in H_{n-1}$ we would have $E(X_n1_A) = 0$ for all $A \in \mathcal{F}_{n-1}$ and hence $E(X_n|\mathcal{F}_{n-1}) = 0$. The next best thing to having $X_0 \in K_0$ is to have

(∗) $X_0 = Y_0 + Z_0 - \theta Z_0$

with $Y_0 \in K_0$ and $Z_0 \in L^2$, for then if we let

$$S_n = \sum_{m=1}^{n} X_m = \sum_{m=1}^{n} \theta^m X_0 \quad \text{and} \quad T_n = \sum_{m=1}^{n} \theta^m Y_0$$

then $S_n = T_n + \theta Z_0 - \theta^{n+1} Z_0$. The $\theta^m Y_0$ are a stationary ergodic martingale difference sequence (ergodicity follows from (1.3) in Chapter 6), so (7.5) implies

$$T_{(n\cdot)}/\sqrt{n} \Rightarrow \sigma B(\cdot) \quad \text{where } \sigma^2 = EY_0^2$$

To get rid of the other term we observe $\theta Z_0/\sqrt{n} \to 0$ a.s. and

$$P\left(\sup_{1 \le m \le n} \theta^{m+1} Z_0 > \epsilon\sqrt{n} \right) \le nP(Z_0 > \epsilon\sqrt{n}) \le \epsilon^{-2} E(Z_0^2; Z_0 > \epsilon\sqrt{n}) \to 0$$

by dominated convergence. To solve (∗) formally, we let

$$Z_0 = \sum_{j=0}^{\infty} E(X_j | \mathcal{F}_{-1})$$

$$\theta Z_0 = \sum_{j=0}^{\infty} E(X_{j+1} | \mathcal{F}_0)$$

$$Y_0 = \sum_{j=0}^{\infty} \{ E(X_j | \mathcal{F}_0) - E(X_j | \mathcal{F}_{-1}) \}$$

and check that

$$Y_0 + Z_0 - \theta Z_0 = E(X_0 | \mathcal{F}_0) = X_0$$

To justify the last calculation we need to know that the series in the definitions of Z_0 and Y_0 converge. Our assumption and shift invariance imply

$$\sum_{j=0}^{\infty} \|E(X_j | \mathcal{F}_{-1})\|_2 < \infty$$

so the triangle inequality implies that the series for Z_0 converges in L^2. Since

$$\|E(X_j | \mathcal{F}_0) - E(X_j | \mathcal{F}_{-1})\|_2 \le \|E(X_j | \mathcal{F}_0)\|_2$$

(the left-hand side is the projection onto $K_0 \subset H_0$), the series for Y_0 also converges in L^2. Putting the pieces together we have shown (7.5) with $\sigma^2 = EY_0^2$. To get the indicated formula for σ^2 observe that conditioning and using Cauchy-Schwarz

$$|EX_0 X_n| = |E(X_0 E(X_n | \mathcal{F}_0))| \le \|X_0\|_2 \|E(X_n | \mathcal{F}_0)\|_2$$

Shift invariance implies $\|E(X_n | \mathcal{F}_0)\|_2 = \|E(X_0 | \mathcal{F}_{-n})\|_2$ so the series converges absolutely.

$$ES_n^2 = \sum_{j=1}^{n} \sum_{k=1}^{n} EX_j X_k = n \, EX_0^2 + 2 \sum_{m=1}^{n-1} (n - m) EX_0 X_m$$

From this it follows easily that

$$n^{-1}ES_n^2 \rightarrow EX_0^2 + 2\sum_{m=1}^{\infty} EX_0X_m$$

To finish the proof let $T_n = \sum_{m=1}^{n} \theta^m Y_0$, observe $\sigma^2 = EY_0^2$ and

$$n^{-1}E(S_n - T_n)^2 = n^{-1}E(\theta Z_0 - \theta^{n+1}Z_0)^2 \leq 4EZ_0^2/n \rightarrow 0$$

since $(a - b)^2 \leq (2a)^2 + (2b)^2$. $\qquad \square$

We turn now to examples. In the first one it is trivial to check the hypothesis of (7.6).

Example 7.1. M dependent sequences. Let X_n, $n \in \mathbf{Z}$, be a stationary sequence with $EX_n = 0$, $EX_n^2 < \infty$, and suppose $\{X_j, j \leq 0\}$ and $\{X_k, k > M\}$ are independent. In this case $E(X_0|\mathcal{F}_{-n}) = 0$ for $n > M$, so (7.5) implies $S_{n,(n\cdot)}/\sqrt{n} \Rightarrow \sigma B(\cdot)$ where

$$\sigma^2 = EX_0^2 + 2\sum_{m=1}^{M} EX_0X_m$$

Remark. For a bare hands approach to this problem see Theorem 7.3.1 in Chung (1974).

EXERCISE 7.2. Consider the special case of Example 7.1 in which ξ_i, $i \in \mathbf{Z}$, are i.i.d. and take values H and T with equal probability. Let $\eta_n = f(\xi_n, \xi_{n+1})$ where $f(H, T) = 1$ and $f(i, j) = 0$ otherwise. $S_n = \eta_1 + \ldots + \eta_n$ counts the number of head runs in $\xi_1, \ldots, \xi_{n+1}$. Apply (7.6) with $\mathcal{F}_n = \sigma(X_m : m \leq n+1)$ to show that there are constants μ and σ so that $(S_n - n\mu)/\sigma n^{1/2} \Rightarrow \chi$. What is the random variable Y_0 constructed in the proof of (7.6) in this case?

Example 7.2. Markov chains. Let ζ_n, $n \in \mathbf{Z}$ be an irreducible Markov chain on a countable state space S in which each ζ_n has the stationary distribution π. Let $X_n = f(\zeta_n)$ where $\sum f(x)\pi(x) = 0$ and $\sum f(x)^2\pi(x) < \infty$. Results in Chapter 6 imply that X_n is an ergodic stationary sequence. If we let $\mathcal{F}_{-n} = \sigma(\zeta_m, m \leq -n)$ then

$$E(X_0|\mathcal{F}_{-n}) = \sum_y p^n(\zeta_{-n}, y)f(y)$$

where $p^n(x, y)$ is the n step transition probability, so

$$\|E(X_0|\mathcal{F}_{-n})\|_2^2 = \sum_x \pi(x) \left(\sum_y p^n(x, y) f(y) \right)^2$$

When f is bounded we can use $\sum f(x)\pi(x) = 0$ to get the following bound

$$\|E(X_0|\mathcal{F}_{-n})\|_2^2 \le \|f\|_\infty \sum_x \pi(x) \|p^n(x, \cdot) - \pi(\cdot)\|$$

where $\|f\|_\infty = \sup |f(x)|$ and $\| \cdot \|$ is the total variation norm.

When S is finite all f are bounded. If the chain is aperiodic, Exercise 5.6 in Chapter 5 implies that

$$\sup_x \|p^n(x, \cdot) - \pi(\cdot)\| \le Ce^{-\epsilon n}$$

and the hypothesis of (7.6) is satisfied. To see that the limiting variance σ^2 may be 0 in this case, consider the modification of Exercise 7.2 in which $X_n = (\xi_n, \xi_{n+1})$, $f(H, T) = 1$, $f(T, H) = -1$ and $f(H, H) = f(T, T) = 0$. In this case $\sum_{m=1}^n f(X_m) \in \{-1, 0, 1\}$ so there is no central limit theorem.

EXERCISE 7.3. Consider the setup of Exercise 7.2 and allow $\int x \, dH_i = \mu_i$ to be different from 0 but assume (without loss of generality) that $\sum \mu_i \pi_i = 0$. Use (7.6) to conclude that $S_{(n \cdot)}/\sqrt{n} \Rightarrow \sigma B(\cdot)$.

Our last example is simple to treat directly, but will help us evaluate the strength of the conditions in (7.6) and in later theorems.

Example 7.3. Moving average process. Suppose

$$X_m = \sum_{k \ge 0} c_k \xi_{m-k} \quad \text{where} \quad \sum_{k \ge 0} c_k^2 < \infty$$

and the ξ_i, $i \in \mathbf{Z}$, are i.i.d. with $E\xi_i = 0$ and $E\xi_i^2 = 1$. If $\mathcal{F}_{-n} = \sigma(\xi_m \; ; \; m \le -n)$ then

$$\|E(X_0|\mathcal{F}_{-n})\|_2 = \left\| \sum_{k \ge n} c_k \xi_{-k} \right\|_2 = \left(\sum_{k \ge n} c_k^2 \right)^{1/2}$$

If for example $c_k = (1 + k)^{-p}$, $\|E(X_0|\mathcal{F}_{-n})\|_2 \sim n^{(1/2)-p}$, and (7.6) applies if $p > 3/2$.

Remark. Theorem 5.3 in Hall and Heyde (1980) shows that

$$\sum_{j \geq 0} \|E(X_j|\mathcal{F}_0) - E(X_j|\mathcal{F}_{-1})\|_2 < \infty$$

is sufficient for a central limit theorem. Using this result shows that $\sum |c_k| < \infty$ is sufficient for the central limit theorem in Example 7.3.

The condition in the improved result is close to the best possible. Suppose the ξ_i take values 1 and -1 with equal probability and let $c_k = (1+k)^{-p}$ where $1/2 < p < 1$. The Lindeberg-Feller theorem can be used to show that $S_n/n^{3/2-p} \Rightarrow \sigma\chi$. To check the normalization (which was wrong in the first edition) note that

$$\sum_{m=1}^{n} X_m = \sum_{j \leq n} a_{n,j}\xi_j$$

If $j \geq 0$ then $a_{n,j} = \sum_{i=0}^{n-j} c_k \approx (n-j)^{1-p}$, so using $0 \leq j \leq n/2$ the variance is at least n^{3-2p}. Further details are left to the reader.

The last three examples show that in many cases it is easy to verify the hypothesis of (7.6) directly. To connect (7.6) with other results in the literature, we will introduce two sufficient conditions phrased in terms of

c. Mixing properties

In each case we will first give an estimate on covariances and then state a central limit theorem. Let

$$\alpha(\mathcal{G}, \mathcal{H}) = \sup\{|P(A \cap B) - P(A)P(B)| : A \in \mathcal{G}, B \in \mathcal{H}\}$$

If $\alpha = 0$, $\mathcal{G}$ and $\mathcal{H}$ are independent so α measures the dependence between the σ-fields.

(7.7) Lemma. Let p, q, $r \in (1, \infty]$ with $1/p + 1/q + 1/r = 1$, and suppose $X \in \mathcal{G}$, $Y \in \mathcal{H}$ have $E|X|^p$, $E|Y|^q < \infty$. Then

$$|EXY - EXEY| \leq 8\|X\|_p\|Y\|_q(\alpha(\mathcal{G}, \mathcal{H}))^{1/r}$$

Here, we interpret $x^0 = 1$ for $x > 0$ and $0^0 = 0$.

Proof If $\alpha = 0$, X and Y are independent and the result is true, so we can suppose $\alpha > 0$. We build up to the result in three steps, starting with the case $r = \infty$.

(a)
$$|EXY - EXEY| \leq 2\|X\|_p\|Y\|_q$$

Proof of (a)   Hölder's inequality ((5.3) in the Appendix) implies $|EXY| \leq \|X\|_p\|Y\|_q$, and Jensen's inequality implies

$$\|X\|_p\|Y\|_q \geq |E|X|E|Y|| \geq |EX\,EY|$$

so the result follows from the triangle inequality.   □

(b) $$|EXY - EX\,EY| \leq 4\|X\|_\infty\|Y\|_\infty \alpha(\mathcal{G},\mathcal{H})$$

Proof of (b)   Let $\eta = \text{sgn}\{E(Y|\mathcal{G}) - EY\} \in \mathcal{G}$. $EXY = E(XE(Y|\mathcal{G}))$, so

$$\begin{aligned}
|EXY - EX\,EY| &= |E(X\{E(Y|\mathcal{G}) - EY\})| \\
&\leq \|X\|_\infty E|E(Y|\mathcal{G}) - EY| \\
&= \|X\|_\infty E(\eta\{E(Y|\mathcal{G}) - EY\}) \\
&= \|X\|_\infty\{E(\eta Y) - E\eta EY\}
\end{aligned}$$

Applying the last result with $X = Y$ and $Y = \eta$ gives

$$|E(Y\eta) - EY\,E\eta| \leq \|Y\|_\infty |E(\zeta\eta) - E\zeta E\eta|$$

where $\zeta = \text{sgn}\{E(\eta|\mathcal{H}) - E\eta\}$. Now $\eta = 1_A - 1_B$ and $\zeta = 1_C - 1_D$ so

$$\begin{aligned}
|E(\zeta\eta) - E\zeta E\eta| &= |P(A \cap C) - P(B \cap C) - P(A \cap D) + P(B \cap D) \\
&\quad - P(A)P(C) + P(B)P(C) + P(A)P(D) - P(B)P(D)| \\
&\leq 4\alpha(\mathcal{G},\mathcal{H})
\end{aligned}$$

Combining the last three displays gives the desired result.   □

(c) $$|EXY - EX\,EY| \leq 6\|X\|_p\|Y\|_\infty \,\alpha(\mathcal{G},\mathcal{H})^{1-1/p}$$

Proof of (c)   Let $C = \alpha^{-1/p}\|X\|_p$, $X_1 = X1_{(|X|\leq C)}$ and $X_2 = X - X_1$

$$\begin{aligned}
|EXY - EX\,EY| &\leq |EX_1Y - EX_1EY| + |EX_2Y - EX_2EY| \\
&\leq 4\alpha C\|Y\|_\infty + 2\|Y\|_\infty E|X_2|
\end{aligned}$$

by (b) and (a). Now

$$E|X_2| \leq C^{-(p-1)}E(|X|^p 1_{(|X|>C)}) \leq C^{-p+1}E|X|^p$$

Combining the last two inequalities and using the definition of C gives

$$|EXY - EX\,EY| \leq 4\alpha^{1-1/p}\|X\|_p\|Y\|_\infty + 2\|Y\|_\infty \alpha^{1-1/p}\|X\|_p^{-p+1+p}$$

which is the desired result. □

Finally to prove (7.7), let $C = \alpha^{-1/q}\|Y\|_q$, $Y_1 = Y \, 1_{(|Y|\leq C)}$, and $Y_2 = Y - Y_1$.

$$|EXY - EXEY| \leq |EXY_1 - EXEY_1| + |EXY_2 - EXEY_2|$$
$$\leq 6C\|X\|_p\alpha^{1-1/p} + 2\|X\|_p\|Y_2\|_\theta$$

where $\theta = (1 - 1/p)^{-1}$ by (c) and (a). Now

$$E|Y_2|^\theta \leq C^{-q+\theta}E(|Y|^q 1_{(|Y|>C)}) \leq C^{-q+\theta}E|Y|^q$$

Taking the $1/\theta$ root of each side and recalling the definition of C

$$\|Y_2\|_\theta \leq C^{-(q-\theta)/\theta}\|Y_2\|_q^{q/\theta} \leq \alpha^{(q-\theta)/\theta q}\|Y_2\|_q$$

so we have

$$|EXY - EXEY| \leq 6\alpha^{-1/q}\|Y\|_q\|X\|_p\alpha^{1-1/p} + 2\|X\|_p\alpha^{1/\theta-1/q}\|Y\|_q^{1/\theta+1/q}$$

proving (7.7). □

Remark. The last proof is from Appendix III of Hall and Heyde (1980). They attribute (b) to Ibragimov (1962) and (c) and (7.7) to Davydov (1968).

Combining (7.6) and (7.7) gives

(7.8) Theorem. Suppose X_n, $n \in \mathbf{Z}$ is an ergodic stationary sequence with $EX_n = 0$, $E|X_0|^{2+\delta} < \infty$. Let $\alpha(n) = \alpha(\mathcal{F}_{-n}, \sigma(X_0))$ where $\mathcal{F}_{-n} = \sigma(X_m : m \leq -n)$ and suppose

$$\sum_{n=1}^{\infty} \alpha(n)^{\delta/2(2+\delta)} < \infty$$

If $S_n = X_1 + \cdots + X_n$, then $S_{(n\cdot)}/\sqrt{n} \Rightarrow \sigma B(\cdot)$ where

$$\sigma^2 = EX_0^2 + 2\sum_{n=1}^{\infty} EX_0 X_n$$

Remark. Let $\bar{\alpha}(n) = \alpha(\mathcal{F}_{-n}, \mathcal{F}_0')$, where $\mathcal{F}_0' = \sigma(X_k, k \geq 0)$. When $\bar{\alpha}(n) \downarrow 0$, the sequence is called **strong mixing**. Rosenblatt (1956) introduced the concept as a condition under which the central limit theorem for stationary sequences could be obtained. Ibragimov (1962) proved $S_n/\sqrt{n} \Rightarrow \sigma\chi$ where $\sigma^2 = \lim_{n\to\infty} ES_n^2/n$ under the assumption

$$\sum_{n=1}^{\infty} \bar{\alpha}(n)^{\delta/(2+\delta)} < \infty$$

See Ibragimov and Linnik (1971) Theorem 18.5.3 or Hall and Heyde (1980) Corollary 5.1 for a proof.

Proof   To use (7.7) to estimate the quantity in (7.6) we begin with

$$(7.9) \qquad \|E(X|\mathcal{F})\|_2 = \sup\{E(XY) : Y \in \mathcal{F}, \|Y\|_2 = 1\}$$

Proof of (7.9)   If $Y \in \mathcal{F}$ with $\|Y\|_2 = 1$ then using a by now familiar property of conditional expectation and the Cauchy-Schwarz inequality

$$EXY = E(E(XY|\mathcal{F})) = E(Y E(X|\mathcal{F})) \le \|E(X|\mathcal{F})\|_2 \|Y\|_2$$

Equality holds when $Y = E(X|\mathcal{F})/\|E(X|\mathcal{F})\|_2$. $\qquad\qquad\square$

Letting $p = 2 + \delta$ and $q = 2$ in (7.7), noticing

$$\frac{1}{r} = 1 - \frac{1}{p} - \frac{1}{q} = 1 - \frac{1}{2+\delta} - \frac{1}{2} = \frac{4 + 2\delta - 2 - (2+\delta)}{(2+\delta)2} = \frac{\delta}{(2+\delta)2}$$

and recalling $EX_0 = 0$, shows that if $Y \in \mathcal{F}_{-n}$

$$|EX_0 Y| \le 8\|X_0\|_{2+\delta} \|Y\|_2 \, \alpha(n)^{\delta/2(2+\delta)}$$

Combining this with (7.9) gives

$$\|E(X_0|\mathcal{F}_{-n})\|_2 \le 8\|X_0\|_{2+\delta} \, \alpha(n)^{\delta/2(2+\delta)}$$

and it follows that the hypotheses of (7.6) are satisfied. $\qquad\qquad\square$

In the M dependent case (Example 7.1), $\alpha(n) = 0$ for $n > M$, so (7.8) applies. As for Markov chains (Example 7.2), in this case

$$\alpha(n) = \sup_{A,B} |P(X_{-n} \in A, X_0 \in B) - \pi(A)\pi(B)|$$

$$= \sup_{A,B} \left| \sum_{x \in A, y \in B} \pi(x)p^n(x,y) - \pi(x)\pi(y) \right|$$

$$\le \sum_x \pi(x)2\|p^n(x,\cdot) - \pi(\cdot)\|$$

so the hypothesis of (7.8) can be checked if we know enough about the rate of convergence to equilibrium.

Finally, to see how good the conditions in (7.8) are we consider the special case of Example 7.3 in which the ξ_i are i.i.d. standard normals. Let

$$\rho(\mathcal{G},\mathcal{H}) = \sup\{\text{corr}(X,Y) : X \in \mathcal{G}, Y \in \mathcal{H}\}$$

where

$$\text{corr}(X,Y) = \frac{EXY - EXEY}{\|X - EX\|_2\|Y - EY\|_2}$$

Clearly $\alpha(\mathcal{G},\mathcal{H}) \leq \rho(\mathcal{G},\mathcal{H})$. Kolmogorov and Rozanov (1960) have shown that when $\mathcal{G}$ and $\mathcal{H}$ are generated by Gaussian random variables then $\rho(\mathcal{G},\mathcal{H}) \leq 2\pi\alpha(\mathcal{G},\mathcal{H})$. They proved this by showing that $\rho(\mathcal{G},\mathcal{H})$ is the angle between $L^2(\mathcal{G})$ and $L^2(\mathcal{H})$. Using the geometric interpretation, we see that if $|c_k|$ is decreasing then $\alpha(n) \leq \bar{\alpha}(n) \leq |c_n|$, so (7.8) requires $\sum |c_n|^{(1/2)-\epsilon} < \infty$ but Ibragimov's result applies if $\sum |c_n|^{1-\epsilon} < \infty$. As Exercise 7.9 (or direct computation shows) the central limit theorem is valid if $\sum |c_n| < \infty$.

Our second mixing concept is more restrictive (and asymmetric) because we divide by $P(B) < 1$. Let

$$\beta(\mathcal{G},\mathcal{H}) = \sup\{P(A|B) - P(A) : A \in \mathcal{G}, B \in \mathcal{H} \text{ with } P(B) > 0\}$$

Clearly $\beta(\mathcal{G},\mathcal{H}) \geq \alpha(\mathcal{G},\mathcal{H})$ and $\beta(\mathcal{G},\mathcal{H}) = 0$ implies that $\mathcal{G}$ and $\mathcal{H}$ are independent. The analogue of (7.7) for β is

(7.10) Lemma. Let $p,q \in (1,\infty)$ with $1/p + 1/q = 1$. Suppose $X \in \mathcal{G}$, and $Y \in \mathcal{H}$ have $E|X|^p$, $E|Y|^q < \infty$. Then

$$|EXY - EXEY| \leq 2\beta(\mathcal{G},\mathcal{H})^{1/p}\|X\|_p\|Y\|_q$$

and the resulting convergence theorem is:

(7.11) Theorem. Suppose X_n, $n \in \mathbf{Z}$ is an ergodic stationary sequence with $EX_n = 0$, $E|X_0|^2 < \infty$. Let $\beta(n) = \alpha(\mathcal{F}_{-n}, \sigma(X_0))$ where $\mathcal{F}_{-n} = \sigma(X_m : m \leq -n)$ and suppose

$$\sum_{n=1}^{\infty} \beta(n)^{1/2} < \infty$$

If $S_n = X_1 + \cdots + X_n$, then $S_{n,(n\cdot)} \Rightarrow \sigma B(\cdot)$ where $\sigma^2 = EX_0^2 + \sum_{n=1}^{\infty} EX_0 X_n$.

Remark. Let $\bar{\beta}(n) = \beta(\mathcal{F}_{-n}, \mathcal{F}_0')$ where $\mathcal{F}_0' = \sigma(X_0, X_1, \ldots)$. If $\bar{\beta}(n) \downarrow 0$ as $n \to \infty$, the sequence X_n is said to be **uniformly mixing**. Billingsley (1968), see Theorem 20.1 on p. 174, gives (7.11) under the assumption $\sum \bar{\beta}(n)^{1/2} < \infty$. A proof of (7.10) can be found on p. 170-171 of his book, and it is a simple exercise to deduce (7.11) from (7.10) using (7.6).

We will not enter into the details of these proofs since they do not give significant results for our examples. In the case of Markov chains (Example 7.2.),

$$\beta(n) = \sup_x \|p^n(x, \cdot) - \pi(\cdot)\|$$

The chain is clearly not uniformly mixing if $\beta(n) = 1$ for all n. Conversely, if $\beta(N) < 1 - \epsilon$ for some N, iterating gives

$$\sup_x \|p^{kN+j}(x, \cdot) - \pi(\cdot)\| \le (1 - \epsilon)^k \quad \text{for } 0 \le j < N$$

so (7.11) applies but by using Example 7.2 one can check the assumptions of (7.6) directly.

Things are even worse for Example 7.3. (Moving average process). It is easy to see that if $c_k \ne 0$ for all k and the distribution of ξ_k is unbounded, $\beta(n) = 1$ for all n. The last result is a little discouraging but there is at least one interesting example:

Example 7.4. The continued fraction transformation (See Exercise 1.6 in Chapter 6) is uniformly mixing with $\bar{\beta}(n) = a\rho^n$ and $\rho < 1$. See Chapter 9 of Lévy (1937). For more on this example see Billingsley (1968), p. 192-194.

*7.8. Empirical Distributions, Brownian Bridge

Let $X_1, X_2, \ldots$ be i.i.d. with distribution F. In Section 1.7 we showed that with probability one the empirical distribution

$$\hat{F}_n(x) = \frac{1}{n} |\{m \le n : X_m \le x\}|$$

converges uniformly to $F(x)$. In this section we will investigate the rate of convergence when F is continuous. We impose this restriction so we can reduce to the case of a uniform distribution on (0,1) by setting $Y_n = F(X_n)$. (See Exercise 1.9 in Chapter 1.) Since $x \to F(x)$ is nondecreasing and continuous and no observations land in intervals of constancy of F, it is easy to see that if we let

$$\hat{G}_n(y) = \frac{1}{n} |\{m \le n : Y_m \le y\}|$$

then

$$\sup_x |\hat{F}_n(x) - F(x)| = \sup_{0 < y < 1} |\hat{G}_n(y) - y|$$

For the rest of the section then, we will assume $Y_1, Y_2, \ldots$ is i.i.d. uniform on $(0,1)$. To be able to apply Donsker's theorem we will transform the problem. Put the observations $Y_1, \ldots, Y_n$ in increasing order: $U_1^n < U_2^n < \ldots < U_n^n$. I claim that

(8.2)
$$\sup_{0 < y < 1} \hat{G}_n(y) - y = \sup_{1 \le m \le n} \frac{m}{n} - U_m^n$$

$$\inf_{0 < y < 1} \hat{G}_n(y) = \inf_{1 \le m \le n} \frac{m-1}{n} - U_m^n$$

since the sup occurs at a jump of $\hat{G}_n$ and the inf right before a jump. We will show that

$$D_n \equiv n^{1/2} \sup_{0 < y < 1} |\hat{G}_n(y) - y|$$

has a limit, so the extra $-1/n$ in the inf does not make any difference.

Our third and final manuever is to give a special construction of the order statistics $U_1^n < U_2^n \ldots < U_n^n$. Let $W_1, W_2, \ldots$ be i.i.d. with $P(W_i > t) = e^{-t}$ and let $Z_n = W_1 + \cdots + W_n$.

(8.3) Lemma. $\{U_k^n : 1 \le k \le n\} \overset{d}{=} \{Z_k/Z_{n+1} : 1 \le k \le n\}$

Proof We change variables $v = r(t)$ where $v_i = t_i/t_{n+1}$ for $i \le n$, $v_{n+1} = t_{n+1}$. The inverse function is

$$s(v) = (v_1 v_{n+1}, \ldots, v_n v_{n+1}, v_{n+1})$$

which has matrix of partial derivatives $\partial s_i/\partial v_j$ given by

$$\begin{pmatrix} v_{n+1} & 0 & \cdots & 0 & v_1 \\ 0 & v_{n+1} & \cdots & 0 & v_2 \\ \vdots & \vdots & \ddots & \vdots & \vdots \\ 0 & 0 & \cdots & v_{n+1} & v_n \\ 0 & 0 & \cdots & 0 & 1 \end{pmatrix}$$

The determinant of this matrix is v_{n+1}^n so if we let $W = (V_1, \ldots, V_{n+1}) = r(Z_1, \ldots, Z_{n+1})$ the change of variables formula implies W has joint density

$$f_W(v_1, \ldots, v_n, v_{n+1}) \doteq \left(\prod_{m=1}^n \lambda e^{-\lambda v_{n+1}(v_m - v_{m-1})} \right) \lambda e^{-\lambda v_{n+1}(1 - v_n)} v_{n+1}^n$$

To find the joint density of $V = (V_1, \ldots, V_n)$ we simplify the preceding formula and integrate out the last coordinate to get

$$f_V(v_1, \ldots, v_n) = \int_0^\infty \lambda^{n+1} v_{n+1}^n e^{-\lambda v_{n+1}} \, dv_{n+1} = n!$$

for $0 < v_1 < v_2 \ldots < v_n < 1$, which is the desired joint density. $\square$

We turn now to the limit law for D_n. As argued above, it suffices to consider

$$D_n' = n^{1/2} \max_{1 \le m \le n} \left| \frac{Z_m}{Z_{n+1}} - \frac{m}{n} \right|$$

$$= \frac{n}{Z_{n+1}} \max_{1 \le m \le n} \left| \frac{Z_m}{n^{1/2}} - \frac{m}{n} \cdot \frac{Z_{n+1}}{n^{1/2}} \right|$$

$$= \frac{n}{Z_{n+1}} \max_{1 \le m \le n} \left| \frac{Z_m - m}{n^{1/2}} - \frac{m}{n} \cdot \frac{Z_{n+1} - n}{n^{1/2}} \right|$$

If we let

$$B_n(t) = \begin{cases} (Z_m - m)/n^{1/2} & \text{if } t = m/n \text{ with } m \in \{0, 1, \ldots, n\} \\ \text{linear} & \text{on } [(m-1)/n, m/n] \end{cases}$$

then

$$D_n' = \frac{n}{Z_{n+1}} \max_{0 \le t \le 1} \left| B_n(t) - t \left\{ B_n(1) + \frac{Z_{n+1} - Z_n}{n^{1/2}} \right\} \right|$$

The strong law of large numbers implies $Z_{n+1}/n \to 1$ a.s., so the first factor will disappear in the limit. To find the limit of the second we observe that Donsker's theorem, (6.6), implies $B_n(\cdot) \Rightarrow B(\cdot)$, a Brownian motion, and computing second moments shows

$$(Z_{n+1} - Z_n)/n^{1/2} \to 0 \text{ in probability}$$

$\psi(\omega) = \max_{0 \le t \le 1} |\omega(t) - t\omega(1)|$ is a continuous function from $C[0,1]$ to $\mathbf{R}$, so it follows from Donsker's theorem that

(8.4) Theorem. $D_n \Rightarrow \max_{0 \le t \le 1} |B_t - tB_1|$, where B_t is a Brownian motion starting at 0.

Remark. Doob (1949) suggested this approach to deriving results of Kolmogorov and Smirnov, which was later justified by Donsker (1952). Our proof follows Breiman (1968).

To identify the distribution of the limit in (8.4), we will first prove

(8.5) $\{B_t - tB_1, 0 \le t \le 1\} \overset{d}{=} \{B_t, 0 \le t \le 1 | B_1 = 0\}$

a process we will denote by B_t^0 and call the **Brownian bridge**. The event $B_1 = 0$ has probability 0, but it is easy to see what the conditional probability should mean. If $0 = t_0 < t_1 < \ldots < t_n < t_{n+1} = 1$, $x_0 = 0$, $x_{n+1} = 0$, and $x_1, \ldots, x_n \in \mathbf{R}$ then

(8.6)
$$P(B(t_1) = x_1, \ldots, B(t_n) = x_n | B(1) = 0)$$

$$= \frac{1}{p_1(0,0)} \prod_{m=1}^{n+1} p_{t_m - t_{m-1}}(x_{m-1}, x_m)$$

where $p_t(x, y) = (2\pi t)^{-1/2} \exp(-(y-x)^2/2t)$.

Proof of (8.5) Formula (8.6) shows that the f.d.d.'s of B_t^0 are multivariate normal and have mean 0. Since $B_t - tB_1$ also has this property, it suffices to show that the covariances are equal. We begin with the easier computation. If $s < t$ then

(8.7) $\qquad E((B_s - sB_1)(B_t - tB_1)) = s - st - st + st = s(1 - t)$

For the other process $P(B_s^0 = x, B_t^0 = y)$ is

$$\frac{\exp(-x^2/2s)}{(2\pi s)^{1/2}} \cdot \frac{\exp(-(y-x)^2/2(t-s))}{(2\pi(t-s))^{1/2}} \cdot \frac{\exp(-y^2/2(1-t))}{(2\pi(1-t))^{1/2}} \cdot (2\pi)^{1/2}$$

$$= (2\pi)^{-1}(s(t-s)(1-t))^{-1/2} \exp(-(ax^2 + 2bxy + cy^2)/2)$$

where

$$a = \frac{1}{s} + \frac{1}{t-s} = \frac{t}{s(t-s)} \qquad b = -\frac{1}{t-s}$$

$$c = \frac{1}{t-s} + \frac{1}{1-t} = \frac{1-s}{(t-s)(1-t)}$$

Recalling the discussion at the end of Section 2.9 and noticing

$$\begin{pmatrix} \frac{t}{s(t-s)} & \frac{-1}{(t-s)} \\ \frac{-1}{(t-s)} & \frac{1-s}{(t-s)(1-t)} \end{pmatrix}^{-1} = \begin{pmatrix} s(1-s) & s(1-t) \\ s(1-t) & t(1-t) \end{pmatrix}$$

(multiply the matrices!) shows (8.5) holds.

Our final step in investigating the limit distribution of D_n is to compute the distribution of $\max_{0 \le t \le 1} |B_t^0|$. To do this we compute $P_x(T_a \wedge T_b > t, B_t \in A)$ where $a < x < b$ and $A \subset (a, b)$. We begin by observing that

(*) $\qquad P_x(T_a \wedge T_b > t, B_t \in A) = P_x(B_t \in A) - P_x(T_a < T_b, T_a < t, B_t \in A)$
$$- P_x(T_b < T_a, T_b < t, B_t \in A)$$

If we let $\rho_a(y) = 2a - y$ be reflection through a and observe that $\{T_a < T_b\}$ is $\mathcal{F}(T_a)$ measurable then it follows from the proof of (4.6) that

$$P_x(T_a < T_b, T_a < t, B_t \in A) = P_x(T_a < T_b, B_t \in \rho_a A)$$

where $\rho_a A = \{\rho_a(y) : y \in A\}$. To get rid of the $T_a < T_b$ we observe that

$$P_x(T_a < T_b, B_t \in \rho_a A) = P_x(B_t \in \rho_a A) - P_x(T_b < T_a, B_t \in \rho_a A)$$

Noticing that $B_t \in \rho_a A$ and $T_b < T_a$ imply $T_b < t$ and using the reflection principle again gives

$$\begin{aligned} P_x(T_b < T_a, B_t \in \rho_a A) &= P_x(T_b < T_a, B_t \in \rho_b \rho_a A) \\ &= P_x(B_t \in \rho_b \rho_a A) - P_x(T_a < T_b, B_t \in \rho_b \rho_a A) \end{aligned}$$

Repeating the last two calculations n more times gives

$$\begin{aligned} P_x(T_a < T_b, B_t \in \rho_a A) &= \sum_{m=0}^{n} P_x(B_t \in \rho_a(\rho_b\rho_a)^m A) - P_x(B_t \in (\rho_b\rho_a)^{m+1} A) \\ &\quad + P_x(T_a < T_b, B_t \in (\rho_b\rho_a)^{n+1} A) \end{aligned}$$

Each pair of reflections pushes A further away from 0 so letting $n \to \infty$ shows

$$P_x(T_a < T_b, B_t \in \rho_a A) = \sum_{m=0}^{\infty} P_x(B_t \in \rho_a(\rho_b\rho_a)^m A) - P_x(B_t \in (\rho_b\rho_a)^{m+1} A)$$

Interchanging the roles of a and b gives

$$P_x(T_b < T_a, B_t \in \rho_b A) = \sum_{m=0}^{\infty} P_x(B_t \in \rho_b(\rho_a\rho_b)^m A) - P_x(B_t \in (\rho_a\rho_b)^{m+1} A)$$

Combining the last two expressions with $(*)$ and using $\rho_c^{-1} = \rho_c$, $(\rho_a\rho_b)^{-1} = \rho_b^{-1}\rho_a^{-1}$ gives

$$P_x(T_a \wedge T_b > t, B_t \in A) = \sum_{m=-\infty}^{\infty} P_x(B_t \in (\rho_b\rho_a)^n A) - P_x(B_t \in \rho_a(\rho_b\rho_a)^n A)$$

To prepare for applications let $A = (u, v)$ where $a < u < v < b$, notice that $\rho_b\rho_a(y) = y + 2(b-a)$ and change variables in the second sum to get

$$P_x(T_a \wedge T_b > t, u < B_t < v) =$$

(8.8)
$$\sum_{n=-\infty}^{\infty} \{ P_x(u + 2n(b-a) < B_t < v + 2n(b-a)) \\ - P_x(2b - v + 2n(b-a) < B_t < 2b - u + 2n(b-a)) \}$$

Letting $u = y - \epsilon$, $v = y + \epsilon$, dividing both sides by 2ϵ, and letting $\epsilon \to 0$ (leaving it to the reader to check that the dominated convergence theorem applies) gives

$$(8.9) \quad P_x(T_a \wedge T_b > t, B_t = y) = \sum_{n=-\infty}^{\infty} P_x(B_t = y + 2n(b-a))$$

$$- P_x(B_t = 2b - y + 2n(b-a))$$

where the probabilities are interpreted as density functions. For a picture of the formula see Figure 7.8.1 and notice that if $x = a$ or b the array of $+$ and $-$ signs is antisymmetric so the answer is 0.

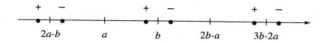

$$\begin{array}{ccccc} + & - & + & - & + & - \\ 2a\text{-}b & a & b & 2b\text{-}a & 3b\text{-}2a \end{array}$$

Figure 7.8.1

Setting $x = y = 0$, $t = 1$, and dividing by $(2\pi)^{-1/2} = P_0(B_1 = 0)$ we get a result for the Brownian bridge B_t^0:

$$(8.10)$$
$$P_0\left(a < \min_{0 \le t \le 1} B_t^0 < \max_{0 \le t \le 1} B_t^0 < b\right)$$
$$= \sum_{n=-\infty}^{\infty} e^{-(2n(b-a))^2/2} - e^{-(2b+2n(b-a))^2/2}$$

Taking $a = -b$ we have

$$(8.11) \qquad P_0\left(\max_{0 \le t \le 1} |B_t^0| < b\right) = 1 + \sum_{m=1}^{\infty} e^{-2m^2 b^2}$$

This formula gives the distribution of the Kolmogorv-Smirnov statistic, which can be used to test if an i.i.d. sequence $X_1, \ldots, X_n$ has distribution F. To do this we transform the data to $F(X_n)$ and look at the maximum discrepancy between the empirical distribution and the uniform. (8.11) tells us the distribution of the error when the X_i have distribution F.

(8.10) gives the joint distribution of the maximum and minimum of Brownian bridge. In theory one can let $a \to -\infty$ in this formula to find the distribution of the maximum but in practice it is easier to start over again.

EXERCISE 8.1. Use Exercise 4.1 and the reasoning that led to (8.10) to conclude

$$P(\max_{0 \le t \le 1} B_t^0 > b) = \exp(-2b^2)$$

*7.9. Laws of the Iterated Logarithm

Our first goal is to show

(9.1) **LIL for Brownian motion.** $\limsup_{t \to \infty} B_t / (2t \log \log t)^{1/2} = 1$ a.s.

Here LIL is short for "law of the iterated logarithm," a name that refers to the $\log \log t$ in the denominator. Once (9.1) is established we can use the Skorokhod respresentation to prove the analogous result for random walks with mean 0 and finite variance. The key to the proof of (9.1) is (4.5).

$$(9.2) \qquad P_0 \left(\max_{0 \le s \le 1} B_s > a \right) = P_0(T_a \le 1) = 2\, P_0(B_1 \ge a)$$

To identify the asymptotic behavior of the right-hand side of (9.2) as $a \to \infty$ we use (1.3) from Chapter 1.

$$(9.3) \qquad \int_x^\infty \exp(-y^2/2)\, dy \le \frac{1}{x} \exp(-x^2/2)$$

$$(9.4) \qquad \int_x^\infty \exp(-y^2/2)\, dy \sim \frac{1}{x} \exp(-x^2/2) \quad \text{as } x \to \infty$$

where $f(x) \sim g(x)$ means $f(x)/g(x) \to 1$ as $x \to \infty$. The last result and Brownian scaling imply that

$$P_0(B_t > (tf(t))^{1/2}) \sim \kappa f(t)^{-1/2} \exp(-f(t)/2)$$

where $\kappa = (2\pi)^{-1/2}$ is a constant that we will try to ignore below. The last result implies that if $\epsilon > 0$ then

$$\sum_{n=1}^\infty P_0(B_n > (nf(n))^{1/2}) \begin{cases} < \infty & \text{when } f(n) = (2+\epsilon)\log n \\ = \infty & \text{when } f(n) = (2-\epsilon)\log n \end{cases}$$

and hence by the Borel-Cantelli lemma that

$$\limsup_{n \to \infty} B_n / (2n \log n)^{1/2} \le 1 \quad \text{a.s.}$$

To replace $\log n$ by $\log \log n$ we have to look along exponentially growing sequences. Let $t_n = \alpha^n$ where $\alpha > 1$.

$$P_0 \left(\max_{t_n \le s \le t_{n+1}} B_s > (t_n f(t_n))^{1/2} \right) \le P_0 \left(\max_{0 \le s \le t_{n+1}} B_s / t_{n+1}^{1/2} > \left(\frac{f(t_n)}{\alpha} \right)^{1/2} \right)$$

$$\le 2\kappa (f(t_n)/\alpha)^{-1/2} \exp(-f(t_n)/2\alpha)$$

by (9.2) and (9.3). If $f(t) = 2\alpha^2 \log\log t$ then

$$\log\log t_n = \log(n\log\alpha) = \log n + \log\log\alpha$$

so $\exp(-f(t_n)/2\alpha) \le C_\alpha n^{-\alpha}$ where C_α is a constant that depends only on α, and hence

$$\sum_{n=1}^{\infty} P_0\left(\max_{t_n \le s \le t_{n+1}} B_s > (t_n f(t_n))^{1/2}\right) < \infty$$

Since $t \to (tf(t))^{1/2}$ is increasing and $\alpha > 1$ is arbitrary, it follows that

(a) $$\limsup B_t/(2t\log\log t)^{1/2} \le 1$$

To prove the other half of (9.1) we again let $t_n = \alpha^n$ but this time α will be large since to get independent events, we will we look at

$$P_0\left(B(t_{n+1}) - B(t_n) > (t_{n+1}f(t_{n+1}))^{1/2}\right) = P_0\left(B_1 > (\beta f(t_{n+1}))^{1/2}\right)$$

where $\beta = t_{n+1}/(t_{n+1} - t_n) = \alpha/(\alpha - 1) > 1$. The last quantity is

$$\ge \frac{\kappa}{2}(\beta f(t_{n+1}))^{-1/2}\exp(-\beta f(t_{n+1})/2)$$

if n is large by (9.4). If $f(t) = (2/\beta^2)\log\log t$ then $\log\log t_n = \log n + \log\log\alpha$ so

$$\exp(-\beta f(t_{n+1})/2) \ge C_\alpha n^{-1/\beta}$$

where C_α is a constant that depends only on α, and hence

$$\sum_{n=1}^{\infty} P_0\left(B(t_{n+1}) - B(t_n) > (t_{n+1}f(t_{n+1}))^{1/2}\right) = \infty$$

Since the events in question are independent, it follows from the second Borel-Cantelli lemma that

(b) $$B(t_{n+1}) - B(t_n) > ((2/\beta^2)t_{n+1}\log\log t_{n+1})^{1/2} \quad \text{i.o.}$$

From (a) we get

(c) $$\limsup_{n\to\infty} B(t_n)/(2t_n\log\log t_n)^{1/2} \le 1$$

Since $t_n = t_{n+1}/\alpha$ and $t \to \log\log t$ is increasing combining (b) and (c), and recalling $\beta = \alpha/(\alpha - 1)$ gives

$$\limsup_{n\to\infty} B(t_{n+1})/(2t_{n+1}\log\log t_{n+1})^{1/2} \ge \frac{\alpha - 1}{\alpha} - \frac{1}{\alpha^{1/2}}$$

Letting $\alpha \to \infty$ now gives the desired lower bound and the proof of (9.1) is complete.

EXERCISE 9.1. Let $t_k = \exp(e^k)$. Show that

$$\limsup_{k \to \infty} B(t_k)/(2t_k \log\log\log t_k)^{1/2} = 1 \quad \text{a.s.}$$

(2.8) implies that $X_t = tB(1/t)$ is a Brownian motion. Changing variables and using (9.1) we conclude

$$(9.5) \qquad \limsup_{t \to 0} |B_t|/(2t \log\log(1/t))^{1/2} = 1 \quad \text{a.s.}$$

To take a closer look at the local behavior of Brownian paths, we note that Blumenthal's 0-1 law (2.6) implies $P_0(B_t < h(t)$ for all t sufficiently small) $\in \{0, 1\}$. h is said to belong to the **upper class** if the probability is 1, the **lower class** if it is 0.

(9.6) Kolmogorov's test. If $h(t) \uparrow$ and $t^{-1/2}h(t) \downarrow$ then h is upper or lower class according as

$$\int_0^1 t^{-3/2} h(t) \exp(-h^2(t)/2t)\, dt \quad \text{converges or diverges}$$

The first proof of this was given by Petrovsky (1935). Recalling (4.1), we see that the integrand is the probability of hitting $h(t)$ at time t. To see what (9.5) says, define $\lg_k(t) = \log(\lg_{k-1}(t))$ for $k \geq 2$ and $t > a_k = \exp(a_{k-1})$ where $\lg_1(t) = \log(t)$ and $a_1 = 0$. A little calculus shows that when $n \geq 4$

$$h(t) = \left(2t \left\{ \lg_2(1/t) + \frac{3}{2}\lg_3(1/t) + \sum_{m=4}^{n-1} \lg_m(1/t) + (1+\epsilon)\lg_n(1/t) \right\} \right)^{1/2}$$

is upper or lower class according as $\epsilon > 0$ or $\epsilon \leq 0$.

Approximating h from above by piecewise constant functions it is easy to show that if the integral in (9.6) converges, $h(t)$ is an upper class function. The proof of the other direction is much more difficult, see Motoo (1959) or Section 4.12 of Itô and McKean (1965).

Turning to random walk, we will prove a result due to Hartman and Wintner (1941):

(9.7) Theorem. If $X_1, X_2, \ldots$ are i.i.d. with $EX_i = 0$ and $EX_i^2 = 1$ then

$$\limsup_{n \to \infty} S_n/(2n \log \log n)^{1/2} = 1$$

Proof   By (5.3) we can write $S_n = B(T_n)$ with $T_n/n \to 1$ a.s. As in the proof of Donsker's theorem this is all we will use in the argument below. (9.7) will follow from (9.1) once we show

(a) $$(S_{[t]} - B_t)/(t \log \log t)^{1/2} \to 0 \quad \text{a.s.}$$

To do this we begin by observing that if $\epsilon > 0$ and $t \geq t_o(\omega)$

(b) $$T_{[t]} \in [t/(1 + \epsilon), t(1 + \epsilon)]$$

To estimate $S_{[t]} - B_t$ we let $M(t) = \sup\{|B(s) - B(t)| : t/(1+\epsilon) \leq s \leq t(1+\epsilon)\}$. To control the last quantity we let $t_k = (1+\epsilon)^k$ and notice that if $t_k \leq t \leq t_{k+1}$

$$M(t) \leq \sup\{|B(s) - B(t)| : t_{k-1} \leq s \leq t_{k+2}\}$$
$$\leq 2 \ \sup\{|B(s) - B(t_{k-1})| : t_{k-1} \leq s \leq t_{k+2}\}$$

Noticing $t_{k+2} - t_{k-1} = \delta t_{k-1}$ where $\delta = (1 + \epsilon)^3 - 1$, scaling implies

$$P\left(\max_{t_{k-1} \leq s \leq t_{k+2}} |B(s) - B(t)| > (3\delta t_{k-1} \log \log t_{k-1})^{1/2} \right)$$
$$= P\left(\max_{0 \leq r \leq 1} |B(r)| > (3 \log \log t_{k-1})^{1/2} \right)$$
$$\leq 2\kappa(3 \log \log t_{k-1})^{-1/2} \exp(-3 \log \log t_{k-1}/2)$$

by a now familiar application of (9.2) and (9.3). Summing over k and using (b) gives

$$\limsup_{t \to \infty} (S_{[t]} - B_t)/(t \log \log t)^{1/2} \leq (3\delta)^{1/2}$$

If we recall $\delta = (1 + \epsilon)^3 - 1$ and let $\epsilon \downarrow 0$, (a) follows and the proof of (9.6) is complete. $\square$

Remark. Since the proof of (9.7) only requires $S_n = B(T_n)$ with $T_n/n \to 1$ a.s., the conclusion generalizes to many of the dependent sequences considered in Section 7.7.

EXERCISE 9.2. Show that if $E|X_i|^\alpha = \infty$ for some $\alpha < 2$ then

$$\limsup_{n \to \infty} |X_n|/n^{1/\alpha} = \infty \quad \text{a.s.}$$

so the law of the iterated logarithm fails.

Strassen (1965) has shown an exact converse. If (9.7) holds then $EX_i = 0$ and $EX_i^2 = 1$. Another one of his contributions to this subject is

(9.8) Strassen's (1964) invariance principle. Let $X_1, X_2, \ldots$ be i.i.d. with $EX_i = 0$ and $EX_i^2 = 1$, let $S_n = X_1 + \cdots + X_n$, and let $S_{(n \cdot)}$ be the usual linear interpolation. The limit set (i.e., the collection of limits of convergent subsequences) of

$$Z_n(\cdot) = (2n \log \log n)^{-1/2} S(n \cdot) \quad \text{for } n \geq 3$$

is $\mathcal{K} = \{f : f(x) = \int_0^x g(y) \, dy \text{ with } \int_0^1 g(y)^2 \, dy \leq 1\}$.

Jensen's inequality implies $f(1)^2 \leq \int_0^1 g(y)^2 \, dy \leq 1$ with equality if and only if $f(t) = t$, so (9.8) contains (9.5) as a special case and provides some information about how the large value of S_n came about.

EXERCISE 9.3. Give a direct proof that under the hypotheses of (9.8) the limit set of $\{S_n/(2n \log \log n)^{1/2}\}$ is $[-1, 1]$.

History Lesson. The LIL for sums of independent *r.v.'s* grew from the early efforts of Hausdorff (1913) and Hardy and Littlewood (1914) to determine the rate of convergence in Borel's theorem about normal numbers (See Example 2.4 in Chapter 6.) The LIL for Bernoulli variables was reached in five steps:

Hausdorff (1913)	$S_n = o(n^{1/2+\epsilon})$
Hardy and Littlewood (1914)	$S_n = O((n \log n)^{1/2})$
Steinhaus (1922)	$\limsup S_n/(2n \log n)^{1/2} \leq 1$
Khintchine (1923)	$S_n = O((n \log \log n)^{1/2})$
Khintchine (1924)	$\limsup S_n/((n/2) \log \log n)^{1/2} = 1$

We have $n/2$ instead of $2n$ in the last result since Bernoulli variables have variance $1/4$. The LIL was proved for bounded independent r.v.'s by Kolmogorov (1929), who did not assume the summands were identically distributed. As mentioned earlier (9.6) is due to Hartman and Wintner (1941). Strassen (1964) gave the proof (of (9.7) and hence (9.6)) using Skorokhod imbedding. The version above follows the treatment in Breiman (1968). For generalizations of (9.7) to martingales and processes with stationary increments see Heyde and Scott (1973), Hall and Heyde (1976), (1980).

There are versions of Kolmogorov's test for random walks. We say that c_n belongs to the upper class $\mathcal{U}$ if $P(S_n > c_n \text{ i.o.}) = 0$, and to the lower class $\mathcal{L}$

otherwise. For Bernoulli variables, P. Lévy (1933) showed

$$(2t \log \log t + a \log \log \log t)^{1/2} \begin{cases} \in \mathcal{U} & \text{if } a > 3 \\ \in \mathcal{L} & \text{if } a \leq 1 \end{cases}$$

Kolmogorov claimed and later Erdös (1942) proved $c_n \in \mathcal{U}, \mathcal{L}$ according as

$$\sum_{n=1}^{\infty} n^{-3/2} c_n \exp(-c_n^2/2n) < \infty \text{ or } = \infty$$

Feller (1943) gave a general version of the test for independent variables without assuming they all had the same distribution. His conditions are stronger than finite variance in the i.i.d. case.

Appendix: Measure Theory

This appendix gives a complete treatment of the results from measure theory that we will need.

A.1. Lebesgue-Stieltjes Measures

To prove the existence of Lebesgue measure (and some related more general measures), we will use the Carathéodary extension theorem, (1.1). To state that result we need several definitions in addition to the ones given in Section 1 of Chapter 1. A collection $\mathcal{A}$ of subsets of Ω is called an **algebra** (or **field**) if $A, B \in \mathcal{A}$ implies A^c and $A \cup B$ are in $\mathcal{A}$. Since $A \cap B = (A^c \cup B^c)^c$, it follows that $A \cap B \in \mathcal{A}$. Obviously a σ-algebra is an algebra. Two cases in which the converse is false are:

Example 1.1. $\Omega = \mathbf{Z} =$ the integers, $\mathcal{A} =$ the collection of $A \subset \mathbf{Z}$ so that A or A^c is finite.

Example 1.2. $\Omega = \mathbf{R}$, $\mathcal{A} =$ the collection of sets of the form

$$\cup_{i=1}^{k}(a_i, b_i] \quad \text{where } -\infty \leq a_i < b_i \leq \infty$$

EXERCISE 1.1. (i) Show that if $\mathcal{F}_1 \subset \mathcal{F}_2 \subset \ldots$ are σ-algebras then $\cup_i \mathcal{F}_i$ is an algebra. (ii) Give an example to show that $\cup_i \mathcal{F}_i$ need not be a σ-algebra.

EXERCISE 1.2. A set $A \subset \{1, 2, \ldots\}$ is said to have **asymptotic density** θ if

$$\lim_{n \to \infty} |A \cap \{1, 2, \ldots, n\}|/n = \theta$$

Let $\mathcal{A}$ be the collection of sets for which the asymptotic density exists. Is $\mathcal{A}$ a σ-algebra? an algebra?

By a **measure on an algebra** $\mathcal{A}$ we mean a set function μ with

(i) $\mu(A) \geq \mu(\emptyset) = 0$ for all $A \in \mathcal{A}$, and

(ii) if $A_i \in \mathcal{A}$ are disjoint *and their union is in* $\mathcal{A}$ then

$$\mu\left(\cup_{i=1}^{\infty} A_i\right) = \sum_{i=1}^{\infty} \mu(A_i)$$

The italicized clause is unnecessary if $\mathcal{A}$ is a σ-algebra so in that case the new definition coincides with the old one. The next exercise generalizes Exercise 1.1 in Chapter 1.

EXERCISE 1.3. We assume that all the sets mentioned are in $\mathcal{A}$.

(i) **monotonicity.** If $A \subset B$ then $\mu(A) \leq \mu(B)$.

(ii) **subadditivity.** If $A \subset \cup_i A_i$ then $\mu(A) \leq \sum_i \mu(A_i)$.

(iii) **continuity from below.** If $A_i \uparrow A$ (i.e., $A_1 \subset A_2, \ldots$ and $\cup_i A_i = A$) then $\mu(A_i) \uparrow \mu(A)$.

(iv) **continuity from above.** If $A_i \downarrow A$ (i.e., $A_1 \supset A_2, \ldots$ and $\cap_i A_i = A$) and $\mu(A_1) < \infty$ then $\mu(A_i) \downarrow \mu(A)$ as $i \uparrow \infty$.

μ is said to be σ-**finite** if there is a sequence of sets $A_n \in \mathcal{A}$, so that $\mu(A_n) < \infty$ and $\cup_n A_n = \Omega$. Letting $A'_1 = A_1$ and for $n \geq 2$

$$A'_n = \cup_{m=1}^{n} A_m \quad \text{or} \quad A'_n = A_n \cap \left(\cap_{m=1}^{n-1} A_m^c \right) \in \mathcal{A}$$

we can without loss of generality assume that $A_n \uparrow \Omega$ or the A_n are disjoint.

(1.1) **Carathéodory's Extension Theorem.** Let μ be a σ-finite measure on an algebra $\mathcal{A}$. Then μ has a unique extension to $\sigma(\mathcal{A}) =$ the smallest σ-algebra containing $\mathcal{A}$.

EXERCISE 1.4. Let $\mathbf{Z} =$ the integers and $\mathcal{A} =$ the collection of subsets so that A or A^c is finite. Let $\mu(A) = 0$ in the first case and $\mu(A) = 1$ in the second. Show that μ has no extension to $\sigma(\mathcal{A})$.

The next section is devoted to the proof of (1.1). To check the hypotheses of (1.1) for Lebesgue measure, we will prove a theorem, (1.3), that will be useful for other examples. To state that result we will need several definitions. A collection S of sets is said to be a **semialgebra** if (i) it is closed under intersection, i.e., $S, T \in S$ implies $S \cap T \in S$, and (ii) if $S \in S$ then S^c is a finite disjoint union of sets in S. An important example of a semialgebra is $\mathcal{R}_o^d =$ the collection of sets of the form

$$(a_1, b_1] \times \cdots \times (a_d, b_d] \subset \mathbf{R}^d \quad \text{where} \quad -\infty \leq a_i < b_i \leq \infty$$

EXERCISE 1.5. Show that $\sigma(\mathcal{R}_o^d) = \mathcal{R}^d$, the Borel subsets of $\mathbf{R}^d$.

(1.2) **Lemma.** If S is a semialgebra then $\bar{S} = \{$finite disjoint unions of sets in $S\}$ is an algebra, called the **algebra generated by** S.

Proof Suppose $A = +_i S_i$ and $B = +_j T_j$ where $+$ denotes disjoint union and we assume the index sets are finite. Then $A \cap B = +_{i,j} S_i \cap T_j \in \bar{S}$. As for complements, if $A = +_i S_i$ then $A^c = \cap_i S_i^c$. The definition of S implies $S_i^c \in \bar{S}$. We have shown that $\bar{S}$ is closed under intersection so it follows by induction that $A^c \in \bar{S}$. $\qquad\qquad\qquad\qquad\qquad\qquad\qquad\qquad\qquad\qquad\qquad\qquad\qquad$ □

Let λ denote Lebesgue measure. The definition gives the values of λ on a semi-algebra S (the half-open intervals). It is easy to see how to extend the definition to the algebra $\bar{S}$ defined in (1.2). We let $\lambda(+_i(a_i, b_i]) = \sum_i(b_i - a_i)$. To assert that λ has an extension to $\sigma(S) = \mathcal{R}$, we have to check that λ is a measure on $\bar{S}$, i.e., if $A \in \bar{S}$ is a countable disjoint union of sets $A_i \in \bar{S}$ then $\lambda(A) = \sum_i \lambda(A_i)$. The next result simplifies that task somewhat.

(1.3) Theorem. Let S be a semialgebra and let μ defined on S have $\mu(\emptyset) = 0$. Suppose (i) if $S \in \mathcal{S}$ is a finite disjoint union of sets $S_i \in \mathcal{S}$ then $\mu(S) = \sum_i \mu(S_i)$, and (ii) if $S_i, S \in \mathcal{S}$ with $S = +_{i \geq 1} S_i$ then $\mu(S) \leq \sum_i \mu(S_i)$. Then μ has a unique extension $\bar{\mu}$ that is a measure on $\bar{S}$ the algebra generated by S. If the extension is σ-finite then by (1.1) there is a unique extension ν that is a measure on $\sigma(S)$.

Remark. In (ii) above, and in what follows, $i \geq 1$ indicates a countable union, while a plain subscript i or j indicates a finite union.

Proof We define $\bar{\mu}$ on $\bar{S}$ by $\bar{\mu}(A) = \sum_i \mu(S_i)$ whenever $A = +_i S_i$. To check that $\bar{\mu}$ is well defined suppose that $A = +_j T_j$ and observe $S_i = +_j(S_i \cap T_j)$ and $T_j = +_i(S_i \cap T_j)$, so (i) implies

$$\sum_i \mu(S_i) = \sum_{i,j} \mu(S_i \cap T_j) = \sum_j \mu(T_j)$$

Our next result takes the first step toward proving that $\bar{\mu}$ is a measure on $\bar{S}$. It includes an extra statement, (b), that will be useful in checking (ii).

(1.4) Lemma. Suppose only that (i) holds.
(a) If $A, B_i \in \bar{S}$ with $A = +_{i=1}^n B_i$ then $\bar{\mu}(A) = \sum_i \bar{\mu}(B_i)$.
(b) If $A, B_i \in \bar{S}$ with $A \subset \cup_{i=1}^n B_i$ then $\bar{\mu}(A) \leq \sum_i \bar{\mu}(B_i)$.

Proof Observe that it follows from the definition that if $A = +_i B_i$ is a finite disjoint union of sets in $\bar{S}$ and $B_i = +_j S_{i,j}$ then

$$\bar{\mu}(A) = \sum_{i,j} \mu(S_{i,j}) = \sum_i \bar{\mu}(B_i)$$

To prove (b) we begin with the case $n = 1$, $B_1 = B$. $B = A + (B \cap A^c)$ and $B \cap A^c \in \bar{S}$ so

$$\bar{\mu}(A) \le \bar{\mu}(A) + \bar{\mu}(B \cap A^c) = \bar{\mu}(B)$$

To handle $n > 1$ now let $F_k = B_1^c \cap \ldots \cap B_{k-1}^c \cap B_k$ and note

$$\cup_i B_i = F_1 + \cdots + F_n$$
$$A = A \cap (\cup_i B_i) = (A \cap F_1) + \cdots + (A \cap F_n)$$

so using (a), (b) with $n = 1$, and (a) again

$$\bar{\mu}(A) = \sum_{k=1}^n \bar{\mu}(A \cap F_k) \le \sum_{k=1}^n \bar{\mu}(F_k) = \bar{\mu}(\cup_i B_i) \qquad \square$$

To extend the additivity property to $A \in \bar{S}$ that are countable disjoint unions $A = +_{i\ge 1} B_i$, where $B_i \in \bar{S}$, we observe that each $B_i = +_j S_{i,j}$ with $S_{i,j} \in S$ and $\sum_{i\ge 1} \bar{\mu}(B_i) = \sum_{i\ge 1} \mu(S_{i,j})$ so replacing the B_i's by $S_{i,j}$'s we can without loss of generality suppose that the $B_i \in S$. Now $A \in \bar{S}$, implies $A = +_j T_j$ (a finite disjoint union) and $T_j = +_{i\ge 1} T_j \cap B_i$, so (ii) implies

$$\mu(T_j) \le \sum_{i\ge 1} \mu(T_j \cap B_i)$$

Summing over j and observing that nonnegative numbers can be summed in any order

$$\bar{\mu}(A) = \sum_j \mu(T_j) \le \sum_{i\ge 1} \sum_j \mu(T_j \cap B_i) = \sum_{i\ge 1} \mu(B_i)$$

the last equality following from (i). To prove the opposite inequality, let $A_n = B_1 + \cdots + B_n$, and $C_n = A \cap A_n^c$. $C_n \in \bar{S}$, since $\bar{S}$ is an algebra, so finite additivity of $\bar{\mu}$ implies

$$\bar{\mu}(A) = \bar{\mu}(B_1) + \cdots + \bar{\mu}(B_n) + \bar{\mu}(C_n) \ge \bar{\mu}(B_1) + \cdots + \bar{\mu}(B_n)$$

and letting $n \to \infty$, $\bar{\mu}(A) \ge \sum_{i\ge 1} \bar{\mu}(B_i)$. $\qquad \square$

With (1.3) established we are ready to prove the existence of Lebesgue measure and a number of other measures.

(1.5) Theorem. Suppose F is (i) nondecreasing and (ii) right continuous, i.e., $F(y) \downarrow F(x)$ when $y \downarrow x$. There is a unique measure μ on $(\mathbf{R}, \mathcal{R})$ with $\mu((a, b]) = F(b) - F(a)$ for all a, b.

Remark. A function F that has properties (i) and (ii) is called a **Stieltjes measure function.** To see the reasons for the two conditions observe that (a) if μ is a measure then $F(b) - F(a) \geq 0$ and (b) part (iv) of Exercise 1.3 implies that if $\mu((a, y]) < \infty$ and $y \downarrow x > a$

$$F(y) - F(a) = \mu((a, y]) \downarrow \mu((a, x]) = F(x) - F(a)$$

Conversely, if μ is a measure on **R** with $\mu((a, b]) < \infty$ when $-\infty < a < b < \infty$ then

$$F(x) = \begin{cases} c + \mu((0, x]) & \text{for } x \geq 0 \\ c - \mu((x, 0]) & \text{for } x < 0 \end{cases}$$

is a function F with $F(b) - F(a) = \mu((a, b])$, and any such function has this form with $c = F(0)$.

Proof Let $\mathcal{S}$ be the semi-algebra of half-open intervals $(a, b]$ with $-\infty \leq a < b \leq \infty$. To define μ on $\mathcal{S}$, we begin by observing that

$$F(\infty) = \lim_{x \uparrow \infty} F(x) \quad \text{and} \quad F(-\infty) = \lim_{x \downarrow -\infty} F(x) \quad \text{exist}$$

and $\mu((a, b]) = F(b) - F(a)$ makes sense for all $-\infty \leq a < b \leq \infty$ since $F(\infty) > -\infty$ and $F(-\infty) < \infty$.

If $(a, b] = +_{i=1}^{n} (a_i, b_i]$ then after relabelling the intervals we must have $a_1 = a$, $b_n = b$, and $a_i = b_{i-1}$ for $2 \leq i \leq n$, so condition (i) in (1.3) holds. To check (ii) suppose first that $-\infty < a < b < \infty$, and $(a, b] \subset \cup_{i \geq 1} (a_i, b_i]$ where (without loss of generality) $-\infty < a_i < b_i < \infty$. Pick $\delta > 0$ so that $F(a + \delta) < F(a) + \epsilon$ and pick η_i so that

$$F(b_i + \eta_i) < F(b_i) + \epsilon 2^{-i}$$

The open intervals $(a_i, b_i + \eta_i)$ cover $[a + \delta, b]$ so there is a finite subcover (α_j, β_j), $1 \leq j \leq J$. Since $(a + \delta, b] \subset \cup_{j=1}^{J} (\alpha_j, \beta_j]$, (b) in (1.4) implies

$$F(b) - F(a + \delta) \leq \sum_{j=1}^{J} F(\beta_j) - F(\alpha_j) \leq \sum_{i=1}^{\infty} (F(b_i + \eta_i) - F(a_i))$$

So by the choice of δ and η_i

$$F(b) - F(a) \leq 2\epsilon + \sum_{i=1}^{\infty} (F(b_i) - F(a_i))$$

and since ϵ is arbitrary we have proved the result in the case $-\infty < a < b < \infty$. To remove the last restriction observe that if $(a, b] \subset \cup_i (a_i, b_i]$ and $(A, B] \subset (a, b]$ has $-\infty < A < B < \infty$ then we have

$$F(B) - F(A) \le \sum_{i=1}^{\infty} (F(b_i) - F(a_i))$$

Since the last result holds for any finite $(A, B] \subset (a, b]$, so the desired result follows.

$$\square$$

Our next goal is to prove a version of (1.5) for $\mathbf{R}^d$. The first step is to introduce the assumptions on the defining function f.

(i) It is nondecreasing, i.e., if $x \le y$ (meaning $x_i \le y_i$ for all i) then $F(x) \le F(y)$.

(ii) F is right continuous, i.e., $\lim_{y \downarrow x} F(y) = F(x)$ (here $y \downarrow x$ means each coordinate $y_i \downarrow x_i$).

To formulate the third and final condition, let

$$A = (a_1, b_1] \times \cdots \times (a_d, b_d]$$
$$V = \{a_1, b_1\} \times \cdots \times \{a_d, b_d\}$$

where $-\infty < a_i < b_i < \infty$. To emphasize that ∞'s are not allowed, we will call A a finite rectangle. Then $V =$ the vertices of the rectangle A. If $v \in V$ let

$$\text{sgn}\,(v) = (-1)^{\# \text{ of } a\text{'s in } v}$$
$$\Delta_A F = \sum_{v \in V} \text{sgn}\,(v) F(v)$$

We will let $\mu(A) = \Delta_A F$ so we must assume

(iii) $\Delta_A F \ge 0$ for all rectangles A.

To see the reason for this definition consider the special case $d = 2$ and then divide one large rectangle into four small ones. For more on this assumption see Section 2.9.

Example 1.3. Suppose $F(x) = \prod_{i=1}^{d} F_i(x)$ where the F_i satisfy (i) and (ii) of (1.5). In this case

$$\Delta_A F = \prod_{i=1}^{d} (F_i(b_i) - F_i(a_i))$$

When $F_i(x) = x$ for all i the resulting measure is Lebesgue measure on $\mathbf{R}^d$.

(1.6) Theorem. Suppose $F : \mathbf{R}^d \to [0,1]$ satisfies (i)–(iii) in Section 2.9. Then there is a unique probability measure μ on $(\mathbf{R}^d, \mathcal{R}^d)$ so that $\mu(A) = \Delta_A F$ for all finite rectangles.

Proof Let $\mathcal{S}$ be the semialgebra of rectangles $A = (a, b]$ where $-\infty \le a_i < b_i \le \infty$. We let $\mu(A) = \Delta_A F$ for all finite rectangles and then use monotonicity to extend the definition to all rectangles.

To check (i), call $A = +_k B_k$ a **regular subdivision** of A if there are sequences $a_i = \alpha_{i,0} < \alpha_{i,1} \dots < \alpha_{i,n_i} = b_i$ so that each rectangle B_k has the form

$$(\alpha_{1,j_1-1}, \alpha_{1,j_1}] \times \cdots \times (\alpha_{d,j_d-1}, \alpha_{d,j_d}] \quad \text{where} \quad 1 \le j_i \le n_i$$

It is easy to see that for regular subdivisions $\lambda(A) = \sum_k \lambda(B_k)$. [First consider the case in which all the endpoints are finite and then take limits to get the general case.] To extend this result to a general finite subdivision $A = +_j A_j$ subdivide further to get a regular one.

The proof of (ii) is almost identical to that in (1.5). To make things easier to write and to bring out the analogies with (1.5), we let

$$(x, y) = (x_1, y_1) \times \cdots \times (x_d, y_d)$$
$$(x, y] = (x_1, y_1] \times \cdots \times (x_d, y_d]$$
$$[x, y] = [x_1, y_1] \times \cdots \times [x_d, y_d]$$

for $x, y \in \mathbf{R}^d$. Suppose first that $-\infty < a < b < \infty$, where the inequalities mean that each component is finite, and suppose $(a, b] \subset \cup_{i \ge 1}(a^i, b^i]$ where (without loss of generality) $-\infty < a^i < b^i < \infty$. Let $\bar{1} = (1, \dots, 1)$, pick $\delta > 0$ so that

$$\mu((a + \delta\bar{1}, b]) < \mu((a, b]) + \epsilon$$

and pick η_i so that

$$\mu((a, b^i + \eta_i\bar{1}]) < \mu((a^i, b^i]) + \epsilon 2^{-i}$$

The open rectangles $(a^i, b^i + \eta_i\bar{1})$ cover $[a + \delta\bar{1}, b]$ so there is a finite subcover (α^j, β^j), $1 \le j \le J$. Since $(a + \delta\bar{1}, b] \subset \cup_{j=1}^{J}(\alpha^j, \beta^j]$, (b) in (1.4) implies

$$\mu([a + \delta\bar{1}, b]) \le \sum_{j=1}^{J} \mu((\alpha^j, \beta^j]) \le \sum_{i=1}^{\infty} \mu((a^i, b^i + \eta_i\bar{1}])$$

So by the choice of δ and η_i

$$\mu((a, b]) \le 2\epsilon + \sum_{i=1}^{\infty} \mu((a^i, b^i])$$

and since ϵ is arbitrary we have proved the result in the case $-\infty < a < b < \infty$. The proof can now be completed exactly as before. $\qquad\square$

A.2. Carathéodary's Extension Theorem

This section is devoted to the proof of (1.1). The proof is slick but rather mysterious. The reader should not worry too much about the details but concentrate on the structure of the proof and the definitions introduced.

Uniqueness. We will prove that the extension is unique before tacking the more difficult problem of proving its existence. The key to our uniqueness proof is Dynkin's $\pi - \lambda$ theorem, a result that we will use many times in the book. As usual, we need a few definitions before we can state the result. $\mathcal{P}$ is said to be a π-**system** if it is closed under intersection, i.e., if $A, B \in \mathcal{P}$ then $A \cap B \in \mathcal{P}$. For example, the collection of rectangles $(a_1, b_1] \times \cdots \times (a_d, b_d]$ is a π-system. $\mathcal{L}$ is said to be a λ-**system** if it satisfies: (i) $\Omega \in \mathcal{L}$. (ii) If $A, B \in \mathcal{L}$ and $A \subset B$ then $B - A \in \mathcal{L}$. (iii) If $A_n \in \mathcal{L}$ and $A_n \uparrow A$ then $A \in \mathcal{L}$. The reader will see in a moment that the next result is just what we need to prove uniqueness of the extension.

(2.1) $\pi - \lambda$ **Theorem.** If $\mathcal{P}$ is a π-system and $\mathcal{L}$ is a λ-system that contains $\mathcal{P}$ then $\sigma(\mathcal{P}) \subset \mathcal{L}$.

Proof: We will show that

(a) if $\ell(\mathcal{P})$ is the smallest λ-system containing $\mathcal{P}$ then $\ell(\mathcal{P})$ is a σ-field.

The desired result follows from (a). To see this note that since $\sigma(\mathcal{P})$ is the smallest σ-field and $\ell(\mathcal{P})$ is the smallest λ-system containing $\mathcal{P}$ we have

$$\sigma(\mathcal{P}) \subset \ell(\mathcal{P}) \subset \mathcal{L}$$

To prove (a) we begin by noting that a λ-system that is closed under intersection is a σ-field since

$$\text{if } A \in \mathcal{L} \text{ then } A^c = \Omega - A \in \mathcal{L}$$
$$A \cup B = (A^c \cap B^c)^c$$
$$\cup_{i=1}^n A_i \uparrow \cup_{i=1}^\infty A_i \text{ as } n \uparrow \infty$$

Thus it is enough to show

(b) $\ell(\mathcal{P})$ is closed under intersection.

To prove (b) we let $\mathcal{G}_A = \{B : A \cap B \in \ell(\mathcal{P})\}$ and prove

(c) if $A \in \ell(\mathcal{P})$ then $\mathcal{G}_A$ is a λ-system.

To check this we note: (i) $\Omega \in \mathcal{G}_A$ since $A \in \ell(\mathcal{P})$.

(ii) if $B, C \in \mathcal{G}_A$ and $B \subset C$ then $A \cap (B - C) = (A \cap B) - (A \cap C) \in \ell(\mathcal{P})$ since $A \cap B, A \cap C \in \ell(\mathcal{P})$ and $\ell(\mathcal{P})$ is a λ-system.

(iii) if $B_n \in \mathcal{G}_A$ and $B_n \uparrow B$ then $A \cap B_n \uparrow A \cap B \in \ell(\mathcal{P})$ since $A \cap B_n \in \ell(\mathcal{P})$ and $\ell(\mathcal{P})$ is a λ-system.

To get from (c) to (b) we note that since $\mathcal{P}$ is a π-system

if $A \in \mathcal{P}$ then $\mathcal{G}_A \supset \mathcal{P}$ and so (c) implies $\mathcal{G}_A \supset \ell(\mathcal{P})$

i.e., if $A \in \mathcal{P}$ and $B \in \ell(\mathcal{P})$ then $A \cap B \in \ell(\mathcal{P})$. Interchanging A and B in the last sentence: if $A \in \ell(\mathcal{P})$ and $B \in \mathcal{P}$ then $A \cap B \in \ell(\mathcal{P})$ but this implies

if $A \in \ell(\mathcal{P})$ then $\mathcal{G}_A \supset \mathcal{P}$ and so (c) implies $\mathcal{G}_A \supset \ell(\mathcal{P})$.

This conclusion implies that if $A, B \in \ell(\mathcal{P})$ then $A \cap B \in \ell(\mathcal{P})$, which proves (b) and completes the proof.   □

To prove that the extension in (1.1) is unique, we will show:

(2.2) Theorem. Let $\mathcal{P}$ be a π-system. If ν_1 and ν_2 are measures (on σ-fields $\mathcal{F}_1$ and $\mathcal{F}_2$) that agree on $\mathcal{P}$ and there is a sequence $A_n \in \mathcal{P}$ with $A_n \uparrow \Omega$ and $\nu_i(A_n) < \infty$, then ν_1 and ν_2 agree on $\sigma(\mathcal{P})$.

Proof   Let $A \in \mathcal{P}$ have $\nu_1(A) = \nu_2(A) < \infty$. Let

$$\mathcal{L} = \{B \in \sigma(\mathcal{P}) : \nu_1(A \cap B) = \nu_2(A \cap B)\}$$

We will now show that $\mathcal{L}$ is a λ-system. Since $A \in \mathcal{P}$, $\nu_1(A) = \nu_2(A)$ and $\Omega \in \mathcal{L}$. If $B, C \in \mathcal{L}$ with $C \subset B$ then

$$\nu_1(A \cap (B - C)) = \nu_1(A \cap B) - \nu_1(A \cap C)$$
$$= \nu_2(A \cap B) - \nu_2(A \cap C) = \nu_2(A \cap (B - C))$$

Here we use the fact that $\nu_i(A) < \infty$ to justify the subtraction. Finally if $B_n \in \mathcal{L}$ and $B_n \uparrow B$, then part (iii) of Exercise 1.3 implies

$$\nu_1(A \cap B) = \lim_{n \to \infty} \nu_1(A \cap B_n) = \lim_{n \to \infty} \nu_2(A \cap B_n) = \nu_2(A \cap B)$$

Since $\mathcal{P}$ is closed under intersection by assumption, the $\pi - \lambda$ theorem implies $\mathcal{L} \supset \sigma(\mathcal{P})$, i.e., if $A \in \mathcal{P}$ with $\nu_1(A) = \nu_2(A) < \infty$ and $B \in \sigma(\mathcal{P})$ then $\nu_1(A \cap B) = \nu_2(A \cap B)$. Letting $A_n \in \mathcal{P}$ with $A_n \uparrow \Omega$, $\nu_1(A_n) = \nu_2(A_n) < \infty$, and using the last result and part (iii) of Exercise 1.3, we have the desired conclusion.

□

EXERCISE 2.1. Give an example of two probability measures $\mu \neq \nu$ on $\mathcal{F} = $ all subsets of $\{1, 2, 3, 4\}$ that agree on a collection of sets $\mathcal{C}$ with $\sigma(\mathcal{C}) = \mathcal{F}$, i.e., the smallest σ-algebra containing $\mathcal{C}$ is $\mathcal{F}$.

Existence. Our next step is to show that a measure (not necessarily σ-finite) defined on an algebra $\mathcal{A}$ has an extension to the σ-algebra generated by $\mathcal{A}$. If $E \subset \Omega$, we let $\mu^*(E) = \inf \sum_i \mu(A_i)$ where the infimum is taken over all sequences from $\mathcal{A}$ so that $E \subset \cup_i A_i$. Intuitively, if ν is a measure that agrees with μ on $\mathcal{A}$ then it follows from part (ii) of Exercise 1.3 that

$$\nu(E) \leq \nu(\cup_i A_i) \leq \sum_i \nu(A_i) = \sum_i \mu(A_i)$$

so $\mu^*(E)$ is an upper bound on the measure of E. Intuitively, the measurable sets are the ones for which the upper bound is tight. Formally, we say that E is **measurable** if

$$(*) \qquad \mu^*(F) = \mu^*(F \cap E) + \mu^*(F \cap E^c) \quad \text{for all sets } F \subset \Omega$$

The last definition is not very intuitive but we will see in the proofs below that it works very well.

It is immediate from the definition that μ^* has the following properties:

(i) **monotonicity.** If $E \subset F$ then $\mu^*(E) \leq \mu^*(F)$.

(ii) **subadditivity.** If $F \subset \cup_i F_i$, a countable union, then $\mu^*(F) \leq \sum_i \mu^*(F_i)$.

Any set function with $\mu^*(\emptyset) = 0$ that satisfies (i) and (ii) is called an **outer measure.** Using (ii) with $F_1 = F \cap E$ and $F_2 = F \cap E^c$ (and $F_i = \emptyset$ otherwise) we see that to prove a set is measurable it is enough to show

$$(*') \qquad \mu^*(F) \geq \mu^*(F \cap E) + \mu^*(F \cap E^c)$$

We begin by showing that our new definition extends the old one.

(2.3) Lemma. If $A \in \mathcal{A}$ then $\mu^*(A) = \mu(A)$ and A is measurable.

Proof   Part (ii) of Exercise 1.3 implies that if $A \subset \cup_i A_i$ then

$$\mu(A) \le \sum_i \mu(A_i)$$

so $\mu(A) \le \mu^*(A)$. Of course we can always take $A_1 = A$ and the other $A_i = \emptyset$ so $\mu^*(A) \le \mu(A)$.

To prove that any $A \in \mathcal{A}$ is measurable, we begin by noting that the inequality is $(*')$ trivial when $\mu^*(F) = \infty$, so we can without loss of generality assume $\mu^*(F) < \infty$. To prove that $(*')$ holds when $E = A$ we observe that since $\mu^*(F) < \infty$ there is a sequence $B_i \in \mathcal{A}$ so that $\cup_i B_i \supset F$ and

$$\sum_i \mu(B_i) \le \mu^*(F) + \epsilon$$

Since μ is additive on $\mathcal{A}$ and $\mu = \mu^*$ on $\mathcal{A}$ we have

$$\mu(B_i) = \mu^*(B_i \cap A) + \mu^*(B_i \cap A^c)$$

Summing over i and using the subadditivity of μ^* gives

$$\mu^*(F) + \epsilon \ge \sum_i \mu^*(B_i \cap A) + \sum_i \mu^*(B_i \cap A^c) \ge \mu^*(F \cap A) + \mu^*(F^c \cap A)$$

which proves the desired result since ϵ is arbitrary. $\square$

(2.4) Lemma. The class $\mathcal{A}^*$ of measurable sets is a σ-field and the restriction of μ^* to $\mathcal{A}^*$ is a measure.

Remark. This result is true for any outer measure.

Proof   It is clear from the definition that

(a) If E is measurable then E^c is.

Our first nontrivial task is to prove:

(b) If E_1 and E_2 are measurable then $E_1 \cup E_2$ and $E_1 \cap E_2$ are.

Proof of (b)   To prove the first conclusion let G be any subset of Ω. Using subadditivity, the measurability of E_2 (let $F = G \cap E_1^c$ in $(*)$) and the measurability of E_1 we get

$$\mu^*(G \cap (E_1 \cup E_2)) + \mu^*(G \cap (E_1^c \cap E_2^c))$$
$$\le \mu^*(G \cap E_1) + \mu^*(G \cap E_1^c \cap E_2) + \mu^*(G \cap E_1^c \cap E_2^c)$$
$$= \mu^*(G \cap E_1) + \mu^*(G \cap E_1^c) = \mu^*(G)$$

To prove that $E_1 \cap E_2$ is measurable, we observe $E_1 \cap E_2 = (E_1^c \cup E_2^c)^c$ and use (a). □

(c) Let $G \subset \Omega$ and $E_1, \ldots, E_n$ be disjoint measurable sets. Then

$$\mu^*(G \cap \cup_{i=1}^n E_i) = \sum_{i=1}^n \mu^*(G \cap E_i)$$

Proof of (c) Let $F_m = \cup_{i \le m} E_i$. E_n is measurable, $F_n \supset E_n$, and $F_{n-1} \cap E_n = \emptyset$ so

$$\mu^*(G \cap F_n) = \mu^*(G \cap F_n \cap E_n) + \mu^*(G \cap F_n \cap E_n^c)$$
$$= \mu^*(G \cap E_n) + \mu^*(G \cap F_{n-1})$$

The desired result follows from this by induction. □

(d) If the sets E_i are measurable then $E = \cup_{i=1}^\infty E_i$ is measurable.

Proof of (d) Let $E_i' = E_i \cap (\cap_{j<i} E_j^c)$. (a) and (b) imply E_i' is measurable so we can suppose without loss of generality that the E_i are pairwise disjoint. Let $F_n = E_1 \cup \ldots \cup E_n$. F_n is measurable by (b) so using monotonicity and (c) we have

$$\mu^*(G) = \mu^*(G \cap F_n) + \mu^*(G \cap F_n^c) \ge \mu^*(G \cap F_n) + \mu^*(G \cap E^c)$$
$$= \sum_{i=1}^n \mu^*(G \cap E_i) + \mu^*(G \cap E^c)$$

Letting $n \to \infty$ and using subadditivity

$$\mu^*(G) \ge \sum_{i=1}^\infty \mu^*(G \cap E_i) + \mu^*(G \cap E^c) \ge \mu^*(G \cap E) + \mu^*(G \cap E^c)$$

which is $(*')$. □

The last step in the proof of (2.4) is

(e) If $E = \cup_i E_i$ where $E_1, E_2, \ldots$ are disjoint and measurable then

$$\mu^*(E) = \sum_{i=1}^\infty \mu^*(E_i)$$

Proof of (e) Let $F_n = E_1 \cup \ldots \cup E_n$. By monotonicity and (c)

$$\mu^*(E) \ge \mu^*(F_n) = \sum_{i=1}^n \mu^*(E_i)$$

Letting $n \to \infty$ now and using subadditivity gives the desired conclusion. □

A.3. Completion, etc.

The proof of (1.1) given in the last section defines an extension to $\mathcal{A}^* \supset \sigma(\mathcal{A})$. Our next goal is to describe the relationship between these two σ-algebras. Let $\mathcal{A}_\sigma$ denote the collection of countable unions of sets in $\mathcal{A}$, and $\mathcal{B}_\delta$ denote the collection of countable intersections of sets in $\mathcal{B}$. Taking $\mathcal{B} = \mathcal{A}_\sigma$ we see that $\mathcal{A}_{\sigma\delta}$ denotes the collection of countable intersections of sets in $\mathcal{A}_\sigma$.

(3.1) **Lemma.** Let E be any set with $\mu^*(E) < \infty$. (i) For any $\epsilon > 0$, there is an $A \in \mathcal{A}_\sigma$ with $A \supset E$ and $\mu^*(A) \leq \mu^*(E) + \epsilon$. (ii) There is a $B \in \mathcal{A}_{\sigma\delta}$ with $B \supset E$ and $\mu^*(B) = \mu^*(E)$.

Proof   By the definition of μ^*, there is a sequence A_i so that $A \equiv \cup_i A_i \supset E$ and $\sum_i \mu(A_i) \leq \mu^*(E) + \epsilon$. The definition of μ^* implies $\mu^*(A) \leq \sum_i \mu(A_i)$, establishing (i). For (ii), let $A_n \in \mathcal{A}_\sigma$ with $A_n \supset E$ and $\mu^*(A_n) \leq \mu^*(E) + 1/n$, and let $B = \cap_n A_n$. Clearly $B \in \mathcal{A}_{\sigma\delta}$, $B \supset E$, and hence by monotonicity, $\mu^*(B) \geq \mu^*(E)$. To prove the other inequality notice that $B \subset A_n$ and hence $\mu^*(B) \leq \mu^*(A_n) \leq \mu^*(E) + 1/n$ for any n. $\square$

EXERCISE 3.1. Let $\mathcal{A}$ be an algebra, μ a measure on $\sigma(\mathcal{A})$, and $B \in \sigma(\mathcal{A})$ with $\mu(B) < \infty$. For any $\epsilon > 0$ there is an $A \in \mathcal{A}$ with $\mu(A \triangle B) < \epsilon$, where $A \triangle B = (A - B) \cup (B - A)$.

(3.2) **Theorem.** Suppose μ is σ-finite on $\mathcal{A}$. $B \in \mathcal{A}^*$ if and only if there is an $A \in \mathcal{A}_{\sigma\delta}$ and a set N with $\mu^*(N) = 0$ so that $B = A - N (= A \cap N^c)$.

Proof   It follows from (2.3) and (2.4) if $A \in \mathcal{A}_{\sigma\delta}$ then $A \in \mathcal{A}^*$. $(*')$ in Section A.2 and monotonicity imply sets with $\mu^*(N) = 0$ are measurable, so using (2.4) again it follows that $A \cap N^c \in \mathcal{A}^*$. To prove the other direction, let Ω_i be a disjoint collection of sets with $\mu(\Omega_i) < \infty$ and $\Omega = \cup_i \Omega_i$. Let $B_i = B \cap \Omega_i$ and use (3.1) to find $A_i^n \in \mathcal{A}_\sigma$ so that $A_i^n \supset B_i$ and $\mu(A_i^n) \leq \mu^*(E_i) + 1/n2^i$. Let $A_n = \cup_i A_i^n$. $B \subset A_n$ and

$$A_n - B \subset \sum_{i=1}^{\infty} (A_i^n - B_i)$$

so by subadditivity

$$\mu^*(A_n - B) \leq \sum_{i=1}^{\infty} \mu^*(A_i^n - B_i) \leq 1/n$$

Since $A_n \in \mathcal{A}_\sigma$, the set $A = \cap_n A_n \in \mathcal{A}_{\sigma\delta}$. Clearly $A \supset B$. Since $N \equiv A - B \subset A_n - B$ for all n, monotonicity implies $\mu^*(N) = 0$ and the proof of (3.2) is complete. $\square$

A measure space $(\Omega, \mathcal{F}, \mu)$ is said to be **complete** if $\mathcal{F}$ contains all subsets of sets of measure 0. In the proof of (3.2) we showed that $(\Omega, \mathcal{A}^*, \mu^*)$ is complete. Our next result shows that $(\Omega, \mathcal{A}^*, \mu^*)$ is the completion of $(\Omega, \sigma(\mathcal{A}), \mu)$.

(3.3) Theorem. If $(\Omega, \mathcal{F}, \mu)$ is a measure space then there is a complete measure space $(\Omega, \bar{\mathcal{F}}, \bar{\mu})$, called the **completion** of $(\Omega, \mathcal{F}, \mu)$, so that: (i) $E \in \bar{\mathcal{F}}$ if and only if $E = A \cup B$ where $A \in \mathcal{F}$ and $B \subset N \in \mathcal{F}$ with $\mu(N) = 0$, (ii) $\bar{\mu}$ agrees with μ on $\mathcal{F}$.

Proof The first step is to check that $\bar{\mathcal{F}}$ is a σ-algebra. If $E_i = A_i \cup B_i$ where $A_i \in \mathcal{F}$ and $B_i \subset N_i$ where $\mu(N_i) = 0$ then $\cup_i A_i \in \mathcal{F}$ and subadditivity implies $\mu(\cup_i N_i) \leq \sum_i \mu(N_i) = 0$, so $\cup_i E_i \in \bar{\mathcal{F}}$. As for complements, if $E = A \cup B$ and $B \subset N$ then $B^c \supset N^c$ so

$$E^c = A^c \cap B^c = (A^c \cap N^c) \cup (A^c \cap B^c \cap N)$$

$A^c \cap N^c$ is in $\mathcal{F}$ and $A^c \cap B^c \cap N \subset N$ so $E^c \in \bar{\mathcal{F}}$.

We define $\bar{\mu}$ in the obvious way: if $E = A \cup B$ where $A \in \mathcal{F}$ and $B \subset N$ where $\mu(N) = 0$ then we let $\bar{\mu}(E) = \mu(A)$. The first thing to show is that $\bar{\mu}$ is well defined, i.e., if $E = A_i \cup B_i$, $i = 1, 2$, are two decompositions then $\mu(A_1) = \mu(A_2)$. Let $A_0 = A_1 \cap A_2$ and $B_0 = B_1 \cup B_2$. $E = A_0 \cup B_0$ is a third decomposition with $A_0 \in \mathcal{F}$ and $B_0 \subset N_1 \cup N_2$, and has the pleasant property that if $i = 1$ or 2

$$\mu(A_0) \leq \mu(A_i) \leq \mu(A_0) + \mu(N_1 \cup N_2) = \mu(A_0)$$

The last detail is to check that $\bar{\mu}$ is measure but that is easy. If $E_i = A_i \cup B_i$ are disjoint then $\cup_i E_i$ can be decomposed as $\cup_i A_i \cup (\cup_i B_i)$ and the $A_i \subset E_i$ are disjoint so

$$\bar{\mu}(\cup_i E_i) = \mu(\cup_i A_i) = \sum_i \mu(A_i) = \sum_i \bar{\mu}(E_i) \qquad \square$$

(1.6) allows us to construct Lebesgue measure λ on $(\mathbf{R}^d, \mathcal{R}^d)$. Using (3.3) we can extend λ to be a measure on $(\mathbf{R}, \bar{\mathcal{R}}^d)$ where $\bar{\mathcal{R}}^d$ is the completion of $\mathcal{R}^d$. Having done this it is natural (if somewhat optimistic) to ask: Are there any sets that are not in $\bar{\mathcal{R}}^d$? The answer is Yes and we will now give an example of a nonmeasurable B in $\mathbf{R}$.

A nonmeasurable subset of $[0,1)$

The key to our construction is the observation that λ is translation invariant: i.e., if $A \in \bar{\mathcal{R}}$ and $x + A = \{x + y : y \in A\}$ then $x + A \in \bar{\mathcal{R}}$ and $\lambda(A) = \lambda(x + A)$.

We say that $x, y \in [0, 1)$ are equivalent and write $x \sim y$ if $x - y$ is a rational number. By the axiom of choice there is a set B that contains exactly one element from each equivalence class. B is our nonmeasurable set. That is,

(3.4) Theorem. $B \notin \bar{\mathcal{R}}$.

Proof   The key is the following

(3.5) Lemma. If $E \subset [0, 1)$ is in $\bar{\mathcal{R}}$, $x \in (0, 1)$, and $x +' E = \{(x + y) \bmod 1 : y \in E\}$ then $\lambda(E) = \lambda(x +' E)$.

Proof   Let $A = E \cap [0, 1 - x)$ and $B = E \cap [1 - x, 1)$. Let $A' = x + A = \{x + y : y \in A\}$ and $B' = x - 1 + B$. $A, B \in \bar{\mathcal{R}}$, so by translation invariance $A', B' \in \bar{\mathcal{R}}$ and $\lambda(A) = \lambda(A')$, $\lambda(B) = \lambda(B')$. Since $A' \subset [x, 1)$ and $B' \subset [0, x)$ are disjoint,

$$\lambda(E) = \lambda(A) + \lambda(B) = \lambda(A') + \lambda(B') = \lambda(x +' E) \qquad \square$$

From (3.5), it follows easily that B is not measurable; for if it were, then $q +' B$, $q \in \mathbf{Q} \cap [0, 1)$ would be a countable disjoint collection of measurable subsets of $[0,1)$ all with the same measure α and having

$$\cup_{q \in \mathbf{Q} \cap [0,1)} (q +' B) = [0, 1)$$

If $\alpha > 0$ then $\lambda([0, 1)) = \infty$, and if $\alpha = 0$ then $\lambda([0, 1)) = 0$. Neither conclusion is compatible with the fact that $\lambda([0, 1)) = 1$ so $B \notin \bar{\mathcal{R}}$. $\qquad \square$

EXERCISE 3.2. Let B be the nonmeasurable set constructed in (3.4). (i) Let $B_q = q +' B$ and show that if $D_q \subset B_q$ is measurable then $\lambda(D_q) = 0$. (ii) Use (i) to conclude that if $A \subset \mathbf{R}$ has $\lambda(A) > 0$, there is a nonmeasurable $S \subset A$.

Letting $B' = B \times [0, 1]^{d-1}$ where B is our nonmeasurable subset of $(0,1)$, we get a nonmeasurable set in $d > 1$. In $d = 3$ there is a much more interesting example, but we need the reader to do some preliminary work. In Euclidean geometry, two subsets of $\mathbf{R}^d$ are said to be **congruent** if one set can be mapped onto the other by translations and rotations. We

Claim. Two congruent measurable sets must have the same Lebesgue measure.

EXERCISE 3.3. Prove the claim in $d = 2$ by showing (i) if B is a rotation of a rectangle A then $\lambda^*(B) = \lambda(A)$. (ii) If C is congruent to D then $\lambda^*(C) = \lambda^*(D)$.

Banach-Tarski Theorem

Banach and Tarski (1924) used the axiom of choice to show that it is possible to partition the sphere $\{x : |x| \le 1\}$ in $\mathbf{R}^3$ into a finite number of sets $A_1, \ldots, A_n$ and find congruent sets $B_1, \ldots, B_n$ whose union is two disjoint spheres of radius 1! Since congruent sets have the same Lebesgue measure, at least one of the sets A_i must be nonmeasurable. The construction relies on the fact that the group generated by rotations in $\mathbf{R}^3$ is not Abelian. Lindenbaum (1926) showed that this cannot be done with any bounded set in $\mathbf{R}^2$. For a popular account of the Banach-Tarski theorem, see French (1988).

Solovay's Theorem

The axiom of choice played an important role in the last two constructions of nonmeasurable sets. Solovay (1970) proved that its use is unavoidable. In his own words, "We show that the existence of a non-Lebesgue measurable set cannot be proved in Zermelo-Frankel set theory if the use of the axiom of choice is disallowed." This should convince the reader that all subsets of $\mathbf{R}^d$ that arise "in practice" are in $\bar{\mathcal{R}}^d$.

A.4. Integration

Let μ be a σ-finite measure on $(\Omega, \mathcal{F})$. In this section we will define $\int f \, d\mu$ for a class of measurable functions. This is a four step procedure:

Step 1. φ is said to be a **simple function** if $\varphi(\omega) = \sum_{i=1}^n a_i 1_{A_i}$ and A_i are disjoint sets with $\mu(A_i) < \infty$. If φ is a simple function we let

$$\int \varphi \, d\mu = \sum_{i=1}^n a_i \mu(A_i)$$

The representation of φ is not unique since we have not supposed that the a_i are distinct. However, it is easy to see that the last definition does not contradict itself.

We will prove the next three conclusions four times, but before we can state them for the first time we need a definition. $\varphi \ge \psi$ μ **almost everywhere** (or $\varphi \ge \psi$ μ-a.e.) means $\mu(\{\omega : \varphi(\omega) < \psi(\omega)\}) = 0$. When there is no doubt about what measure we are referring to we drop the μ.

(4.1) Lemma. Let φ and ψ be simple functions.
   (i) If $\varphi \ge 0$ a.e. then $\int \varphi \, d\mu \ge 0$.

(ii) For any $a \in \mathbf{R}$, $\int a\varphi \, d\mu = a \int \varphi \, d\mu$.

(iii) $\int \varphi + \psi \, d\mu = \int \varphi \, d\mu + \int \psi \, d\mu$.

Proof   (i) and (ii) are immediate consequences of the definition. To prove (iii) suppose

$$\varphi = \sum_{i=1}^{m} a_i 1_{A_i} \quad \text{and} \quad \psi = \sum_{j=1}^{n} b_j 1_{B_j}$$

To make the supports of the two functions the same we let $A_0 = \cup_i B_i - \cup_i A_i$, let $B_0 = \cup_i A_i - \cup_i B_i$, and let $a_0 = b_0 = 0$. Now

$$\varphi + \psi = \sum_{i=0}^{m} \sum_{j=0}^{n} (a_i + b_j) 1_{(A_i \cap B_j)}$$

and the $A_i \cap B_j$ are pairwise disjoint, so

$$\int (\varphi + \psi) \, d\mu = \sum_{i=0}^{m} \sum_{j=0}^{n} (a_i + b_j) \mu(A_i \cap B_j)$$

$$= \sum_{i=0}^{m} \sum_{j=0}^{n} a_i \mu(A_i \cap B_j) + \sum_{j=0}^{n} \sum_{i=0}^{m} b_j \mu(A_i \cap B_j)$$

$$= \sum_{i=0}^{m} a_i \, \mu(A_i) + \sum_{j=0}^{n} b_j \, \mu(B_j) = \int \varphi \, d\mu + \int \psi \, d\mu$$

In the next to last step we used $A_i = +_j (A_i \cap B_j)$ and $B_j = +_i (A_i \cap B_j)$, where $+$ denotes a disjoint union. $\qquad\square$

We will prove (i)–(iii) three more times as we generalize our integral. See (4.3), (4.5), and (4.7). As a consequence of (i)-(iii) we get three more useful properties. To keep from repeating their proofs, which do not change, we will prove

(4.2) Lemma. If (i) and (iii) hold then we have:

(iv) If $\varphi \le \psi$ a.e. then $\int \varphi \, d\mu \le \int \psi \, d\mu$.

(v) If $\varphi = \psi$ a.e. then $\int \varphi \, d\mu = \int \psi \, d\mu$.

If, in addition, (ii) holds when $a = -1$ we have

(vi) $|\int \varphi \, d\mu| \le \int |\varphi| \, d\mu$

Proof   By (iii), $\int \psi \, d\mu = \int \varphi \, d\mu + \int (\psi - \varphi) \, d\mu$ and the second integral is ≥ 0 by (i), so (iv) holds. $\varphi = \psi$ a.e. implies $\varphi \le \psi$ a.e. and $\psi \le \varphi$ a.e. so (v) follows from two applications of (iv). To prove (vi) now notice that $\varphi \le |\varphi|$ so (iv) implies $\int \varphi \, d\mu \leq \int |\varphi| \, d\mu$. $-\varphi \leq |\varphi|$ so (iv) and (ii) imply $-\int \varphi \, d\mu \leq \int |\varphi| \, d\mu$. Since $|y| = \max(y, -y)$, the result follows. $\qquad\square$

Step 2. Let E be a set with $\mu(E) < \infty$ and let f be a bounded function that vanishes on E^c. To define the integral of f we observe that if φ, ψ are simple functions that have $\varphi \leq f \leq \psi$ then we want to have

$$\int \varphi \, d\mu \leq \int f \, d\mu \leq \int \psi \, d\mu$$

so we let

$(*)$ $$\int f \, d\mu = \sup_{\varphi \leq f} \int \varphi \, d\mu = \inf_{\psi \geq f} \int \psi \, d\mu$$

Here and for the rest of Step 2, we assume that φ and ψ vanish on E^c. To justify the definition in $(*)$ we have to prove that the sup and inf are equal. It follows from (iv) in (4.2) that

$$\sup_{\varphi \leq f} \int \varphi \, d\mu \leq \inf_{\psi \geq f} \int \psi \, d\mu$$

To prove the other inequality, suppose $|f| \leq M$ and let

$$E_k = \left\{ x \in E : \frac{kM}{n} \geq f(x) > \frac{(k-1)M}{n} \right\} \quad \text{for } -n \leq k \leq n$$

$$\psi_n(x) = \sum_{k=-n}^{n} \frac{kM}{n} 1_{E_k} \qquad \varphi_n(x) = \sum_{k=-n}^{n} \frac{(k-1)M}{n} 1_{E_k}$$

By definition $\psi_n(x) - \varphi_n(x) = (M/n) 1_E$ so

$$\int \psi_n(x) - \varphi_n(x) \, d\mu = \frac{M}{n} \mu(E)$$

Since $\varphi_n(x) \leq f(x) \leq \psi_n(x)$ it follows from (iii) in (4.1) that

$$\sup_{\varphi \leq f} \int \varphi \, d\mu \geq \int \varphi_n \, d\mu = -\frac{M}{n} \mu(E) + \int \psi_n \, d\mu$$

$$\geq -\frac{M}{n} \mu(E) + \inf_{\psi \geq f} \int \psi \, d\mu$$

The last inequality holds for all n, so the proof of $(*)$ is complete. $\qquad\square$

(4.3) Lemma. Let E be a set with $\mu(E) < \infty$. If f and g are bounded functions that vanish on E^c then:

(i) If $f \geq 0$ a.e. then $\int f \, d\mu \geq 0$.

(ii) For any $a \in \mathbf{R}$, $\int af \, d\mu = a \int f \, d\mu$.

(iii) $\int f + g \, d\mu = \int f \, d\mu + \int g \, d\mu$.

(iv) If $g \leq f$ a.e. then $\int g \, d\mu \leq \int f \, d\mu$.

(v) If $g = f$ a.e. then $\int g \, d\mu = \int f \, d\mu$.

(vi) $|\int f \, d\mu| \leq \int |f| \, d\mu$

Proof Since we can take $\varphi \equiv 0$, (i) is clear from the definition. To prove (ii) we observe that if $a > 0$ for then $a\varphi \leq af$ if and only if $\varphi \leq f$ so

$$\int af \, d\mu = \sup_{\varphi \leq f} \int a\varphi \, d\mu = \sup_{\varphi \leq f} a \int \varphi \, d\mu = a \sup_{\varphi \leq f} \int \varphi \, d\mu = a \int f \, d\mu$$

For $a < 0$ we observe that $a\varphi \leq af$ if and only if $\varphi \geq f$ so

$$\int af \, d\mu = \sup_{\varphi \geq f} \int a\varphi \, d\mu = \sup_{\varphi \geq f} a \int \varphi \, d\mu = a \inf_{\varphi \geq f} \int \varphi \, d\mu = a \int f \, d\mu$$

To prove (iii) we observe that if $\psi_1 \geq f$ and $\psi_2 \geq g$ then $\psi_1 + \psi_2 \geq f + g$ so

$$\inf_{\psi \geq f+g} \int \psi \, d\mu \leq \inf_{\psi_1 \geq f, \psi_2 \geq g} \int \psi_1 + \psi_2 \, d\mu$$

Using linearity for simple functions it follows that

$$\int f + g \, d\mu = \inf_{\psi \geq f+g} \int \psi \, d\mu$$
$$\leq \inf_{\psi_1 \geq f, \psi_2 \geq g} \int \psi_1 \, d\mu + \int \psi_2 \, d\mu = \int f \, d\mu + \int g \, d\mu$$

To prove the other inequality observe that the last conclusion applied to $-f$ and $-g$ and (ii) imply

$$-\int f + g \, d\mu \leq -\int f \, d\mu - \int g \, d\mu$$

(iv)–(vi) follow from (i)–(iii) by (4.2). □

Notation. We define the integral of f over the set E:

$$\int_E f \, d\mu \equiv \int f \cdot 1_E \, d\mu$$

Step 3. If $f \geq 0$ then we let

$$\int f \, d\mu = \sup \left\{ \int h \, d\mu : 0 \leq h \leq f, h \text{ is bounded and } \mu(\{x : h(x) > 0\}) < \infty \right\}$$

The last definition is nice since it is clear that this is well defined. The next result will help us compute the value of the integral.

(4.4) Lemma. Let $E_n \uparrow \Omega$ have $\mu(E_n) < \infty$ and let $a \wedge b = \min(a, b)$. Then

$$\int_{E_n} f \wedge n \, d\mu \uparrow \int f \, d\mu \quad \text{as } n \uparrow \infty$$

Proof It is clear that from (iv) in (4.3) that the left-hand side increases as n does. Since $h = (f \wedge n)1_{E_n}$ is a possibility in the sup, each term is smaller than the integral on the right. To prove that the limit is $\int f \, d\mu$ observe that if $0 \leq h \leq f$, $h \leq M$, and $\mu(\{x : h(x) > 0\}) < \infty$ then for $n \geq M$ using $h \leq M$, (iv), and (iii),

$$\int_{E_n} f \wedge n \, d\mu \geq \int_{E_n} h \, d\mu = \int h \, d\mu - \int_{E_n^c} h \, d\mu$$

Now $0 \leq \int_{E_n^c} h \, d\mu \leq M\mu(E_n^c \cap \{x : h(x) > 0\}) \to 0$ as $n \to \infty$, so

$$\liminf_{n \to \infty} \int_{E_n} f \wedge n \, d\mu \geq \int h \, d\mu$$

which proves the desired result since h is an arbitrary member of the the class that defines the integral of f. □

(4.5) Lemma. Suppose $f, g \geq 0$.
   (i) $\int f \, d\mu \geq 0$
   (ii) If $a > 0$ then $\int af \, d\mu = a \int f \, d\mu$.
   (iii) $\int f + g \, d\mu = \int f \, d\mu + \int g \, d\mu$
   (iv) If $0 \leq g \leq f$ a.e. then $\int g \, d\mu \leq \int f \, d\mu$.
   (v) If $0 \leq g = f$ a.e. then $\int g \, d\mu = \int f \, d\mu$.

Proof (i) is trivial from the definition. (ii) is clear since when $a > 0$, $ah \leq af$ if and only if $h \leq f$ and we have $\int ah \, d\mu = a \int h \, du$ for h in the defining class. For (iii) we observe that if $f \geq h$ and $g \geq k$ then $f + g \geq h + k$ so taking the sup over h and k in the defining classes for f and g gives

$$\int f + g \, d\mu \geq \int f \, d\mu + \int g \, d\mu$$

To prove the other direction we observe $(a + b) \wedge n \le (a \wedge n) + (b \wedge n)$ so (iv) from (4.2) and (iii) from (4.3) imply

$$\int_{E_n} (f + g) \wedge n \, d\mu \le \int_{E_n} f \wedge n \, d\mu + \int_{E_n} g \wedge n \, d\mu$$

Letting $n \to \infty$ and using (4.4) gives (iii). As before (iv) and (v) follow from (i), (iii), and (4.2).   □

EXERCISE 4.1. Show that if $f \ge 0$ and $\int f \, d\mu = 0$ then $f = 0$ a.e.

EXERCISE 4.2. Let $f \ge 0$ and $E_{n,m} = \{x : m/2^n \le f(x) < (m + 1)/2^n\}$. As $n \uparrow \infty$

$$\sum_{m=1}^{\infty} \frac{m}{2^n} \mu(E_{n,m}) \uparrow \int f \, d\mu$$

Step 4. We say f is integrable if $\int |f| \, d\mu < \infty$. Let

$$f^+(x) = f(x) \vee 0 \quad \text{and} \quad f^-(x) = (-f(x)) \vee 0$$

where $a \vee b = \max(a, b)$. Clearly,

$$f(x) = f^+(x) - f^-(x) \quad \text{and} \quad |f(x)| = f^+(x) + f^-(x)$$

We define the integral of f by

$$\int f \, d\mu = \int f^+ \, d\mu - \int f^- \, d\mu$$

The right-hand side is well defined since $f^+, f^- \le |f|$ and we have (iv) in (4.5). For the final time we will prove our six properties. To do this it is useful to know:

(4.6) **Lemma.** If $f = f_1 - f_2$ where $f_1, f_2 \ge 0$ and $\int f_i \, d\mu < \infty$ then

$$\int f \, d\mu = \int f_1 \, d\mu - \int f_2 \, d\mu$$

Proof $f_1 + f^- = f_2 + f^+$ and all four functions are ≥ 0 so by (iii) of (4.5)

$$\int f_1 \, d\mu + \int f^- \, d\mu = \int f_1 + f^- \, d\mu = \int f_2 + f^+ \, d\mu = \int f_2 \, d\mu + \int f^+ \, d\mu$$

Rearranging gives the desired conclusion. □

(4.7) Theorem. Suppose f and g are integrable.
    (i) If $f \geq 0$ a.e. then $\int f \, d\mu \geq 0$.
    (ii) For all $a \in \mathbf{R}$, $\int af \, d\mu = a \int f \, d\mu$.
    (iii) $\int f + g \, d\mu = \int f \, d\mu + \int g \, d\mu$
    (iv) If $g \leq f$ a.e. then $\int g \, d\mu \leq \int f \, d\mu$.
    (v) If $g = f$ a.e. then $\int g \, d\mu = \int f \, d\mu$.
    (vi) $|\int f \, d\mu| \leq \int |f| \, d\mu$

Proof (i) is trivial. (ii) is clear since if $a > 0$ then $(af)^+ = a(f^+)$, and so on. To prove (iii) observe that $f + g = (f^+ + g^+) - (f^- + g^-)$, so using (4.6) and (4.5)

$$\int f + g \, d\mu = \int f^+ + g^+ \, d\mu - \int f^- + g^- \, d\mu$$

$$= \int f^+ \, d\mu + \int g^+ \, d\mu - \int f^- \, d\mu - \int g^- \, d\mu$$

As usual (iv)–(vi) follow from (i)–(iii) and (4.2). □

Notation for special cases:

(a) When $(\Omega, \mathcal{F}, \mu) = (\mathbf{R}^d, \mathcal{R}^d, \lambda)$ we write $\int f(x) \, dx$ for $\int f \, d\lambda$.

(b) When $(\Omega, \mathcal{F}, \mu) = (\mathbf{R}, \mathcal{R}, \lambda)$ and $E = [a, b]$ we write $\int_a^b f(x) \, dx$ for $\int_E f \, d\lambda$.

(c) When $(\Omega, \mathcal{F}, \mu) = (\mathbf{R}, \mathcal{R}, \mu)$ with $\mu((a, b]) = G(b) - G(a)$ for $a < b$ we write $\int f(x) \, dG(x)$ for $\int f \, d\mu$.

(d) When Ω is a countable set, $\mathcal{F} =$ all subsets of Ω, and μ is counting measure we write $\sum_{i \in \Omega} f(i)$ for $\int f \, d\mu$.

We mention example (d) primarily to indicate that results for sums follow from those for integrals.

For the rest of this section we will consider the case $(\Omega, \mathcal{F}, \mu) = (\mathbf{R}, \mathcal{R}, \lambda)$.

Littlewood's principles

Speaking of the theory of functions of a real variable Littlewood (1944) said

"The extent of knowledge required is nothing like so great as is sometimes supposed. There are three principles, roughly expressable in the following terms:

1. Every measurable set is roughly a finite union of intervals.

2. Every measurable function is almost continuous.

3. Every convergent sequence of measurable functions is almost uniformly convergent.

Most of the results of the theory are fairly intuitive applications of the theory and the student armed with them should be equal to most occasions when real variable theory is called for."

Exercise 3.1 above and Exercise 2.8 in Chapter 1 give versions of the first and third principles. The next two exercise develops a version of the second.

EXERCISE 4.3. Let g be an integrable function on $\mathbf{R}$ and $\epsilon > 0$. (i) Use the definition of the integral to conclude there is a simple function $\varphi = \sum_k b_k 1_{A_k}$ with $\int |g - \varphi| \, dx < \epsilon$. (ii) Use Exercise 3.1 to approximate the A_k by finite unions of intervals to get a **step function**

$$q = \sum_{j=1}^{k} c_j 1_{(a_{j-1}, jm)}$$

with $a_0 < a_1 < \ldots < a_k$, so that $\int |\varphi - q| < \epsilon$. (iii) Round the corners of q to get a continuous function r so that $\int |q - r| \, dx < \epsilon$.

EXERCISE 4.4. Prove the **Riemann-Lebesgue lemma**. If g is integrable then

$$\lim_{n \to \infty} \int g(x) \cos nx \, dx = 0$$

Hint: If g is a step function this is easy. Now use Exercise 4.3.

* Riemann Integration

Our treatment of the Lebesgue integral would not be complete if we did not prove the classic theorem of Lebesgue that identifes the functions for which the Riemann integral exists. Let $-\infty < a < b < \infty$. A subdivision σ of $[a, b]$ is a finite sequence $a = x_0 < x_1 \ldots < x_n = b$. Given a subdivision σ, we define the

upper Riemann sum $U(\sigma) = \sum_{i=1}^{n} (x_{i+1} - x_i) \sup\{f(y) : y \in [x_{i-1}, x_i]\}$

lower Riemann sum $L(\sigma) = \sum_{i=1}^{n} (x_{i+1} - x_i) \inf\{f(y) : y \in [x_{i-1}, x_i]\}$

We say that f is **Riemann integrable on $[a, b]$ in the liberal sense** if

$$\infty > \inf_{\sigma} U(\sigma) = \sup_{\sigma} L(\sigma) > -\infty$$

The function $q(x)$ that is 1 if x is irrational and 0 if x is rational is the classic example of a function that is not Riemann integrable on $[0,1]$ in the liberal sense but is Lebesgue integrable on $[0,1]$. ($q1_{[0,1]}$ is a simple function!) The next result gives a necessary condition for Riemann integrability.

(4.8) Theorem. If f is Riemann integrable on $[a,b]$ in the liberal sense, then f is bounded and continuous a.e. on $[a,b]$.

Proof If f is unbounded above then $U(\sigma) = \infty$ for all subdivisions. Likewise if f is unbounded below $L(\sigma) = -\infty$ for all subdivisions. Thus f must be bounded. To prove that it must be continuous a.e., we begin by letting

$$u_n(x) = \sup\{f(y) : |x - y| < 2^{-n} \text{ and } y \in [a,b]\}$$
$$v_n(x) = \inf\{f(y) : |x - y| < 2^{-n} \text{ and } y \in [a,b]\}$$

Exercise 2.6 in Chapter 1 implies u_n and v_n are measurable. Let

$$f^0 = \lim_{n\to\infty} u_n \quad \text{and} \quad f_0 = \lim_{n\to\infty} v_n$$

$f^0(x) \geq f_0(x)$ with equality if and only if f is continuous at x. Given a subdivision σ,

$$\sup\{f(y) : y \in [x_{i-1}, x_i]\} \geq f^0(x) \quad \text{for } x \in (x_{i-1}, x_i)$$

so $U(\sigma) \geq \int_{[a,b]} f^0 \, dx$, the (Lebesgue) integral existing since f^0 is bounded and measurable. Similar reasoning shows that any lower Riemann sum has $\int_{[a,b]} f_0 dx \geq L(\sigma)$, so if f is Riemann integrable in the liberal sense $\int_{[a,b]} f^0 - f_0 \, dx = 0$ and it follows from Exercise 4.1 that $f^0 = f_0$ a.e. □

To state a converse to (4.8) we need two definitions. The **mesh** of a subdivision $= \sup(x_i - x_{i-1})$. f is said to be **Riemann integrable on $[a,b]$ in the strict sense** if for any sequence of subdivisions with mesh $\to 0$, $U(\sigma_n) - L(\sigma_n) \to 0$.

(4.9) Theorem. If f is bounded and continuous a.e. on $[a,b]$ then f is Riemann integrable on $[a,b]$ in the strict sense.

Proof We need a little more theory before you can give a simple proof of this. See Exercise 5.6.

EXERCISE 4.5. Give examples to show that for a function f defined on **R** neither statement implies the other. (a) f is continuous a.e. (b) There is a continuous function g so that $f = g$ a.e.

EXERCISE 4.6. Let $(\Omega, \mathcal{F}, \mu)$ be a finite measure space and let f be a function with $|f| < M$. Given a sequence of subdivisions $-M = x_0^n < x_1^n < \ldots < x_n^n = M$ define the

upper Lebesgue sum $\bar{U}(\sigma_n) = \sum_{m=1}^{n} x_m^n \mu(\{\omega : f(\omega) \in [x_{m-1}^n, x_m^n)\})$

lower Lebesgue sum $\bar{L}(\sigma_n) = \sum_{m=1}^{n} x_{m-1}^n \mu(\{\omega : f(\omega) \in [x_{m-1}^n, x_m^n)\})$

Show that if $\text{mesh}(\sigma_n) \to 0$ then $\bar{U}(\sigma_n), \bar{L}(\sigma_n) \to \int f \, d\mu$. In short, in Riemann integration we subdivide the domain and in Lebesgue integration we subdivide the range.

A.5. Properties of the Integral

In this section we will develop properties of the integral defined in the last section. Our first result generalizes (vi) from (4.7).

(5.1) Jensen's inequality. Suppose φ is convex, that is,

$$\lambda \varphi(x) + (1 - \lambda)\varphi(y) \geq \varphi(\lambda x + (1 - \lambda)y)$$

for all $\lambda \in (0, 1)$ and $x, y \in \mathbf{R}$. If μ is a probability measure, i.e., $\mu(\mathbf{R}) = 1$ and f and $\varphi(f)$ are integrable then

$$\varphi\left(\int f \, d\mu\right) \leq \int \varphi(f) \, d\mu$$

Proof   Let $c = \int f \, d\mu$ and let $\ell(x) = ax + b$ be a linear function that has $\ell(c) = \varphi(c)$ and $\varphi(x) \geq \ell(x)$. To see that such a function exists recall that convexity implies

$$\lim_{h \downarrow 0} \frac{\varphi(c) - \varphi(c - h)}{h} \leq \lim_{h \downarrow 0} \frac{\varphi(c + h) - \varphi(c)}{h}$$

(The limits exist since the sequences are monotone.) If we let a be any number between the two limits and let $\ell(x) = a(x - c) + \varphi(c)$ then ℓ has the desired properties. With the existence of ℓ established, the rest is easy. (iv) in (5.1) implies

$$\int \varphi(f) \, d\mu \geq \int (af + b) \, d\mu = a \int f \, d\mu + b = \ell\left(\int f \, d\mu\right) = \varphi\left(\int f \, d\mu\right)$$

since $c = \int f \, d\mu$ and $\ell(c) = \varphi(c)$. $\qquad\qquad\qquad\qquad\qquad\qquad\qquad\qquad$ $\square$

Let $\|f\|_p = (\int |f|^p \, d\mu)^{1/p}$ for $1 \leq p < \infty$, and notice $\|cf\|_p = |c| \cdot \|f\|_p$ for any real number c.

(5.2) Hölder's inequality. If $p, q \in (1, \infty)$ with $1/p + 1/q = 1$ then

$$\int |fg| \, d\mu \leq \|f\|_p \|g\|_q$$

Proof If $\|f\|_p$ or $\|g\|_q = 0$ then $|fg| = 0$ a.e., so it suffices to prove the result when $\|f\|_p$ and $\|g\|_q > 0$ or by dividing both sides by $\|f\|_p\|g\|_q$, when $\|f\|_p = \|g\|_q = 1$. Fix $y \geq 0$ and let

$$\varphi(x) = x^p/p + y^q/q - xy \quad \text{for} \quad x \geq 0$$
$$\varphi'(x) = x^{p-1} - y \quad \text{and} \quad \varphi''(x) = (p-1)x^{p-2}$$

so φ has a minimum at $x_o = y^{1/(p-1)}$. $x_o^p = y^{p/(p-1)} = y^q$ and $q = p/(p-1)$ so

$$\varphi(x_o) = y^q(1/p + 1/q) - y^{1/(p-1)}y = 0$$

Since x_o is the minimum, it follows that $xy \leq x^p/p + y^q/q$. Letting $x = |f|$, $y = |g|$, and integrating

$$\int |fg| \, d\mu \leq \frac{1}{p} + \frac{1}{q} = 1 = \|f\|_p\|g\|_q \qquad\qquad\qquad \square$$

Remark. The special case $p = q = 2$ is called the **Cauchy-Schwarz inequality**. one can give a direct proof of the result in this case by observing that for any θ

$$0 \leq \int (f + \theta g)^2 \, d\mu = \int f^2 \, d\mu + \theta \left(2 \int fg \, d\mu \right) + \theta^2 \left(\int g^2 \, d\mu \right)$$

so the quadratic $a\theta^2 + b\theta + c$ on the right-hand side has at most one real root. Recalling the formula for the roots of a quadratic

$$\frac{-b \pm \sqrt{b^2 - 4ac}}{2a}$$

we see $b^2 - 4ac \leq 0$ which is the desired result.

EXERCISE 5.1. Let $\|f\|_\infty = \inf\{M : \mu(\{x : |f(x)| > M\}) = 0\}$. Prove that

$$\int |fg|\,d\mu \le \|f\|_1 \|g\|_\infty$$

EXERCISE 5.2. Show that if μ is a probability measure then

$$\|f\|_\infty = \lim_{p\to\infty} \|f\|_p$$

EXERCISE 5.3. **Minkowski's inequality.** (i) Suppose $p \in (1,\infty)$. The inequality $|f + g|^p \le 2^p(|f|^p + |g|^p)$ shows that if $\|f\|_p$ and $\|g\|_p$ are $< \infty$ then $\|f + g\|_p < \infty$. Apply Hölder's inequality to $|f||f + g|^{p-1}$ and $|g||f + g|^{p-1}$ to show $\|f + g\|_p \le \|f\|_p + \|g\|_p$. (ii) Show that the last result remains true when $p = 1$ or $p = \infty$.

Our next goal is to give conditions that guarantee

$$\lim_{n\to\infty} \int f_n\,d\mu = \int \left(\lim_{n\to\infty} f_n\right) d\mu$$

First we need a definition. We say that $f_n \to f$ **in measure**, i.e., for any $\epsilon > 0$, $\mu(\{x : |f_n(x) - f(x)| > \epsilon\}) \to 0$ as $n \to \infty$. This is a weaker assumption than $f_n \to f$ a.e., but the next result is easier to prove in the greater generality.

(5.3) **Bounded convergence theorem.** Let E be a set with $\mu(E) < \infty$. Suppose f_n vanishes on E^c, $|f_n(x)| \le M$, and $f_n \to f$ in measure. Then

$$\int f\,d\mu = \lim_{n\to\infty} \int f_n\,d\mu$$

Example 5.1. The functions $f_n(x) = 1_{[n,n+1)}(x)$, on $\mathbf{R}$ equipped with the Borel sets $\mathcal{R}$ and Lebesgue measure λ, show that the conclusion of (5.4) does not hold when $\mu(E) = \infty$.

Proof Let $\epsilon > 0$, $G_n = \{x : |f_n(x) - f(x)| < \epsilon\}$ and $B_n = E - G_n$. Using (iii) and (iv) from (4.7),

$$\left|\int f\,d\mu - \int f_n\,d\mu\right| = \left|\int (f - f_n)\,d\mu\right| \le \int |f - f_n|\,d\mu$$

$$= \int_{G_n} |f - f_n|\,d\mu + \int_{B_n} |f - f_n|\,d\mu$$

$$\le \epsilon\mu(E) + 2M\mu(B_n)$$

$f_n \to f$ in measure implies $\mu(B_n) \to 0$. $\epsilon > 0$ is arbitrary and $\mu(E) < \infty$ so the proof is complete. $\quad\square$

EXERCISE 5.4. Use (5.3) to prove (4.9). Hint: given a subdivision σ let

$$f^\sigma(x) = \sup\{f(y) : y \in [x_{i-1}, x_i]\} \quad \text{for } x \in (x_{i-1}, x_i)$$

so that $\int_{[a,b]} f^\sigma(x)dx = U(\sigma)$.

(5.4) Fatou's lemma. If $f_n \geq 0$ then

$$\liminf_{n\to\infty} \int f_n \, d\mu \geq \int \left(\liminf_{n\to\infty} f_n\right) d\mu$$

Example 5.2. Example 5.1 shows that we may have strict inequality in (5.4). The functions $f_n(x) = n1_{(0,1/n]}(x)$ on $(0,1)$ equipped with the Borel sets and Lebesgue measure show that this can happen on a space of finite measure.

Proof   Let $g_n(x) = \inf_{m\geq n} f_m(x)$. $f_n(x) \geq g_n(x)$ and as $n \uparrow \infty$,

$$g_n(x) \uparrow g(x) = \liminf_{n\to\infty} f_n(x)$$

Since $\int f_n \, d\mu \geq \int g_n \, d\mu$, it suffices then to show that

$$\liminf_{n\to\infty} \int g_n \, d\mu \geq \int g \, d\mu$$

Let $E_m \uparrow \Omega$ be sets of finite measure. Since $g_n \geq 0$ and for fixed m

$$(g_n \wedge m) \cdot 1_{E_m} \to (g \wedge m) \cdot 1_{E_m} \quad \text{a.e.}$$

the bounded convergence theorem (5.3) implies

$$\liminf_{n\to\infty} \int g_n \, d\mu \geq \int_{E_m} g_n \wedge m \, d\mu \to \int_{E_m} g \wedge m \, d\mu$$

Taking the sup over m and using (4.4) gives the desired result. $\quad\square$

(5.5) Monotone convergence theorem. If $f_n \geq 0$ and $f_n \uparrow f$ then

$$\int f_n \, d\mu \uparrow \int f \, d\mu$$

Proof Fatou's lemma, (5.4), implies $\liminf \int f_n \, d\mu \geq \int f \, d\mu$. On the other hand $f_n \leq f$ implies $\limsup \int f_n \, d\mu \leq \int f \, d\mu$. □

EXERCISE 5.5. If $g_n \uparrow g$ and $\int g_1^- \, d\mu < \infty$ then $\int g_n \, d\mu \uparrow \int g \, d\mu$.

EXERCISE 5.6. If $g_m \geq 0$ then $\int \sum_{m=0}^{\infty} g_m \, d\mu = \sum_{m=0}^{\infty} \int g_m \, d\mu$.

EXERCISE 5.7. Let $f \geq 0$. (i) Show that $\int f \wedge n \, d\mu \uparrow \int f \, d\mu$ as $n \to \infty$. (ii) Use (i) to conclude that if g is integrable and $\epsilon > 0$ then we can pick $\delta > 0$ so that $\mu(A) < \delta$ implies $\int_A |g| d\mu < \epsilon$.

(5.6) **Dominated convergence theorem.** If $f_n \to f$ a.e., $|f_n| \leq g$ for all n, and g is integrable, then $\int f_n \, d\mu \to \int f \, d\mu$.

Proof $f_n + g \geq 0$ so Fatou's lemma implies

$$\liminf_{n \to \infty} \int f_n + g \, d\mu \geq \int f + g \, d\mu$$

Subtracting $\int g \, d\mu$ from both sides gives

$$\liminf_{n \to \infty} \int f_n \, d\mu \geq \int f \, d\mu$$

Applying the last result to $-f_n$ we get

$$\limsup_{n \to \infty} \int f_n \, d\mu \leq \int f \, d\mu$$

and the proof is complete. □

EXERCISE 5.8. If f is integrable and E_n are disjoint sets with union E then

$$\sum_{m=0}^{\infty} \int_{E_m} f \, d\mu = \int_E f \, d\mu$$

So if $f \geq 0$ then $\nu(E) = \int_E f \, d\mu$ defines a measure.

EXERCISE 5.9. Show that if f is integrable on $[a, b]$, $g(x) = \int_{[a,x]} f(y) \, dy$ is continuous on (a, b).

EXERCISE 5.10. Show that if f has $\|f\|_p = (\int |f|^p d\mu)^{1/p} < \infty$, then there are simple functions φ_n so that $\|\varphi_n - f\|_p \to 0$.

EXERCISE 5.11. Show that if $\sum_n \int |f_n| d\mu < \infty$ then $\sum_n \int f_n \, d\mu = \int \sum_n f_n \, d\mu$.

A.6. Product Measures, Fubini's Theorem

Let $(X, \mathcal{A}, \mu_1)$ and $(Y, \mathcal{B}, \mu_2)$ be two σ-finite measure spaces. Let

$$\Omega = X \times Y = \{(x, y) : x \in X, y \in Y\}$$
$$\mathcal{S} = \{A \times B : A \in \mathcal{A}, B \in \mathcal{B}\}$$

Sets in $\mathcal{S}$ are called **rectangles**. It is easy to see that $\mathcal{S}$ is a semi-algebra:

$$(A \times B) \cap (C \times D) = (A \cap C) \times (B \cap D)$$
$$(A \times B)^c = (A^c \times B) \cup (A \times B^c) \cup (A^c \times B^c)$$

Let $\mathcal{F} = \mathcal{A} \times \mathcal{B}$ be the σ-algebra generated by $\mathcal{S}$.

(6.1) Theorem. There is a unique measure μ on $\mathcal{F}$ with

$$\mu(A \times B) = \mu_1(A)\mu_2(B)$$

Notation. μ is often denoted by $\mu_1 \times \mu_2$.

Proof   By (1.3) it is enough to show that if $A \times B = +_i(A_i \times B_i)$ then

$$\mu(A \times B) = \sum_i \mu(A_i \times B_i)$$

For each $x \in A$, let $I(x) = \{i : x \in A_i\}$. $B = +_{i \in I(x)} B_i$ so

$$1_A(x)\mu_2(B) = \sum_i 1_{A_i}(x)\mu_2(B_i)$$

Integrating with respect to μ_1 and using Exercise 5.6 gives

$$\mu_1(A)\mu_2(B) = \sum_i \mu_1(A_i)\mu_2(B_i)$$

which proves (6.1). $\square$

EXERCISE 6.1. Let $\mathcal{A}_o \subset \mathcal{A}$ and $\mathcal{B}_o \subset \mathcal{B}$ be semi-algebras with $\sigma(\mathcal{A}_o) = \mathcal{A}$ and $\sigma(\mathcal{B}_o) = \mathcal{B}$. Given a measure μ_1 on $\mathcal{A}$ and a measure μ_2 on $\mathcal{B}$, there is a unique measure μ on $\mathcal{A} \times \mathcal{B}$ that has $\mu(A \times B) = \mu_1(A)\mu_2(B)$ for $A \in \mathcal{A}_o$ and $B \in \mathcal{B}_o$. The point of this exercise is that we can define Lebesgue measure on $\mathbf{R}^2$ by the requirement that $\lambda((a, b] \times (c, d]) = (b - a)(d - c)$.

Using (6.1) and induction it follows that if $(\Omega_i, \mathcal{F}_i, \mu_i)$, $i = 1, \ldots, n$, are σ-finite measure spaces, and $\Omega = \Omega_1 \times \cdots \times \Omega_n$ there is a unique measure μ on the σ-algebra $\mathcal{F}$ generated by sets of the form $A_1 \times \cdots \times A_n$, $A_i \in \mathcal{F}_i$, that has

$$\mu(A_1 \times \cdots \times A_n) = \prod_{m=1}^{n} \mu_m(A_m)$$

When $(\Omega_i, \mathcal{F}_i, \mu_i) = (\mathbf{R}, \mathcal{R}, \lambda)$ for all i, the result is Lebesgue measure on the Borel subsets of n dimensional Euclidean space $\mathbf{R}^n$.

Returning to the case in which $(\Omega, \mathcal{F}, \mu)$ is the product of two measure spaces, $(X, \mathcal{A}, \mu)$ and $(Y, \mathcal{B}, \nu)$, our next goal is to prove

(6.2) Fubini's theorem. If $f \geq 0$ or $\int |f| \, d\mu < \infty$ then

$$(*) \quad \int_X \int_Y f(x, y) \, \mu_2(dy) \, \mu_1(dx) = \int_{X \times Y} f \, d\mu = \int_Y \int_X f(x, y) \, \mu_1(dx) \, \mu_2(dy)$$

Proof We will prove only the first equality, since the second one is similar. Two technical things that need to be proved before we can assert that the first integral makes sense are:

When x is fixed, $y \to f(x, y)$ is $\mathcal{B}$ measurable.

$x \to \int_Y f(x, y)\mu_2(dy)$ is $\mathcal{A}$ measurable.

We begin with the case $f = 1_E$. Let $E_x = \{y : (x, y) \in E\}$ be the **cross section** at x.

(6.3) Lemma. If $E \in \mathcal{F}$ then $E_x \in \mathcal{B}$.

Proof $(E^c)_x = (E_x)^c$ and $(\cup_i E_i)_x = \cup_i (E_i)_x$ so if $\mathcal{E}$ is the collection of sets E for which $E_x \in \mathcal{B}$, then $\mathcal{E}$ is a σ-algebra. Since $\mathcal{E}$ contains the rectangles the result follows. $\square$

(6.4) Lemma. If $E \in \mathcal{F}$ then $g(x) \equiv \mu_2(E_x)$ is $\mathcal{A}$ measurable and

$$\int_X g \, d\mu_1 = \mu(E)$$

Notice that it is not obvious that the collection of sets for which the conclusion is true is a σ-algebra since $\mu(E_1 \cup E_2) = \mu(E_1) + \mu(E_2) - \mu(E_1 \cap E_2)$. Dynkin's $\pi - \lambda$ theorem (2.1) was tailor made for situations like this.

Proof of (6.4) If conclusions hold for E_n and $E_n \uparrow E$ then (2.5) in Chapter 1 and the monotone convergence theorem imply that they hold for E. Since μ_1 and μ_2 are σ-finite, it is enough then to prove the result for $E \subset A \times B$ with $\mu_1(A) < \infty$ and $\mu_2(B) < \infty$, or taking $\Omega = A \times B$ we can suppose without loss of generality that $\mu(\Omega) < \infty$. Let $\mathcal{L}$ be the collection of sets E for which the conclusions hold. We will now check that $\mathcal{L}$ is a λ-system. Property (i) of a λ-system is trivial. (iii) follows from the first sentence in the proof. To check (ii) we observe that

$$\mu_2((A - B)_x) = \mu_2(A_x - B_x) = \mu_2(A_x) - \mu_2(B_x)$$

and integrating over x gives the second conclusion. Since $\mathcal{L}$ contains the rectangles, a π-system that generates $\mathcal{F}$, the desired result follows from the $\pi - \lambda$ theorem. □

We are now ready to prove (6.2) by verifying it in four increasingly more general special cases.

CASE 1. If $E \in \mathcal{F}$ and $f = 1_E$ then $(*)$ follows from (6.4).

CASE 2. Since each integral is linear in f, it follows that $(*)$ holds for simple functions.

CASE 3. Now if $f \geq 0$ and we let $f_n(x) = ([2^n f(x)]/2^n) \wedge n$, where $[x] = $ the largest integer $\leq x$, then the f_n are simple and $f_n \uparrow f$ so it follows from the monotone convergence theorem that $(*)$ holds for all $f \geq 0$.

CASE 4. The general case now follows by writing $f(x) = f(x)^+ - f(x)^-$ and applying Case 3 to f^+, f^-, and $|f|$. □

To illustrate why the various hypotheses of (6.2) are needed we will now give some examples where the conclusion fails.

Example 6.1. Let $X = Y = \{1, 2, \ldots\}$ with $\mathcal{A} = \mathcal{B} = $ all subsets and $\mu_1 = \mu_2 = $ counting measure. For $m \geq 1$ let $f(m, m) = 1$ and $f(m+1, m) = -1$, and let $f(m, n) = 0$ otherwise. We claim that

$$\sum_m \sum_n f(m, n) = 1 \quad \text{but} \quad \sum_n \sum_m f(m, n) = 0$$

A picture is worth several dozen words:

$$
\begin{matrix}
\vdots & \vdots & \vdots & \vdots & \vdots \\
0 & 0 & 0 & 1 & \cdots \\
\uparrow\;\; 0 & 0 & 1 & -1 & \cdots \\
n\;\; 0 & 1 & -1 & 0 & \cdots \\
1 & -1 & 0 & 0 & \cdots \\
& m & \rightarrow & &
\end{matrix}
$$

In words, if we sum the columns first the first one gives us a 1 and the others 0, while if we sum the rows each one gives us a 0.

Example 6.2. Let $X = (0,1)$, $Y = (1,\infty)$, both equipped with the Borel sets and Lebesgue measure. Let $f(x,y) = e^{-xy} - 2e^{-2xy}$.

$$
\int_0^1 \int_1^\infty f(x,y)\,dy\,dx = \int_0^1 x^{-1}(e^{-x} - e^{-2x})\,dx > 0
$$

$$
\int_1^\infty \int_0^1 f(x,y)\,dx\,dy = \int_1^\infty y^{-1}(e^{-2y} - e^{-y})\,dy < 0
$$

The next example indicates why μ_1 and μ_2 must be σ-finite.

Example 6.3. Let $X = (0,1)$ with $\mathcal{A} = $ the Borel sets and $\mu_1 = $ Lebesgue measure. Let $Y = (0,1)$ with $\mathcal{B} = $ all subsets and $\mu_2 = $ counting measure. Let $f(x,y) = 1$ if $x = y$ and 0 otherwise

$$
\int_Y f(x,y)\,\mu_2(dy) = 1 \quad \text{for all } x \text{ so} \quad \int_X \int_Y f(x,y)\,\mu_2(dy)\,\mu_1(dx) = 1
$$

$$
\int_X f(x,y)\,\mu_1(dx) = 0 \quad \text{for all } y \text{ so} \quad \int_Y \int_X f(x,y)\,\mu_1(dy)\,\mu_2(dx) = 0
$$

Our last example shows that measurability is important, or maybe that some of the axioms of set theory are not as innocent as they seem.

Example 6.4. By the axiom of choice and the continuum hypothesis one can define an order relation $<'$ on $(0,1)$ so that $\{x : x <' y\}$ is countable for each y. Let $X = Y = (0,1)$, let $\mathcal{A} = \mathcal{B} = $ the Borel sets and $\mu_1 = \mu_2 = $ Lebesgue measure. Let $f(x,y) = 1$ if $x <' y$, $= 0$ otherwise. Since $\{x : x <' y\}$ and $\{y : x <' y\}^c$ are countable

$$
\int_X f(x,y)\,\mu_1(dx) = 0 \quad \text{for all } y
$$

$$
\int_Y f(x,y)\,\mu_2(dy) = 1 \quad \text{for all } x
$$

We turn now to applications of (6.2).

EXERCISE 6.2. If $\int_X \int_Y |f(x,y)|\mu_2(dy)\mu_1(dx) < \infty$ then

$$\int_X \int_Y f(x,y)\mu_2(dy)\mu_1(dx) = \int_{X \times Y} f \, d(\mu_1 \times \mu_2) = \int_Y \int_X f(x,y)\mu_1(dx)\mu_2(dy)$$

Example 6.5. Let $X = \{1,2,\ldots\}$, $\mathcal{A} = $ all subsets of X, and $\mu_1 = $ counting measure. If $\sum_n \int |f_n| d\mu < \infty$ then $\sum_n \int f_n \, d\mu = \int \sum_n f_n \, d\mu$.

EXERCISE 6.3. Let $g \geq 0$ be a measurable function on $(X, \mathcal{A}, \mu)$. Use (6.2) to conclude that

$$\int_X g \, d\mu = (\mu \times \lambda)(\{(x,y) : 0 \leq y < g(x)\}) = \int_0^\infty \mu(\{x : g(x) > y\}) \, dy$$

EXERCISE 6.4. Let F, G be Stieltjes measure functions and let μ, ν be the corresponding measures on $(\mathbf{R}, \mathcal{R})$. Show that

(i) $\int_{(a,b]} \{F(y) - F(a)\} dG(y) = (\mu \times \nu)(\{(x,y) : a < x \leq y \leq b\})$

(ii) $$\int_{(a,b]} F(y) \, dG(y) + \int_{(a,b]} G(y) \, dF(y)$$
$$= F(b)G(b) - F(a)G(a) + \sum_{x \in (a,b]} \mu(\{x\})\nu(\{x\})$$

To see the second term is needed let $F(x) = G(x) = 1_{[0,\infty)}(x)$ and $a < 0 < b$.

(iii) If $F = G$ is continuous then $\int_{(a,b]} 2F(y)dF(y) = F^2(b) - F^2(a)$.

EXERCISE 6.5. Let μ be a finite measure on $\mathbf{R}$ and $F(x) = \mu((-\infty, x])$. Show that

$$\int (F(x+c) - F(x)) \, dx = c\mu(\mathbf{R})$$

EXERCISE 6.6. Show that $e^{-xy} \sin x$ is integrable in the strip $0 < x < a$, $0 < y$. Perform the double integral in the two orders to get:

$$\int_0^a \frac{\sin x}{x} \, dx = \frac{\pi}{2} - (\cos a) \int_0^\infty \frac{e^{-ay}}{1+y^2} \, dy - (\sin a) \int_0^\infty \frac{ye^{-ay}}{1+y^2} \, dy$$

and replace $1 + y^2$ by 1 to conclude $\left| \int_0^a (\sin x)/x \, dx - (\pi/2) \right| \leq 2/a$ for $a \geq 1$.

A.7. Kolmogorov's Extension Theorem

To construct some of the basic objects of study in probability theory we will need an existence theorem for measures on infinite product spaces. Let $\mathbf{N} = \{1, 2, \ldots\}$ and

$$\mathbf{R}^{\mathbf{N}} = \{(\omega_1, \omega_2, \ldots) : \omega_i \in \mathbf{R}\}$$

We equip $\mathbf{R}^{\mathbf{N}}$ with the product σ-algebra $\mathcal{R}^{\mathbf{N}}$, which is generated by the **finite dimensional rectangles** = sets of the form $\{\omega : \omega_i \in (a_i, b_i]$ for $i = 1, \ldots, n\}$, where $-\infty \le a_i < b_i \le \infty$.

(7.1) Kolmogorov's extension theorem. Suppose we are given probability measures μ_n on $(\mathbf{R}^n, \mathcal{R}^n)$ that are consistent, that is,

$$\mu_{n+1}((a_1, b_1] \times \ldots \times (a_n, b_n] \times \mathbf{R}) = \mu_n((a_1, b_1] \times \ldots \times (a_n, b_n])$$

Then there is a unique probability measure P on $(\mathbf{R}^{\mathbf{N}}, \mathcal{R}^{\mathbf{N}})$ with

$$(*) \qquad P(\omega : \omega_i \in (a_i, b_i], 1 \le i \le n) = \mu_n((a_1, b_1] \times \ldots \times (a_n, b_n])$$

An important example of a consistent sequence of measures is

Example 7.1. Let $F_1, F_2, \ldots$ be distribution functions and let μ_n be the measure on $\mathbf{R}^n$ with

$$\mu_n((a_1, b_1] \times \ldots \times (a_n, b_n]) = \prod_{m=1}^n (F_m(b_m) - F_m(a_m))$$

In this case if we let $X_n(\omega) = \omega_n$ then the X_n are independent and X_n has distribution F_n.

Proof of (7.1)   Let $\mathcal{S}$ be the sets of the form $\{\omega : \omega_i \in (a_i, b_i], 1 \le i \le n\}$, and use $(*)$ to define P on $\mathcal{S}$. $\mathcal{S}$ is a semialgebra, so by (1.3) it is enough to show that if $A \in \mathcal{S}$ is a disjoint union of $A_i \in \mathcal{S}$ then $P(A) \le \sum_i P(A_i)$. If the union is finite then all the A_i are determined by the values of a finite number of coordinates and the conclusion follows from results in Section A.6.

Suppose now that the union is infinite. Let $\mathcal{A} = \{$ finite disjoint unions of sets in $\mathcal{S}\}$ be the algebra generated by $\mathcal{S}$. Since $\mathcal{A}$ is an algebra (by (1.2))

$$B_n \equiv A - \cup_{i=1}^n A_i$$

is a finite disjoint union of rectangles, and by the result for finite unions

$$P(A) = \sum_{i=1}^n P(A_i) + P(B_n)$$

It suffices then to show

(7.2) Lemma. If $B_n \in \mathcal{A}$ and $B_n \downarrow \emptyset$ then $P(B_n) \downarrow 0$.

Proof Suppose $P(B_n) \downarrow \delta > 0$. By repeating sets in the sequence we can suppose

$$B_n = \cup_{k=1}^{K_n} \{\omega : \omega_i \in (a_i^k, b_i^k], 1 \le i \le n\} \quad \text{where} \quad -\infty \le a_i^k < b_i^k \le \infty$$

The strategy of the proof is to approximate the B_n from within by compact rectangles with almost the same probability and then use a diagonal argument to show that $\cap_n B_n \ne \emptyset$. There is a set $C_n \subset B_n$ of the form

$$C_n = \cup_{k=1}^{K_n} \{\omega : \omega_i \in [\bar{a}_i^k, \bar{b}_i^k], 1 \le i \le n\} \quad \text{with} \quad -\infty < \bar{a}_k^i < \bar{b}_k^i < \infty$$

that has $P(B_n - C_n) \le \delta/2^{n+1}$. Let $D_n = \cap_{m=1}^n C_m$.

$$P(B_n - D_n) \le \sum_{m=1}^n P(B_m - C_m) \le \delta/2$$

so $P(D_n) \downarrow$ a limit $\ge \delta/2$. Now there are sets C_n^*, $D_n^* \subset \mathbf{R}^n$ so that

$$C_n = \{\omega : (\omega_1, \ldots, \omega_n) \in C_n^*\} \quad \text{and} \quad D_n = \{\omega : (\omega_1, \ldots, \omega_n) \in D_n^*\}$$

Note that

$$C_n = C_n^* \times \mathbf{R} \times \mathbf{R} \times \ldots \quad \text{and} \quad D_n = D_n^* \times \mathbf{R} \times \mathbf{R} \times \ldots$$

so C_n and C_n^* (and D_n and D_n^*) are closely related but $C_n \subset \Omega$ and $C_n^* \subset \mathbf{R}^n$. C_n^* is a finite union of closed rectangles so

$$D_n^* = C_n^* \cap_{m=1}^{n-1} (C_m^* \times \mathbf{R}^{n-m})$$

is a compact set. For each m let $\omega_m \in D_m$. $D_m \subset D_1$ so $\omega_{m,1}$ (i.e., the first coordinate of ω_m) is in D_1^* Since D_1^* is compact we can pick a subsequence $m(1, j) \ge j$ so that as $j \to \infty$

$$\omega_{m(1,j),1} \to \text{a limit } \theta_1$$

For $m \ge 2$, $D_m \subset D_2$ and hence $(\omega_{m,1}, \omega_{m,2}) \in D_2^*$. Since D_2^* is compact we can pick a subsequence of the previous subsequence (i.e., $m(2, j) = m(1, i_j)$) with $i_j \ge j$) so that as $j \to \infty$

$$\omega_{m(2,j),2} \to \text{a limit } \theta_2$$

Continuing in this way we define $m(k,j)$ a subsequence of $m(k-1,j)$ so that as $j \to \infty$

$$\omega_{m(k,j),k} \to \text{ a limit } \theta_k$$

Let $\omega_i' = \omega_{m(i,i)}$. ω_i' is a subsequence of all the subsequences so $\omega_{i,k}' \to \theta_k$ for all k. Now $\omega_{i,1}' \in D_1^*$ for all $i \geq 1$ and D_1^* is closed so $\theta_1 \in D_1^*$. Turning to the second set, $(\omega_{i,1}', \omega_{i,2}') \in D_2^*$ for $i \geq 2$ and D_2^* is closed so $(\theta_1, \theta_2) \in D_2^*$. Repeating the last argument we conclude that $(\theta_1, \ldots, \theta_k) \in D_k^*$ for all k, so $\omega = (\theta_1, \theta_2, \ldots) \in D_k$ (no star here since we are now talking about subsets of Ω) for all k and

$$\emptyset \neq \cap_k D_k \subset \cap_k B_k$$

a contradiction which proves the desired result. $\square$

A.8. Radon-Nikodym Theorem

In this section we prove the Radon-Nikodym theorem. To develop that result we begin with a topic that at first may appear to be unrelated. Let $(\Omega, \mathcal{F})$ be a measurable space. α is said to be a **signed measure** on $(\Omega, \mathcal{F})$ if (i) α takes values in $(-\infty, \infty]$, (ii) $\alpha(\emptyset) = 0$, and (iii) if $E = +_i E_i$ is a disjoint union then $\alpha(E) = \sum_i \alpha(E_i)$, in the following sense:

If $\alpha(E) < \infty$, the sum converges absolutely and $= \alpha(E)$.

If $\alpha(E) = \infty$, then $\sum_i \alpha(E_i)^- < \infty$ and $\sum_i \alpha(E_i)^+ = \infty$.

Clearly, a signed measure cannot be allowed to take both the values ∞ and $-\infty$ since then $\alpha(A) + \alpha(B)$ might not make sense. In most formulations a signed measure is allowed to take values in either $(-\infty, \infty]$ or $[-\infty, \infty)$. We will ignore the second possibility to simplify statements later. As usual, we turn to examples to help explain the definition.

Example 8.1. Let μ be a measure, f be a function with $\int f^- \, d\mu < \infty$, and let $\alpha(A) = \int_A f \, d\mu$. Exercise 5.8 implies that α is a signed measure.

Example 8.2. Let μ_1 and μ_2 be measures with $\mu_2(\Omega) < \infty$, and let $\alpha(A) = \mu_1(A) - \mu_2(A)$.

The Jordan decomposition ((8.4) below) will show that Example 8.2 is the general case. To derive that result we begin with two definitions. A set A is **positive** if every measurable $B \subset A$ has $\alpha(B) \geq 0$. A set A is **negative** if every measurable $B \subset A$ has $\alpha(B) \leq 0$.

EXERCISE 8.1. In Example 8.1, A is positive if and only if $\mu(A \cap \{x : f(x) < 0\}) = 0$.

(8.1) **Lemma.** (i) Every measurable subset of a positive set is positive. (ii) If the sets A_n are positive then $A = \cup_n A_n$ is also positive.

Proof (i) is trivial. To prove (ii) observe that

$$B_n = A_n \cap \left(\cap_{m=1}^{n-1} A_m^c \right) \subset A_n$$

are positive, disjoint, and $\cup_n B_n = \cup_n A_n$. Let $E \subset A$ be measurable, and let $E_n = E \cap B_n$. $\alpha(E_n) \geq 0$ since B_n is positive, so $\alpha(E) = \sum_n \alpha(E_n) \geq 0$. □

The conclusions in (8.1) remain valid if positive is replaced by negative. The next result is the key to the proof of (8.3).

(8.2) **Lemma.** Let E be a measurable set with $\alpha(E) < 0$. Then there is a negative set $F \subset E$ with $\alpha(F) < 0$.

Proof If E is negative this is true. If not, let n_1 be the smallest positive integer so that there is an $E_1 \subset E$ with $\alpha(E_1) \geq 1/n_1$. Let $k \geq 2$. If $F_k = E - (E_1 \cup \ldots \cup E_{k-1})$ is negative, we are done. If not, we continue the construction letting n_k be the smallest positive integer so that there is an $E_k \subset F_k$ with $\alpha(E_k) \geq 1/n_k$. If the construction does not stop for any $k < \infty$, let

$$F = \cap_k F_k = E - (\cup_k E_k)$$

Since $0 > \alpha(E) > -\infty$ and $\alpha(E_k) \geq 0$, it follows from the definition of signed measure that

$$\alpha(E) = \alpha(F) + \sum_{k=1}^{\infty} \alpha(E_k)$$

$\alpha(F) \leq \alpha(E) < 0$, and the sum is finite. From the last observation and the construction it follows that F can have no subset G with $\alpha(G) > 0$, for then $\alpha(G) \geq 1/N$ for some N and we would have a contradiction. □

(8.3) **Hahn decompositon.** Let α be a signed measure. Then there is a positive set A and a negative set B so that $\Omega = A \cup B$ and $A \cap B = \emptyset$.

Proof Let $c = \inf\{\alpha(B) : B \text{ is negative}\} \leq 0$. Let B_i be negative sets with $\alpha(B_i) \downarrow c$. Let $B = \cup_i B_i$. By (8.1), B is negative, so by the definition of c, $\alpha(B) \geq c$. To prove $\alpha(B) \leq c$ we observe that $\alpha(B) = \alpha(B_i) + \alpha(B - B_i) \leq \alpha(B_i)$ since B is negative, and let $i \to \infty$. The last two inequalities show that

$\alpha(B) = c$ and it follows from our definition of a signed measure that $c > -\infty$. Let $A = B^c$. To show A is positive, observe that if A contains a set with $\alpha(E) < 0$ then by (8.2), it contains a negative set F with $\alpha(F) < 0$, but then $B \cup F$ would be a negative set that has $\alpha(B \cup F) = \alpha(B) + \alpha(F) < c$, a contradiction. $\qquad \square$

The Hahn decompositon is not unique. In Example 8.1, A can be any set with

$$\{x : f(x) > 0\} \subset A \subset \{x : f(x) \geq 0\} \quad \text{a.e.}$$

where $B \subset C$ a.e. means $\mu(B \cap C^c) = 0$. The last example is typical of the general situation. Suppose $\Omega = A_1 \cup B_1 = A_2 \cup B_2$ are two Hahn decompositions. $A_2 \cap B_1$ is positive and negative so it is a **null set**: all its subsets have measure 0. Similarly $A_1 \cap B_2$ is a null set.

Two measures μ_1 and μ_2 are said to be **mutually singular** if there is a set A with $\mu_1(A) = 0$ and $\mu_2(A^c) = 0$. In this case, we also say μ_1 is **singular with respect to** μ_2 and write $\mu_1 \perp \mu_2$.

EXERCISE 8.2. Show that the uniform distribution on the Cantor set (Example 1.8 in Chapter 1) is singular with respect to Lebesgue measure.

(8.4) Jordan decompositon. Let α be a signed measure. There are mutually singular measures α_+ and α_- so that $\alpha = \alpha_+ - \alpha_-$. Moreover, there is only one such pair.

Proof Let $\Omega = A \cup B$ be a Hahn decomposition. Let

$$\alpha_+(E) = \alpha(E \cap A) \quad \text{and} \quad \alpha_-(E) = -\alpha(E \cap B)$$

Since A is positive and B is negative, α_+ and α_- are measures. $\alpha_+(A^c) = 0$ and $\alpha_-(A) = 0$, so they are mutually singular. To prove uniqueness suppose $\alpha = \nu_1 - \nu_2$ and D is a set with $\nu_1(D) = 0$ and $\nu_2(D^c) = 0$. If we set $C = D^c$, then $\Omega = C \cup D$ is a Hahn decomposition, and it follows from the choice of D that

$$\nu_1(E) = \alpha(C \cap E) \quad \text{and} \quad \nu_2(E) = -\alpha(D \cap E)$$

Our uniqueness result for the Hahn decomposition shows that $A \cap D = A \cap C^c$ and $B \cap C = A^c \cap C$ are null sets, so $\alpha(E \cap C) = \alpha(E \cap (A \cup C)) = \alpha(E \cap A)$ and $\nu_1 = \alpha_+$. $\qquad \square$

EXERCISE 8.3. Show that $\alpha_+(E) = \sup\{\alpha(F) : F \subset E\}$.

Remark. Let α be a **finite signed measure** (i.e., one that does not take the value ∞ or $-\infty$) on $(\mathbf{R}, \mathcal{R})$. Let $\alpha = \alpha_+ - \alpha_-$ be its Jordan decomposition. Let

$A(x) = \alpha((-\infty, x])$, $F(x) = \alpha_+((-\infty, x])$, and $G(x) = \alpha_-((-\infty, x])$. $A(x) = F(x) - G(x)$ so the distribution function for a finite signed measure can be written as a difference of two bounded increasing functions. It follows from Example 8.2 that the converse is also true. Let $|\alpha| = \alpha^+ + \alpha^-$. $|\alpha|$ is called the **total variation** of α, since in this example $|\alpha|((a, b])$ is the total variation of A over $(a, b]$ as defined in analysis textbooks. See, for example, Royden (1988) p. 103. We exclude the left endpoint of the interval since a jump there makes no contribution to the total variation on $[a, b]$ but it does appear in $|\alpha|$.

Our third and final decomposition is:

(8.5) Lebesgue decomposition. Let μ, ν be σ-finite measures. ν can be written as $\nu_r + \nu_s$ where ν_s is singular with respect to μ and

$$\nu_r(E) = \int_E g \, d\mu$$

Proof By decomposing $\Omega = +_i \Omega_i$ we can suppose without loss of generality that μ and ν are finite measures. Let $\mathcal{G}$ be the set of $g \geq 0$ so that $\int_E g \, d\mu \leq \nu(E)$ for all E.

(a) If $g, h \in \mathcal{G}$ then $g \vee h \in \mathcal{G}$.

Proof of (a) Let $A = \{g > h\}$, $B = \{g \leq h\}$.

$$\int_E g \vee h \, d\mu = \int_{E \cap A} g \, d\mu + \int_{E \cap B} h \, d\mu \leq \nu(E \cap A) + \nu(E \cap B) = \nu(E)$$

Let $\kappa = \sup\{\int g \, d\mu : g \in \mathcal{G}\} \leq \nu(\Omega) < \infty$. Pick g_n so that $\int g_n \, d\mu > \kappa - 1/n$ and let $h_n = g_1 \vee \ldots \vee g_n$. By (a), $h_n \in \mathcal{G}$. As $n \uparrow \infty$, $h_n \uparrow h$. The definition of κ, the monotone convergence theorem, and the choice of g_n imply that

$$\kappa \geq \int h \, d\mu = \lim_{n \to \infty} \int h_n \, d\mu \geq \lim_{n \to \infty} \int g_n \, d\mu = \kappa$$

Let $\nu_r(E) = \int_E h \, d\mu$ and $\nu_s(E) = \nu(E) - \nu_r(E)$. The last detail is to show:

(b) ν_s is singular with respect to μ.

Proof of (b) Let $\epsilon > 0$ and let $\Omega = A_\epsilon \cup B_\epsilon$ be a Hahn decomposition for $\nu_s - \epsilon\mu$. Using the definition of ν_r and then the fact that A_ϵ is positive for $\nu_s - \epsilon\mu$ (so $\epsilon\mu(A_\epsilon \cap E) \leq \nu_s(A_\epsilon \cap E)$)

$$\int_E (h + \epsilon 1_{A_\epsilon}) \, d\mu = \nu_r(E) + \epsilon\mu(A_\epsilon \cap E) \leq \nu(E)$$

This holds for all E, so $k = h + \epsilon 1_{A_\epsilon} \in \mathcal{G}$. It follows that $\mu(A_\epsilon) = 0$ for if not then $\int k \, d\mu > \kappa$ a contradiction. Letting $A = \cup_n A_{1/n}$, we have $\mu(A) = 0$. To see that $\nu_s(A^c) = 0$ observe that if $\nu_s(A^c) > 0$ then $(\nu_s - \epsilon\mu)(A^c) > 0$ for small ϵ, a contradiction since $A^c \subset B_\epsilon$, a negative set. $\square$

EXERCISE 8.4. Prove that the Lebesgue decomposition is unique. Note that you can suppose without loss of generality that μ and ν are finite.

We are finally ready for the main business of the section. We say a measure ν is **absolutely continuous with respect to** μ (and write $\nu \ll \mu$) if $\mu(A) = 0$ implies that $\nu(A) = 0$.

EXERCISE 8.5. If $\mu_1 \ll \mu_2$ and $\mu_2 \perp \nu$ then $\mu_1 \perp \nu$.

(8.6) **Radon-Nikodym theorem.** If μ, ν are σ-finite measure and ν is absolutely continuous with respect to μ then there is a $g \geq 0$ so that $\nu(E) = \int_E g \, d\mu$. If h is another such function then $g = h$ μ a.e.

Proof Let $\nu = \nu_r + \nu_s$ be any Lebesgue decomposition. Let A be chosen so that $\nu_s(A^c) = 0$ and $\mu(A) = 0$. Since $\nu \ll \mu$, $0 = \nu(A) \geq \nu_s(A)$ and $\nu_s \equiv 0$. To prove uniqueness observe that if $\int_E g \, d\mu = \int_E h \, d\mu$ for all E, then letting $E \subset \{g > h, g \leq n\}$ be any subset of finite measure we conclude $\mu(g > h, g \leq n) = 0$ for all n, so $\mu(g > h) = 0$, and similarly $\mu(g < h) = 0$. $\square$

Example 8.3. (8.6) may fail if μ is not σ-finite. Let $(\Omega, \mathcal{F}) = (\mathbf{R}, \mathcal{R})$, $\mu = $ counting measure and $\nu = $ Lebesgue measure.

The function g whose existence is proved in (8.6) is often denoted $d\nu/d\mu$. This notation suggests the following properties, whose proofs are left to the reader.

EXERCISE 8.6. If $\nu_1, \nu_2 \ll \mu$ then $\nu_1 + \nu_2 \ll \mu$

$$d(\nu_1 + \nu_2)/d\mu = d\nu_1/d\mu + d\nu_2/d\mu$$

EXERCISE 8.7. If $\nu \ll \mu$ and $f \geq 0$ then $\int f \, d\nu = \int f \frac{d\nu}{d\mu} \, d\mu$.

EXERCISE 8.8. If $\pi \ll \nu \ll \mu$ then $d\pi/d\mu = (d\pi/d\nu) \cdot (d\nu/d\mu)$.

EXERCISE 8.9. If $\nu \ll \mu$ and $\mu \ll \nu$ then $d\mu/d\nu = (d\nu/d\mu)^{-1}$.

A.9. Differentiating Under the Integral

At several places in the text we need to interchange differentiate inside a sum or an integral. This section is devoted to results which can be used to justify those computations.

(9.1) Theorem. Let $(S, \mathcal{S}, \mu)$ be a measure space. Let f be a complex valued function defined on $\mathbf{R} \times S$. Let $\delta > 0$ and suppose that for $x \in (y - \delta, y + \delta)$ we have

(i) $u(x) = \int_S f(x, s) \, \mu(ds)$ with $\int_S |f(x, s)| \, \mu(ds) < \infty$

(ii) for fixed s, $\partial f / \partial x (x, s)$ exists and is a continuous function of x,

(iii) $v(x) = \int_S \frac{\partial f}{\partial x}(x, s) \, \mu(ds)$ is continuous at $x = y$,

and (iv) $\int_S \int_{-\delta}^{\delta} \left| \frac{\partial f}{\partial x}(y + \theta, s) \right| d\theta \, \mu(ds) < \infty$

then $u'(y) = v(y)$.

Proof Letting $|h| \leq \delta$ and using (i), (ii), (iv), and Fubini's theorem in the form given in Exercise 6.2 we have

$$u(y + h) - u(y) = \int_S f(y + h, s) - f(y, s) \, \mu(ds)$$

$$= \int_S \int_0^h \frac{\partial f}{\partial x}(y + \theta, s) \, d\theta \, \mu(ds)$$

$$= \int_0^h \int_S \frac{\partial f}{\partial x}(y + \theta, s) \, \mu(ds) \, d\theta$$

The last equation implies

$$\frac{u(y + h) - u(y)}{h} = \frac{1}{h} \int_0^h v(y + \theta) \, d\theta$$

Since v is continuous at y by (iii), letting $h \to 0$ gives the desired result. $\square$

Example 9.1. For a result in Section 2.3 we need to know that we can differentiate under the integral sign in

$$u(x) = \int \cos(xs) e^{-s^2/2} \, ds$$

For convenience, we have dropped a factor $(2\pi)^{-1/2}$ and changed variables to match (9.1). Clearly (i) and (ii) hold. The dominated convergence theorem implies (iii)

$$x \to \int -s\sin(sx)e^{-s^2/2}\,ds$$

is a continuous. For (iv) we note

$$\int \left|\frac{\partial f}{\partial x}(x,s)\right|\,ds = \int |s|e^{-s^2/2}\,ds < \infty$$

and the value does not depend on x so (iv) holds.

To prepare for the next example, we introduce

(9.2) Lemma. For (iii) and (iv) to hold, it is sufficient that we have (ii) and

(iii')
$$\int_S \sup_{\theta\in[-\delta,\delta]} \left|\frac{\partial f}{\partial x}(y+\theta,s)\right|\,\mu(ds) < \infty$$

Proof Since

$$\int_{-\delta}^{\delta} \left|\frac{\partial f}{\partial x}(y+\theta,s)\right|\,d\theta \le 2\delta \sup_{\theta\in[-\delta,\delta]}\left|\frac{\partial f}{\partial x}(y+\theta,s)\right|$$

it is clear that (iv) holds. To check (iii) we note that

$$|v(x) - v(y)| \le \int_S \left|\frac{\partial f}{\partial x}(x,s) - \frac{\partial f}{\partial x}(y,s)\right|\mu(ds)$$

(ii) implies that the integrand $\to 0$ as $x \to y$. The desired result follows from (iii') and the dominated convergence theorem. $\qquad\square$

To indicate the usefulness of the new result we prove.

(9.3) Theorem. If $\varphi(\theta) = Ee^{\theta Z} < \infty$ for $\theta \in [-\epsilon, \epsilon]$ then $\varphi'(0) = EZ$.

Proof Here θ plays the role of x and we take μ to be the distribution of Z. Let $\delta = \epsilon/2$. $f(x,s) = e^{xs} \ge 0$ so (i) holds by assumption. $\partial f/\partial x = se^{xs}$ is clearly a continuous function so (ii) holds. To check (iii') we note that there is a constant C so that if $x \in (-\delta, \delta)$ then $|s|e^{xs} \le C\,(e^{-\epsilon s} + e^{\epsilon s})$. $\qquad\square$

Taking $S = \mathbf{Z}$ with $\mathcal{S} = $ all subsets of S, and μ is counting measure in (9.1) and using (9.2) gives the following.

(9.4) Corollary. Let $\delta > 0$. Suppose that for $x \in (y - \delta, y + \delta)$ we have

(i) $u(x) = \sum_{n=1}^{\infty} f_n(x)$ with $\sum_{n=1}^{\infty} |f_n(x)| < \infty$

(ii) for each n, $f_n'(x)$ exists and is a continuous function of x,

and (iii) $\sum_{n=1}^{\infty} \sup_{\theta \in (-\delta, \delta)} |f_n'(y + \theta)| < \infty$

then $u'(x) = v(x)$.

Example 9.2. We want to show that if $p \in (0, 1)$ then

$$\left(\sum_{n=1}^{\infty} (1 - p)^n \right)' = \sum_{n=1}^{\infty} n(1 - p)^{n-1}$$

Let $f_n(x) = (1 - x)^n$, $y = p$, and pick δ so that $[y - \delta, y + \delta] \subset (0, 1)$. Clearly
(i) $\sum_{n=1}^{\infty} |(1 - x)^n| < \infty$ and (ii) $f_n'(x) = n(1 - x)^{n-1}$ is continuous for $x \in [y - \delta, y + \delta]$. To check (iii) we note that if we let $2\eta = y - \delta$ then there is a constant C so that if $x \in [y - \delta, y + \delta]$ and $n \geq 1$ then

$$n(1 - x)^{n-1} = \frac{n(1 - x)^{n-1}}{(1 - \eta)^{n-1}} \cdot (1 - \eta)^{n-1} \leq C(1 - \eta)^{n-1} \qquad \square$$

References

O. Adelman (1985) Brownian motion never increases: a new proof of a result of Dvoretsky, Erdös, and Kakutani. *Israel J. Math.* 50, 189-192

D. Aldous and P. Diaconis (1986) Shuffling cards and stopping times. *Amer. Math. Monthly.* 93, 333-348

P.H. Algoet and T.M. Cover (1988) A sandwich proof of the Shannon McMillan Breiman theorem. *Ann. Probab.* 16, 899-909

D.V. Anosov (1963) Ergodic properties of geodesic flows on closed Riemannian manifolds of negative curvature. *Soviet Math. Doklady* 4, 1153-1156

D.V. Anosov (1967) Geodesic flows on compact Riemannian manifolds of negative curvature. *Proceedings of the Steklov. Inst. of Math*, No. 90

K. Athreya and P. Ney (1972) *Branching Processes.* Springer-Verlag, New York

K. Athreya and P. Ney (1978) A new approach to the limit theory of recurrent Markov chains. *Trans. AMS.* 245, 493-501

K. Athreya, D. McDonald, and P. Ney (1972) Coupling and the renewal theorem. *Amer. Math. Monthly.* 85, 809-814

L. Bachelier (1900) Théorie de la spéculation. *Ann. Sci. École Norm. Sup.* 17, 21-86

R. Ballerini and S. Resnick (1985) Records from improving populations. *J. Appl. Probab.* 22, 487-502

R. Ballerini and S. Resnick (1987) Records in the presence of a linear trend. *Adv. Appl. Probab.* 19, 801-828

S. Banach and A. Tarski (1924) Sur la décomposition des ensembles de points en parties respectivements congruent. *Fund. Math.* 6, 244-277

O. Barndorff-Nielsen (1961) On the rate of growth of partial maxima of a sequence of independent and identically distributed random variables. *Math. Scand.* 9, 383-394

L.E. Baum and P. Billingsley (1966) Asymptotic distributions for the coupon collector's problem. *Ann. Math. Statist.* 36, 1835-1839

F. Benford (1938) The law of anomalous numbers. *Proc. Amer. Phil. Soc.* 78, 552-572

J.D. Biggins (1977) Chernoff's theorem in branching random walk. *J. Appl. Probab.* 14, 630-636

J.D. Biggins (1978) The asymptotic shape of branching random walk. *Adv. in Appl. Probab.* 10, 62-84

J.D. Biggins (1979) Growth rates in branching random walk. *Z. Warsch. verw. Gebiete.* 48, 17-34

P. Billingsley (1961) The Lindeberg-Lévy theorem for martingales. *Proc. AMS* 12, 788-792

P. Billingsley (1968) *Weak convergence of probability measures.* John Wiley and Sons, New York

P. Billingsley (1979) *Probability and measure.* John Wiley and Sons, New York

G.D. Birkhoff (1931) Proof of the ergodic theorem. *Proc. Nat. Acad. Sci.* 17, 656-660

D. Blackwell and D. Freedman (1964) The tail σ-field of a Markov chain and a theorem of Orey. *Ann. Math. Statist.* 35, 1291-1295

R.M. Blumenthal and R.K. Getoor (1968) *Markov processes and their potential theory.* Academic Press, New York

E. Borel (1909) Les probabilités dénombrables et leur applications arithmét -iques. *Rend. Circ. Mat. Palermo.* 27, 247-271

L. Breiman (1957) The individual ergodic theorem of ergodic theory. *Ann. Math. Statist.* 28 (1957), 809-811, Correction 31, 809-810

L. Breiman (1968) *Probability.* Addison-Wesley, Reading, MA

K. Burdzy (1990) On the non-increase of Brownian motion. *Ann. Probab.* 18, 978-980

Y.S. Chow and H. Teicher (1988) *Probability theory: independence, interchangeability, martingales, second edition.* Springer-Verlag, New York

K.L. Chung (1961) A note on the ergodic theorem of information theory. *Ann. Math. Statist.* 32, 612-614

K.L. Chung (1974) *A Course in Probability Theory*, second edition. Academic Press, New York.

K.L. Chung, P. Erdös, and T. Sirao (1959) On the Lipschitz's condition for Brownian motion. *J. Math. Soc. Japan.* 11, 263-274

K.L. Chung and W.H.J. Fuchs (1951) On the distribution of values of sums of independent random variables. *Memoirs of the AMS*, No. 6

V. Chvátal and D. Sankoff (1975) Longest common subsequences of two random sequences. *J. Appl. Probab.* 12, 306-315

Z. Ciesielski (1961) Hölder condition for realizations of Gaussian processes. *Trans. AMS.* 99, 403-413

J. Cohen, H. Kesten, and C. Newman (1985) *Random matrices and their applications.* AMS Contemporary Math. 50, Providence, RI

J.T. Cox and R. Durrett (1981) Limit theorems for percolation processes with necessary and sufficient conditions. *Ann. Probab.* 9, 583-603

B. Davis (1983) On Brownian slow points. *Z. Warsch. verw. Gebiete.* 64, 359-367

Y. A. Davydov (1968) Convergence of distributions generated by stationary stochastic processes. *Theory Probab. Appl.* 13, 691-696

Y. Derriennic (1983) Une théorème ergodique presque sous additif. *Ann. Probab.* 11, 669-677

P. Diaconis and D. Freedman (1980) Finite exchangeable sequences. *Ann. Prob.* 8, 745-764

J. Dieudonné (1948) Sur la théorème de Lebesgue-Nikodym, II. *Ann. Univ. Grenoble* 23, 25-53

M. Donsker (1951) An invariance principle for certain probability limit theorems. *Memoirs of the AMS*, No. 6

M. Donsker (1952) Justification and extension of Doob's heurisitc approach to the Kolmogorov-Smirnov theorems. *Ann. Math. Statist.* 23, 277-281

J.L. Doob (1949) A heuristic approach to the Kolmogorov-Smirnov theorems. *Ann. Math. Statist.* 20, 393-403

J.L. Doob (1953) *Stochastic Processes.* John Wiley and Sons, New York

P. Doyle and L. Snell (1984) *Random walks and electrical networks.* Carus Monograph, Math. Assoc. of America

L.E. Dubins (1968) On a theorem of Skorkhod. *Ann. Math. Statist.* 39, 2094-2097

L.E. Dubins and D.A. Freedman (1965) A sharper form of the Borel-Cantelli lemma and the strong law. *Ann. Math. Statist.* 36, 800-807

L.E. Dubins and D.A. Freedman (1979) Exchangeable processes need not be distributed mixtures of independent and identically distributed random variables. *Z. Warsch. verw. Gebiete.* 48, 115-132

R. Durrett (1984) *Brownian motion and martingales in analysis.* Wadsworth Pub. Co., Pacific Grove, CA

R. Durrett and S. Resnick (1978) Functional limit theorems for dependent random variables. *Ann. Probab.* 6, 829-846

A. Dvoretsky (1972) Asymptotic normality for sums of dependent random variables. *Proc. 6th Berkeley Symp.*, Vol. II, 513-535

A. Dvoretsky and P. Erdös (1950) Some problems on random walk in space. *Proc. 2nd Berkeley Symp.*, 353-367

A. Dvoretsky, P. Erdös, and S. Kakutani (1961) Nonincrease everywhere of the Brownian motion process. *Proc. 4th Berkeley Symp.*, Vol. II, 103-116

E.B. Dynkin (1955) Continuous one-dimensional Markov processes. *Dokl. Akad. Nauk. SSR* 105, 405-408

E.B. Dynkin (1961) Some limit theorems for sums of independent random variables with infinite mathematical expectation. *Selected Translations in Math. Stat. and Prob.* Vol. 1, 171-189

E.B. Dynkin (1965) *Markov processes.* Springer-Verlag, New York

P. Erdös (1942) On the law of the iterated logarithm. *Ann. Math.* 43, 419-436

P. Erdös and M. Kac (1946) On certain limit theorems of the theory of probability. *Bull. AMS.* 52, 292-302

P. Erdös and M. Kac (1947) On the number of positive sums of independent random variables. *Bull. AMS.* 53, 1011-1020

P. Erdös and B. Szerkeres (1935) A combinatorial problem in geometry. *Compositio Math.* 2, 463-470

N. Etemadi (1981) An elementary proof of the strong law of large numbers. *Z. Warsch. verw. Gebiete.* 55, 119-122

A.M. Faden (1985) The existence of regular conditional probabilities: necessary and sufficient conditions. *Ann. Probab.* 13, 288-298

M. Fekete (1923) Über die Verteilung der Wurzeln bei gewissen algebraischen Gleichungen mit ganzzahligen Koeffizienten. *Math. Z.* 17, 228-249

W. Feller (1943) The general form of the so-called law of the iterated logarithm. *Trans. AMS.* 54, 373-402

W. Feller (1946) A limit theorem for random variables with infinite moments. *Amer. J. Math.* 68, 257-262

W. Feller (1961) A simple proof of renewal theorems. *Comm. Pure Appl. Math.* 14, 285-293

W. Feller (1968) *An introduction to probability theory and its applications, Vol. I*, third edition. John Wiley and Sons, New York

W. Feller (1971) *An introduction to probability theory and its applications, Vol. II*, second edition. John Wiley and Sons, New York

S.R. Foguel (1969) *The ergodic theory of Markov processes.* Van Nostrand, New York

D. Freedman (1965) Bernard Friedman's urn. *Ann. Math. Statist.* 36, 956-970

D. Freedman (1971a) *Brownian motion and diffusion.* Originally published by Holden Day, San Francisco, CA. Second edition by Springer-Verlag, New York

D. Freedman (1971b) *Markov chains.* Originally published by Holden Day, San Francisco, CA. Second edition by Springer-Verlag, New York

D. Freedman (1980) A mixture of independent and identically distributed random variables need not admit a regular conditional probability given the exchangeable σ-field. *Z. Warsch. verw. Gebiete* 51, 239-248

R.M. French (1988) The Banach-Tarski theorem. *Math. Intelligencer* 10, No. 4, 21-28

B. Friedman (1949) A simple urn model. *Comm. Pure Appl. Math.* 2, 59-70

H. Furstenburg (1970) Random walks in discrete subgroups of Lie Groups. In *Advances in Probability* edited by P.E. Ney.

H. Furstenburg and H. Kesten (1960) Products of random matrices. *Ann. Math. Statist.* 31, 451-469

A. Garsia (1965) A simple proof of E. Hopf's maximal ergodic theorem. *J. Math. Mech.* 14, 381-382

M.L. Glasser and I.J. Zucker (1977) Extended Watson integrals for the cubic lattice. *Proc. Nat. Acad. Sci.* 74, 1800-1801

B.V. Gnedenko (1943) Sur la distribution limité du terme maximum d'une série aléatoire. *Ann. Math.* 44, 423-453

B.V. Gnedenko and A.V. Kolmogorov (1954) *Limit distributions for sums of independent random variables.* Addison-Wesley, Reading, MA

M.I. Gordin (1969) The central limit theorem for stationary processes. *Soviet Math. Doklady.* 10, 1174-1176

D.R. Grey (1989) Persistent random walks may have arbitrarily large tails. *Adv. Appl. Probab.* 21, 229-230

D. Griffeath and T.M. Liggett (1982) Critical phenomena for Spitzer's reversible nearest particle systems. *Ann. Probab.* 10, 881-895

P. Hall (1982) *Rates of convergence in the central limit theorem.* Pitman Pub. Co., Boston, MA

P. Hall and C.C. Heyde (1976) On a unified approach to the law of the iterated logarithm for martingales. *Bull. Austral. Math. Soc.* 14, 435-447

P. Hall and C.C. Heyde (1980) *Martingale limit theory and its application.* Academic Press, New York

P.R. Halmos (1950) *Measure theory.* Van Nostrand, New York

P.R. Halmos (1956) *Lectures on ergodic theory.* Chelsea Pub. Co., New York

J.M. Hammersley (1970) A few seedlings of research. *Proc. 6th Berkeley Symp.*, Vol. I, 345-394

G.H. Hardy and J.E. Littlewood (1914) Some problems of Diophantine approximation. *Acta Math.* 37, 155-239

G.H. Hardy and E.M. Wright (1959) *An introduction to the theory of numbers*, fourth edition. Oxford University Press, London

T.E. Harris (1956) The existence of stationary measures for certain Markov processes. *Proc. 3rd Berkeley Symp.*, Vol. II, 113-124

P. Hartman and A. Wintner (1941) On the law of the iterated logarithm. *Amer. J. Math.* 63, 169-176

F. Hausdorff (1913) *Gründzuge der Mengenlehre.* Viet, Leipzig

E. Hewitt and L.J. Savage (1956) Symmetric measures on Cartesian products. *Trans. AMS.* 80, 470-501

E. Hewitt and K. Stromberg (1965) *Real and abstract analysis.* Springer-Verlag, New York

C.C. Heyde (1963) On a property of the lognormal distribution. *J. Royal. Stat. Soc. B.* 29, 392-393

C.C. Heyde (1967) On the influence of moments on the rate of convergence to the normal distribution. *Z. Warsch. verw. Gebiete.* 8, 12-18

C.C. Heyde and D.J. Scott (1973) Invariance principles for the law of the iterated logarithm for martingales and for processes with stationary increments. *Ann. Probab.* 1, 428-436

P. Hoel, S. Port, and C. Stone (1972) *Introduction to Stochastic Processes.* Houghton Mifflin, Boston, MA

T. Hida (1980) *Brownian motion.* Springer-Verlag, New York

J.L. Hodges, Jr. and L. Le Cam (1960) The Poisson approximation to the binomial distribution. *Ann. Math. Statist.* 31, 737-740

G. Hunt (1956) Some theorems concerning Brownian motion. *Trans. AMS* 81, 294-319

I.A. Ibragimov (1962) Some limit theorems for stationary processes. *Theory Probab. Appl.* 7, 349-382

I.A. Ibragimov (1963) A central limit theorem for a class of dependent random variables. *Theory Probab. Appl.* 8, 83-89

I.A. Ibragimov and Y. V. Linnik (1971) *Independent and stationary sequences of random variables.* Wolters-Noordhoff, Groningen

H. Ishitani (1977) A central limit theorem for the subadditive process and its application to products of random matrices. *RIMS, Kyoto.* 12, 565-575

K. Itô and H.P. McKean (1965). *Diffusion processes and their sample paths.* Springer-Verlag, New York

N.C. Jain and S. Orey (1969) On the range of random walk. *Israel J. Math.* 6, 373-380

N.C. Jain and W.E. Pruitt (1971) The range of random walk. *Proc. 6th Berkeley Symp.*, Vol. III, 31-50

M. Kac (1947a) Brownian motion and the theory of random walk. *Amer. Math. Monthly.* 54, 369-391

M. Kac (1947b) On the notion of recurrence in discrete stochastic processes. *Bull. AMS.* 53, 1002-1010

M. Kac (1949) On deviations between theoretical and empirical distribution functions. *Proc. Nat. Acad. Sci.* 35, 252-257

M. Kac (1959) *Statistical independence in probability, analysis, and number theory.* Carus Monographs, Math. Assoc. of America

S. Karlin and H.M. Taylor (1975) *A first course in stochastic processes*, second edition. Academic Press, New York

Y. Katznelson and B. Weiss (1982) A simple proof of some ergodic theorems. *Israel J. Math.* 42, 291-296

E. Keeler and J. Spencer (1975) Optimal doubling in backgammon. *Operations Research.* 23, 1063-1071

H. Kesten (1986) Aspects of first passage percolation. In *École d'été de probabilités de Saint-Flour XIV.* Lecture Notes in Math 1180, Springer-Verlag, New York

H. Kesten (1987) Percolation theory and first passage percolation. *Ann. Probab.* 15, 1231-1271

A. Khintchine (1923) Über dyadische Brüche. *Math. Z.* 18, 109-116

A. Khintchine (1923) Über einen Satz der Wahrscheinlichkeitsrechnung. *Fund. Math.* 6, 9-20

J. Kielson and D.M.G. Wishart (1964) A central limit theorem for processes defined on a Markov chain. *Proc. Camb. Phil. Soc.* 60, 547-567

J.F.C. Kingman (1968) The ergodic theory of subadditive processes. *J. Roy. Stat. Soc. B* 30, 499-510

J.F.C. Kingman (1973) Subadditive ergodic theory. *Ann. Probab.* 1, 883-909

J.F.C. Kingman (1975) The first birth problem for age dependent branching processes. *Ann. Probab.* 3, 790-801

A.N. Kolmogorov and Y.A. Rozanov (1964) On strong mixing conditions for stationary Gaussian processes. *Theory Probab. Appl.* 5, 204-208

A.N. Kolmogorov (1929) Über das Gesetz des iterierten Logarithmus. *Math. Ann.* 101, 126-135

K. Kondo and T. Hara (1987) Critical exponent of susceptibility for a general class of ferromagnets in $d > 4$ dimensions. *J. Math. Phys.* 28, 1206-1208

U. Krengel (1985) *Ergodic theorems.* deGruyter, New York

A. Lasota and M. MacKay (1985) *Probabilistic properties of deterministic systems.* Cambridge Univ. Press, London

J.F. LeGall (1985) Un théorème central limite pour le nombre de points visités par une marche aléatoire plane récurrente. *C.R. Acad. Sci. Paris.* 300, 505-508

J.F. LeGall (1986a) Propriétés d'intersection des marches aléatoires, I. Convergence vers le temps local d'intersection. *Comm. Math. Phys.* 104, 471-507

J.F. LeGall (1986b) Propriétés d'intersection des marches aléatoires, II. Étude de cas critiques. *Comm. Math. Phys.* 104, 509-528

R. Leipnik (1981) The lognormal law and strong non-uniqueness of the moment problem. *Theory Probab. Appl.* 26, 850-852

S. Leventhal (1988) A proof of Liggett's version of the subadditive ergodic theorem. *Proc. AMS.* 102, 169-173

P. Lévy (1931) Sur un théorème de M. Khintchine. *Bull. Sci. Math* (2), 55, 145-160

P. Lévy (1937) *Théorie de l'addition des variables aléatoires.* Gauthier-Villars, Paris

P. Lévy (1939) Sur certains processus stochastiques homogènes. *Compositio Math.* 7, 283-339

P. Lévy (1948) *Processus stochastiques et mouvement Brownien.* Gauthier-Villars, Paris

T.M. Liggett (1985) An improved subadditive ergodic theorem. *Ann. Probab.* 13, 1279-1285

A. Lindenbaum (1926) Contributions à l'étude de l'espace metrique. *Fund. Math.* 8, 209-222

T. Lindvall (1977) A probabilistic proof of Blackwell's renewal theorem. *Ann. Probab.* 5, 482-485

T. Lindvall (1979) On coupling of discrete renewal processes. *Z. Warsch. verw. Gebiete.* 48, 57-70

J.E. Littlewood (1944) *Lectures on the theory of functions.* Oxford U. Press, London

B.F. Logan and L.A. Shepp (1977) A variational problem for random Young tableaux. *Adv. in Math.* 26, 206-222

T. Lyons (1983) A simple criterion for transience of a Markov chain. *Ann. Probab.* 11, 393-402

H.P. McKean (1969) *Stochastic integrals.* Academic Press, New York

B. McMillan (1953) The basic theorems of information theory. *Ann. Math. Statist.* 24, 196-219

M. Motoo (1959) Proof of the law of the iterated logarithm through diffusion equation. *Ann. Inst. Stat. Math.* 10, 21-28

S. Newcomb (1881) Note on the frequency of use of the different digits in natural numbers. *Amer. J. Math.* 4, 39-40

J. Neveu (1965) *Mathematical foundations of the calculus of probabilities.* Holden-Day, San Francisco, CA

J. Neveu (1975) *Discrete parameter martingales.* North Holland, Amsterdam

G. O'Brien (1974) Limit theorems for sums of chain dependent processes. *J. Appl. Probab.* 11, 582-587

D. Ornstein (1969) Random walks. *Trans. AMS.* 138, 1-60

V.I. Oseledec (1968) A multiplicative ergodic theorem. Lyapunov characteristic numbers for synmaical systems. *Trans. Moscow Math. Soc.* 19, 197-231

R.E.A.C. Paley and N. Wiener (1934) *Fourier transforms in the complex domain.* Amer. Math. Soc. Colloq. Pub. XIX

R.E.A.C. Paley, N. Wiener and A. Zygmund (1933) Notes on random functions. *Math. Z.* 37, 647-668

I. Petrovski (1935) Zur ersten Randwertaufgabe der Wärmeleitungsgleichung. *Compositio Math.* 1, 383-419

E.J.G. Pitman (1956) On derivatives of characteristic functions at the origin. *Ann. Math. Statist.* 27, 1156-1160

S.C. Port and C.J. Stone (1969) Potential theory of random walks on abelian groups. *Acta Math.* 122, 19-114

M.S. Ragunathan (1979) A proof of Oseledec's multiplicative ergodic theorem. *Israel J. Math.* 32, 356-362

R. Raimi (1976) The first digit problem. *Amer. Math. Monthly.* 83, 521-538

S. Resnick (1987) *Extreme values, regular variation, and point processes.* Springer-Verlag, New York

D. Revuz (1984) *Markov chains*, second edition. North Holland, Amsterdam

D.H. Root (1969) The existence of certain stopping times on Brownian motion. *Ann. Math. Statist.* 40, 715-718

M. Rosenblatt (1956) A central limit theorem and a strong mixing condition. *Proc. Nat. Acad. Sci.* 42, 43-47

H. Royden (1988) *Real analysis*, third edition. McMillan, New York

D. Ruelle (1979) Ergodic theory of differentiable dynamical systems. *IHES Pub. Math.* 50, 275-306

C. Ryll-Nardzewski (1951) On the ergodic theorems, II. *Studia Math.* 12, 74-79

L.J. Savage (1972) *The foundations of statistics*, second edition. Dover, New York

D.J. Scott (1973) Central limit theorems for martingales and for processes with stationary independent increments using a Skorokhod representation approach. *Adv. Appl. Probab.* 5, 119-137

C. Shannon (1948) A mathematical theory of communication. *Bell Systems Tech. J.* 27, 379-423

L.A. Shepp (1964) Recurrent random walks may take arbitrarily large steps. *Bull. AMS.* 70, 540-542

S. Sheu (1986) Representing a distribution by stopping Brownian motion: Root's construction. *Bull. Austral. Math. Soc.* 34, 427-431

A. Skorokhod (1965) *Studies in the theory of random processes.* Originally published by Addison Wesley, Reading, MA. Second edition (1982) Dover, New York

N.V. Smirnov (1949) Limit distributions for the terms of a variational series. *AMS Transl. Series.* 1, No. 67

R. Smythe and J.C. Wierman (1978) *First passage percolation on the square lattice.* Lecture Notes in Math 671, Springer-Verlag, New York

R.M. Solovay (1970) A model of set theory in which every set of reals is Lebesgue measurable. *Ann. Math.* 92, 1-56

E. Sparre-Andersen and B. Jessen (1984) On the introduction of measures in infinite product spaces. *Danske Vid. Selsk. Mat.-Fys. Medd.* 25, No. 4

F. Spitzer (1964) *Principles of random walk.* Van Nostrand, Princeton, NJ

J.M. Steele (1989) Kingman's subadditive ergodic theorem. *Ann. Inst. H. Poincaré* 25, 93-98

C. Stein (1987) *Approximate computation of expectations.* IMS Lecture Notes Vol. 7

H. Steinhaus (1922) Les probabilités denombrables et leur rapport à la theorie de la mesure. *Fund. Math.* 4, 286-310

C.J. Stone (1969) On the potential operator for one dimensional recurrent random walks. *Trans. AMS.* 136, 427-445

J. Stoyanov (1987) *Counterexamples in probability.* John Wiley and Sons, New York

V. Strassen (1964) An invariance principle for the law of the iterated logarithm. *Z. Warsch. verw. Gebiete* 3, 211-226

V. Strassen (1965) A converse to the law of the iterated logarithm. *Z. Warsch. verw. Gebiete.* 4, 265-268

V. Strassen (1967) Almost sure behavior of the sums of independent random variables and martingales. *Proc. 5th Berkeley Symp.*, Vol. II, 315-343

H. Thorisson (1987) A complete coupling proof of Blackwell's renewal theorem. *Stoch. Proc. Appl.* 26, 87-97

H. Trotter (1958) A property of Brownian motion paths. *Illinois J. Math.* 2, 425-433

P. van Beek (1972) An application of Fourier methods to the problem of sharpening the Berry-Esseen inequality. *Z. Warsch. verw. Gebiete.* 23, 187-196

A.M. Vershik and S.V. Kerov (1977) Asymptotic behavior of the Plancherel measure of the symmetric group and the limit form of random Young tableau. *Dokl. Akad. Nauk SSR* 233, 1024-1027

H. Wegner (1973) On consistency of probability measures. *Z. Warsch. verw. Gebiete* 27, 335-338

L. Weiss (1955) The stochastic convergence of a function of sample successive differences. *Ann. Math. Statist.* 26, 532-536

N. Wiener (1923) Differential space. *J. Math. Phys.* 2, 131-174

K. Yosida and S. Kakutani (1939) Birkhoff's ergodic theorem and the maximal ergodic theorem. *Proc. Imp. Acad. Tokyo* 15, 165-168

A. Zygmund (1947) *Trigonometric series.* Cambridge U. Press, London

Notation

N	natural numbers $1, 2, \ldots$
Z	integers
Q	rational numbers
R	real numbers
C	complex numbers

Real numbers

$[x]$	integer part of x, the largest integer $n \le x$
$x \wedge y$	minimum of x and y
$x \vee y$	maximum of x and y
x^+	positive part, $x \vee 0$
x^-	negative part, $(-x) \vee 0$
$\operatorname{sgn}(x)$	the sign of x, 1 if $x > 0$, -1 if $x < 0$, 0 if $x = 0$
$x_n \to x$	$\lim_{n \to \infty} x_n = x$
$a_n \uparrow$	$a_1 \le a_2 \le \cdots$
$a_n \downarrow a$	$a_1 \ge a_2 \ge \ldots$ and $a_n \to a$
$\sim$	asymptotically, $a_n \sim b_n$ means $a_n/b_n \to 1$ as $n \to \infty$
O	$f(t)$ is $O(t^2)$ as $t \to 0$ means $\limsup_{t \to 0} |f(t)|/t^2 < \infty$
o	$f(t)$ is $o(t)$ as $t \to 0$ means $f(t)/t \to 0$ as $t \to 0$

Complex numbers, $z = a + bi$

$\bar{z}$	complex conjugate, $= a - bi$
$|z|$	modulus, $= (a^2 + b^2)^{1/2}$
$\operatorname{Re} z$	real part of z, $= a$
$\operatorname{Im} z$	imaginary part of z, $= b$

Vectors

$\|x\|$	the length of x, $(x_1^2 + \ldots + x_d^2)^{1/2}$
$\|x\|_1$	the L^1 norm, $\|x_1\| + \cdots + \|x_d\|$
$x \cdot y$	dot product, $x_1 y_1 + \ldots + x_d y_d$

Sets in Euclidean Space

$\bar{A}$	closure of A
A^o	interior of A
∂A	boundary of A, $= \bar{A} - A^o$
$B(x, r)$	ball of radius r with center at x, $\{y : \|x - y\| < r\}$
$\partial B(x, r)$	boundary of $B(x, r)$, $\{y : \|x - y\| = r\}$
$x + A$	$\{x + y : y \in A\}$
rA	$\{rx : x \in A\}$

Set Theory

$\emptyset$	empty set
$A \cup B$	the union of A and B
$A \cap B$	the intersection of A and B
$A + B$	disjoint union, i.e., $A + B = A \cup B$ and $A \cap B = \emptyset$
A^c	the complement of A
$A - B$	difference, $A \cap (B^c)$
$A \Delta B$	symmetric difference, $(A - B) \cup (B - A)$
$\limsup A_n$	$\cap_{m \geq 1} (\cup_{n \geq m} A_n)$, points in infinitely many A_n
$\liminf A_n$	$\cup_{m \geq 1} (\cap_{n \geq m} A_n)$, points in all but finitely many A_n
$A_n \downarrow A$	$A_1 \supset A_2 \supset \ldots$ and $\cap_n A_n = A$
$A_n \uparrow A$	$A_1 \subset A_2 \subset \ldots$ and $\cup_n A_n = A$

Probability

$\mathcal{R}^d$	Borel subsets of $\mathbf{R}^d$
$\|A\|$	Lebesgue measure of A
1_A	indicator function, $= 1$ on A, $= 0$ on A^c
$X \in \mathcal{F}$	X is measurable with respect to $\mathcal{F}$, see Section 1.2

$\sigma(\ldots)$	σ-field generated by $\ldots$
$\sigma(\mathcal{C})$	the smallest σ-field containing all the sets in $\mathcal{C}$
$\sigma(X)$	the smallest σ-field $\mathcal{G}$ so that $X \in \mathcal{G}$
$\mathcal{F}_n \uparrow \mathcal{F}_\infty$	$\mathcal{F}_1 \subset \mathcal{F}_2 \subset \ldots, \sigma(\cup \mathcal{F}_n) = \mathcal{F}_\infty$
$\mathcal{F}_n \downarrow \mathcal{F}_\infty$	$\mathcal{F}_1 \supset \mathcal{F}_2 \supset \ldots, \cap \mathcal{F}_n = \mathcal{F}_\infty$
EX	expected value of X, see Section 1.3
$\|X\|_p$	$(E|X|^p)^{1/p}$
$\text{var}(X)$	the variance of X, $= E(X - EX)^2$
$L^2(\mathcal{F})$	$\{X : X \in \mathcal{F}, EX^2 < \infty\}$
$E(X|\mathcal{F})$	conditional expectation of X given $\mathcal{F}$, see Section 4.1
$P(A|\mathcal{F})$	$E(X|\mathcal{F})$ when $X = 1_A$
χ	random variable with a standard normal distribution
$\mathcal{N}(x)$	$P(\chi \le x)$, normal distribution function
$X =_d Y$	X and Y have the same distribution
$\Rightarrow$	converges weakly, see Section 2.2

Abbreviations

a.e.	almost everywhere
a.s.	almost surely
ch.f.	characteristic function
CLT	central limit theorem
d.f.	distribution function
f.d.d.'s	finite dimensional distributions
g.c.d.	greatest common divisor
i.i.d.	independent and identically distributed
i.o.	infinitely often
LIL	law of the iterated logarithm
l.s.c.	lower semicontinuous
MCT	monotone class theorem
r.c.d.	regular conditional distribution
r.v.	random variable
u.s.c.	upper semicontinuous
w.r.t.	with respect to

Table of the Normal Distribution

$$\Phi(x) = \int_{-\infty}^{x} \frac{1}{\sqrt{2\pi}} e^{-y^2/2} \, dy$$

To illustrate the use of the table: $\Phi(0.36) = 0.6406$, $\Phi(1.34) = 0.9099$

	0	1	2	3	4	5	6	7	8	9
0.0	0.5000	0.5040	0.5080	0.5120	0.5160	0.5199	0.5239	0.5279	0.5319	0.5359
0.1	0.5398	0.5438	0.5478	0.5517	0.5557	0.5596	0.5636	0.5675	0.5714	0.5753
0.2	0.5793	0.5832	0.5871	0.5910	0.5948	0.5987	0.6026	0.6064	0.6103	0.6141
0.3	0.6179	0.6217	0.6255	0.6293	0.6331	0.6368	0.6406	0.6443	0.6480	0.6517
0.4	0.6554	0.6591	0.6628	0.6664	0.6700	0.6736	0.6772	0.6808	0.6844	0.6879
0.5	0.6915	0.6950	0.6985	0.7019	0.7054	0.7088	0.7123	0.7157	0.7190	0.7224
0.6	0.7257	0.7291	0.7324	0.7357	0.7389	0.7422	0.7454	0.7486	0.7517	0.7549
0.7	0.7580	0.7611	0.7642	0.7673	0.7703	0.7734	0.7764	0.7793	0.7823	0.7852
0.8	0.7881	0.7910	0.7939	0.7967	0.7995	0.8023	0.8051	0.8078	0.8106	0.8133
0.9	0.8159	0.8186	0.8212	0.8238	0.8264	0.8289	0.8315	0.8340	0.8365	0.8389
1.0	0.8413	0.8438	0.8461	0.8485	0.8508	0.8531	0.8554	0.8577	0.8599	0.8621
1.1	0.8643	0.8665	0.8686	0.8708	0.8729	0.8749	0.8770	0.8790	0.8810	0.8830
1.2	0.8849	0.8869	0.8888	0.8907	0.8925	0.8943	0.8962	0.8980	0.8997	0.9015
1.3	0.9032	0.9049	0.9066	0.9082	0.9099	0.9115	0.9131	0.9147	0.9162	0.9177
1.4	0.9192	0.9207	0.9222	0.9236	0.9251	0.9265	0.9279	0.9292	0.9306	0.9319
1.5	0.9332	0.9345	0.9357	0.9370	0.9382	0.9394	0.9406	0.9418	0.9429	0.9441
1.6	0.9452	0.9463	0.9474	0.9484	0.9495	0.9505	0.9515	0.9525	0.9535	0.9545
1.7	0.9554	0.9564	0.9573	0.9582	0.9591	0.9599	0.9608	0.9616	0.9625	0.9633
1.8	0.9641	0.9649	0.9656	0.9664	0.9671	0.9678	0.9686	0.9693	0.9699	0.9706
1.9	0.9713	0.9719	0.9726	0.9732	0.9738	0.9744	0.9750	0.9756	0.9761	0.9767
2.0	0.9772	0.9778	0.9783	0.9788	0.9793	0.9798	0.9803	0.9808	0.9812	0.9817
2.1	0.9821	0.9826	0.9830	0.9834	0.9838	0.9842	0.9846	0.9850	0.9854	0.9857
2.2	0.9861	0.9864	0.9868	0.9871	0.9875	0.9878	0.9881	0.9884	0.9887	0.9890
2.3	0.9893	0.9896	0.9898	0.9901	0.9904	0.9906	0.9909	0.9911	0.9913	0.9916
2.4	0.9918	0.9920	0.9922	0.9924	0.9927	0.9929	0.9931	0.9932	0.9934	0.9936
2.5	0.9938	0.9940	0.9941	0.9943	0.9945	0.9946	0.9948	0.9949	0.9951	0.9952
2.6	0.9953	0.9955	0.9956	0.9957	0.9959	0.9960	0.9961	0.9962	0.9963	0.9964
2.7	0.9965	0.9966	0.9967	0.9968	0.9969	0.9970	0.9971	0.9972	0.9973	0.9974
2.8	0.9974	0.9975	0.9976	0.9977	0.9977	0.9978	0.9979	0.9979	0.9980	0.9981
2.9	0.9981	0.9982	0.9982	0.9983	0.9984	0.9984	0.9985	0.9985	0.9986	0.9986
3.0	0.9986	0.9987	0.9987	0.9988	0.9988	0.9989	0.9989	0.9989	0.9990	0.9990

Index